Freight Systems 2018

AA005162

Transportation Research Record: Journal of the Transportation Research Board Volume 2672, Issue 9

ISBN: 978-1-5108-8178-5

Printed from e-media with permission by:

Curran Associates, Inc.
57 Morehouse Lane
Red Hook, NY 12571

Some format issues inherent in the e-media version may also appear in this print version.

Published by Sage Publications USA

ISBN (Print) 978-1-5108-8178-5
ISBN 2018 Printed Set (All Issues) 978-1-5108-7735-1

Printed by Curran Associates, Inc. (2019)

Additional copies of this publication are available from:

Curran Associates, Inc.
57 Morehouse Lane
Red Hook, NY 12571 USA
Phone: 845-758-0400
Fax: 845-758-2633
Email: curran@proceedings.com
Web: www.proceedings.com

Volume 2672 Number 9 December 2018

TRR
JOURNAL OF THE TRANSPORTATION RESEARCH BOARD

Contents

Freight Systems

Articles

Evolving Supply Chains and Local Freight Flows: A Geographic Information System Analysis of Minnesota Cereal Grain Movement 1
Travis Fried, Lee Munnich, Thomas Horan, and Brian Hilton

Using Data from a State Travel Demand Model to Develop a Multi-Criteria Framework for Transload Facility Location Planning 12
Magdalena I. Asborno and Sarah Hernandez

Development of a Cost Estimation Framework for Potential Transload Facilities 24
Sadie Smith, Andrew Braham, Sarah Hernandez, and John Kent

Joint Deployment of Quay Cranes and Yard Cranes in Container Terminals at a Tactical Level 35
Lingxiao Wu and Shuaian Wang

Integrated Scheduling of Rail-Mounted Gantry Cranes, Internal Trucks and Reach Stackers in Railway Operation Area of Container Terminal 47
Baicheng Yan, Xiaoning Zhu, Li Wang, and Yimei Chang

Fuel Consumption Optimization Model for the Multi-Period Inventory Routing Problem 59
Chun Cheng, Mingyao Qi, and Louis-Martin Rousseau

Supplier Evaluation Model in Freight Activity Microsimulation Estimator 70
Seyed Mehdi Mahmoudifard, Ramin Shabanpour, Nima Golshani, Kiana Mohammadian, and Abolfazl Mohammadian

Dimensionality Reduction to Reveal Urban Truck Driver Activity Patterns 81
Fangping Lu, Fang Zhao, and Lynette Cheah

Advanced Woody Biomass Logistics for Co-Firing in Existing Coal Power Plant: Case Study of the Great Lakes States 93
Sangpil Ko and Pasi Lautala

Freight Mode Choice: A Regret Minimization and Utility Maximization Based Hybrid Model 107
Nowreen Keya, Sabreena Anowar, and Naveen Eluru

Weigh Station Impact on Truck Travel Time Reliability: Results and Findings from a Field Study and a Simulation Experiment 120
M. Shoaib Samandar, Billy M. Williams, and Ishtiak Ahmed

Reducing Unmet Demand and Spoilage in Cut Rose Logistics: Modeling and Control of Fast Moving Perishable Goods 130
Xiao Lin, Rudy R. Negenborn, Mark B. Duinkerken, and Gabriel Lodewijks

Evaluation of Commercial Truck Parking Detection for Rest Areas 141
Wei Sun, Ethan Stoop, and Scott S. Washburn

Comprehensive Exploratory Analysis of Truck Route Choice Diversity in Florida 152
Trang D. Luong, Divyakant Tahlyan, and Abdul R. Pinjari

Evaluating the Impact of Real-Time Mobility and Travel Time Reliability Information on Truck Drivers' Routing Decisions 164
Xiaoqiang Kong, William L. Eisele, Yunlong Zhang, and Daren B. H. Cline

Delivery Process for an Office Building in the Seattle Central Business District 173
Haena Kim, Linda Ng Boyle, and Anne Goodchild

The Scope for Pavement Porters: Addressing the Challenges of Last-Mile Parcel Delivery in London 184
Julian Allen, Tolga Bektas, Tom Cherrett, Oliver Bates, Adrian Friday, Fraser McLeod, Maja Piecyk, Marzena Piotrowska, ThuBa Nguyen, and Sarah Wise

Potential of Implementing Urban Freight Strategies in the Accommodation and Food Services Sector 194
Iván Sánchez-Díaz

Parking Conditions for Residential Delivery in New York City: A Case Study Analysis 204
Pierre Escand, Quanquan Chen, and Alison Conway

Testing the "Freight Landscape" Concept for Paris 216
Takanori Sakai, Adrien Beziat, Adeline Heitz, and Laetitia Dablanc

Statistical Distributions of Pavement Damage Associated with Overweight Vehicles: Methodology and Case Study 229
Dario Batioja-Alvarez, Seyed-Farzan Kazemi, Elie Y. Hajj, Raj V. Siddharthan, and Adam J. T. Hand

A Comparative Analysis of Truck Parking Travel Diary Data 242
Caroline Boris and Rebecca Brewster

Train-Level and Car-Level Modeling of Hazardous Materials Release in Railroad Incidents 249
Amirfarrokh Iranitalab and Aemal Khattak

Developing an Empirical Pipeline and Rail Crude Oil Mode Split and Route Assignment Model 261
Adam Morrison, Chris Bachmann, and Frank Saccomanno

---

* Membership as of February 2019.

TRB

TRANSPORTATION RESEARCH BOARD

CORE SPONSORS

Transportation Departments of the 50 States and the District of Columbia

Federal Government

U.S. Department of Transportation
- Federal Aviation Administration (FAA)
- Federal Highway Administration
- Federal Motor Carrier Safety Administration (FMCSA)
- Federal Railroad Administration (FRA)
- Federal Transit Administration
- National Highway Traffic Safety Administration (NHTSA)
- Office of the Assistant Secretary of Transportation for Research and Technology (OST-R)

U.S. Air Force Civil Engineer Center (AFCEC)
U.S. Coast Guard (USCG)
U.S. Department of Energy
U.S. Department of the Interior (DOI)
U.S. Environmental Protection Agency (EPA)

Nongovernmental Organizations

American Public Transportation Association (APTA)
Association of American Railroads (AAR)
California Air Resources Board

The National Academies of
SCIENCES · ENGINEERING · MEDICINE

The **National Academy of Sciences** was established in 1863 by an Act of Congress, signed by President Lincoln, as a private, nongovernmental institution to advise the nation on issues related to science and technology. Members are elected by their peers for outstanding contributions to research. Dr. Marcia McNutt is president.

The **National Academy of Engineering** was established in 1964 under the charter of the National Academy of Sciences to bring the practices of engineering to advising the nation. Members are elected by their peers for extraordinary contributions to engineering. Dr. C. D. Mote, Jr., is president.

The **National Academy of Medicine** (formerly the Institute of Medicine) was established in 1970 under the charter of the National Academy of Sciences to advise the nation on medical and health issues. Members are elected by their peers for distinguished contributions to medicine and health. Dr. Victor J. Dzau is president.

The three Academies work together as the National Academies of Sciences, Engineering, and Medicine to provide independent, objective analysis and advice to the nation and conduct other activities to solve complex problems and inform public policy decisions. The Academies also encourage education and research, recognize outstanding contributions to knowledge, and increase public understanding in matters of science, engineering, and medicine.

Learn more about the National Academies of Sciences, Engineering, and Medicine at **www.national-academies.org.**

The **Transportation Research Board** is one of seven major programs of the National Academies of Sciences, Engineering, and Medicine. The mission of the Transportation Research Board is to increase the benefits that transportation contributes to society by providing leadership in transportation innovation and progress through research and information exchange, conducted within a setting that is objective, interdisciplinary, and multimodal. The Board's varied activities annually engage about 7,000 engineers, scientists, and other transportation researchers and practitioners from the public and private sectors and academia, all of whom contribute their expertise in the public interest. The program is supported by state transportation departments, federal agencies including the component administrations of the U.S. Department of Transportation, and other organizations and individuals interested in the development of transportation.

Learn more about the Transportation Research Board at **www.TRB.org**.

TRR
JOURNAL OF THE TRANSPORTATION RESEARCH BOARD

Article

Transportation Research Record
2018, Vol. 2672(9) 1–11

DOI: 10.1177/0361198118759952
journals.sagepub.com/home/trr
SAGE

Evolving Supply Chains and Local Freight Flows: A Geographic Information System Analysis of Minnesota Cereal Grain Movement

Travis Fried[1], Lee Munnich[1], Thomas Horan[2], and Brian Hilton[3]

Abstract
In Minnesota, technological and economic shifts in the grain supply chain have altered the way grain producers and sellers navigate their local freight network. In particular, many producers have been increasing their personal trucking capacity and taking longer trips to intermodal and domestic market options. This logistical reshaping of local grain supply chains pressure transportation officials to reconsider the consequences for road infrastructure and congested freight corridors. Studies are discussing the potential of disaggregated commodity flow survey (CFS) data as a critical tool in understanding small-scale freight movement and informing infrastructural investment decisions. Utilizing ArcGIS's Network Analyst and Hot spot tools to analyze inter-county grain trucking, our study effectively differentiates highly active freight corridors. The model is used to further inform an ongoing infrastructure development project in the Twin Cities metro area by contextualizing road usage within the economic framework of the grain supply chain. However, this study finds CFS data alone fails to account for shifting supply chain conditions, and their consequent impact on the road network. Employing United States Department of Agriculture crop production and cropland data, this study additionally builds an original, computational model that simulates corn producer shipment reaction to market price competition within two key grain-producing counties. Results visualize how producers, during spot months, may be incentivized to haul longer distances to more competitive markets—especially emerging biofuel industries. This lesson proves crucial for state and local transportation officials who wish to identify freight infrastructure development opportunities that invigorate and accommodate growth in Minnesota's expanding agricultural industry cluster.

Like the state's grain milling roots, cereal grain today remains a central driver of the Minnesota economy. Last year, Minnesota produced nearly 55 million tons of corn and soybeans, making the state the third largest producer of grain and oilseed in the country. This ranking does nothing to highlight powerful Minnesota industries that rely on grain such as biofuel, livestock, dairy, and food processing. In fact, grain moving to and between these markets contributes to nearly 28% of all freight volume on the road, the highest share of any freight commodity in the state. Therefore, ensuring safe and efficient grain movement not only sustains the livelihoods of 340,000 state residents who work in Minnesota's growing agricultural sector but also crucially reinforces the state's economic health and competitiveness.

With both grain production and, consequently, grain operators' reliance on rural and urban roadways projected to grow, so too are concerns for accelerated pavement degradation and shipment delays due to highway bottlenecking, traffic congestion, limited intermodal access, and outdated seasonal load-weight restrictions (*1*). In response, Minnesota transportation officials have implemented freight network planning in strategic, statewide infrastructure development initiatives. Last year, the Minnesota Department of Transportation (MnDOT) reissued their *Statewide Freight System Plan*, outlining the importance of efficient freight transportation across statewide industry clusters and proposing various strategies to meet infrastructure demands of principal commercial corridors. Among the strategies discussed, adapting freight network resiliency and redundancy to evolving supply chain operations assumed an emphasized importance. Rapid advancements in geographic information system (GIS) technology have allowed for the efficient, inexpensive handling of complex commodity flow data necessary for evaluating freight

[1]Humphrey School of Public Affairs, University of Minnesota, MN
[2]University of Redlands, Redlands, CA
[3]Claremont Graduate University, Claremont, CA

Corresponding Author:
Address correspondence to Lee Munnich: lmunnich@umn.edu

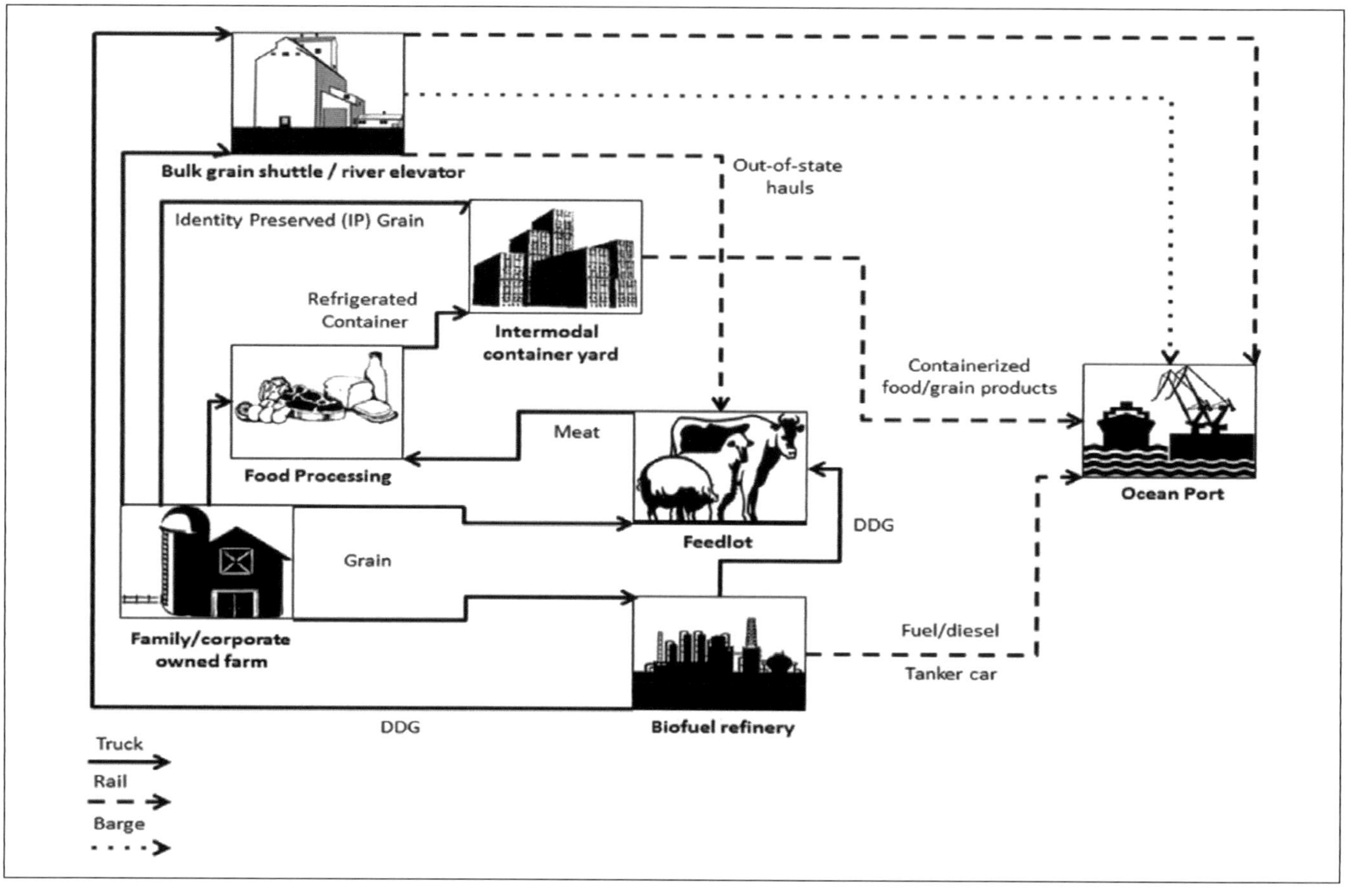

Figure 1. Simplified illustration of a farm-generated grain export chain with value-added, bulk, and containerized grain products.

impact on local transportation networks. These new technologies task Minnesota transportation officials to look toward accessible industry data and innovative spatial tools that effectively model grain flow patterns across key trucking corridors and shifting supply chain conditions.

Evolving Grain Supply Chain

In the past two decades, the grain supply chain experienced a number of political, technological, and market innovations that have directly influenced the way grain producers and wholesalers navigate their local freight networks. Traditionally in production states, trucking moves cereal grain—especially corn, comprising two-thirds of all grain movements—to domestic market sites (see Figure 1). Major rail companies had a substantial hand in solidifying the truck's role as a preliminary and intermediary connector within the supply chain. Federal deregulation policies since the 1980's Staggers Rail Act has granted major railroad companies greater economic freedom in consolidating less profitable shortline rails and adjusting rate incentives for bulk, primarily agricultural, shippers (*2*). This effect has larger, Class-I rail-served elevators incentivized to increase loading-capacity to serve 100–150 cars ("shuttles") at a time, often outcompeting smaller country elevators. In North Dakota, for instance, less than a dozen shuttle grain elevators handled two-thirds of the state's crop production (*3*).

Simultaneously, increasing international demand for grain-derived products, like biofuel and livestock, has stimulated greater competition among domestic market options (*4*). The United States Department of Agriculture (USDA) cites that domestic usage for corn more than doubled between 1998 and 2013. Ethanol in particular is a major industry player in the consumption of locally sourced grain. According to the Minnesota Biofuels Association, last year Minnesota refineries processed 5.54 billion bushels of corn—roughly a third of the state's total corn supply—and produced 3.6 million tons of dried distillers' grain (DDG), a production byproduct and common supplement for animal feed. Though newer plants are being equipped with rail-serving capabilities, most refineries rely on steady shipments of corn feedstock carried via truck. Thus, rail and barge have become increasingly relegated to long-haul, bulk shipping to export facilities on the coasts; meanwhile, trucks remain the principal link connecting production sites to subterminal elevators and domestic terminal markets.

Both phenomena greet similar effects—grain operators are upping truck fleets and hauling longer distances to meet more competitive buyers (*4*), in some instances, driving up to 100 miles (*5*). Trucks are increasingly occupying a greater modal share of cereal grain transportation at the expense of rail and barge, which have experienced a steady decline in modal usage. In the case of U.S. corn, the proportion hauled by truck increased by 22% between 1998 and 2013, while barge and rail modal usage dropped 10% and 12%, respectively (*6*). In other words, today's evolving grain supply chain has roads progressively assuming an exaggerated responsibility for connecting farmers to the marketplace.

Literature Review

Idealizing a Freight Flow Model

TRB's SHRP2 highlights the growing need for fine-tuned freight flow models to inform transportation investment decisions (*7*). The "ideal" freight flow model must a) account for freight operator behavior, b) span multiple shipping modes, c) integrate both commodity and passenger movements, and d) be capable of measuring supply chain shifts, such as those caused by policy change or technology advancement (*8*). Although a number of freight model methodologies have been conceptualized (see *7*), Wisetjindawat et al. explain that efforts to simulate freight movement tend to adhere broadly to two major categories, *trip-* and *commodity-based* approaches (*9*). Trip-based approaches estimate vehicle trips at a geographically aggregated level, relying on empirical data such as employment and manufacturer information. Results simulate broader freight movement between region centers or major market terminals, or both, as origin–destination pairs (O-D). However, in aggregating empirical freight data, trip-based approaches compromise the ability to integrate behavioral modeling and are thus limited when evaluating supply chain evolution.

When freight operators discuss supply chain evolution they are referring to producer behavior adapting to changing consumer demands. To account for these changes, commodity-based models estimate freight movement by volume, relying on more direct economic variables such as production and consumption. In international studies, commodity flow between producers and consumers can be disaggregated to a "micro" level, especially in urban areas, and better reflect changing freight operator behavior (*9–11*).

In the U.S., however, limited freight data availability has severely constrained efforts to develop an ideal, unified methodology for commodity flow modeling (*7*). The commodity flow survey (CFS) produced by the FHWA's Freight Analysis Framework (FAF4) is the leading data source for commodity-based modeling. CFS offers up-to-date freight movement data between domestic and export origins/destinations; however, data are geographically limited to interregional flows between states or major metropolitan areas. As a result, U.S. freight flow models intent on depicting more localized network movements require sophisticated data integration and disaggregation techniques, occasionally employing exhaustive industry surveys, proprietary datasets, and expensive transportation modeling packages (*12–14*).

This complexity elevates the importance of GIS as an integrative tool capable of effectively managing, visualizing, and analyzing intricate commodity flow data (*15*). Furthermore, rapid advancements in open spatial software and data call forth the authors of this study to contribute to the building of a more flexible commodity-based model methodology that,

- Identifies key transportation corridors in need of strategic freight network investment;
- Simulates shifts in supply chain logistics and measures consequent impacts on local infrastructure; and
- Relies on readily available, inexpensive commodity flow data while taking advantage of state-of-the-art GIS technology and techniques.

Results have the potential to direct future planning and policy considerations for state and local freight planners wishing to invigorate and accommodate growth in Minnesota's expanding agricultural industry cluster.

GIS Analysis of Grain Movement

List and Turnquist (*16*) first introduced GIS in freight flow modeling to fulfill the need for a robust, spatial data management system capable of handling multivariate O-D matrixes, incorporating vehicle class types, time, and distance. Using a trip-based, linear programming model to estimate least-distance links between zonal centroids, their study estimates truck flow across key highway corridors around the New York metro area (for early examples of GIS freight modeling in the U.S., see *15, 17, 18*). At the time of these studies, ArcGIS Network Analyst toolbox—now an instrumental tool for transportation researchers—was in its infancy. More recent studies on grain flow modeling take advantage of the toolbox's powerful vehicle routing problem (VRP) tool to draw least-cost paths between O-D pairs across transportation infrastructure. Gleim and Nolan explore the impact of the dissolution of the Canadian Wheat Board on rail-based wheat movements to Canadian export ports (*19*). Using time as the primary impedance in VRP, they take a trip-based approach to model approximate policy scenarios and their consequent influence on O-D routing. Though all these studies visualize the extent of freight network usage to varying degrees of detail, their approaches assume a uniform decision-making entity in regards to freight distribution. In reality, it is distinct industry operators competing within shifting supply chain conditions that determine commodity movement.

In the case of grain, these operators are producers and wholesalers. Whereas later Canadian studies incorporate rail rates and freight penalties to model shipper-agent competition

(*20*), the U.S. grain supply chain provides their own unique challenges when it comes to simulating shifting grain producer behaviors and their impact on road infrastructure. Early U.S. efforts to map these behaviors began with Ellis and Jessup's study on grain flows in rural eastern Washington (*21*). Their study measured the impacts of the Snake River drawdown on intermodal access and highway grain flows. Employing an "optimized transportation" approach, they used local survey results to calculate trucking-cost coefficients and assign least-cost trips to elevator locations. In disaggregating commodity flows to a more localized level, their study measured grain shipper response to changing market climates.

North Dakota State University's Upper Great Plains Transportation Institute (UGPTI) has performed a number of studies tying grain movement to the state's economic and transportation development (*3, 22, 23*). Their analysis of agriculture supply chain operations sets the bar in terms of both the GIS methods and language used for mapping grain flow. Mitra et al. (*22*) uses USDA's cropland raster data layer to estimate grain freight generation around the state. They derive fixed and variable trucking costs and axle/payload configurations to develop a freight distribution and trip generation model. In later studies, they introduce grain bid pricing to estimate commodity flow reaction to spatial competition between specific grain markets, primarily biofuel refineries (*23*). Rather than minimized distance or time, simulated grain volume navigated road networks to maximize profit. Though the reports' implications are discussed later, their results lay the foundation for modeling the economic impact of shifting supply chain conditions.

However, UGPTI's studies rely on extensive, statewide manufacturer surveys. These studies also made use of proprietary and at times inflexible transportation optimization packages such as TransCAD, GAMS, and CUBE. Furthermore, our research makes use of Esri ArcGIS Network and Spatial Analyst toolboxes to model grain flow. Our first approach employed a proprietary CFS-derived data set to evaluate inter-county cereal grain movement. Although revealing of broader interregional flow and potential corridors of high, grain-related freight bottlenecking, a pure network analyst approach using disaggregated CFS data is insufficient for understanding how changing grain operator behavior affects freight flow.

To capture the impact of grain operator behavior on the roadways, a more novel approach to deriving and modeling grain freight data is needed. The initial phase of this study began with a series of seven interviews with grain producers around Minnesota, as well as various focus meetings with grain freight operators and state transportation officials. Throughout the year, grain producers and wholesalers negotiate between changing grain basis contracts and operation costs to determine when and where to sell or store grain. Therefore, this study also employs original geocomputational modeling to follow corn as it flows from specific farm locations to its most competitive market option in two key corn-producing counties. The model is based on both current and predicted future grain price data, Mitra et al.'s distance-cost algorithm (*22*), and OpenStreetMap's (OSM) distance-matrix API. Freight generation sites are derived from USDA's cropland data layer. The South Dakota DOT identifies the cropland data layer as a powerful tool that identifies common commodity production sites around the country (*24*). Since data availability is a generally acknowledged obstacle to disaggregated commodity-based freight modeling, the study explores the merits and challenges that both approaches and datasets present for Minnesota freight planners and transportation officials seeking to inform strategic freight network investment decisions.

Methodology

Network Analyst Approach

The study takes a network analyst approach to model inter-county cereal grain volume flow and evaluate its potential in identifying freight bottlenecks and global supply chain shifts. CFS data is provided by Quetica, LLC (see Table 1), a Minnesota-based supply chain management company that most recently used commodity flow analysis to optimize freight network planning for Iowa's DOT (*25*). Quetica's data disaggregates CFS on a county level and represents larger interregional movements rather than farm-to-market movements. The sample dataset includes cereal grain shipment weight via truck mode for all Minnesota O-D and Minnesota destination counties in 2014. "Cereal grain" is designated by its standard classification of transported goods code, which includes wheat, corn, sorghum and barley, among others (*26*). Preliminary analysis showed 4,970 O-D linkages shared between 71 Minnesota counties.

O-D points were derived from county centroids then joined to the CFS dataset. Using Esri's Streets Edge network dataset, O-D points were loaded into ArcGIS VRP with origins as facilities and destinations as incidents. Distance was given primary impedance and unpaved roads were allowed to ensure access to all incidents. The solution produced unique route lines between all O-D pairs (7,569 lines). The next step required a spatial join to Esri's underlying road network to identify route names and aggregate shipment values on overlapping lines. The results produced a road network capable of visualizing inter-county cereal grain flow.

Optimized Market Approach

This study employs an additional approach to explore the integration of more publicly available software and datasets into freight flow modeling. The optimized market approach models the network impact of producer reaction to market competition. Data is derived from USDA's National Accident Sampling System service, with GIS shapefiles provided by

Table 1. Outline of Software, Geo-Tools, and Data Tools Used in Both the Network Analyst and Optimized Market Approach

Approach	Model type	Software	Geo-tools used	Data sources
Network analyst	Commodity-based	ESRI ArcGIS	VRP/Closest facility analysis	Minnesota Department of Transportation (2016). GIS data products
			Hot spot analysis	Quetica's county-level CFS data
			Ordinary least squares	Minnesota Pollution Control Agency (2016). Feedlots in Minnesota
				Minnesota Dept. of Agriculture (2014). Ethanol plants, Minnesota
				ReferenceUSA (2017). Businesses with NAICS 424510
Optimized market	Commodity-based	Jupyter Notebook	Python 3	U.S. Dept. of Agriculture (2014). Cropland data layer
			OpenStreetMap routing API	U.S. Dept. of Agriculture (2014). Minnesota corn production, by county
		ESRI ArcGIS	VRP/Closest facility analysis	AgWeb (2017). Cash grain bids*

*AgWeb is an open resource that allows users to freely look up grain price data based on zipcode proximity. More complete and historical grain market data can be purchased from GeoGrain (http://geograin.com).

sources accessible on Minnesota's Geospatial Commons. The study sampled two corn-producing counties—Swift County in western Minnesota, Dodge County in the south—during two separate months. April, the time of the study, used current prices, or "spot", whereas October relied on predicted "future" prices calculated from the contract basis. Market data were derived from AgWeb's Cash Grain Bids tool and elevator websites. Market locations identified by the standardized industry code for Grain and Field Bean Merchant Wholesalers (NAICS 424510) were gathered from ReferenceUSA's company inventory. A new data set was created by joining the markets' spatial and price data. Markets within a 50-mi buffer of the test county were included in the routing problem as destinations.

Origins were generated using USDA's 2014 cropland layer, with all corn-classified pixels converted to polygons. The resulting vector layer was queried to remove shapes with an area less than the mean. This process reduced data size without interfering with general cornfield distribution. Polygon area geometries were calculated, converted to point features via their centroids, and then spatially joined to their containing county polygons. County corn production was proportionally distributed to each origin site based on corresponding field size. South Dakota describes a similar procedure for deriving production values from field size (*24*). April's and October's production were calculated using the state's monthly shares of corn sales for 2014.

Both counties' spatial datasets were reintegrated into a Python-based linear programming model. An algorithm was designed to determine optimized O-D pairs given maximum profit. Distance between origin sites and market destinations was computed using OSM's distance-matrix API, an open-sourced application. O-D distances were passed into a fixed-transportation cost function derived from Mitra et al.'s 2007 survey on agricultural freight flows. The function calculates expense from computed O-D distance (x), assuming trucks travel at a constant 60 mph and are unlimited by time delays and road load-weight restrictions. Optimized profit is calculated to be the maximum difference between incomes gained by a single generation site (origin o) via all possible markets (destination d), based on month m (see Table 2).

The fixed expense equation calculates both loaded and unloaded round-trip costs based on the type of truck used in delivery. The optimized profit variable relies on two conditional income calculations. If the input month is April (spot) a basic price per ton calculation is conducted. Otherwise, the month represents a future price, such as October 2017, and the month's contract basis is used to calculate a forecasted price ("futures") for grain (basis = spot price – futures) (*27*). The program outputs two comma-separated values containing spatial and thematic data for separate origins (all origin sites to all destination markets) and destinations (unique market destinations only). The two files were formatted in-program for import into the ArcGIS VRP.

Results

Identifying Grain-Related Bottlenecks

In 2015, MnDOT invested US$70 million in their expanded Corridors of Commerce project to improve the flow of commodities and passenger vehicles around key interregional highways and intermodal access roads around the Twin Cities metro area. With 18.3 million tons of grain moving through highways located in the seven-county metro area, roughly 285,000 full truckloads, our results reveal how

Table 2. Truck-Cost Transportation Matrix and Optimized O-D Distance-Profit Calculation

Truck Type	Proportion used (%)	Cost (Per-ton mile)	Payload cap. (tons)	Tare weight (tons)
Two-axle single unit	9	0.052	11.25	4.75
Three-axle single unit	34	0.041	14.80	8.20
Four-axle single unit	14	0.029	21.25	8.75
Five-axle semi	43	0.029	26.6	13.40

$$\sum_{t=1}^{t} Expense_{o,d} = \frac{\left(x_{o,d} * Cost_t * \left(W_o^m + N_t * Tare_t\right)\right) + \left(x_{o,d} * Cosst * N_t * Tare_t\right)}{Proportion_t}$$

$$Optimized\ profit_o^m = MAX\left(Income_{o,d}^m - Expense_{o,d}\ \forall\ d\right)$$

where:

t = truck type

W = origin corn weight (tons)

N = total number of trucks$_t$ = INT $((W_o^m / payload_t) + 1)$

$$Income_o^{m=apr.} = W_o^{m=apr} * Price_d^{per\ ton}$$

$$Income_o^{m=oct.} = W_o^{m=oct} * \left(Price_d^{per\ ton} - Basis_d^{m=oct}\right)$$

cereal grain represents an important strategic consideration for freight network improvement initiatives in both the Metro and Greater Minnesota Area. An average total of 154 million tons of grain travel between Minnesota counties. Metro highways are responsible for moving 12% of all grain freight, mainly moving northwest to southeast down I-94, I-494, and I-35 (see Figure 2). Identifying corridors more prone to grain-related bottlenecking, such as these, is crucial for informing transportation officials of potential development opportunities.

Grain-related bottlenecks are determined using Hot spot analysis (Getis-Ord-Gi*). ArcGIS Hot spot tool determines the level of significance between the spatial clustering of variables. The output assigns each feature with a GI-bin integer ranging between −3 and 3, where −3 indicates a high significance for weak clustering ("cold spot"), 0 no significant clustering, and 3 a high for strong clustering ("hot spot"). Route usage—total O-D links shared by a route segment—is inputted to reveal hot spot route segments that show significantly high inter-county usage relative to all other route segments. Given the topological nature of road networks, special consideration is given to conceptualizing spatial relationships. The Gi* test was conceptualized using a fixed-distance band with a reduced threshold radius to minimize the influence of neighboring, non-shared routes while maximizing the influence of connecting segments. The analysis then queried for route segments with a GI-bin of 3 (99% p-value confidence) and grain shipment weights 1 *SD* above the mean. Thus, this study defines grain-specific bottlenecks as routes measuring a powerful level of significance for high inter-county usage and carrying above average grain volume.

Table 3 provides a sample of routes that meet the bottleneck criteria. The hot spot matrix successfully determines highly active cereal grain corridors. For instance, 6.9 million tons of grain travel down US-71 between I-94 and I-90. Where US-71 intersects with US-212 (W Lincoln Ave) holds the maximum value in terms of volume (over 13 million tons) and is among the highest in terms of O-D link usage. Interestingly, a number of interviewees referred to this route segment as a highly used intersection connecting producers to Wilmar to the north, Windom to the south, and the Twin Cities to the east. Although revealing of grain movement, results do little to show interaction between freight and passenger traffic. Strategic investment decisions to improve grain network resiliency should target corridors in which grain flow has a sizeable impact on commercial and passenger activity. As a result, further validation is needed to determine the impacts of grain movement on active commercial corridors. Future efforts should take advantage of the American Transportation Research Institute's (ATRI) truck GPS data, utilized in MnDOT's 2014 study, to explore how cereal grain movement influences general freight truck performance (*28*).

Case Study: Highway 13 Corridor

Located in Savage, MN, Highway (HWY) 13 is a critical corridor for grain traffic en route to river ports along the Minnesota River. Lack of capacity for heavy commercial trucks, passenger delays, and growing accidents have officials investing in development projects for HWY 13 between US-169 and I-35 (See Figure 3). Among the plans outlined in 2013 are, improving interchanges and widening road access to HWY 13, installing traffic signals at key connector roads to river elevators, road reconstruction, optimizing signal timing, improving sidewalks, and extending high-use county roads to major highway connectors (*28*). Given that the

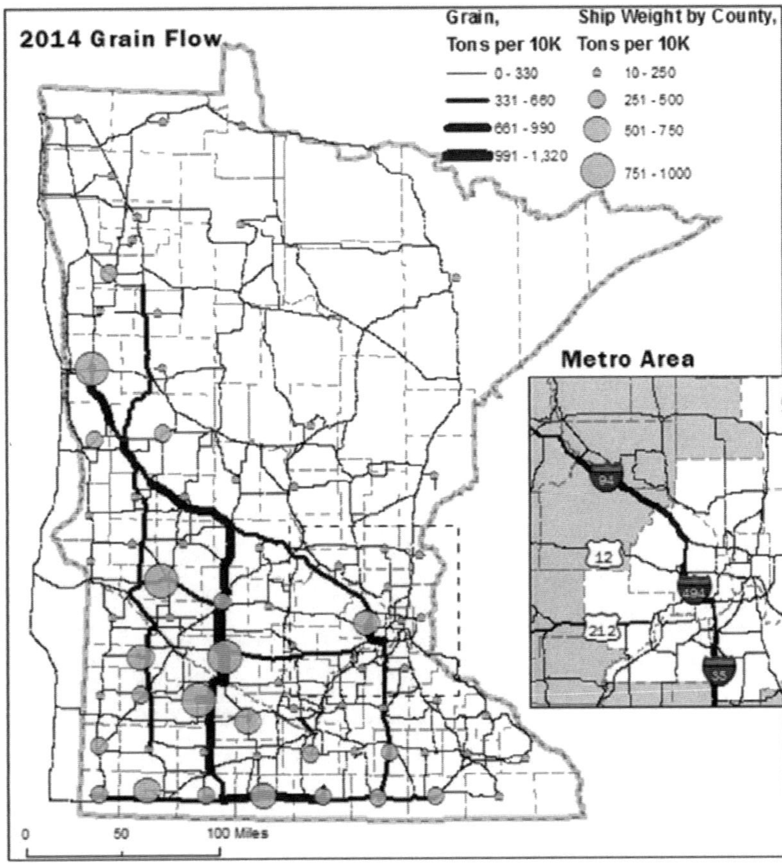

Figure 2. Network analyst results.

Table 3. Sample Hot spot Matrix Identifying Grain-Related Bottlenecks Based on County Route Usage and Average Shipment Weights. Includes Average Usage, Shipment Weights, and Cumulative Route Length As Well As Variables of Significance

Route name	Avg. usage	GI-bin	Route length (MI)	Avg. weight (tons per 10K)	*P*-value	*Z*-score
W Lincoln Ave (US-212/US-71)	429	3	2	922	0.0000	12.78
1ST ST S (US-71)	353	3	3	907	0.0087	2.36
E Bridge St (US-71)	219	3	2	880	0.0006	3.30
US-71	319	3	136	762	0.0000	6.02
I-494 E	506	3	7	681	0.0000	10.29
I-494 S	449	3	16	642	0.0000	7.08
MN-9	178	2	3	621	0.0250	2.24
I-90 W	322	3	36	558	0.0004	2.37
MN-4	180	3	3	556	0.0077	2.59
I-94 E	577	3	151	544	0.0000	7.45
I-35W S	316	3	6	500	0.0000	5.58

HWY 13 corridor represents an important intermodal cereal grain linkage, detailed grain flow modeling can inform and legitimize investment decisions.

This study replicates the network analyst approach to map inter-county flow to the ports of Savage. Figure 3 illustrates inter-county grain flow to destination ports. Savage lies in the northern end of Scott County, with four grain markets located just off HWY 13 along the Minnesota River. Since these locations represent the only market option in Scott County, it is assumed that all grain flowing to Scott County terminates at one of the four river ports. The 2014 network simulates the routes taken by over 3,600 truckloads, carrying

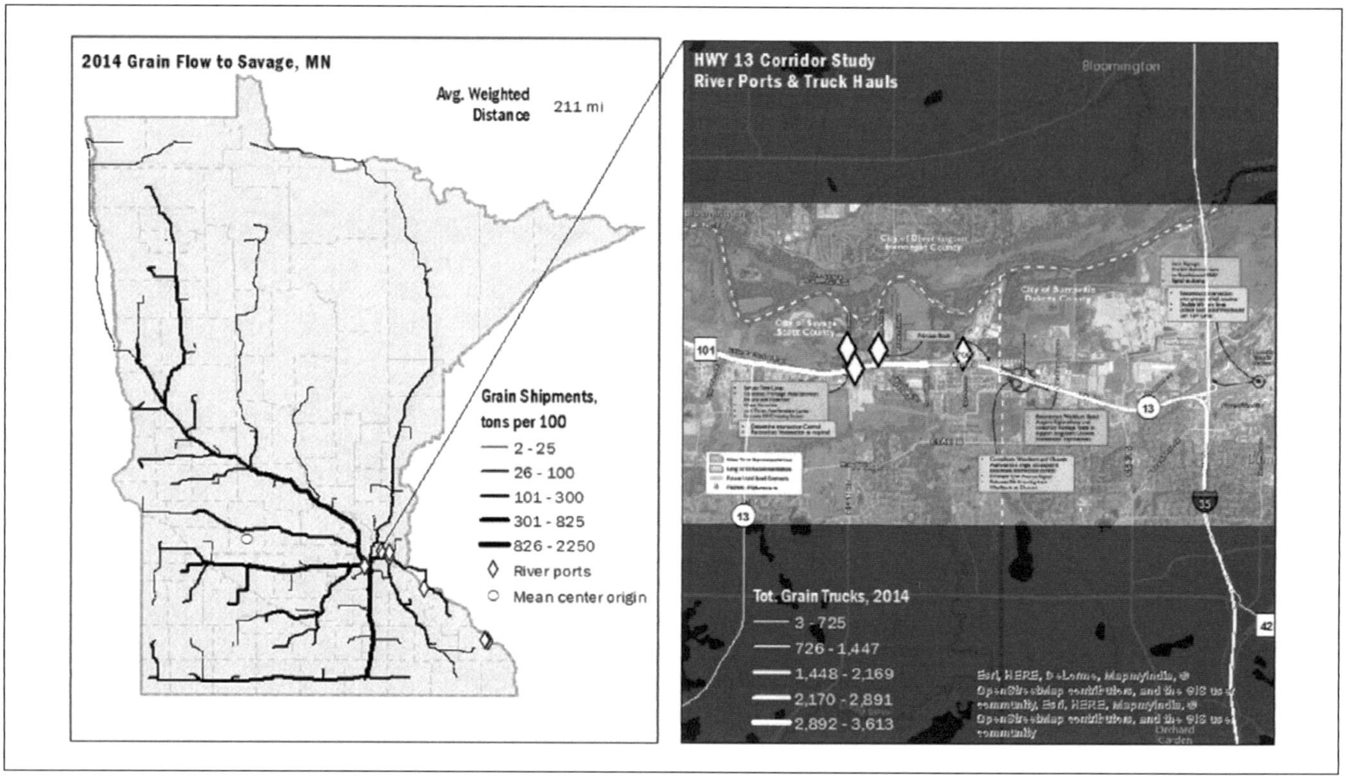

Figure 3. Analyzing grain flow of HWY 13 corridor near ports of Savage.

roughly 225,000 tons of cereal grain. Savage is the destination for over 70 counties, with trucks traveling a weighted mean distance of 211 mi. US-212 and US-169 feed into County HWY 101 entering the ports from the northwest and carrying over 150,000 tons; I-35N, feeding into HWY 13 from the south, moves nearly a third of total grain truck traffic. The map on the left focuses in on the HWY 13 corridor overlaid with a georeferenced image of proposed infrastructure plans in 2012. The diagram juxtaposes 2012 project construction plans in relation to grain flow.

Shifting Producer Behavior and its Freight Flow Impact

To better explore the relationship between grain movement and supply chain demand, this study designed an additional commodity-based model that derived freight generation from USDA's cropland data layer and took a market optimized approach to simulate freight distribution. Whereas this study's network analyst approach captures generalized grain movement between terminal markets, the market optimized approach models producer behavior given the grain bids of spatially competitive buyers.

Most evident in Figure 4 is that spot prices draw producers to farther, more competitive market options. In the case of Dodge, an ethanol plant offering US\$3.37/bu, the highest offer within 50 mi, simulated producers hauling roughly 44,000 tons an average distance of 55 route miles. The river port in Red Wing held the second highest offer and attracted the remaining 15% of April's yield 45 mi. Meanwhile, Granite Falls' ethanol plant drew all of Swift County's April corn yields, averaging 39 mi. Granite Falls proved to be more downstream competitive than a neighboring shuttle elevator offering US\$0.04/bu more, twice the distance away. Granite Falls also proved more profitable than a competing, similarly sized ethanol plant located within Swift County. Spot price modeling supports the notion that producers forgo local market options in favor of more distant, competitive ones. In the counties selected, the strongest markets proved to be ethanol plants, with secondary competition from high-capacity elevators such as those that serve river barges and shuttle trains.

Opposite trends were noted when inputting October's basis contract. Transportation costs became a more sensitive, limiting factor with producers not driving over 17 mi in any county. Additionally, there was no relationship between flow and market types; distance remained the more appropriate determinant in producer behavior when considering October futures. One explanation is that basis prices reflect more historical, economic trends, whereas market operators often adjust spot prices to meet instantaneous, just-in-time demand. As a result, basis prices are relatively stable over time—with little fluctuation between markets—and spot prices more volatile and more closely mirroring the evolving competitive

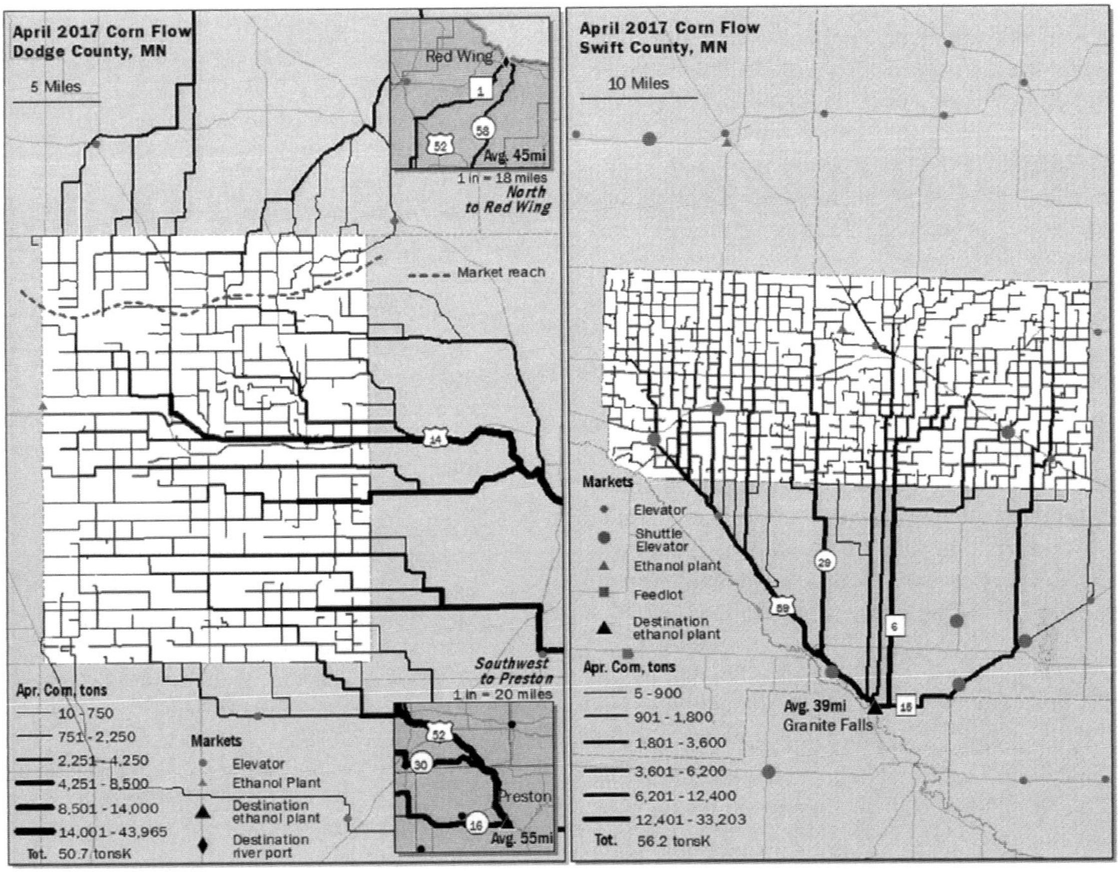

Figure 4. Optimized market results for April corn flows in Dodge County and Swift County.

market climate. Expanded market data would allow for more months to be tested and allow future iterations of the model to explore this relationship. Another explanation is that the initial test employed an overly simplified and inconsistent conception of basis pricing. Mitra et al. (*23*) derives their own basis using price data, transportation costs, demand and production variables, and market type. A more locally sensitive basis allows for predictive grain flow modeling beyond the spot month. Though important for future improvements of this study's commodity-based algorithm, broader implications regarding October grain flow modeling remain limited in this phase of the study.

Conclusion

A Focus on Grain-Derived, Value-added Supply Chains

The optimized market approach effectively captures grain operator behavior during spot months and visualizes the market impact on producers' navigation of their local road networks. Considering that ethanol plants have a strong mediating effect on downstream grain market pricing (*29*), this model suggests that the emergence of biofuel has an especially substantial effect on producer behavior and grain trucking movement. Results mirror Mitra et al.'s findings of increased grain truck flow to biofuel facilities, which equated to US$4.13 million of additional pavement maintenance costs for load-restricted roads within the North Dakota study area (*23*).

Meanwhile, Minnesota district engineers and grain freight operators are expressing a number of concerns about the effect of feedlot growth, and international demand for specialized bean products. In the case of the former, expanding dairy farm operations in southern Minnesota has the potential to greatly affect grain feed flow, adding to the wear-and-tear of rural roads, bridges, and major highway connectors. Additionally, increasing East Asian demand for identity preserved grain—namely soybeans—is raising concerns for empty container availability and the lack of rail portals to California ports (*30*). Emerging markets have a substantial impact on spatial competition between grain buyers and their consequent attraction of producer freight. Especially for south Minnesota district and county engineers, modeling relationships between newer grain markets and operator behavior provides invaluable insight for managing freight infrastructure assets.

Expanding Multimodal Services

Rail and waterway infrastructure is not only a central feature of the agricultural economy but also a substantial mediator of roadway usage. Roughly 40% of Minnesota corn is destined for international ports, over two-thirds of which travel down the Mississippi River via barge (*1*). These river ports—like shuttle elevators—represent an attractive market option, not only in southeastern Minnesota, but for grain operators throughout the state (see Figure 3). With the expansion of the Panama Canal expecting to increase barge traffic along the Mississippi River, freight planners are beginning to renew discussions on optimizing road access to river ports (e.g., HWY 13) and improving aging waterway infrastructure. Meanwhile, expanding shortline rail services to smaller, more isolated elevators has been a long-discussed topic. Stimulating multimodal competition has the potential of lowering long-haul trucking between satellite elevators and larger intermodal facilities. Adding multimodal functionality to grain flow models can simulate economic results of strategic investments in improved rail and waterway infrastructure.

Revisiting Seasonal Load-Weight Restrictions

What is already an old topic for many producers, some local infrastructure engineers are beginning to discuss the possibilities of reducing <10-ton load-weight restrictions on state and rural roads. Increasing trucking capacities, grain yields, and market completion means route diversions caused by weight restrictions are taking a bigger toll on freight costs. Mitra et al.'s North Dakota study found that maintenance and improvement costs from removing all load-weight restrictions exceeded cost savings from improved agricultural freight flow. However, they found that small-scale weight limit reductions on connector roads to more dominant grain market areas proved cost-effective for reducing transportation prices (*22*). Although the current model omits the impact of load-weight restrictions on grain flow, this study's approach can be utilized to model corridor hot spots and producer behavior to determine high-use, weight-restricted routes most affected by market competition.

Toward Strategic Analysis of Grain Movements

Grain movements will continue to be a significant driver of freight volume on Minnesota highways. This analysis has demonstrated the dynamic nature of this demand and how industry and market changes can affect the spatial distribution of bottlenecks and other transportation impacts. Additionally, this study addresses the considerable "gap" in unified freight and agricultural modeling methodology, as outlined by SHRP2 and South Dakota DOT, by exploring the potential of disaggregated grain flow modeling using open-source spatial computing tools and publicly available data (*7, 24*). Though preliminary, this approach effectively models grain operator behavior reacting to evolving market conditions on a micro level; however, further datasets and methodologies must be employed to validate and expand implications of these results across variant timespans, geographic extents, policy scenarios, passenger-commercial vehicle interactions, and modal options.

Moreover, the issue needs to be considered an economic competiveness issue, not just a transportation issue. It is important to consider economic data relating to industry clusters to obtain a more complete picture of freight flow dynamics.

References

1. Minnesota Department of Transportation. *Statewide Freight System Plan*. May 2016. http://www.dot.state.mn.us/planning/freightplan. Accessed May 19, 2017.
2. Wilson, W. W., and W. W. Wilson. Deregulation, Rate Incentives, and Efficiency in the Railroad Market. *Transportation After Deregulation*, Vol. 6, 2001, pp. 1–24.
3. Vachal, K., and K. J. Button. Impact of Shuttle Rates on Local Grain Flows. *International Journal of Transport Economics*, Vol. 30, No. 1, 2003, pp. 91–102.
4. U.S. Department of Agriculture. *Study of Rural Transportation Issues*. April 2010. https://www.ams.usda.gov/sites/default/files/media/RTIFullReport.pdf. Accessed May 19, 2017.
5. Tolliver, D., M. Berwick, and K. Vachal. Farm-to-Market Transportation Patterns and Truck Use in the Northern Plains. Prepared for Bureau of Transportation Statistics and U.S. Department of Transportation, 2005.
6. U.S. Department of Agriculture. *Transportation of U.S. Grains: A Modal Share Analysis 1978-2013 Update*. June 2015. https://www.ams.usda.gov/sites/default/files/media/ModalJune2015.pdf. Accessed May 19, 2017.
7. Chase, K. M., P. Anater, and T. Phelan. *SHRP2 Report S2-C20-RR-1: Freight Demand Modeling and Data Improvement*. Transportation Research Board of the National Academies, Washington, D.C., 2013.
8. Chow, J. Y. J., C. H. Yang, and A. C. Regan. State-of-the art of Freight Forecast Modeling: Lessons Learned and the Road Ahead. *Transportation*, Vol. 37, 2010, pp. 1011–1030.
9. Wisetjindawat, W., K. Sano, S. Matsumoto, and P. Raothanachonkun. Micro-Simulation Model for Modeling Freight Agents Interactions in Urban Freight Movement. Presented at 83rd Annual Meeting of the Transportation Research Board, Washington D.C., 2006.
10. Boerkamps, J. H. K., A. J. van Binsbergen, and P. K. L. Bovy. Modeling Behavioral Aspects of Urban Freight Movement in Supply Chains. *Transportation Research Record: Journal of the Transportation Research Board*, 2000. 1725: 17–25.
11. de Jong, G., and M. Ben-Akiva. A Micro-Simulation Model of Shipment Size and Transport Chain Choice. *Transportation Research Part B: Methodological*, Vol. 41, No. 9, 2007, pp. 921–1078.
12. Southworth, F. Filling Gaps in the U.S. Commodity Flow Picture: Using the CFS with Other Data Sources. Prepared for the 2005 Commodity Flow Survey Users' Conference, Boston, Mass., 2005.

13. Mitra, S., and D. Tolliver. Framework for Modeling Statewide Freight Movement Using Publicly Available Data. *Transportation Research Forum*, Vol. 48, No. 2, 2009, pp. 83–102.
14. Giuliano, G., P. Gordon, J. Park, L. Wang, and Q. Pan. Estimating Freight Flows for Metropolitan Area Highway Networks Using Secondary Data Sources. Prepared for Transportation Research Circular E-C088: Commodity Flow Survey Conference, 2010.
15. Southworth, F., and B. E. Peterson. Intermodal and International Freight Network Modeling. *Transportation Research C: Emerging Technologies*, Vol. 8, 2000, pp. 146–166.
16. List, G. F., and M. A. Turnquist. Estimating Truck Travel Patterns in Urban Areas. *Transportation Research Record: Journal of the Transportation Research Board*, 1994. 1430: 1–9.
17. Alam, M., and E. Fekpe. Application of Geographic Information Systems Technology in Freight Data Analysis: Case Study of I-90/I-94 Corridor Analysis. *Transportation Research Record: Journal of the Transportation Research Board*, 1998. 1625: 173–183.
18. Gordon, P., and Q. Pan. Assembling and Processing Freight Shipment Data: Developing a GIS-Based Origin-Destination Matrix for Southern California Freight Flows. Prepared for METRANS Transportation Center, 2001.
19. Gleim, S., and J. Nolan. Canada's Grain Handling and Transportation System: A GIS-based Evaluation of Potential Policy Changes. *Journal of the Transportation Research Forum*, Vol. 54, No. 3, 2015, pp. 99–111.
20. Lawrence, R., J. Nolan, and R. Schoney. Simulating Contestability in Freight Transportation: The Canadian Grain Handling and Transportation System, *Journal of Transport Economics and Policy*, Vol. 50, No. 4, 2016, pp. 325–349.
21. Jessup, E. L., J. Ellis, and K. Casavant. A GIS Commodity Flow Model for Transportation Policy Analysis: A Case Study of the Impacts of a Snake River Drawdown. Prepared for the Eastern Washington Intermodal Transportation Study, 1999.
22. Mitra, S., D. Tolliver, A. Varma, and A. Dybing. Analyzing Effects of Spring Highway Load Restrictions on North Dakota's Agricultural Freight Flows. *Transportation Research Record: Journal of the Transportation Research Board*, 2007. 2008: 92–99.
23. Mitra, S., D. Tolliver, and A. Dybing. Impacts of Ethanol on Highway Networks. *The Journal of Transport and Land Use*, Vol. 4, No. 1, 2011, pp. 71–81.
24. Cambridge Systematics Inc. *Agricultural Freight Data Improvement Study SD2014-09*. Prepared for South Dakota Department of Transportation, 2016.
25. Quetica, LLC. Iowa Statewide Freight Transportation Network Optimization Strategy. 2016. http://quetica.com/portfolio/iowa-statewide-freight-transportation-network-optimization-strategy/. Accessed May 19, 2017.
26. U.S. Census Bureau. 2007 Commodity Flow Survey Standard Classification of Transported Goods (SCTG), 2007. https://www.census.gov/svsd/www/cfsdat/cfs071200.pdf. Accessed May 19, 2007.
27. Baldwin, E. D. Understanding and Using Basis-Grains. N.d. http://rnrinag.uwagec.org/. Accessed May 19, 2017.
28. Liao, C. Using GPS Data to Identify Traffic Bottlenecks for Trucks. Prepared for Minnesota Department of Transportation, 2014. http://www.dot.state.mn.us/research/TS/2014/201414TS.pdf. Accessed May 19, 2017.
29. McNew, K., and D. Griffith. Measuring the Impact of Ethanol Plants on Local Grain Prices. *Review of Agricultural Economics*, Vol. 27, No. 2, 2005, pp. 164–180.
30. Vachal, K. Marketing U.S. Grain and Oilseed by Container. *Prepared for Upper Great Plains Transportation Institute*, 2014.

The Standing Committee on Agricultural Transportation (AT030) peer-reviewed this paper (18-03357).

Article

TRR
JOURNAL OF THE TRANSPORTATION RESEARCH BOARD

Transportation Research Record
2018, Vol. 2672(9) 12–23

DOI: 10.1177/0361198118772699
journals.sagepub.com/home/trr
SAGE

Using Data from a State Travel Demand Model to Develop a Multi-Criteria Framework for Transload Facility Location Planning

Magdalena I. Asborno[1] **and Sarah Hernandez**[1]

Abstract

The majority of freight is transported within the U.S. by road. However, the use of alternative modes, such as rail and barge, is associated with lower transportation and infrastructure maintenance costs, release of highway capacity, increased safety, and lower emissions. Thus, there is a latent opportunity for shippers and consumers to benefit from modal shift. In this context, strategically located freight-transfer facilities to improve rail and barge access is key. Moreover, for states with lower commodity tonnages and access to short-line rail and navigable waterways, transload facilities have significant potential to shift freight to underutilized modes. This paper develops a multi-criteria assessment framework to identify strategic locations for transload facilities at the state level. Using a statewide travel demand model (STDM) as the main data source, this framework provides a sketch-planning tool to support decision-making for state Departments of Transportation and economic development agencies. The multi-criteria quantify four measures of facility potential: (a) interaction with the transportation network, (b) amount of freight transported between major freight routes, (c) spatial aggregation, and (d) directionality aggregation. Each criterion is estimated and combined at the county level to produce a multi-criteria score, which defines a county's potential to support transload movements. Using this score, counties are ranked, and facilities prioritized. The framework is applied to Arkansas and validated using the STDM for base (2010) and forecast (2040) years.

In the U.S. the vast majority of freight is transported by road (*1*), for several reasons: (a) lack of knowledge of alternatives, (b) risk associated with alternatives, (c) shortage of resources to explore alternatives, and (d) the absence of transportation infrastructure and services limiting modal accessibility (*2, 3*). However, several benefits can be realized by shippers and consumers by shifting freight to more efficient modes, such as rail or water, or adopting a multimodal scheme. Use of rail and barge is associated with lower transportation and infrastructure maintenance costs, release of highway capacity, increased safety, and lower emissions (*4–6*). Given the reasons why road trucking remains the predominant mode, this paper focuses on provision of accessibility to other modes as a means to promote modal shift. The provision of strategically located freight-transfer facilities to improve rail and barge access and ensure efficient, safe, and resilient freight movement is key (*7*). From a business perspective, the number and location of freight facilities in the transportation network has a direct impact on the cost of the final product, and a positive effect on the ability of a region to attract industries and trigger economic growth (*8*).

The type and size of a freight-transfer facility suitable to a region depends on the characteristics of regional freight, which is shaped by shippers, the transportation network, and the type, quantity, and shipment distances of commodities (*9*). In particular, transload facilities are freight-transfer facilities where non-containerized freight is transferred from one mode to another. Transload facilities are not distribution centers, where goods brought in by truck are later shipped out by truck. Unlike intermodal facilities, which handle containerized cargo, transload facilities serve bulk, dimensional, equipment, or warehouse

[1]Department of Civil Engineering, University of Arkansas, Fayetteville, AR

Corresponding Author:
Address correspondence to Magdalena I. Asborno:
miasborn@email.uark.edu

products (*9, 10*). Examples of bulk products are grain, aggregate, and cement; dimensional goods include lumber, steel coils, and pipes; equipment includes farm and earthwork equipment; and warehouse goods include paper, plywood panels, or non-refrigerated/refrigerated foods (*9*). While feasible intermodal facilities require 100,000 carloads traveling 2,000 miles annually (*8*), feasible transload facilities operate at lower capacities of 1,500 annual carloads (*9*). Transload facilities, therefore, are more attractive to regions with relatively smaller amounts of freight and are the focus of this paper.

The decision-making process to find suitable locations for freight facilities first examines current and future needs, followed by network modeling, location screening, field validation, cost modeling, and site selection (*8*). The first step adopts a planning framework to identify how current and future needs can be addressed by the proposed project. In this context, private companies rely on experience, market knowledge, and proprietary business data to identify needs. For public agencies, this data is usually not available and other sources must be found.

Statewide travel demand models (STDMs) with freight components are one such source of facility planning data. Due to recent legislation such as the Moving Ahead for Progress in the 21st Century Act (MAP-21) and the Fixing America's Surface Transportation Act (FAST) (*9, 10*), many states have developed STDMs with freight components to evaluate infrastructure needs and identify future system deficiencies. There is valuable data within these models that can be mined for multimodal freight planning. This paper explores the use of STDMs to identify the main characteristics of regional freight, namely, commodity volume and origin–destination (OD) patterns. Using this data, this paper outlines a multi-criteria, sketch-planning framework to support the first step in state-level facility location planning. This approach is particularly useful to state agencies, as it leverages data that is already available through their STDMs. However, the framework is flexible as it could use alternative freight forecasting data sources. The framework evaluates and ranks the potential for each county within a state to serve as a location for a transload facility using four criteria:

(1) Interaction with the existing transportation network,
(2) Ton-distance transported along major shipping lanes,
(3) Spatial aggregation of commodities that use the major shipping lanes, and
(4) Origin–destination directionality aggregation (proximity of a county's markets).

A planning horizon of 20 years or more is utilized for facilities involving a significant investment in infrastructure (*8*), such as transload facilities. Consequently, this paper adopts 2040 as the future scenario.

The remainder of this paper is organized as follows. The background section presents commodities and distance thresholds for transload potential, and an introduction to STDMs and multi-criteria decision-making. An explanation of the framework methodology follows. Next, the framework is calibrated and validated using Arkansas data. Lastly, key findings from the case study are summarized.

Background

Commodities with Transload Potential

Of the many factors influencing freight mode choice, the physical attributes of goods and spatial distribution of shipments are central in determining if a commodity has potential to use a transload facility (*12*). To identify commodities with transload potential, data from the Freight Analysis Framework (FAF) is used to compare state mode shares with national averages. For commodities where the percentage of tons transported by truck is higher than the national average, an opportunity for mode shift may exist. However, this opportunity is not exhaustive. Where the percentage of tonnage transportation by barge and rail in a state is higher than the national average, there may still be opportunity to further increase barge and rail shares.

Mode share distributions by commodity for Arkansas for shipments made in 2015 are shown in Table 1 (*1*). For consistency, the standard classification of transported goods commodity codes used by FAF have been aggregated to the commodity groups adopted by the Arkansas STDM (AR-STDM) (*13, 14*). Table 1 shows that the share of chemicals moved by truck in Arkansas is higher than the national average (73.8% statewide versus 64.6% nationwide), while the share moved by water is 0.0% statewide and 8.7% national average. This indicates the potential to shift chemicals to water by providing transload opportunities in the counties with high chemical tonnage production and/or attraction.

Distance Threshold and Transload Potential

Building upon the factors that influence mode selection, total logistics cost per mode has a direct relationship with transportation distance (*12*). Thus, transport distance is important in determining the feasibility of a transload facility within a county. A review of previous studies revealed a wide range of distance thresholds at which a certain mode is more advantageous for a commodity. These thresholds are study specific, depending on product value and density (*15, 16*). As the scope of this paper covers a broad range of commodities, the

Table 1. Comparison of State and National Mode Shares by Commodity Type

Commodity (AR-STDM Categorization)		Statewide mode shares (% of tons, 2015) Truck	Rail	Water	National average mode shares (% of tons, 2015) Truck	Rail	Water
1	Farm products	86.0	12.0	1.9	82.4	14.8	2.8
2	Mining	87.9	6.3	5.9	50.3	24.5	25.1
3	Coal	1.1	98.9	0.0	25.2	68.1	6.8
4	Nonmetallic minerals	88.2	11.8	0.0	91.5	5.3	3.2
5	Food	92.8	5.7	1.6	91.5	7.5	1.1
6	Consumer manufacturing	89.8	1.0	9.2	97.5	1.4	1.1
7	Non-durable manufacturing	93.7	5.5	0.8	76.1	23.3	0.6
8	Lumber	97.0	3.0	0.1	95.8	3.9	0.3
9	Durable manufacturing	93.4	6.1	0.5	94.6	4.4	1.0
10	Paper	84.3	15.7	0.0	84.6	14.7	0.7
11	Chemicals	73.8	26.2	0.0	64.6	26.8	8.7
12	Petroleum	98.1	1.2	0.7	79.3	4.5	16.2
13	Clay, concrete, glass	97.8	2.0	0.2	95.3	3.7	1.0
14	Primary metal	58.8	32.8	8.3	84.8	13.5	1.7
15	Miscellaneous mixed	89.0	4.7	6.4	95.6	3.4	1.0

Note: Shaded row indicates commodity group with high transload potential (where for "Truck" the national average share is lower than the state share).

specific thresholds identified in studies such as (*15, 16*) may not be applicable. The FHWA provides a more general definition for short-haul freight as movements up to 200 miles (*15*), the distance under which truck and rail do not compete (*12*). Thus, a 200-mile distance threshold is adopted in this paper. A regionally-defined, commodity-specific, and mode-specific distance threshold could be adopted in the future.

Statewide Travel Demand Models

STDMs are used for long-term travel forecasting and mostly follow a sequential four-step approach: trip generation, trip distribution, mode choice, and route assignment (*17*). The zoning system for STDMs typically follows census boundaries aggregated to traffic analysis zones (TAZs) and/or counties. While zones within the state can be disaggregated to TAZs, external zones tend to be much larger areas consistent with the FAF zones. In most cases, statewide models have two distinct components, a passenger model and a freight model, which are combined before route assignment.

For the freight component, it is common to use commodity-based forecasting models (*18*). These are analogous to trip based passenger models with the replacement of person trips with commodity tons. Commodity data sources include publicly available data like the Commodity Flow Survey (CFS) or FAF, and/or proprietary sources such as Transearch (*19*). Freight forecasting by commodity occurs in the first step of the model (trip generation) by estimating production and attraction of freight (in annual tons) within each zone. Annual tonnage by commodity is then distributed across origin–destination (OD) pairs. Next, annual tonnages by OD pair are disaggregated by mode, generating OD matrices with annual tonnages of freight per mode, per commodity. STDMs may include truck, rail, water, air, and intermodal (a combination of modes, not containerized transport) modes.

This paper uses STDM commodity OD matrices to obtain tonnages by OD and commodity for the base and forecast years. The use of STDM data to locate potential freight facilities is beneficial for state planners as it leverages an existing resource and requires no additional investment in proprietary data or surveys.

Multi-Criteria Decision-Making

Since this paper focuses on prioritization or ordering of the full set of location alternatives, we opted to adopt a multi-criteria decision-making (MCDM) framework instead of an optimization method. The objective of optimization methods is to reach an absolute solution to a problem. Depending on the optimization method, the solution can be unique, or it can consist of a set of Pareto-optimal solutions (as in the case of multi-objective optimization). Pareto-optimal solutions are equally desirable, since the full set of alternatives cannot be uniquely ordered (*20*). MCDM is used to prioritize differing and conflicting alternatives, with the objective to identify the best choice among them, based on multiple criteria being evaluated (*21*). Transportation investment alternatives have been evaluated using multi-criteria analysis (*22, 23*).

Criteria can be quantitative or qualitative measures converted to a quantifiable scale. Each criterion is combined to produce a "score" for that alternative. The combination can include equal ranking of criteria or a weighting scheme. Since the relative importance of conflicting criteria depends on the specific context of the alternatives being ranked, researchers have introduced weighting scenarios (*24, 25*). Chow et al. (*21*) applied MCDM to evaluate feasible and sustainable transit options using criteria weights obtained through stakeholder interviews. Conversely, Nijkamp et al. (*26*) created a MCDM framework for land use and transport modeling without criterion weights. For this paper, which is general as it considers all commodities and is applicable to differing regions and users, a systematic weighting system is not adopted. However, this is a topic for future exploration.

A benefit of MCDM is that it combines graphical tools with an analytical framework to appeal to decision-makers of varied backgrounds. For visualization, each independent criterion and the combined multi-criteria score can be shown on a series of thematic maps (Figures 2 and 3), and in radial graphs (Figure 4). Visualizations help decision-makers locate and contextualize each proposed location against existing facilities and the transportation network.

Methodology

According to NCFRP Report 13, the essential element when planning the location of a freight facility is the ability to access key markets, which is intrinsically related to the connectivity of the transportation network (*8*). Thus, the MCDM framework includes transportation-related factors that positively contribute to market accessibility. These criteria are:

1. **Interaction (I_c):** Interaction with transportation network and mode options;
2. **Ton-distance (T_c):** Ton-distance transported between major freight routes;
3. **Spatial Aggregation (A_c):** Spatial aggregation potential of commodities to share one facility; and
4. **Directionality (D_c):** Origin–destination directionality, evaluating the proximity of markets served from each county.

Freight traveling distances shorter than 200 miles or passing-through is not considered in the calculations, since it is unlikely to use a transload facility within the state. Each criterion receives a score between 0 and 1. The scoring for all four categories is combined for each alternative facility location,

$$S_c = a \times (I_c + T_c + A_c + D_c) \qquad (1)$$

where $a = \begin{cases} 1 \text{ if } I_c \neq 0 \\ 0 \text{ if } I_c = 0 \end{cases}$

S_c = overall framework scoring of county c
I_c = Interaction scoring of county c
T_c = Ton-distance scoring of county c
A_c = Spatial Aggregation scoring of county c
D_c = Directionality scoring of county c

Each county serves as an alternative location for a potential transload facility. Counties are prioritized from highest to lowest S_c. Counties with higher scores have a higher priority to locate a transload facility. Evaluation is made at county level primarily because this framework serves as a sketch-planning tool for initial stages of decision-making. In later stages, further spatial disaggregation is possible given data availability. Secondly, required trip distribution data per commodity obtained from STDMs may not be available at spatial levels finer than counties. The framework flowchart is shown in Figure 1. Each criterion is described below.

Criterion 1. Interaction with Transportation Network (I_c)

I_c captures access to the railways, navigable waterways, and major highways within a county:

$$I_c = \begin{cases} w \times W_c + r \times R_c + h \times H_c \\ 0 \quad \text{if } W_c = 0 \text{ and } R_c = 0 \end{cases} \qquad (2)$$

where I_c = Interaction scoring of county c
W_c = 1 if county c contains navigable waterways and 0 otherwise
R_c = 1 if county c contains railways and 0 otherwise
H_c = 1 if county c contains major highways and 0 otherwise

Counties with all modes present have a maximum score of 1.00. This framework only measures the presence of a mode and not access to that mode, for example, there does not have to be a port or railyard in the county to receive a positive score, because the transload facility serves as the modal access.

The weights w, r, and h for water, rail, and highway access, respectively, represent the relative importance of each mode. These weights are determined during framework calibration. Scores of I_c = 0 and S_c = 0 are assigned to counties that do not have railways or waterways, since those counties are not suitable to house a transload facility. "Major Highway" refers to roadways categorized as Interstates or Other Freeways & Expressways by the FHWA (functional class 1 or 2) (*27*). It is assumed that the external zone trip-end has access to the same modes to facilitate (un)loading.

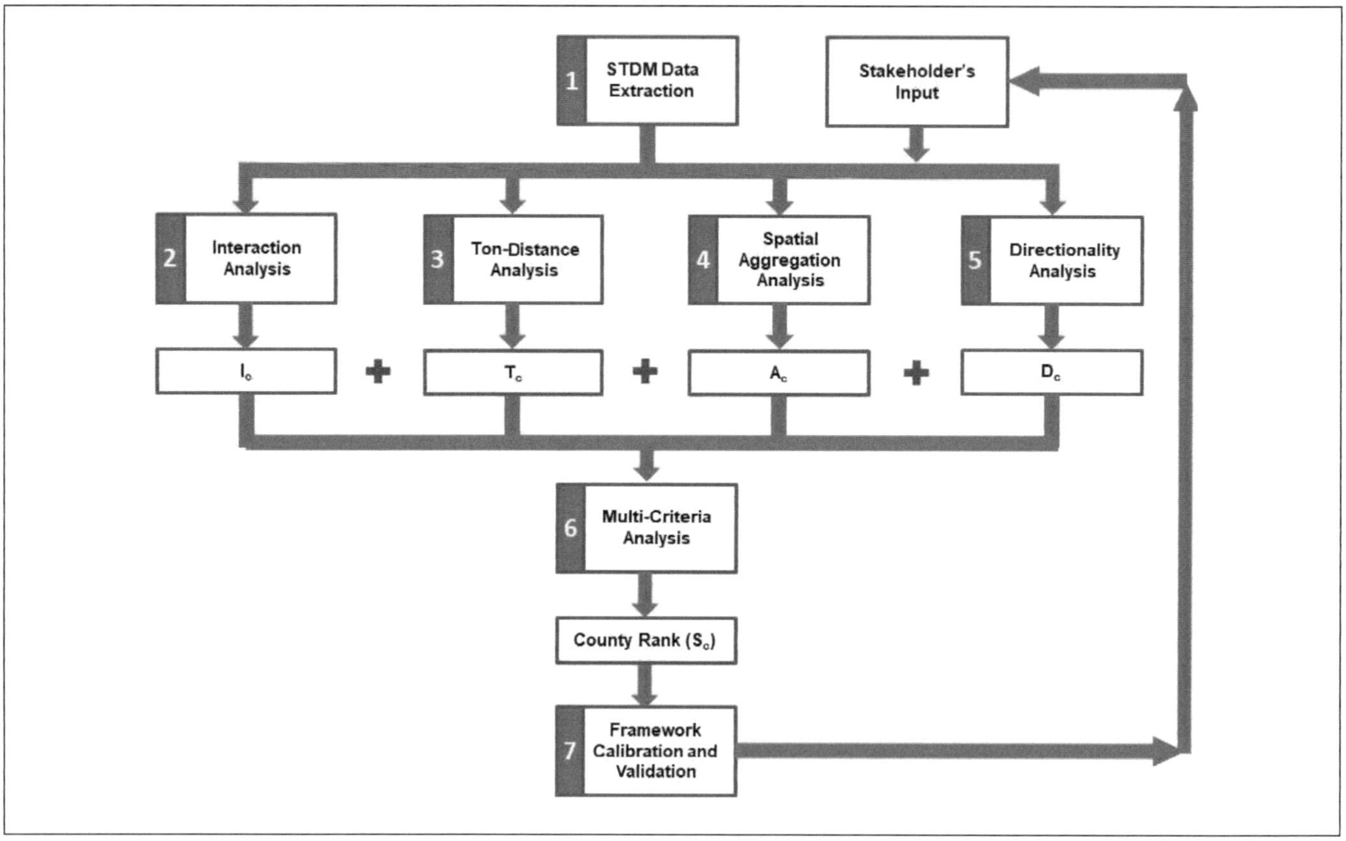

Figure 1. Framework flowchart.

Criterion 2. Ton-distance Transported between Major Freight Routes (T_c)

For T_c, the following are extracted from the STDM: distances between each OD pair $\left(d^{OD}\right)$, the annual tons (t) by truck (tr) for the base year (b) for each commodity (g) for each OD pair $\left(t^{OD}_{tr,b,g}\right)$, and the difference in annual tons between the forecast year and the base year (f) by all modes (a) for each commodity for each OD pair $\left(t^{OD}_{a,f,g}\right)$. Data from the base year captures existing conditions while data from the forecast year considers anticipated facility capacity needs over the planning horizon. The premise of this study is that non-truck modes are more efficient and thus a potential facility should be planned to encourage use of non-truck modes. Therefore, for the base year, only freight currently shipped by truck would potentially use a transload facility to shift commodities to rail or water. For the future scenario, we consider the forecasted total tonnage by all modes to capture the possibility of expanding existing facilities that already serve as transload from truck to rail or water. The use of forecasted tonnage by all modes, which is produced by trip generation, the first step of the STDM (and not by the mode choice model), alleviates the concern that a transload decision can affect the forecasted mode-split in a county. Forecasted tonnages, then, are not contingent on future mode decisions. Base and future year tonnages are combined to calculate total tonnage of commodity flow between each OD pair, for example, $t^{OD}_{g} = t^{OD}_{tr,b,g} + t^{OD}_{a,f,g}$.

After calculating the ton-distance for each OD pair for each commodity $\left(m^{OD}_{g} = d^{OD} \times t^{OD}_{g}\right)$, the top commodity OD pair (OD') based on ton-distance is identified for each county (c) $\left(M^{OD'}_{c,g}\right)$. Next, the top-$p$ counties are identified by ranking $M^{OD'}_{c,g}$. Focusing on the counties that have highest ton-distance for a single commodity transported between a single OD pair captures the key markets corresponding to each state. The value of p is determined through calibration. The county with maximum $M^{OD'}_{c,g}$ receives the highest possible score for this category, 1.00. The remaining p-1 counties are scored as per their ton-distance, normalized by the maximum. Counties outside the top-p ranking receive a score of zero. T_c scoring is shown in the following equation:

$$T_c = \begin{cases} \frac{M_{c,g}^{OD'}}{\max_{i=1...C}(M_{i,g}^{OD'})} & \text{for } rank_{i=1...C}\left(M_{c,g}^{OD'}\right) \geq p \\ 0 & \text{otherwise} \end{cases} \quad (3)$$

where T_c = Ton-distance scoring for county c; and

$M_{c,g}^{OD'}$ = highest ton-distance for county c, serving OD pair OD' and commodity g, for example, $\max_{j \in N}\left(m_{c,j}^{OD'}\right)$ where N is the set of OD'-g combinations.

Criterion 3. Spatial Aggregation Potential (A_c)

A_c accounts for multiple commodities sharing the same transload facility. The ability to serve more than one commodity along the same shipping lane provides flexibility and minimizes business risks. The spatial aggregation potential of a county is measured as the total ton-distance of all commodities transported between its main OD pair. A county's main OD pair is the OD pair with maximum total ton-distance of all commodities ($\sum_{j=1}^{G} m_{c,j}^{OD'}$). The county with maximum total ton-distance is given a score of 1.00; the scores of all other counties are normalized by the maximum:

$$A_c = \frac{\max_{OD'} \sum_{j=1}^{G} m_{c,j}^{OD'}}{\max_{C}\left(\max_{OD'} \sum_{j=1}^{G} m_{c,j}^{OD'}\right)} \quad (4)$$

where A_c = Spatial Aggregation scoring for county c,

$m_{C,j}^{OD'}$ = ton-distance for OD pair OD', with origin or destination in county c, and all commodities ($j = 1 \ldots G$).

Criterion 4. Origin–Destination Directionality Aggregation (D_c)

D_c measures the proximity of markets served by the same county, capturing the ability of a facility to combine shipments orientated along the same shipment lane. For this paper, shipment lane is loosely defined as the inter-cardinal direction of the OD pair. Directionality is defined as the angle between the line connecting an OD pair and due north measured clockwise from north (i.e., the azimuth). Directionality is aggregated to 45-degree sections centered in the eight inter-cardinal directions (i.e., NNE, NEE, SEE, etc.). Aggregation is necessary due to the abstract nature of the zoning system used in STDMs, where multi-region external zones are represented by centroids resulting in OD directions that are quite abstract.

For each county, the standard deviation of the tons transported within 45-degree regions is used to capture directional aggregation potential. Higher standard deviation means that most of the tons are traveling to/from similar directions, for example, the narrowest angle from/to a county. Thus, the county with highest standard deviation receives the highest score, 1.00; while the county with lowest standard deviation receives the lowest score, 0.00. All remaining counties are scored as per the relative standard deviation according to

$$D_c = 1 - \frac{\max_{i=1...C}(s_i) - (s_c)}{\max_{i=1...C}(s_i) - \min_{i=1...C}(s_i)} \quad (5)$$

where D_c = Directionality scoring of county c,

s_c = standard deviation of $X_{c,k}$, calculated according to Equation 7,

$X_{c,k}$ = freight (in weight) transported within each 45-degree segment, to/from county c, calculated according to Equation 7.

$$s_c = \sqrt{\frac{\sum_{k=1}^{N}\left(X_{c,k} - \overline{X_{c,k}}\right)^2}{N}} \quad (6)$$

$$X_{c,k} = \sum_{k}^{k+1} t_{c,k}^{OD} \quad (7)$$

where k represents each of the 45-degree directionality segments, $k = 1 \ldots 8$

$t_{c,k}^{OD}$ = freight (in weight) of all commodities transported between OD pair OD, to/from county c.

Case Study: Arkansas

This section presents the results of applying the framework to Arkansas, and a discussion of the results.

Study Area and Data

Arkansas has 2,662 miles of active rail lines, 1,683 miles of which are Class I railroads and the remainder are 23 short-line, Class III railroads (*28*). There are five commercially navigable rivers in Arkansas totaling 1,000 miles (*29*). The Mississippi river constitutes the state's eastern boundary. Arkansas highway network consists of 16,444 miles of state highways, 8,447 of which constitute the Arkansas Primary Highway Network. Figure 5 depicts Arkansas's multimodal transportation network and existing transload facilities.

The freight component of the AR-STDM is a direct commodity, four-step model based on Transearch data (*13*). Notably, Transearch data has been used to assess the potential for modal substitution in previous studies (*30*). Commodities are aggregated into 15 groups (Table 1). Although freight trip generation is estimated at the TAZ level in the AR-STDM, trip distribution is estimated at the county level (*31*). Therefore, the case study is applied at county level to OD pairs consisting of 75 counties and 481 external FAF zones (*32*).

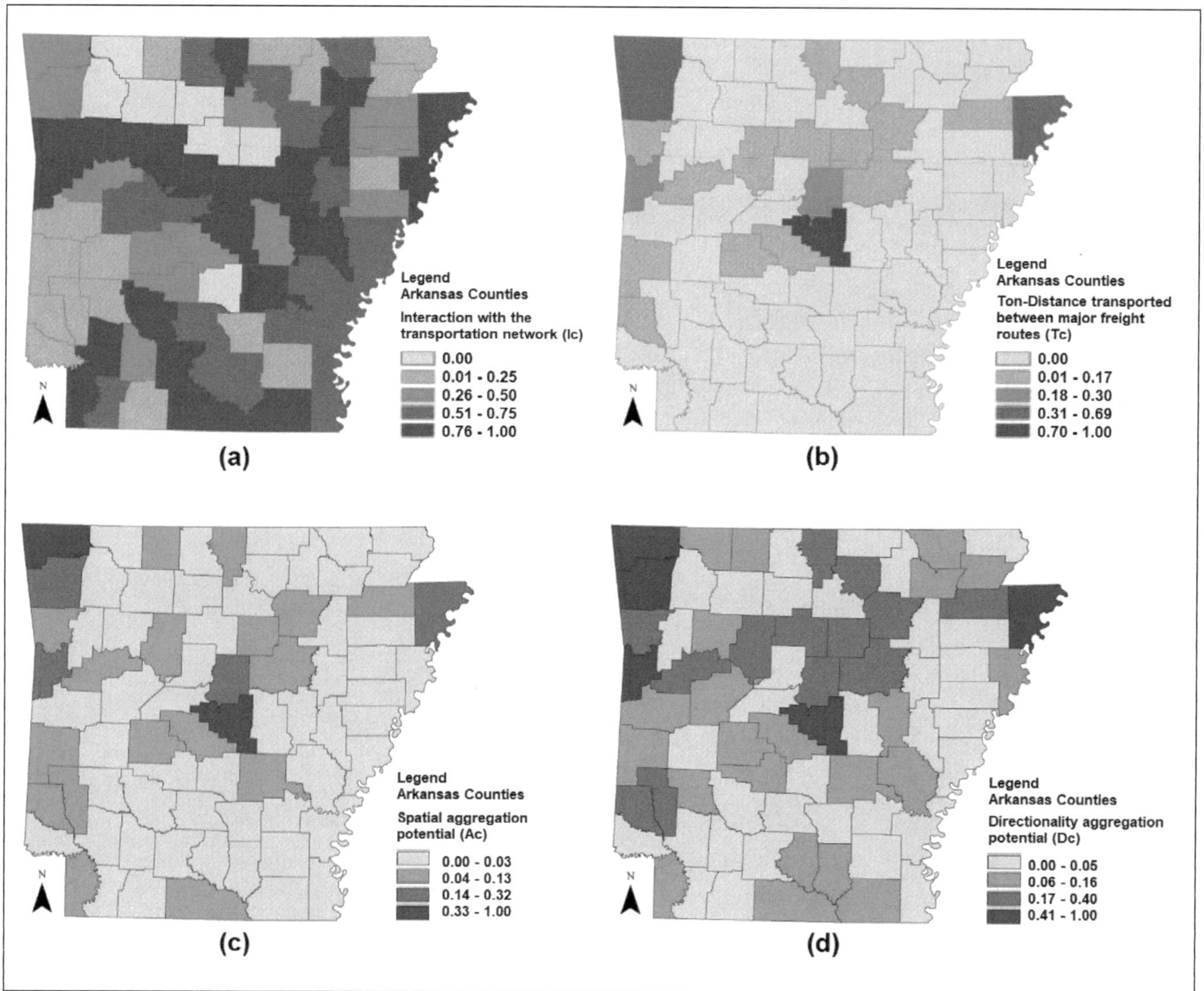

Figure 2. Intermediate results of framework scoring: (a) Interaction (I_c); (b) Ton-distance (T_c); (c) Spatial aggregation (A_c); (d) Directionality (D_c).

Based on stakeholder input and the AR-STDM model output, shipments of coal to Arkansas serve power-generation plants, and are already moved exclusively by rail. Thus, we exclude coal from the set of commodity groups, since there is no significant potential for trans-loading multiple commodity types or for serving multiple customers.

Framework Calibration

The weights w, r, and h of water, rail, and highway access, respectively, are calculated for the case study using the 2010 data. First, the weights are initialized to $1/n$, where n is the number of modes. For calibration, the framework is evaluated with these values and the relative ranking of each county is compared with the location of existing transload facilities in the state. We observe that counties near navigable rivers have transload facilities, but score low on I_c and S_c. Thus, weighting factors w, r, and h are modified accordingly, and the model is recalibrated. This iterative process is repeated until changes in weighting factors do not improve the relative ranking of each county compared with the location of existing facilities. The weighting factors adopted for this case study are

$$w = 0.50 \quad r = 0.25 \quad h = 0.25$$

Similarly, the relative ranking position (p) used to define the top OD pairs by ton-distance used in Criterion

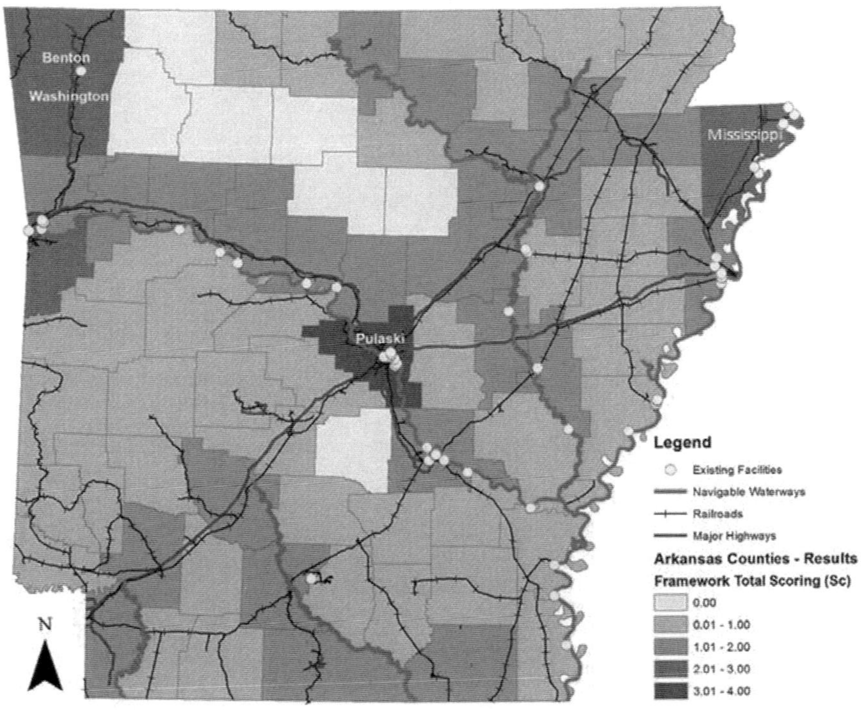

Figure 3. Final results of framework scoring.

2 requires calibration. Through the same process of comparing current locations with those resulting from the framework, a value of $p = 20$ is found to be appropriate for this case study.

Results

The Arkansas counties with the highest scores are Pulaski, Mississippi, and Benton. Based on actual and future freight scenarios, these counties are the strongest candidates to locate a freight transload site to facilitate access to key markets. Figure 2 shows county scores for each of the four criteria. The total score of each county is shown in Figure 3. Considering the total scoring range, 47 counties are in the lowest quartile (scores, 0.00–0.99), 23 counties are in the second quartile (1.00–1.99), 4 counties are in the third quartile (2.00–2.99), and only one county is in the highest quartile (3.00–4.00). Figure 4 depicts the radial graph for select counties. The size of the diamond represents the relative ranking of the county based on the four criteria. This provides decision-makers with easily digestible comparisons of transload potential among counties.

Framework Validation

The methodology is validated by applying the framework to the AR-STDM 2010 scenario. Counties with transload potential, as defined by the framework, are compared with the location of existing freight transload facilities (*33*). The percentage of existing facilities within the upper quartile of ranked counties is used as the performance metric. The validation results indicate that 65% of existing facilities are located within the upper quartile of ranked counties. In addition, the highest-scoring county is the one with the highest number of existing facilities (Pulaski), and the 30 lowest ranked counties do not house any facility. Results are shown in Figure 5.

Figure 5 shows that a few counties with low scores (i.e., not expected to house transload facilities) currently have transload facilities. Most of these facilities are located on rivers along the state boundary. Phillips County has six existing freight facilities, while its overall scoring is relatively low. This could be explained by recent efforts made by Phillips to attract new business, which include incentives that resulted in the development of Helena Harbor (*34–36*). The strategic location of this port serves a region that spans Arkansas and Mississippi. Since the proposed approach only uses data pertaining to shipments with origin or destination in Arkansas, we are unable to capture tonnage originating in a bordering state that is drayed to Arkansas for a transload. An analysis considering data from adjacent states would be necessary to capture this dynamic.

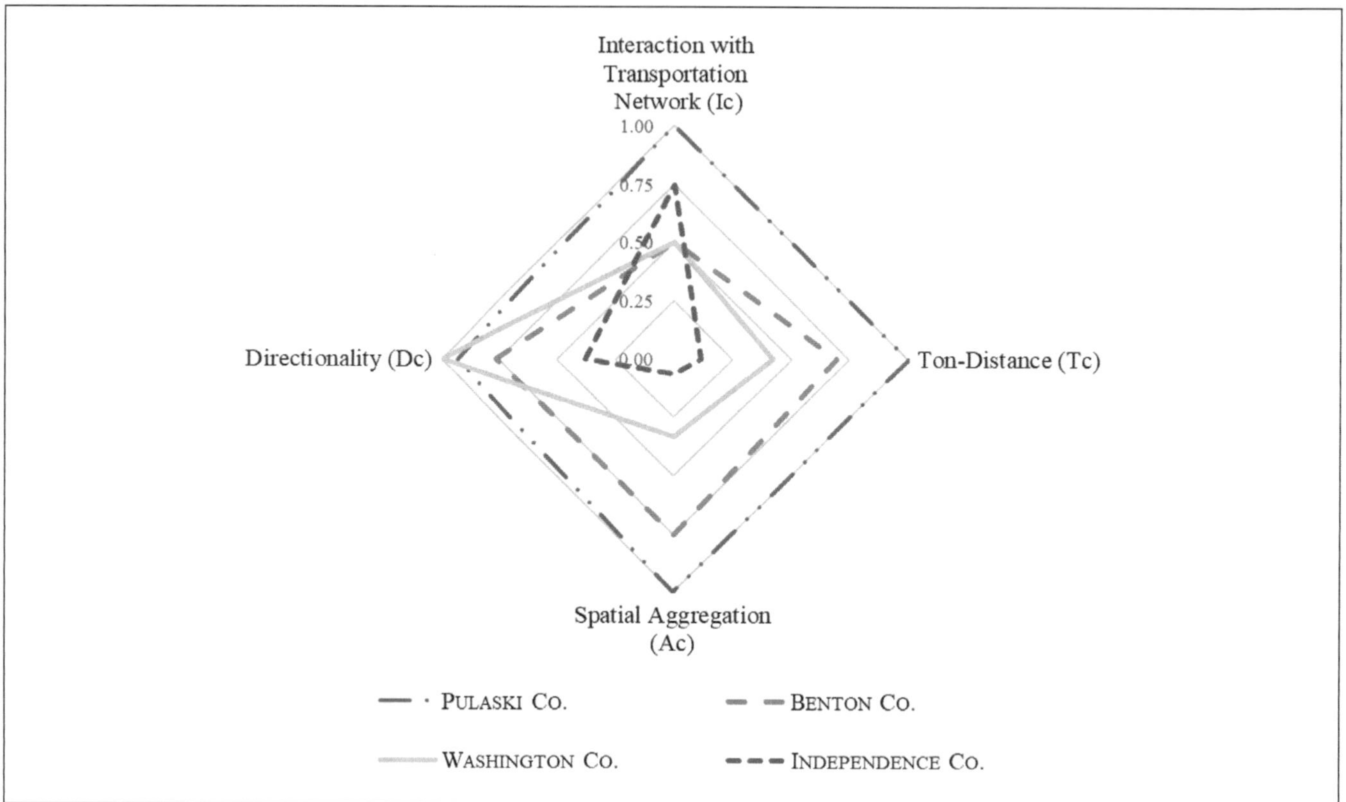

Figure 4. MCDM radial graph for location of transload facilities.

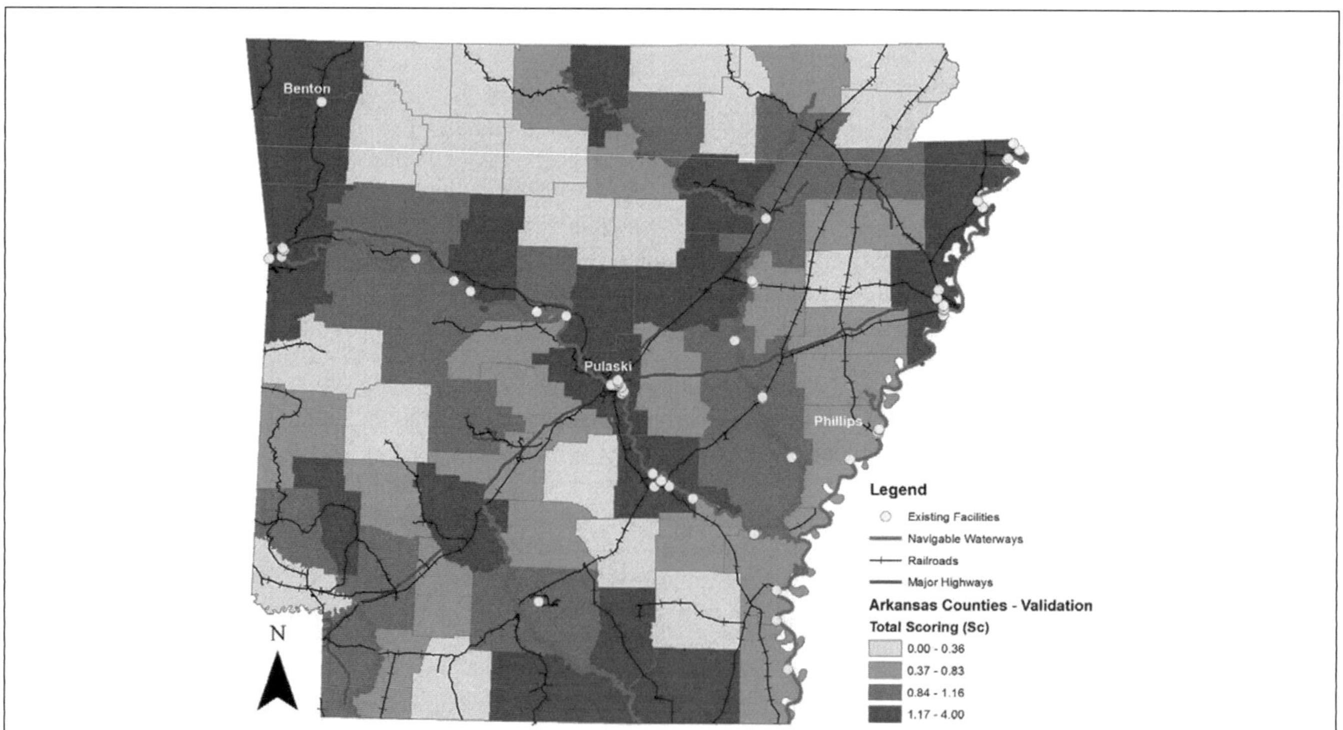

Figure 5. Framework validation results.

Discussion of Case Study Findings

The framework scoring results prioritize Arkansas counties as per their relative suitability to host a freight transload facility. There is only one county, Pulaski, ranking in the highest quartile (score 3.00–4.00). As noted in Figure 3, Pulaski is centrally located in Arkansas, and has access to multimodal transportation, allowing many shipping lanes to be served. In contrast, counties without access to the multimodal transportation network receive the lowest possible score, and are not suitable to house a facility.

To demonstrate the effect of the assumed distance threshold (e.g., 200 miles) on the location ranking, a sensitivity analysis was performed. Results indicate no difference in the ranking of facility locations given a 200-mile or 500-mile threshold assumption. The sensitivity analysis demonstrates the robustness of the model for the range of distance thresholds identified in previous studies, likely due to the minimal differences in mode share that exist below 500 miles (*1*).

When analyzing the criterion scores (Figure 2), we observe that the counties that do not have access to a multimodal transportation network score very low in other criteria. This may indicate that investing in transportation infrastructure in these counties (to support potential future investment in a freight transload facility) may not be the best choice, unless it is accompanied by a comprehensive set of policies to generate regional economic development. We also note that the ranking according to the interaction criteria alone results in a very different set of counties than when all criteria are considered. This exemplifies the necessity to apply a multi-criteria analysis. If only one criteria had been considered, important conflicting alternatives may have been left out. For example, if interaction with transportation network had been the only criteria considered, Benton County would rank 24th instead of 2nd.

Some limitations of this framework are exposed by the validation exercise. The most notable may be the relatively poor scoring of counties located along state boundaries, which may be handling freight from adjacent states. This limitation may arise because the STDM data does not consider the movement of freight that does not have an origin or destination in the state. Freight with its ultimate origin or destination in a bordering state may be drayed by truck across the border to a transload facility in the state of interest. Data from a national (e.g., FAF) or sub-national freight model could be implemented into the framework to account for this limitation. This "bordering" effect could be an added opportunity for a facility because of the potential to attract out-of-state customers. For example, Benton, Washington, Ashley, and Union counties are located along the state boundary, are connected by waterway and/or railway, and are currently served by only one or no facilities. The combination of all these factors is easily visualized for decision-makers in a single image (Figure 3), an advantage of the MCDM framework.

Moreover, Benton and Washington's proximity and conditions (as evidenced by the framework results) indicate that there is a strong case to combine the freight of both counties by strategically siting a single facility. This case is further supported by the radial graph (Figure 4) showing that the spatial aggregation of Benton and the directionality potential of Washington are complementary, making the location of a transload facility in northwest Arkansas even more attractive than if each of these two counties were considered individually.

The results obtained are highly dependent on the STDM. A limitation of this framework is the commodity aggregation defined by the AR-STDM. If disaggregated commodity data were available (tons of commodities transported between the modeled OD pairs, at STCC two-digit level, indicating the origin and destination zone in each case), a more detailed analysis for specific commodities could be made. Further, a sound freight model (i.e., a model that is based on representative data, considers proper spatial generators, is subject to calibration and validation, and contains justifiable predictions of future freight flows) would lead to a more realistic prioritization of potential freight facility locations. In particular, the use of Transearch as the data source for the STDM imposes limitations. These limitations occur because of the way Transearch handles shipments from external zones with intermediate stops at warehouses or distribution centers located out-of-state. Transearch records the first portion of the trip (from the origin to the distribution center) in the national database, while the second portion of the trip (from the distribution center to its destination in Arkansas) is recorded individually in the state database (*12*). This makes it impossible to identify the actual origin of some commodity flows. Transearch reports all movements made from distribution centers as a unique STCC commodity code (*13*), which the AR-STDM incorporates into a single commodity group. Lastly, many commodities are not reported in Transearch, such as construction, retail, refined petroleum, municipal solid waste and farm-based agriculture shipments (*13*), some of which may have transload potential.

Conclusions

STDM data is used in this paper to prioritize counties as per their suitability to house a freight transload facility, based on four criteria: (1) interaction with transportation network, (2) ton-distance of freight transported between major OD pairs, (3) spatial aggregation of commodities potential, and (4) OD directionality aggregation. In this context, this paper makes the following contributions: (a) an original transferable framework, (b) a data-driven method to rank and prioritize facility locations from a statewide perspective, and (c) the presentation of GIS visualization tools.

The proposed framework serves as a flexible and transferable tool for state planning officials by leveraging resources already available to most planning agencies (e.g., STDMs or, more generally, freight tonnage forecasts). The framework is flexible in that it can be adapted to specific commodities, perhaps those that comprise a key industry for a state. The framework could also be implemented using forecast data other than the STDM. For example, private companies could apply the framework to business operations data. Moreover, the framework could be generalized to include seaports and airports, provided that additional data on the amount and type of commodities shipped is available.

The driver for this work is the need to examine, from a statewide level, where freight-transfer facilities might have the most impact. Almost every region in the state has an intermodal freight authority or economic development agency that recommends its region as the best place to locate new facilities. The state Department of Transportation needs a way to determine which location would be feasible. This study provides a data-driven method to rank and prioritize facilities. For example, the case study indicated three of the 75 counties in which a transload facility would be preferable. These locations could be incorporated into planning considerations by regional transportation agencies, economic development commissions, or private companies.

The results of the case study are presented in radial graphs and GIS maps to help decision-makers contextualize, compare, and contrast alternatives considered in the MCDM framework. This paper presents thematic GIS maps overlapping the framework scoring results with the location of existing facilities and the transportation network. Overall framework and individual category scorings are presented. A simple observation of these images revealed which pairs of adjacent counties would be better candidates to house freight facility locations when considered together, rather than individually.

The sketch-planning framework presented here supports and justifies decision-making processes for the location of freight facilities at the early planning stages. Further steps to identify specific TAZs and sites within a county are being developed. The framework could be adapted to incorporate factors relevant to advanced stages of decision-making, such as conditions of labor and the employment market, overall regional cost, local regulations and permits, taxation, and incentives. Next, real estate tools would be used to evaluate and compare site-specific characteristics such as site suitability, utilities, and cost. To improve this framework further, methods of determining weighting factors for each criterion could be explored, and the list of commodities could be refined according to their transload potential.

Acknowledgments

The authors thank Andrew Braham and John Kent for their assistance, and the Arkansas Department of Transportation for sponsoring this paper under the "Locating Transload Facilities to Ease Highway Congestion and Safeguard the Environment" project.

References

1. Federal Highway Administration. *Freight Analysis Framework Data Tabulation Tool (FAF4)*. U.S. Department of Transportation, Washington, D.C. http://faf.ornl.gov/fafweb/Extraction1.aspx. Accessed June 06, 2017.
2. The Tioga Group. *Goods Movement Truck and Rail Study*. Project 99-130. Southern California Association of Governments, Los Angeles, California, 2003.
3. Rodrigue, J.P., C. Comtois, and B. Slack. *The Geography of Transport Systems, 4th ed.* https://people.hofstra.edu/geotrans/eng/ch3en/conc3en/ch3c1en.html. Accessed November 11, 2017.
4. Bhamidipati, C. and M. Demetsky. Evaluation of Impacts of Intermodal Terminals on the Highway System. *Transportation Research Record: Journal of the Transportation Research Board*, 2009. 2097: 97–108.
5. Bryan, J., G. Weisbrod, and C.D. Martland. Rail Freight as a Means of Reducing Roadway Congestion: Feasibility Considerations for Transportation Planning. *Transportation Research Record: Journal of the Transportation Research Board*, 2007. 2008: 75–83.
6. Nachtmann, H., O. Boudhoum, and F. Oztanriseven. *Regional Economic Impact Study for the McClellan Kerr Arkansas River Navigation System*. Maritime Transportation Research and Education Center, Mack-Blackwell Transportation Center, Fayetteville, AR, 2015.
7. National Freight Strategic Plan. *Draft for Public Comment*. U.S. Department of Transportation, Washington, D.C. https://www.transportation.gov/freight/NFSP. Accessed July 7, 2017.
8. Steele, C. W. and D. Hodge. *NCFRP Report 13: Background Research Material for Freight Facility Location Selection: A Guide for Public Officials*. Transportation Research Board of the National Academies, Washington, D.C., 2011.
9. Thompson, D. Transloads: Freight Movement Efficiencies in the Next Decade. Presented at the ASME/ASCE/IEEE 2012 Joint Rail Conference, Philadelphia, Pennsylvania, April 17–19, 2012.
10. Federal Highway Administration. *Moving Ahead for Progress in the 21st Century Act (MAP-21). A Summary of Highway Provisions*. U.S. Department of Transportation, Washington, D.C. https://www.fhwa.dot.gov/MAP21/summaryinfo.cfm. Accessed July 2, 2017.

11. Federal Highway Administration. *Fixing America's Surface Transportation Act. A Summary of Highway Provisions*. U.S. Department of Transportation, Washington, D.C. https://www.fhwa.dot.gov/fastact/summary.cfm. Accessed July 2, 2017.
12. The Center for Urban Transportation Research at the University of South Florida. *Analysis of Freight Movement Mode Choice Factors*. Florida Department of Transportation, Tallahassee, FL, 2002.
13. Alliance Transportation Group and Cambridge Systematics. *Arkansas Statewide Travel Demand Model Phase I: Freight Model*. Arkansas Department of Transportation, Little Rock, AR, 2011.
14. Horowitz, A., C. Cohen, and R. Pendyala. *NCHRP Report 606: Forecasting Statewide Freight Toolkit*. Transportation Research Board of the National Academies, Washington, D.C., 2008.
15. Federal Highway Administration. *Comprehensive Truck Size and Weight Study. Volume II*. U.S. Department of Transportation, Washington, D.C., 2000.
16. Gonzales, D., E.M. Searcy, and S.D. Eksioglu. *Cost Analysis for High-Volume and Long-Haul Transportation of Densified Biomass Feedstock. Transportation Research Part A: Policy and Practice*, Vol. 49, 2013, pp. 48–61.
17. Ortúzar, J.dD., and L.G. Willumsen. *Modelling Transport*, 4th ed. John Wiley Sons, West Sussex, U.K., 2011.
18. Cambridge Systematics and GeoStats. *NCFRP Report 8: Freight-Demand Modeling to Support Public-Sector Decision Making*. Transportation Research Board of the National Academies, Washington, D.C., 2010.
19. Horowitz, A. *NCHRP Synthesis 358: Statewide Travel Forecasting Models*. Transportation Research Board of the National Academies, Washington, D.C., 2006.
20. Branke, J., K. Deb, K. Miettinen, and R. Słowiński. *Multiobjective Optimization: Interactive and Evolutionary Approaches*. Springer-Verlag Berlin, Heidelberg, 2008.
21. Chow, Hernandez, Bhagat, and McNally. Multicriteria Sustainability Assessment in Transport Planning for Recreational Travel. *International Journal of Sustainable Transportation*, Vol. 8, No. 2, 2013, pp. 151–175.
22. Jakimavičius, , and Burinskiene. Assessment of Vilnius City Development Scenarios Based on Transport System Modelling and Multicriteria Analysis. *Journal of Civil Engineering and Management*, Vol. 15, No. 4, 2009, pp. 361–368.
23. Tsambouolas, Yiotis, and Mikroudis. A Method for Multicriteria Analysis in Transportation Infrastructure Investments. *International Journal of Transport Economics*, Vol. 34, No. 1, 2007, pp. 113–131.
24. Guy and Urli. Port Selection and Multicriteria Analysis: An Application to the Montreal-New York Alternative. *Maritime Economics & Logistics*, Vol. 8, No. 2, 2006, pp. 169–186.
25. Tsamboulas, D.A. and P. Moraiti. Identification of Potential Target Locations and Attractiveness Assessment Due to Terrorism in the Freight Transport. *Journal of Transportation Security*, Vol. 1, No. 3, 2008, pp. 189–207.
26. Nijkamp, P., M.T. Borzacchiello, B. Ciuffo, and F. Torrieri. Sustainable Urban Land Use and Transportation Planning: A Cognitive Decision Support System for the Naples Metropolitan Area. *International Journal of Sustainable Transportation*, Vol. 1, No. 2, 2007, pp. 91–114.
27. Federal Highway Administration. *Highway Functional Classification Concepts, Criteria and Procedures*. U.S. Department of Transportation, Washington, D.C., 2013.
28. *2015 Arkansas State Rail Plan. Final Report*. Arkansas Department of Transportation, Little Rock, AR, 2016.
29. Arkansas Waterways Commission. *Rivers*. State of Arkansas, Little Rock, AR. http://www.waterways.arkansas.gov/rivers. Accessed June 9, 2017.
30. Aultman–Hall, L., B. Johnson, and B. Aldridge. Assessing Potential for Modal Substitution from Statewide Freight Commodity Flow Data. *Transportation Research Record: Journal of the Transportation Research Board,* 2000. 1719: 10–16.
31. Alliance Transportation Group and Cambridge Systematics. *Arkansas Statewide Travel Demand Model Phase II: Freight Model*. Arkansas Department of Transportation, 2015.
32. Alliance Transportation Group and Cambridge Systematics. *Arkansas Statewide Travel Demand Model Phase II: Zone Structure Report*. Arkansas Department of Transportation, 2015.
33. Smith, S., S. Hernandez, and A. Braham. Developing a Statewide Transload Facility Inventory: Challenges and Lessons Learned Presented at 96th Annual Meeting of the Transportation Research Board, Washington, D.C., 2017.
34. *Strategic Community plan 2005-2010*. Phillips County, Arkansas, 2005.
35. *Phillips County and Your Business*. Phillips County Chamber of Commerce. https://www.phillipscountychamber.org/your-business/. Accessed July 19, 2017
36. *Helena Harbor*. http://helenaharbor.com. Accessed July 19, 2017.

The Standing Committee on Intermodal Freight Transport (AT045) peer-reviewed this paper (18-02670).

Article

TRR
JOURNAL OF THE TRANSPORTATION RESEARCH BOARD

Transportation Research Record
2018, Vol. 2672(9) 24–34

DOI: 10.1177/0361198118774690
journals.sagepub.com/home/trr
SAGE

Development of a Cost Estimation Framework for Potential Transload Facilities

Sadie Smith[1], Andrew Braham[1], Sarah Hernandez[1], and John Kent[2]

Abstract

As the cost of transportation continues to rise and there is a growing push for a more environmentally friendly transportation network, optimizing mode distributions becomes an attractive solution. One way to optimize mode distributions is through strategically located freight transfer facilities, such as transload facilities. While there are many benefits to this type of facility, such as the emissions savings or pavement damage reductions garnered by shifting commodity tonnage onto alternative modes of transport, it is also essential that transportation planners understand the costs associated with building transload facilities. Unfortunately, literature does not provide an adequately disaggregated and scalable cost estimation approach that could be applied to various configurations and types of transload facilities. In this research, a cost estimation framework was created to determine the basic cost of transload facilities by type using unit costs from a construction cost database, equipment costs from local dealers, the projected commodity tonnage, design recommendations from literature, and survey responses from local facilities. A case study based upon proposed facilities in Arkansas was completed to illustrate the effectiveness of this methodology. While there is currently no construction design for these facilities, this framework yielded costs consistent with those expected. A key finding was that storage costs could account for up to 81% of a transload facility's costs. Overall, this cost framework is believed to balance general scalability with accuracy well to provide reasonable cost estimations for constructing new or expanded facilities.

As freight transportation costs continue to increase and the demand for environmentally friendly transportation options grows, transportation logistics and planning agencies have sought innovative solutions. Shifting freight away from trucks to more efficient modes of transportation addresses the issues of increasing fuel prices and higher costs of maintaining infrastructure, as well as concerns regarding environmental degradation. Barges and trains are often more cost-effective for long haul shipments and are responsible for fewer emissions and less damage to highway infrastructure as freight is moved off roadways onto railroads and rivers (*1*, *2*); however, trucking remains the dominate mode due to high accessibility. Therefore, to incentivize efficient mode choices, conveniently located freight transfer facilities, including intermodal rail (container on freight car, COFC) and barge (container on barge, COB) terminals, marine ports, or bulk-transfer (transload) facilities, are necessary. In providing these facilities and offering a wide array of transportation options, economic development programs aimed at attracting new industries to a state or region can be strengthened.

Regional freight and transportation network characteristics significantly influence the type of freight transfer facility that would be best in a given region and promote optimal mode distributions. The availability and convergence of highways, rail networks, and waterways play a pivotal role in determining the location of a potential facility (*3*). In addition, the type, quantity, and shipment distances of commodities being moved contribute to the type and size of the facility (*4*).

Transload facilities, in particular, are suitable for regions with bulk, warehouse, or dimensional freight commodities, and for regions with lower volumes of freight than would be needed to support an intermodal facility (*4*, *5*). A transload facility is defined as a "receiving and distributing facility for lumber, grain, concrete, petroleum, aggregates, and other such bulk products" that provides access to multiple transportation modes (*3*). An important distinction for this research is the difference between transload facilities and intermodal facilities. Transload facilities handle bulk products that are not dependent on any sort of volume or weight dimensions, whereas intermodal facilities handle containerized goods (*6*).

[1]Department of Civil Engineering, University of Arkansas, Fayetteville, AR
[2]Department of Supply Chain Management, University of Arkansas, Fayetteville, AR

Corresponding Author:
Address correspondence to Sadie Smith: ses007@uark.edu

Intelligent siting of transload facilities could better leverage a state's multi-modal transportation network by shifting freight from truck to barge or rail and tapping into latent demand for short line and regional rail and waterways. Many regional intermodal agencies and economic development commissions wish to attract private development of transload facilities as a means to spur the local economy. From their perspective it would be beneficial to provide private developers with cost estimates of a proposed transload facility construction or expansion as part of a marketing package to attract private interest. From a statewide perspective, state departments of transportation and metropolitan planning organizations should prioritize transload facility projects proposed by regional authorities. Such prioritization might include a cost-benefit analysis for which initial cost estimates would be needed. Although transload operators are typically private companies or individuals that have a good handle on facility costs, initial cost estimates may not be available to public agencies (*4*).

Transload facilities can have a wide array of configurations or sizes, therefore, a generalized cost estimation framework should be scalable so that it can be applied to any facility scenario. To develop a complete and accurate cost estimate for transload facilities, multiple dimensions are considered in the framework, including modes of transportation serviced, category of commodities handled, and storage needs. This paper describes the development of this cost estimation framework and the assumptions made, it applies the framework to a unique case study in Arkansas, and conducts sensitivity analysis on key model assumptions.

Background

Freight transfer facilities play a key role in multi-modal connectivity by providing access to efficient alternative freight modes. Over the last 30 years, multi-modal freight transportation has grown rapidly and is often considered the fastest growing segment of transportation (*7*). Mode shifts away from truck have the potential to reduce pavement damage, roadway congestion, and emissions. In general, one barge can move the same quantity of freight as 16 rail cars or 70 trucks, thus creating potential to reduce vehicles on the highway network (*8*). As a result, transportation planners have begun to closely examine the role of freight transfer facilities within a multi-modal transportation network (*1*).

The type of facility and required infrastructure is determined by the accessible modes of transportation, commodities handled, and storage needs. This paper focuses on transload facilities as they play a significant role in the multi-modal movement of freight that is common to more rural, agricultural states. Studies show that feasible transload facilities handle 1,500 annual rail carloads compared with 100,000 carloads needed for a feasible intermodal facility (*4*). The modes of transportation served dictate certain features of the facility. For instance, if a facility services rail, rail spurs and some length of track are required, with the ideal rail capacity including two tracks with enough length to accommodate one train of 100 cars at 60 ft/car (*9*). A facility handling truck traffic, on the other hand, would require loading docks and paved truck bays, each sized 55 ft × 11 ft (*9*). The type, quantity, and shipping distance of a commodity dictates how it is handled, the mode of transport available, and storage requirements. In regards to storage, the type, space required, and length of time the commodity will be stored are important considerations. Modes of transportation, commodities, and storage also influence the equipment necessary to ensure efficient operation of the facility.

Transload facilities are seen by transportation planners and economic development officials as suitable and potentially cost-effective means to encourage mode shift. Regardless of the decision process to determine the actual location of the new or expanded facility, estimation of facility construction costs should be used to rank competing location alternatives. Unfortunately, the existing body of literature does not provide a generalized cost framework that can be applied to facilities of various types, sizes, commodities, and regions. Thompson describes in detail facility design components with specific reference to a facility in the city of Davenport (*4*). Thompson lists the total facility area of approximately 12.5 acres but does not discuss costs of individual infrastructure elements (*4*). Similarly, Bhamidipati and Demetsky present a case study in Virginia for a freight facility of approximately 40 acres, but there is no discussion of facility cost (*10*). The Upper Great Plains Transportation Institute conducted a study evaluating the feasibility of a "logistics center," or intermodal facility, including the development of an economic engineering model to simulate costs for an intermodal facility (*11*). This cost information was neither disaggregated nor robust enough to be generalized to various configurations of transload facilities. Williams Associates-Engineers presented disaggregated cost information for barge transload facilities (*12*). However, these costs are based on estimates from local contractors and are limited to barge components, not including truck, rail, or storage requirements (*12*).

While the research mentioned above provides general guidance on transload facility design considerations including the size of a facility and relative magnitudes of their total construction cost, it would be difficult to apply this information to facilities of a different size, configuration, mode combination, or commodity type. Additionally, in each of these cases, some preliminary facility design existed. This paper discusses a framework to create basic construction cost estimations for potential transload facilities using data from existing literature, survey responses, and stakeholder interviews. The methodology provides robust and accurate construction cost estimates, without requiring a facility design, which is not available at the initial stage of site selection. The framework serves as a tool that would allow transportation planners and economic developers to provide a basic construction cost estimate for a new or expanded transload facility prior to the construction design phase. Such estimates could be used by public sector decision makers to make high-level comparisons of alternative transload facility investment projects and decide on appropriate subsidies to spur investment, for example.

Methodology

A three-stage approach is adopted to estimate the cost of a transload facility expansion or new construction:

(1) Identify transload facility components (infrastructure and equipment) based on size, modes, and types of commodities handled,
(2) Estimate component costs using construction cost estimates,
(3) Determine total facility construction cost.

Identification of Transload Facility Components

The major groups of transload facility costs were identified as (*4*):

(a) Site Costs: site acquisition and land preparation.
(b) Building Costs: basic facility infrastructure, for example, parking lots, offices, storage, and so forth.
(c) Mode Access Costs: mode infrastructure, for example, truck bays, rail spurs, barge docks, and so on.
(d) Equipment Costs: transload equipment, for example, forklifts, conveyors, and so forth.

Each of these major groups was further divided into more specific components for which unit costs were determined. Components were selected based upon facility design information found in literature, interviews with local facility operators, and responses to the online survey sent to existing facilities in Arkansas (*4*, *9*, *13*).

Estimation of Component Costs

The facility components (Tables 1–3) were selected from a material, construction, and equipment cost database (*14*, *15*). This database is comprised of national averages updated annually and is often used by contractors to provide accurate estimates for project costs. For this research, the unit costs selected included labor and materials as well as general overhead and profit to provide the most complete picture of potential costs associated with each item. For the majority of these cost items, an appropriately aggregated item was listed in the database. An example of this is the aggregated square footage costs used for the office and storage warehouse, which is assumed to include things like the foundation, utilities, construction materials, and so forth. The unit cost databases and unit cost items used to determine facility component costs were validated through interviews with industry representatives familiar with this type of cost estimate.

There were no aggregated cost items for barge berth. Therefore, the items included for barge berth cost were selected based upon design recommendations found in Ernst and Runge-Schmidt's *Planning and Design of Ports and Marine Terminals* (*16*). Considering that transload terminals are common on the inland waterway network, these berths are designed as open-piled structures as this is the most widely used design (*16*). These structures typically include a concrete deck supported by steel or concrete piles with some slope stabilizer for the bank beneath the deck (*16*). Comparable cost items were then selected from the cost database for use in the cost estimate. In order to verify the items selected for the barge berth, the cost of constructing a berth with a 60-ft × 60-ft deck was compared with the cost of a similar pile-supported concrete dock using the unit price listed in the U.S. Army Corps of Engineers final report on barge landing system design alternatives for various locations in Alaska (*17*). Using the cost database, the construction price of this berth with a 60-ft × 60-ft deck was found to be US$543,267, compared with US$990,000 determined using the U.S. Army Corps of Engineers' prices. While this difference may initially seem significant, when considering the total cost of a transload facility with waterfront access, which could be about US$20 million, a difference of US$446,733 is only 2% of the total facility cost. Additionally, it is worth mentioning that prices in Alaska are often higher due to the need to import many materials. Finally, all of the salient costs of items in Table 2 are of the same order as those found by Williams Associates-Engineers for a Maryland port (*12*). Therefore, it was decided that the items selected from the cost database were appropriate for inclusion in this cost estimate.

This database did not have the purchase prices for equipment however, only rental prices. Thus, local equipment dealers were contacted to provide equipment purchase costs.

Determination of Facility Construction Cost

Once the facility components and unit costs were selected, a procedure was developed for a complete cost estimation for a proposed transload facility. It is important to again note that at this stage in the decision-making process of choosing a facility location, no construction design exists for the facilities. Rather, only very general facility characteristics have been outlined for the facilities. For example, the proposed facility may be characterized to handle a certain type and quantity of a commodity and provide access to barge or rail, or both. Therefore, a number of assumptions were made in order to produce a final construction cost estimate without specific quantity requirements.

Table 1. Transload Facility Site Preparation and Building Components

	Description	Unit
Site preparation components		
Site clearing	Remove trees and stumps up to 6" by cut and chip and stump haul away	1 acre
	Strip topsoil and stockpile, 6-in. deep by skid steer	Square yard
Site earthwork	Gravel cut and fill, 80-hp dozer and compact, 50-ft haul, 6-in. lift, 2 passes	Cubic yard
	Excavate sand and gravel, ½-yd^3 backhoe, two 8-yd^{3}dump trucks, 2 MRT	Cubic yard
	Earth cut and fill, 80-hp dozer and compactor, 50-ft haul, 4-in. lift, 2 passes	Cubic yard
	Excavate common earth, ½-yd^3 backhoe, two 8-yd^{3}dump trucks, 1 MRT	Cubic yard
	Clay cut & fill, 80-hp dozer and compactor, 50ft haul, 4in. lifts, 2 passes	Cubic yard
	Excavate clay, ½-yd^3 backhoe, two 8-yd^3 dump trucks, 2 mi. round trip	Cubic yard
	Load and haul rock, 1-1/2-yd^3 track loader, six 8-yd^3 trucks, 1 MRT	Cubic yard
	Excavate sandy clay/loam, ½-yd^3 backhoe, two 8-yd^{3}dump trucks, 1 MRT	Cubic yard
Building components		
Enclosed storage	Warehouses and storage buildings	Square foot
Silo	Steel, factory fab., 30,000 gall capacity, painted, economy	Each
Covered storage	Metal canopies, aluminum prefinished, 12 ft × 40 ft	Each
Storage tank	Steel, ground level, ht./diam. less than 1; not including foundation, 100,000 gall	Each
Paved storage	Asphalt concrete paving, parking lots and driveways; 6-ft stone base, 2 in.-binder course, 1-ft topping	Square foot
Unpaved storage	Crushed 1-1/2-ft stone base, compacted to 8-in. deep	Square yard
Employee parking	Asphalt concrete paving, parking lots and driveways; 6-ft stone base, 2-in. binder course, 2-in. topping	Square foot
Office	Office building	Square foot

Table 2. Transload Facility Mode Access Components

Mode	Component	Description	Unit
Truck	Loading dock	Dock boards, 60 in. × 60 in., aluminum, 15,000 lb capacity	Each
		Dock leveler, hinged for trucks, 10-ton capacity, 6 ft × 8 ft	Each
		Shelters, fabric, for truck or train	Each
	Pavement under loaded trucks	Asphalt concrete paving, parking lots and driveways; 6-in. stone base, 3-in. binder course, 2-in. topping	Square foot
	Truck scales	Truck scales, digital, electric, 60-ft × 10-ft platform	Each
		Concrete foundation pit, 70-ft × 10-ft platform, 40 yd^3 required	Each
Rail	Railroad sidings	Wood ties and ballast, 100-lb new rail	Linear foot
	Railroad turnouts	Turnout, #8 complete, w/rails, plates, bars, frog, switch point, timbers, and ballast to 6 in. below bottom of ties	Each
Barge	Dredging	Hydraulic method, pumped 1000 ft to shore dump maximum	B.C.Y.
	Elevated slab	Elevated slab (4,000 psi), one-way beam and slab, 125 psf sup. Load, 15-ft span	Cubic yard
	Concrete caisson	Concrete caissons for marine construction; cased shafts, 140 to 175 ton capacity, 19-in. diameter, 40-ft depth	Vertical linear foot
	Concrete revetment	Concrete revetment matt 8 ft × 20 ft × 4-1/2 in., excluding site prep, includes all labor, material and equipment for installation	Each
	Gravel base course	Crushed 1-1/2-in. stone base, compacted to 8-in. deep	Square yard

Table 3. Transload Facility Equipment Costs

Component	Approximate unit price (US$)[a]
Air compressor	$45,000
Auger	$3,500
Back hoe	$90,000
C hook	$11,000
Car puller[b]	
Containment pans	$1,500
Conveyor	$15,000
Crane	$300,000–$1,900,000; Avg.=$1,100,000
Excavator	$210,000
Fork lift	$77,500
Pneumatic lift	$3,500
Pallet jacks	$700
Ramp-portable	$2,600
Roll clamp	$2,000
Wheel loader	$152,500

[a]Cost estimates for equipment are highly dependent on type of freight and should be verified for local needs.
[b]Cost estimates for the indicated items were unavailable.

Facility Infrastructure. Quantity estimates were based largely on the annual tonnage of commodities projected for a forecast year. Annual tonnage forecasts were converted to an equivalent number of annual trucks using truck payload factors. Modal conversion factors from (*8*) were used to determine equivalent modal volumes, i.e., 70 truckloads can be transported by 16 carloads and one barge load. This simple conversion could be enhanced using commodity-specific mode conversion factors available from the Commodity Flow Survey or a region-specific commodity flow database (*5*). For the purposes of this paper, a general conversion approach was deemed appropriate given the high-level cost estimates desired.

It was assumed that 3% of the commodity tonnage in the region would use the transload facility. This factor represents the tonnage that would shift from truck to rail or barge. The assumed transload percent, e.g., 3%, should be altered by project stakeholders for specific projects. For the purposes of the case study presented later, an assumed 3% shift delivers reasonable annual tonnage volumes (*4*).

The expected number of daily trucks, rail cars, and barges to serve the assumed tonnage was used to determine the number of loading docks, truck parking spots, length of rail track, number of berths, and the storage requirements. The procedure and assumptions used to determine these quantities are as follows.

- Truck loading docks were determined using the assumption that trucks would be at the facility no longer than 2 h due to the potential of incurring detention charges and a 12-h work day (*18*).
- The amount of parking space required for trucks was based upon the number of daily trucks expected and by assuming truck spots are 55 ft × 11 ft (*9*).
- The length of rail track was based upon the daily number of rail cars expected and the ideal rail capacity recommendations of 100 rail cars, which is one train, over two tracks (*9*).
- To account for rail car storage, the number of daily rail cars was multiplied by the typical ratio of 3:2 for storage to working track (*9*).
- The number of berths required was based upon the number of barges expected per day.
- Storage requirements were found by multiplying the daily number of trucks by the volume of a 53-ft truck trailer and 30 days based upon the assumption that long-term storage of commodities would typically be one month (*19*, *20*).
- The type of storage required for each commodity was determined based upon interviews with local facility operators and a discussion of handling and storage (*4*).
- Three different storage scenarios were determined for each facility in which three different percentages of the commodity tonnage was assumed to require storage, for example, 10%, 50%, and 90%. This was adjusted depending on the amount of land available on the proposed sites for facilities.

Projected commodity tonnages were used to determine mode access infrastructure and storage quantities as follows. To find the daily number of trucks, loading docks, barge berths, and railcar spots required, the annual number of trucks, railcars and barges was multiplied by the assumed transload percent and divided by the number of working days, assumed to be 254 days per year. For the mode access items, such as loading docks and railcar spots, the ratio of the amount of time for which the trucks and railcars are assumed to remain at the facility are multiplied by the daily number of trucks/barges/railcars. Storage quantities are calculated by multiplying the volume of a standard truck, for example, WB-67, by the assumed 30 days for which the commodity is stored, the number of daily trucks, and the percentage of commodity requiring storage.

Google Earth was used to measure average quantities of office space, employee parking, and truck driveway for seven identified facilities (*13*). These facilities were selected because the facility capacity was known through survey responses and these facilities represented a range of facility sizes.

Facility Equipment. A final consideration in completing the facility construction cost estimate is selecting the equipment required to move commodities served at that facility. Using general facility characteristic data available at this stage in the facility location planning process, it is possible to identify the types of commodities that a facility might handle. For example, in Arkansas, typical transloaded commodities included crushed stone, sand, gravel, food products, metals, feed products, manufactured products, chemicals, waste, and

Table 4. Equipment Required to Handle Commodities

Equipment	Commodity					
	Crushed stone, sand, gravel	Food products	Metals	Feed products	Chemicals	Waste
Air compressor	X		X	X	X	
Auger				X		
Back hoe						X
C hook			X			
Car puller	X	X	X	X	X	
Containment pans					X	
Conveyor	X			X	X	X
Crane	X		X	X		X
Excavator	X		X	X		X
Fork lift		X	X		X	X
Lift			X			
Pallet jacks		X	X			
Ramp-portable		X	X			
Roll clamp			X			
Wheel loader	X		X	X		X

Note: Blank cells indicates that commodity requires list equipment item.

lumber. For this set of commodities, the specific equipment needs were determined via a previously conducted online survey (*13*). Survey responses linked commodity type to equipment needed to handle that equipment, as shown in Table 4. While this table provides an example for specific commodities listed in the top row, it could be expanded to a complete set of bulk commodities using similar survey approaches or site visits.

Total Construction Cost Estimate. A cost estimate can be prepared for a proposed facility by combining the costs associated with components listed in Tables 1–3 with quantity estimations based upon projected commodity tonnage and assumed transload percent and the equipment requirements (Table 4) for the commodities of interest. Site preparation, office infrastructure, and equipment required are combined with the mode access and storage items to create the total facility cost estimate.

Case Study: Arkansas

The Arkansas Department of Transportation along with the Arkansas Economic Development Commission (AR EDC) jointly sponsored a research project to determine the potential market and location of new transload facilities in Arkansas. The goal of the project was to prioritize transload facilities based on potential to shift freight to more efficient modes, protect highway infrastructure, and attract industry to the state. Five facilities were proposed: two of these facilities would provide transfer from truck to rail and three of the facilities would provide transfer from truck to barge and rail. The projected number of annual trucks, approximate facility storage area, and total facility costs are shown for the five proposed locations in Table 5.

Comparing the projected number of annual trucks, it is evident that there is a significant range from the 8,514 projected for the Jefferson County area to the 515,566 in Pulaski County. Although a direct comparison to existing transload sites is difficult due to a lack of information in the literature regarding the projected tonnage of commodities expected, the storage area and the facility costs are of the same magnitude as those seen in literature (*4*, *10*, *17*).

To illustrate the use of the cost estimation framework described above, the procedure followed to estimate the cost of a single facility in Pulaski County is presented.

Transload Facility in Pulaski County, Arkansas

This facility was identified as an optimal location for a new transload facility based on a statewide analysis of commodity flows forecasted for 2040. This facility would transload nonmetallic minerals, durable manufacturing, primary metal, and secondary and miscellaneous mixed commodities. An available plot of land was selected from the AR EDC's Building and Sites Database (*21*) that provides a searchable database of available sites. Figure 1 displays an aerial image of the potential site in Pulaski County. This location provides access to rail, water, and the highway. This particular site is limited to 6.5 acres, a smaller acreage than comparable facilities; however, due to convenient access to water and rail it is a desirable location (*2*, *10*).

Table 6 displays the projected number of annual trucks, equivalent number of barges and railcars, loading dock, barge berth, and railcar spot requirements. The number of

Table 5. Proposed Transload Facilities in Arkansas

Facility location	Projected annual trucks	Approximate storage area (acres)	Total facility cost (US$)
Benton/Washington County	209,804	8	$26,891,858
Jefferson County	8,514	4	$9,293,312
Pulaski County	515,566	6	$11,887,697
Hot Spring County	19,477	5	$8,341,055
Crawford/Sebastian County	9,937	4	$9,206,577

Figure 1. Potential location of Pulaski County facility (image from maps.google.com).

daily trucks was found by multiplying the total annual number of trucks by the 3% assumed to be transloaded, divided by 254 working days per year. The number of loading docks required was found by multiplying this number of daily trucks by the ratio of 2 h that the truck is expected to be at the facility to the 12-h workday. Using a similar process and the equivalent amounts for barge and rail car, it was determined that one berth and storage for at least 28 railcars would be required. The type of storage for each commodity type was identified from the survey results (*13*).

The site earthwork quantity was found by summing the area of all other construction, and the amount of site clearing was approximated by measuring the area of trees that would need to be removed, using Google Earth. The storage quantities presented in Table 7 are based upon the assumption that 4.5% of the commodity tonnage entering the facility would require storage. This particular percentage corresponds to complete use of the 6.5 acres available for this site with 6 acres used for storage and 0.5 acres used for office and other non-storage purposes. Each commodity shown in Table 6 requires a different type of storage and thus corresponds to a different unit cost and area calculation. A more disaggregated evaluation of the types of commodities included in these larger commodity groups, information obtained from local facility operators, and the discussions found in literature (*4*) all contributed to the selection of storage requirements for each commodity group. For instance, because durable manufacturing may include weather-sensitive products such as furniture, machinery, and electrical equipment, these items would require enclosed storage. Nonmetallic minerals, on the other hand, include aggregate, which are construction materials, which are not weather sensitive and could be off-loaded onto unpaved storage. Primary metal was determined to require covered storage, and secondary and miscellaneous mixed was assigned to paved storage based upon weather sensitivity and a look at the disaggregated commodities.

From the capacity requirements found using the projected commodity tonnage, the unit prices found in the cost database (Tables 1 and 2) and the commodity equipment requirements (Tables 3 and 4), the cost estimate shown in Table 7 was created. Overall, the total cost of the facility was estimated to be US$11,606,635.

Table 6. Projected Facility Capacity

Commodity	2040 annual trucks	Barge equivalent	Rail car equivalent	Daily trucks	Loading docks	Barge berths	Railcar spots
Nonmetallic minerals	254,482	3,635	58,167	31	6	1	14
Durable manufacturing	55,866	798	12,769	7	2	1	4
Primary metal	105,717	1,510	24,164	13	3	1	6
Secondary and misc. mixed	99,501	1,421	22,743	12	2	1	6
Total	515,566	7,364	117,843	63	1w	4	30

Table 7. Pulaski County Facility Cost Estimate

Group	Component	Unit	Quantity	Extended cost (US$)
Site costs	Site clearing	1 acre	0.50	$3,545
	Site earthwork	Cubic yard	30,860	$296,566
Building costs	Enclosed storage	Square foot	37,328	$4,218,008
	Covered storage	Square foot	69,600	$2,515,388
	Paved storage	Square foot	63,990	$168,934
	Unpaved storage	Square yard	6,123	$70,103
	Employee parking	Square foot	1,920	$5,990
	Office	Square foot	896	$157,696
Truck access	Loading dock	Each	11	$139,150
	Pavement under loaded trucks	Square foot	48,905	$174,591
	Truck scales	Each	1	$71,700
Rail access	Railroad sidings	Linear foot	8,520	$1,686,960
	Railroad turnouts	Each	2	$126,000
Barge access	Dredging	B.C.Y.	560	$9,604
	Elevated slab	Cubic yard	267	$293,333
	Concrete caisson	Vertical linear foot	1,600	$119,200
	Concrete revetment	Each	31.5	$116,550
	Gravel base course	Square yard	400	$4,580
Equipment	Type			Estimated price
	Fork lift			$77,500
	Conveyor			$15,000
	Excavator			$210,000
	Ramp			$2,600
	Roll clamp			$2,000
	Air compressor			$45,000
	C hook			$11,000
	Wheel loader			$152,500
	Pallet jacks			$700
	Back hoe			$90,000
	Crane			$1,100,000
	Lift			$3,500
Total estimated price	$11,887,697			

Assuming that 4.5% of the commodity tonnage would require storage, the total cost of storage is US$6,972,431, which is 60% of the total cost of the facility. In order to explore the impact of quantity of tonnage requiring storage, Figure 2 displays the three different scenarios for which storage quantities were calculated. The three scenarios represent (i) 50%, (ii) 100%, and (iii) 200% (e.g., doubling of the site are) of the proposed site being developed. The quantity of storage required to allow 100% of the total commodity tonnage to be stored was found then scaled to fit within the constraints of the available site resulting in 2.25%, 4.5%, and 9% of the commodity tonnage being stored, respective to scenarios (i–iii). Comparing the total storage cost, it is evident that there is a large variance between these three scenarios. For the 2.25% storage scenario, the storage cost would be 41% of the total cost, and for the 9% storage scenario, storage would be 74% of the total cost. In future iterations of this framework, significant improvement should be

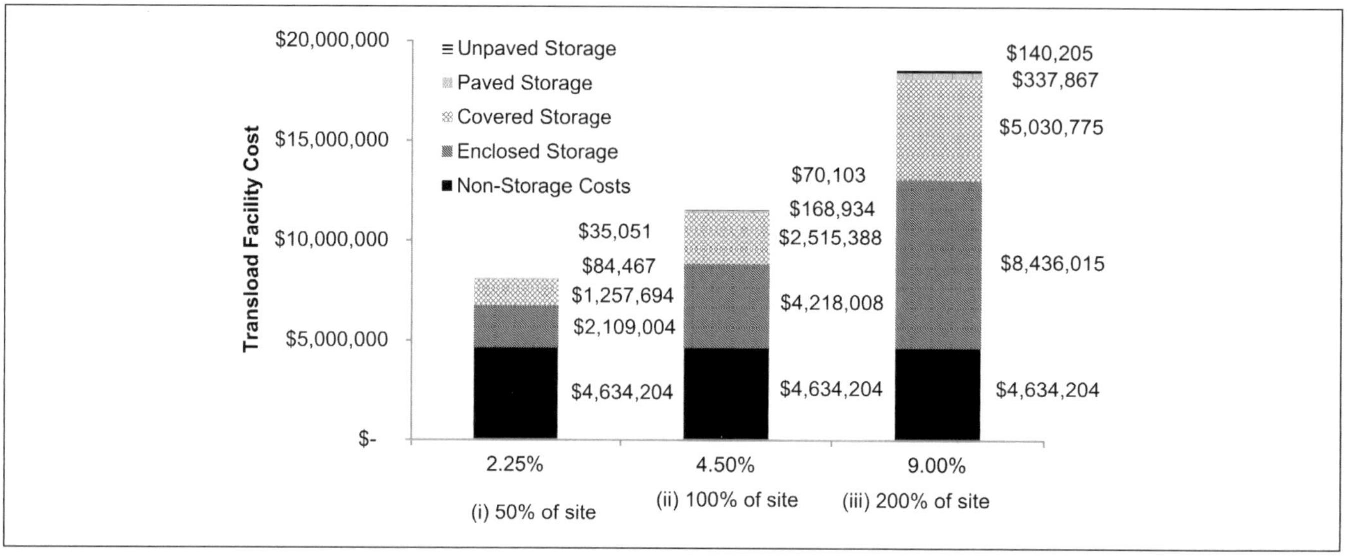

Figure 2. Facility storage scenarios.

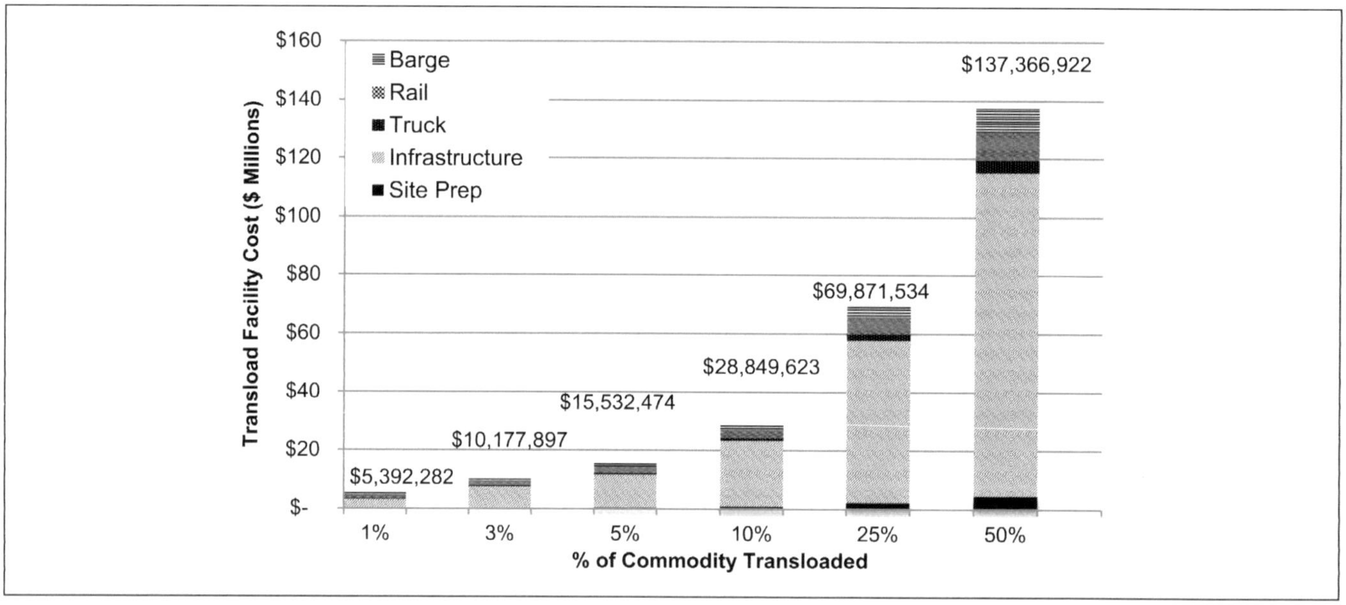

Figure 3. Facility cost sensitivity analysis.

made to better estimate the storage requirement for each commodity type.

As shown in Figure 2, storage costs are a significant portion of the total estimated construction cost. The storage requirements are based on assumptions of the (a) percent of commodity being transloaded (e.g., 3%), (b) percent of the commodity being stored (e.g., 10%, 50%, and 90%), and (c) duration of the storage period (e.g., 30 days). Of these, percent of commodity being transloaded has the largest effect on overall facility costs as it mandates the capacity of the facility. In order to evaluate the influence of the assumption of the percent of commodity being transloaded, a sensitivity analysis was completed by altering the percent transloaded, holding constant all other assumed values, and restricting the storage to 4.5% of the tonnage. The results of this analysis are shown in Figure 3.

As the percentage of commodity being transloaded approaches 100%, infrastructure (primarily storage), approaches a maximum of 81% of the total facility cost. However, the percentages both rail and barge costs contribute to the total facility cost generally decrease as the percentage of commodity being transloaded increases. This is likely because, as transloaded commodity tonnage increases, barge and rail costs do not increase at as great a rate as infrastructure costs increase. Truck costs consistently comprise 3 to 4% of the total facility cost. From this analysis, it is evident that incorporating stakeholder feedback to further refine this key assumption will influence the accuracy of this cost estimate.

Conclusion

The framework presented in this paper can be used to estimate transload facility cost by type using unit costs from a construction cost database, equipment costs from local dealers, the projected commodity tonnage, design recommendations from literature, and survey responses from local facilities. This can be used by economic development commissions to prepare marketing materials to attract private developers to a region. For a state and local agencies, in addition to determining the benefits of mode shift (e.g., emissions reductions, better use of highway capacity, etc.), understanding costs associated with constructing transload facilities is necessary to prioritize alternatives proposed by different regional authorities. Moreover, initial cost estimates could be used to determine the viability of a developer's proposal or inform public policy on investment decisions regarding subsidies for needed facilities. Unfortunately, the existing body of literature did not provide an adequately disaggregated and scalable cost estimation approach to suit these needs. However, this paper fills that gap.

This framework is scalable to differing transload facility characteristics yet accurate enough to give decision makers reasonable cost estimates for constructing new or expanded facilities. In order to illustrate the usefulness of this methodology, a case study of proposed facilities in Arkansas was presented. There is currently no construction design for these facilities; however, this framework yielded costs consistent with those found in the literature. Of particular interest, the storage costs of facilities were a significant cost of the overall facility. One case study showed that increasing the facility storage from 2.25% to 9.00% of the commodity tonnage would increase the cost of the facility from approximately US$8 million to US$18 million. An additional sensitivity analysis showed that up to 81% of a transload facility's costs could potentially be for storage.

Improvements to the framework include: (a) use of commodity-specific payload factors; (b) more accurate methods of estimating the number of berths, loading docks, length of rail track, and storage; and (c) the addition of operational costs. The number of loading docks, berths, railcar spots, and quantity of storage area were all determined based upon the assumed number of daily trucks for a facility that is dependent on payload factors. Payload factors based on vehicle type and commodity would improve the estimated number of trucks needed to serve a facility of a specified size. Maintenance and operational costs including local labor costs could be added to the framework to allow for life cycle analysis. Future research could include cost estimates for intermodal, containerized transport facilities, such as COFC and COB.

Acknowledgments

The authors thank the Arkansas Department of Transportation for their financial support of this project under TRC1608, "Locating Transload Facilities to Ease Highway Congestion and Safeguard the Environment."

Author Contributions

The authors confirm contribution to the paper as follows: study conception and design: S.S., A.B., S.H., and J.K.; data collection: S.S., A.B., S.H., and J.K.; analysis and interpretation of results: S.S., A.B., S.H., and J.K.; draft manuscript preparation: S.S., A.B., S.H., and J.K. All authors reviewed the results and approved the final version of the manuscript.

References

1. Bryan, J., G. Weisbrod, and C. Martland. *NCHRP Report 586: Rail Freight Solutions to Roadway Congestion—Final Report and Guidebook*. Transportation Research Board of the National Academies, Washington, D.C., 2007.
2. Bomba, M., and R. Harrison. Feasibility of a Container-on-Barge Network along the Texas Gulf Coast. *Transportation Research Record: Journal of the Transportation Research Board*, 2002. 1782: 23–30.
3. Steele, C.W., and D. Hodge; Fitzgerald and Halliday, Inc., Halcow, Inc., and Resource Systems Group, Inc. *Background Research Material for Freight Facility Location Selection: A Guide for Public Officials*. NCFRP Report 13, National Academies Press, 2011.
4. Thompson, D. *Transloads- Freight Movement Efficiencies in the Next Decade*. Presented at ASME/ASCE/IEEE 2012 Joint Rail Conference, Philadelphia, Pennsylvania, 2012.
5. Bureau of Transportation Statistics (BTS), US Department of Transportation, Commodity Flow Survey. http://www.rita.dot.gov/bts/sites/rita.dot.gov.bts/files/publications/commodity_flow_survey/index.html. Accessed May 4, 2015.
6. Jones, B., R. Cassady, and R. Bowden. *Developing a Standard Definition of Intermodal Transportation*. National Center for Intermodal Transportation Final Research Project, Transportation Law Journal, 2000.
7. Ozment, J. *Demand for Intermodal Transportation in Arkansas*. Walton College of Business, University of Arkansas, Arkansas, 2001.
8. Kruse, C. J., A. Protopapas, and L. E. Olson. *A Modal Comparison of Domestic Freight Transportation Effects on the General Public: 2001-2009*. Final Report for the National Waterways Foundation, Texas A&M Transportation Institute, 2012.
9. Chimka, J. *Multimodal Transport and Transload Facilities in Arkansas*. Maritime Transportation Research and Education Center, University of Arkansas, Arkansas, 2015.
10. Bhamidipati, C., and M. Demetsky. *Framework for Evaluation of System Impacts of Intermodal Terminals using Commodity Flow Data*. Research Report no. UVACTS-14-5-111. University of Virginia Center for Transportation Studies, Charlotteseville, VA, 2008.
11. Upper Great Plains Transportation Institute. *Feasibility of a Logistics Center Including Container/Trailer Intermodal Transportation in the Fargo/Moorhead Area*. North Dakota State University, 2007.

12. Williams Associates-Engineers. *Barge Transload Facility—Soybean Site Location Analysis*. MPA-MA Project 013-2004-02. Maryland Port Administration, 2004.
13. Smith, S., S. Hernandez, and A. Braham. *Developing a Statewide Transload Facility Inventory: Challenges and Lessons Learned*. Paper No. 17-05477, TRB 2017 Annual Meeting Compendium of Papers, Washington, D.C., 2017.
14. RSMeans. *Building Construction Costs with RSMeans Data*, Norwell, MA, Gordian, 2014.
15. RSMeans. *Heavy Construction Costs with RSMeans Data*, Norwell, MA, Gordian, 2017.
16. Agerschou, H., I. Ernst, T. Ghoos, H. Jensen, O. Korsgaard, J. Land, J. McKay, T. Oumeraci, H. Petersen, J. Runge-Schmidt, L. Svendsen, and L. Hanne. *Planning and Design of Ports and Marine Terminals*, 2nd ed. ICE Publishing, London, England, 2004.
17. URS Corporation. *Alaska Barge Landing System Assessment and Design, Various Locations Statewide, Alaska Phase 2*. U.S. Army Engineer District, Alaska, 2010.
18. Farrell, L., S. Soccolich, and R. Hanowski. *Assessing Daily Driving and Working Hours Within the Context of Hours-of-service Regulations*. Virginia Tech Transportation Institute, Linköping, 2016.
19. HA Logistics, Inc. *Trailer Specifications*. http://www.halogistics.com/resources/trailer_specs.php. Accessed July 19, 2017.
20. Union Pacific Distribution Services. *UPDS Metals Pricing*. UPDS Program for Odessa, TX, 2017.
21. Arkansas Site Selection Center. *Entergy Corporation*. http://www.arkansassiteselection.com/About.aspx. Accessed July 19, 2017.

The Standing Committee on Intermodal Freight Transport (AT045) peer-reviewed this paper (18-03659).

Article

TRR
JOURNAL OF THE TRANSPORTATION RESEARCH BOARD

Transportation Research Record
2018, Vol. 2672(9) 35–46

DOI: 10.1177/0361198118780881
journals.sagepub.com/home/trr
SAGE

Joint Deployment of Quay Cranes and Yard Cranes in Container Terminals at a Tactical Level

Lingxiao Wu[1] and Shuaian Wang[2,3]

Abstract

This paper discusses tactical joint quay crane (QC) and yard crane (YC) deployment in container terminals. The deployments of QCs and YCs are critical for the efficiency of container terminals. Although they are closely intertwined, the deployments of QCs and YCs are usually sequential. This paper proposes a mixed-integer programming model for the joint deployment of QCs and YCs in container terminals. The objective of the model is to minimize the weighted vessel turnaround time and the weighted delayed workload for external truck service in yard blocks, both of great importance for a container terminal but rarely considered together in the literature. This paper proves that the studied problem is NP-hard in the strong sense. Case studies demonstrate that the proposed model can obtain better solutions than the sequential method. This paper also investigates the most effective combinations of QCs and YCs for a container terminal at various demand levels.

Containerized maritime transportation has grown steadily in the past few decades. According to the UN Conference on Trade and Development (UNCTAD) (*1*), the total containerized trade in 2015 was estimated to be 175 million 20ft equivalent units. Container terminals are the heart of international container transportation, providing the linkage between inland and seaborne transportation. With the annual volume of containerized trade growing rapidly, container terminals are confronting an immense increase in service demand. Besides, competition among container terminals, especially geographically close ones, has become fiercer than ever. To deal with the increased demand and win competitive advantages, port managers are working to improve terminal efficiency.

The efficiency of container terminals is largely determined by the use of port resources. The most important equipment used in port operations includes quay cranes (QCs) and yard cranes (YCs), which are both gantry cranes. QCs are located on the quay side and are used to load and discharge containers into and from vessels. YCs are used on the yard side to stack and retrieve containers stored in the blocks. Both the QCs and the YCs play very important roles in container terminals, and more often than not they are bottlenecks in container handling. In addition, because both QCs and YCs are very expensive and have high maintenance costs, it is impossible to get all the berths and blocks fully equipped with QCs and YCs. (Fully equipped means a berth has the maximum number of QCs that can work simultaneously for a vessel, or a block has the largest allowable number of YCs.) Moreover, the workload in berths and blocks varies from time to time. To make better use of limited resources and to ensure the smooth flow of containers, container terminals should deploy QCs and YCs in an efficient way.

The QC deployment problem (QCDP) and the YC deployment problem (YCDP) are intertwined. On the one hand, the efficiency of QCs directly determines the workload of YCs in each block in the yard. On the other hand, the work of YCs in the yard has a significant impact on the speed of ship handling in berths. To make overall improvements in terminal efficiency, port managers should work out an integrated scheme that handles the QCDP and the YCDP simultaneously. Failure to handle either of these two problems efficiently may result in (i) congestion in the yard, (ii) idleness of QCs or YCs, and (iii) delays of vessels and external container trucks.

[1]Department of Logistics and Maritime Studies, The Hong Kong Polytechnic University, Hung Hom, Hong Kong
[2]The Hong Kong Polytechnic University Shenzhen Research Institute, Shenzhen, China
[3]Department of Logistics and Maritime Studies, The Hong Kong Polytechnic University, Hung Hom, Hong Kong

Corresponding Author:
Address correspondence to Shuaian Wang: wangshuaian@gmail.com

Terminal operation planning can be divided into operational and tactical levels (*2*). Tactical planning of container terminals considers the periodic calling of liner vessels. For terminal operators, tactical planning works as the foundation for operational planning. This paper studies the tactical joint QC and YCDP (TJCDP). Studies in the literature have focused on the tactical planning of container terminals, with most focused on the tactical berth allocation problem (TBAP) (*3*), the tactical yard template problem (TYTP) (*4*), or the integration of the two (*5*, *6*). These studies are closely related to the problem considered here. While the TBAP and the TYAP provide input data for the TJCDP, the TJCDP gives an implemental guarantee for the two problems.

To address the needs of container terminals, the TJCDP is considered in this paper, whose contribution is threefold. First, a mixed-integer programming model for the TJCDP is proposed and the model is further linearized. Second, the considered problem is proved to be NP-hard in the strong sense. Third, a series of application cases are solved by the model and discussed in detail, and the results demonstrate the validity of the model.

Literature Review

Efforts have been devoted to operational problems in container terminals. Most discussed are (i) the berth allocation problem (BAP) (*7*, *8*), (ii) the QC assignment problem (QCAP) and the QC scheduling problem (QCSP) (*9*, *10*), (iii) the yard template problem (YTP) (*4*), and (iv) the YCDP and the YC scheduling problem (YCSP) (*11*, *12*). This research can be seen as an extension of these studies, especially the studies on QCAP and YCDP.

The QCAP has usually been considered together with the BAP. Meisel and Bierwirth were among the first to integrate the BAP and the QCAP (*13*). They developed a heuristic algorithm to solve the integrated problem. Blazewicz et al. formulated an integrated problem of berth allocation and QC assignment as a moldable task scheduling problem and solved the problem by a heuristic algorithm (*14*). Some studies have considered the QCAP and the QCSP simultaneously. Daganzo studied the QC assignment and scheduling problems for multiple container ships with the objective of minimizing the delay of vessels (*9*). An exact method and a heuristic method were developed to solve small-size instances and large ones, respectively.

Studies that focused on the YCDP have included Zhang et al., who studied a dynamic YCDP, aiming to minimize the delay of the workload in all blocks (*11*). They proposed a mixed-integer programming model for the problem, and the model was then solved by Lagrangian relaxation. Cheung et al. proposed a mixed-integer programming model for the YCDP (*15*). A Lagrangian decomposition solution procedure and a successive piecewise-linear approximation method were developed to solve instances with various sizes. The YCDP was investigated by Linn and Zhang, who proposed a heuristic method that they proved could deliver near-optimal solutions (*16*). Jin et al. integrated the space allocation problem and the YCDP (*17*). They proposed an integer programming model with the objective of minimizing the YC operating cost and the YC interblock movement cost. An efficient heuristic algorithm was developed to solve the problem.

Although the QCAP and the YCDP have received great attention in the literature for more than 20 years, to the best of the authors' knowledge there have been no studies considering the joint deployment problem of QCs and YCs.

Problem Description

A typical container terminal, as shown in Figure 1, is composed of two parts—the quay and the yard. The quay of a container terminal is divided into independent segments called berths where containers are loaded onto or discharged from vessels. The yard of a container terminal is an area reserved for storing containers. A large-scale yard may be divided into blocks, which are separated rectangular places where containers are stacked side by side and one on top of another. Three types of containers are used in a container terminal. An inbound container is first discharged from a vessel by a QC, then transported by an internal truck (IT) from the berth to the designated block, then stacked by a YC into the block and stored in the block until it is picked up by a YC to load onto an external truck (XT) that carries the container to an inland consignee. An outbound container undergoes a reverse flow. An XT first brings a container into the designated block. A YC then discharges the container from the XT and places it into the block. When the vessel for transporting the container arrives, the container is carried to the relevant berth by an IT and then loaded by a QC onto the vessel. The third kind of container is the transshipment container, which is transshipped from one vessel to another. After a vessel calls at the berth, a transshipment container is first discharged and then transported to its designated block, using the same procedure as with an inbound container. When the vessel for transshipment arrives, the container is then transported from the block to the relevant berth and loaded onto the vessel, using the same procedure as with an outbound container.

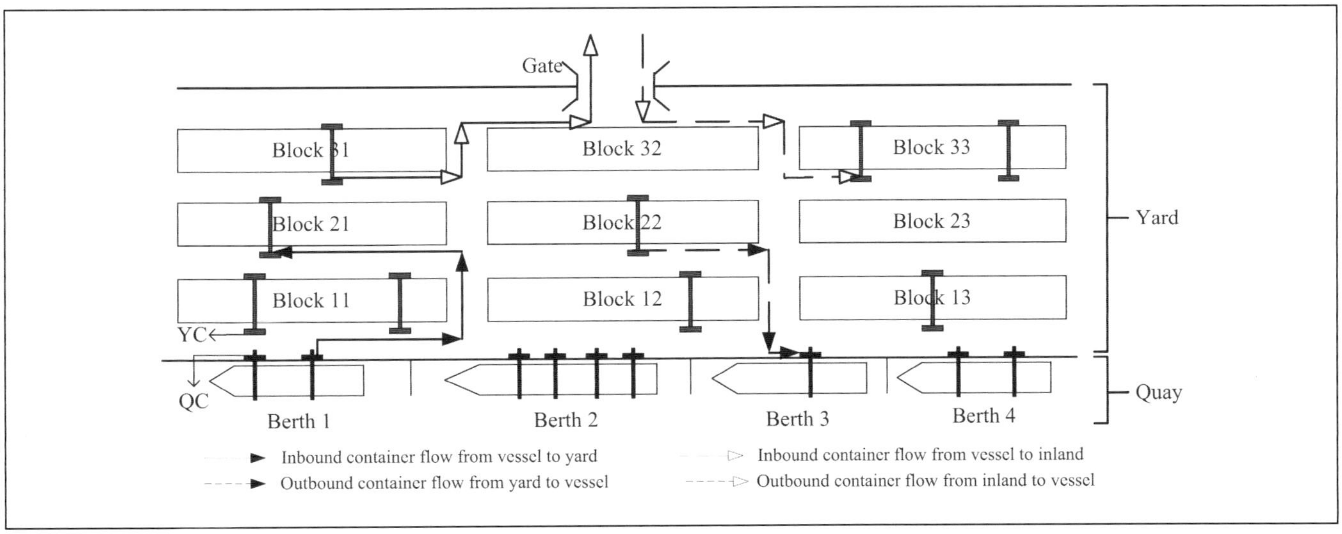

Figure 1. Container terminal layout.

In terminal operation planning, a working day is normally divided into several 4- or 6-hour work shifts. At the tactical level, operation plans are normally made by taking a work shift as a planning unit.

Tactical Quay Crane Deployment

Vessels visit container terminals on a fixed, periodic schedule. Before handling starts, a container ship should first moor in a particular berth. In practice, port managers work out a tactical berth allocation plan to designate the berth and the time window for each vessel in a planning horizon for loading and discharging containers. Such a plan is settled according to the estimated time of arrival, the draft of the vessel, and the blocks assigned to the vessel.

It is of great importance to ensure that the handling of each vessel can be completed in time, because the delay in the handling of a vessel may not only cause extra expenses of the port (like demurrage), but also can lead to delays of subsequent vessels that call at the same berth. In addition, shorter turnaround time is always encouraged by liner companies and helps improve customer satisfaction. The workload of vessel handling is defined in QC work shifts, that is, the number of work shifts needed by one QC to finish handling a vessel. Therefore, for a certain vessel, more assigned QCs mean less turnaround time.

It is a common rule in container terminals that QCs should remain in the same berth within one shift and can only be moved between berths between two shifts. Therefore, given the berth plan of all ships in a planning horizon, container terminals should decide how to deploy QCs dynamically to fully exploit their productivity and to ensure ships are handled within the allowable time.

Tactical Yard Crane Deployment

The yard serves as a buffer zone between sea transportation and inland transportation. A tactical yard management plan assigns for vessels in a planning horizon the blocks to store containers that are to be loaded onto the vessel and the blocks to store containers that are to be discharged from the vessel. When a vessel moors in a berth, containers to be loaded onto or discharged from the vessel are transported by ITs between the corresponding blocks and the berth. Besides, in negotiation with the liner companies or the consignees, port managers also need to work out a plan that schedules the timetable of certain XTs to arrive at designated blocks for discharging or loading containers.

YCs play a critical role in yard operations, serving both the ITs and the XTs. The workload (measured by YC work shifts) for YCs in a block within a work shift is composed of two parts: (i) the workload related to moving containers in the block that are to be loaded or discharged by QCs onto or from vessels in the work shift, and (ii) the workload related to moving containers from or onto XTs in the work shift.

Most YCs are tire mounted and can move freely in the yard, and as shown in Figure 2, two types of YC movements between blocks are identified. One is the movement between two blocks in the same row (e.g., movement from block 11 to block 12). Another is the movement between two blocks in separate rows (e.g., movement from block 11 to block 21). Because any movement of YCs can cause a loss in YC productivity and possible congestion in the yard, such movements

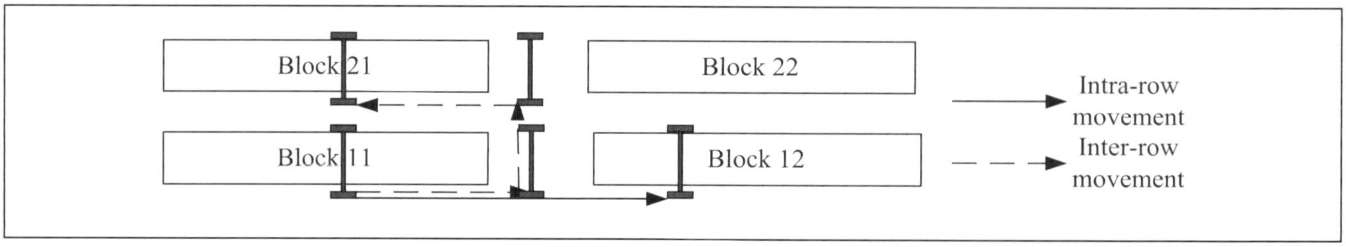

Figure 2. Yard crane movements.

should not happen frequently. A common rule is that intra-row movements of YCs are implemented only between work shifts; and inter-row movements, which cause more traffic congestion and productivity losses, can only be implemented once in a day (usually during the midnight work shift) (*11*). Moreover, to reduce interference between two working YCs, the number of YCs in a block is limited to no more than two at any time.

Given the projected workload in each block in each work shift, port managers should decide the number of YCs in each block in each work shift with the aim of ensuring smooth flows of ITs and XTs, which are critical for reducing vessel turnaround time and XT delays and improving customer satisfaction.

Tactical Joint Quay Crane and Yard Crane Deployment Problem

Operations of QCs and YCs are closely interconnected. Ship handling processes involve inbound and outbound flows of containers and therefore call for cooperation between YCs and QCs. Therefore, for a balanced use of the QCs and YCs and to ensure a smooth flow of containers, it is of great importance for the port managers to deploy QCs and YCs in an integrated manner.

Considering vessels are heterogeneous in size, load capacity, and lateness, when making decisions, terminal managers assign a particular weight for each vessel. Similarly, the weight for delayed XT service is also assigned according to its relative importance.

The TJCDP in container terminals can be formally stated as follows: Given the berth allocation plan, the yard management plan, and the XT transportation plan for a planning horizon, the port managers make the following three decisions jointly:

- The number of QCs deployed in each berth in each work shift.
- The number of YCs deployed in each row of blocks in each working day.
- The number of YCs in each block in each work shift.

The considered objective is to minimize the summation of the weighed turnaround time of each vessel and the weighed delayed workload for XT service in blocks in a planning horizon.

Further, according to industrial practices, in this paper the following assumptions are made for the TJCDP:

A1. All the YCs are identical in terms of working efficiency.

A2. Movements of YCs between two rows happen at most once in the midnight shift of a working day.

A3. At most two YCs can work in parallel in a block at any time.

A4. Containers are not allowed to move among blocks.

A5. All the QCs are identical in terms of working efficiency.

A6. There may be lower and upper bounds for the number of QCs that can work simultaneously on a vessel. The minimum number of cranes to be assigned for handling a vessel can be specified in the contract between the corresponding terminal authority and shipping company. Further, for safety and efficiency considerations, the number of QCs that can work simultaneously on a vessel should be less than an upper bound.

A7. Efficiency losses of individual QCs and YCs as a result of crane interference are not considered. This assumption implies that the handling efficiencies of QCs and YCs working on the same vessel or block are proportional to the number of cranes assigned. This assumption may not be true in practice, but it is used to simplify the analysis.

A8. Deployments of QCs and YCs should remain unchanged within one work shift.

A9. Each berth can handle at most one vessel, and each handling vessel occupies only one berth at any time.

A10. Sufficient ITs carry containers between the berths and the blocks, and no crane idleness may be caused by delayed arrivals of ITs.

A11. Direct transshipments between two berthing ships in the terminal are not allowed.

Table 1. Notations Used in Formulations

Indices	
k	Index for vessels in a planning horizon, arranging in an alphabetical order
i	Index for berths, arranging in an alphabetical order
j	Index for blocks, arranging in an alphabetical order
r	Index for block rows, arranging in an alphabetical order
d	Index for working days, arranging in an alphabetical order
t	Index for work shifts, arranging in an alphabetical order
Sets	
V	Set of vessels
V_i^I	Set of vessels that moor in berth i
V_t^T	Set of vessels whose handling time windows cover work shift t
V_j^J	Set of vessels to which block j is assigned to hold containers that are to be loaded onto or discharged from them
I	Set of berths
J	Set of blocks
R	Set of rows for blocks
J_r^R	Set of blocks in row r
D	Set of work days
T	Set of work shifts
T_d^D	Set of work shifts in day d
T_k^V	Set of work shifts within the time window set for handling vessel k
Z_+	Set of non-negative integers
Parameters	
s_k	Berthing time (work shift) of vessel k; s_k is the smallest element in T_k
b_k	Berth assigned to vessel k
$q_k^{\min}$	Minimum number of QCs agreed to serve vessel k at any time
$q_k^{\max}$	Maximum number of QCs allowed to serve vessel k at any time
w_k^Q	Quay crane workload (in QC work shifts) for handling vessel k
w_{jk}^Y	Yard crane workload (in YC work shifts) in block j for handling vessel k
w_{jt}^E	Yard crane workload (in YC work shifts) in block j for serving XTs in work shift t
H	Number of QCs
G	Number of YCs
ω_1^k	Weight assigned to the turnaround time of vessel k in the objective
ω_2	Weight assigned to the delayed workload of YCs for serving XTs in the objective
M	A constant large enough
M_k^1	A constant set equal to $\min\{q_k^{\min}, H\}$
M_k^2	A constant set equal to $\min\{q_k^{\max}, H\}$
Decision variables	
α_k^t	1 if vessel k is being handled in work shift t and 0, otherwise
c_k	Completion time (work shift) of handling vessel k
x_i^t	Number of QCs assigned to berth i in work shift t
y_j^t	Number of YCs assigned to block j in work shift t
z_r^d	Number of YCs assigned to row r in working day d
v_k^t	Number of QCs assigned to handle vessel k in work shift t
u_j^t	Delayed workload (in YC work shifts) of YCs in block j for serving XTs in work shift t, where the initial delayed workload u_j^0 is defined as 0 for each j

Model Formulation

In this section, two mixed-integer programming models for the TJCDP are presented. Notations used in the models are introduced by Table 1.

A Non-Linear Model

The first mathematical model (M1) for the TJCDP is presented as follows:

$$[\mathbf{M1}]\ \min F = \sum_{k \in V} \omega_1^k (c_k - s_k) + \omega_2 \sum_{t \in T} \sum_{j \in J} u_j^t \tag{1}$$

subject to:

$$\sum_{i \in I} x_i^t \leq H, \quad \forall t \in T \tag{2}$$

$$\sum_{k \in V_i^I \cap V_t^T} \alpha_k^t \leq 1, \ \forall i \in I, \forall t \in T \tag{3}$$

$$x_i^t - q_k^{\min} \geq (\alpha_k^t - 1)M, \quad \forall i \in I, \forall k \in V_i^I, \forall t \in T_k^V \quad (4)$$

$$x_i^t - q_k^{\max} \leq (1 - \alpha_k^t)M, \quad \forall i \in I, \forall k \in V_i^I, \forall t \in T_k^V \quad (5)$$

$$\sum_{t \in T_k^V} \alpha_k^t x_{i=b_k}^t \geq w_k^Q, \quad \forall k \in V \quad (6)$$

$$y_j^t \leq 2, \quad \forall t \in T, \forall j \in J \quad (7)$$

$$\sum_{j \in J_r^R} y_j^t \leq z_r^d, \quad \forall r \in R, \forall d \in D, \forall t \in T_d^D \quad (8)$$

$$\sum_{r \in R} z_r^d \leq G, \quad \forall d \in D \quad (9)$$

$$y_j^t \geq \sum_{k \in V_t^T \cap V_j^J} \frac{w_{jk}^Y}{w_k^Q} \alpha_k^t x_{i=b_k}^t, \quad \forall t \in T, \forall j \in J \quad (10)$$

$$c_k \geq t\alpha_k^t, \quad \forall k \in V, \forall t \in T_k^V \quad (11)$$

$$u_j^t \geq u_j^{t-1} + w_{jt}^E + \sum_{k \in V_t^T \cap V_j^J} \frac{w_{jk}^Y}{w_k^Q} \alpha_k^t x_{i=b_k}^t - y_j^t, \quad \forall t \in T, \forall j \in J \quad (12)$$

$$\alpha_k^t \in \{0, 1\}, \quad \forall k \in V, \forall t \in T_k^V \quad (13)$$

$$x_i^t \in Z+, \quad \forall t \in T, \forall i \in I \quad (14)$$

$$y_j^t \in Z+, \quad \forall t \in T, \forall j \in J \quad (15)$$

$$z_r^d \in Z+, \quad \forall d \in D, \forall r \in R \quad (16)$$

$$u_j^t \geq 0, \quad \forall t \in T, \forall j \in J \quad (17)$$

The objective function (1) minimizes the summation of the weighted turnaround time of each ship and the weighted delayed workload of YCs for serving XTs. In the objective function, $c_k - s_k$ calculates the actual handling time of vessel k, and $\sum_{t \in T}\sum_{j \in J} u_j^t$ calculates the total delayed YC workload for XT services. By minimizing the objective function, a deployment strategy is sought that encourages shorter vessel turnaround time and less delayed XT services in blocks. This objective function also gives terminal managers the freedom to assign different priorities to different vessels and the delayed YC workload for XT services. Constraint (2) ensures the total number of QCs located at all berths does not exceed the total available quantity in each work shift. Constraint (3) ensures that at any time, at most one vessel can be handled in a berth. Constraints (4) and (5) define the lower bound and upper bound of the number of QCs that can work simultaneously for each vessel in each work shift. Constraint (6) enforces that workload for QCs to handle a vessel can be completed within the allowable time window. Constraint (7) ensures that the number of YCs in a block does not exceed two in each work shift. Constraint (8) ensures that the total number of YCs located in all blocks in a row does not exceed the number of YCs assigned to the row in any work shift within a working day. Constraint (9) defines the upper bound of the total number of YCs in all blocks. Constraint (10) ensures that in each work shift, workload related to vessel handling in each block can be completed. In the right part of the constraint, $\frac{w_{jk}^Y}{w_k^Q} \alpha_k^t x_{i=b_k}^t$ calculates the proportion of workload for QCs to handle the vessel k in work shift t. Accordingly, the overall workload in block j for handling vessels in shift t is $\sum_{k \in V_t^T \cap V_j^J} \frac{w_{jk}^Y}{w_k^Q} \alpha_k^t x_{i=b_k}^t$. We assume YCs in a yard must be able to complete workload related to vessel handling, since otherwise, QCs deployed to handle certain vessels may be left idle if they have to wait for the corresponding yard work to get ready. Constraint (11) calculates the completion time of vessel handling. Constraint (12) calculates delayed workload of YCs in each block for serving XTs, which is obtained by adding the delayed workload of the block in the current work shift to the delayed workload of the block in the last work shift. As the workload related to vessel handling is given priority, YC capacity in each block available for XT services in the current work shift are obtained by deducting the workload related to vessel handling $\sum_{k \in V_t^T \cap V_j^J} \frac{w_{jk}^Y}{w_k^Q} \alpha_k^t x_{i=b_k}^t$ from the total YC capacity y_j^t. Thereby, delayed workload in the current work shift can be calculated by $w_{jt}^E + \sum_{k \in V_t^T \cap V_j^J} \frac{w_{jk}^Y}{w_k^Q} \alpha_k^t x_{i=b_k}^t - y_j^t$. Constraint (13) defines binary variables. Constraints (14)–(16) ensure the corresponding variables to be non-negative integers. Constraint (17) ensures u_j^t to be non-negative.

A Strengthened Linear Model

Constraints (6), (10) and (12) in M1 involve multiplication among decision variables, making the proposed model non-linear in nature. However, popular optimization software like LINGO and CPLEX are unable to solve, or can only obtain less satisfactory solutions for non-linear models. In view of this, a linear model (M2) is provided for the studied problem where $\alpha_k^t x_i^t$ in M1 is replaced by v_k^t. In addition, constraints (4) and (5) in M1 are strengthened by constraints (19) and (20) in M2 by assigning tighter bounds.

The second model (M2) is presented as follows:

$$[\mathbf{M2}] \min F = \sum_{k \in V} \omega_1^k (c_k - s_k) + \omega_2 \sum_{t \in T}\sum_{j \in J} u_j^t \quad (1)$$

subject to:

$$\sum_{k\in V_t^T} v_k^t \le H, \ \forall t \in T \tag{18}$$

$$v_k^t - q_k^{\min} \ge (\alpha_k^t - 1)M_k^1, \ \forall k \in V, \forall t \in T_k^V \tag{19}$$

$$v_k^t - q_k^{\max} \le (1 - \alpha_k^t)M_k^2, \ \forall k \in V, \forall t \in T_k^V \tag{20}$$

$$v_k^t \le \alpha_k^t M_k^2, \ \forall k \in V, \forall t \in T_k^V \tag{21}$$

$$\sum_{t\in T_k^V} v_k^t \ge w_k^Q, \ \forall k \in V \tag{22}$$

$$y_j^t \ge \sum_{i\in I} \sum_{k\in V_t^T\cap V_i^I\cap V_j^J} \frac{w_{jk}^Y}{w_k^Q} v_k^t, \ \forall t \in T, \forall j \in J \tag{23}$$

$$u_j^t \ge u_j^{t-1} + w_{jt}^E + \sum_{i\in I} \sum_{k\in V_t^T\cap V_i^I\cap V_j^J} \frac{w_{jk}^Y}{w_k^Q} v_k^t - y_j^t, \ \forall t \in T, \forall j \in J \tag{24}$$

$$v_k^t \in Z+, \ \forall k \in V, \forall t \in T_k^V \tag{25}$$

and constraints (3), (7)–(9), (11), (13) and (15)–(17).

Constraint (18) imposes a capacity limitation for available QCs that can work simultaneously. Constraints (19) and (20) ensure that when a vessel is being handled, the number of QCs assigned for it should be limited in an allowable range. Constraint (21) enforces v_k^t to be 0 when vessel k is not being handled. Constraint (22) ensures each vessel can be handled within the allowable time window. Constraints (23) and (24) have the same functions like constraint (10) and (12) in M1. Constraint (25) defines v_k^t to be non-negative and integral.

Hardness of the Problem

This section demonstrates that the TJCDP is NP-hard in the strong sense. To do this, we show that the decision version of the TJCDP is strongly NP-hard. That is, given the berth allocation plan and the yard management plan, it cannot be determined in polynomial time or even in pseudo-polynomial time whether the objective value F is no greater than a given constant λ unless P = NP.

We prove the NP-hardness of the studied problem by reducing a well-known strongly NP-hard problem—the multi-dimensional bin packing problem (MDBPP)—into a decision version of the TJCDP. The decision version of the MDBPP can be stated as follows:

There is a set $V$of items, each of which has volumes in J dimensions, and for item k, the jth dimensional volume is w_{kj}. There is also a set T of J-dimensional bins, each of which has the capacity of σ_j in the jth dimension. Let V_t denote the subset of items packed in bin t, then the MDBPP is to decide whether there is a packing strategy such that the following two conditions hold:

$$\sum_{k\in V_t} w_{kj} \le \sigma_j, \ \forall t \in T, \forall j \in \{1, 2, ..., J\} \tag{26}$$

$$\bigcup_{t\in T} V_t = V \tag{27}$$

Theorem 1: The decision version of the TJCDP is strongly NP-hard.

Proof: We only need to prove that an MDBPP can be transformed into a TJCDP in polynomial time. The transformation can be done as follows. A TJCDP instance corresponding to an arbitrary MDBPP instance has a set $V$of vessels to be handled in a container terminal within a planning horizon $[1, \widehat{T}]$, where $\widehat{T} = |T|$. Suppose the terminal has J blocks and there are infinite numbers of QCs, YCs and berths in the terminal. The handling time windows T_k^V for each vessel k is set equal to the planning horizon $[1, \widehat{T}]$, and the allowable handling QC number is set as $q_k^{\min} = q_k^{\max} = 2$. The workload of QC for handling vessel k (w_k^Q) is set equal to 2 QC work shifts, and workload of YC in block related to handling vessel k (w_{jk}^Y) is set as $\frac{2w_{kj}}{\Delta}$ YC work shifts, where $\Delta = \max_{j\in J}\{\sigma_j\}$. In addition, the workload for handling XT containers w_{jt}^E is assumed to be $2 - \frac{2\sigma_j}{\Delta}$ YC work shifts for block j in each work shift. Moreover, we set the weights for all vessels, ω_1^k s to be 0, the weight for the delayed workload of YCs for serving XTs ω_2 to be 1 and the constant value λ is set to be 0. Therefore, the decision version of the TJCDP is to determine whether there exists a deployment plan such that all the involved vessels can complete handling within the planning horizon and no delay happens for the YC workload related to XT services.

Let V_t denote the subset of vessels handled in work shift t. Observe that:

(i) The above transformation can be done in polynomial time.

(ii) All vessels require exactly one work shift to handle owing to the QC assignment and QC workload settings.

(iii) Since the number of YCs is assumed to be infinite, the handling capacity of all blocks in any work shift reaches the maximum of two YC work shifts.

(iv) A workload of $\frac{2w_{kj}}{\Delta}$ for YC is required by vessel k in block j in the work shift when it is being handled, and therefore, a total of $2 - \sum_{k\in V_t} \frac{2w_{kj}}{\Delta}$ capacity is left for handling XT services in block j in work shift t.

(v) If $F \leq \lambda$ then there must exist a handling plan such that the following conditions hold:

$$2 - \sum_{k \in V_t} \frac{2w_{kj}}{\Delta} \geq 2 - \frac{2\sigma_j}{\Delta}, \quad \forall t \in T, \forall j \in J \qquad (28)$$

$$\bigcup_{t \in T} V_t = V \qquad (29)$$

Obviously, Equation 28 is equivalent to Equation 26.

Therefore, it follows easily now that a TJCDP will generate an objective value of λ or less if and only if the MDBPP has a feasible solution. Thus, an MDBPP can be solved by solving a TJCDP problem.

It is worth noting that even a specialized sub-problem of MDBPP, the bin packing problem which involves only one dimension is already NP-Hard in the strong sense. It can be naturally obtained that even when only one block is involved in the TJCDP, the problem is already strongly NP-hard. Moreover, we can also prove in a similar way that the TJCDP is strongly NP-hard by assuming the block handling capacity to be infinite but the number of available QCs to be limited.

Model Application

This section describes a series of case studies that investigate how the integration of QC and YC deployments can affect the overall performance of a container terminal. These cases are designed based on practical situations. To understand the effectiveness of the joint deployment policy, these cases are solved by model M2 and the obtained results are compared with the results delivered by the method that deploys QCs and YCs in a sequential way.

For a better observation, optimal solutions should be examined. Thereby, in view of the complexity of the problem, the authors have deliberately chosen to use small test cases which can be well solved by some state-of-the-art mixed-integer software to optimum.

Experimental Settings

In the studied cases, there are 6 to 15 vessels to be handled at a container terminal within 1 week's time. A typical working day is divided into six work shifts, each containing 4 hours. There are three berths and eight blocks located in two rows. The numbers of QCs and YCs are 6 and 8, respectively. Three classes of vessels call at the terminal periodically, and detailed information of them is given in Table 2. It is assumed that these vessels arrive at the terminal randomly, and for each vessel, the designated blocks are also generated in a random way. Based on these data, a berth allocation plan is then generated accordingly. In addition, it is assumed that external trucks arrive at the terminal randomly. Therefore, the required YC workload for XT services (in YC work shifts) for each block in each work shift is randomly generated within the range [0, 0.9]. Besides, the weight of delayed workload for XT services is set as 1.

Moreover, the objective function F in M2 is divided into two parts, namely $F1 = \sum_{k \in V} \omega_1^k (c_k - s_k)$ representing vessel handling cost and $F2 = \omega_2 \sum_{t \in T} \sum_{j \in J} u_j^t$ representing the cost for delayed YC workload for XT services.

The effectiveness of the joint deployment strategy is analyzed by studying the differences between results obtained by M2 and those obtained by the sequential deployment policy. In order to capture the procedures of a sequential deployment, two new models are proposed (M3 and M4). Model M3 is used to optimize QC deployment, and model M4 is formulated to deploy YCs based on the QC deployment pattern obtained in M3.

Before presenting the two models, several new notations are introduced. First, since in M3, deployment pattern of YCs is fixed and required as an input, a parameter b_j representing the number of deployed YC in block j is introduced. In all cases, the b_j s are set in a way to ensure a most balanced distribution of a given number of YCs among all blocks. Moreover, considering that there is a possibility that some vessels cannot finish handling within the given time window under certain YC deployment patterns, the handling time window of each vessel is relaxed by extending the end of the time window to $\widetilde{T}(\widetilde{T} \geq |T|)$ to ensure feasibility of M3. The extended handling time window for vessel k is denoted as $\widetilde{T}_k^V$. It is worth noting that in practice, such infeasibility caused by fixed YC deployment can be solved by rescheduling the berth allocation plan, which is, however, out of the scope of this paper.

The QC deployment model (M3) is given as follows:

$$[\mathbf{M3}] \min F_1 \qquad (30)$$

Table 2. Parameters for Vessels Used in Test Cases

Class	Ratio	Handling time	Used QC	QC workload	YC workload	Dedicated blocks	Weight in objective
Feeder	1/3	2–4	1–3	2–5	2–10	2–6	1–3
Medium	1/3	3–5	2–4	6–14	6–28	3–8	4–6
Jumbo	1/3	4–6	3–6	15–20	15–40	4–8	7–9

Table 3. Result of the Cases

		M2			M3 + M4			
Case	Vessel	F1[a1]	F2[a2]	Time(s)	F1[b1]	F2[b2]	Time(s)	Gap (%)
1	6	31	21.53	2.70	76	10.35	0.44	39.17
2	7	47	29.66	147.89	68	21.59	0.43	14.43
3	8	83	34.55	3.80	112	24.18	0.68	13.68
4	9	53	20.93	3.14	110	18.25	0.43	42.36
5	10	88	33.86	12.44	169	31.94	1.28	39.36
6	11	124	32.26	21.70	172	30.99	1.75	23.02
7	12	108	67.54	37.16	193	45.90	1.31	26.52
8	13	127	67.79	193.84	213	73.72	2.76	32.06
9	14	115	54.36	120.84	241	42.11	5.91	40.18
10	15	89	33.37	16.88	140	29.45	1.33	27.78
Average		86.50	39.58	56.04	149.40	32.85	1.63	29.86

Note: Gap is calculated by $100 \bullet (b1 + b2 - a1 - a2)/(b1 + b2)$. The "Time" columns report the computational times of CPLEX.

subject to:

$$\sum_{t \in \bar{T}_k^V} v_k^t \geq w_k^Q, \quad \forall k \in V \tag{31}$$

$$\sum_{k \in V_t^T \cap V_j^J} \frac{w_{jk}^Y}{w_k^Q} v_k^t \leq b_j, \quad \forall t \in T, \forall j \in J \tag{32}$$

and constraints (3),(11), (13), (18)–(21), (25).

Constraint (31) ensures enough QC capacity is scheduled for vessel handling. Constraint (32) enforces yard handling capacity limitation.

The YCDP is formulated in model M4:

$$[\mathbf{M4}] \min F_2 \tag{33}$$

subject to: constraints (7)–(9), (15)–(17), (23) and (24).

Note that in constraint (23) and (24), variable v_k^t is fixed and obtained by solving model M3.

Computational Results

The 10 cases are solved by M2, M3, and M4 and the results are shown in Table 3.

As shown in Table 3, for all of the 10 cases, M2 delivers better solutions than M3 and M4. The minimum optimality gap between the two deployment methods is 13.68% for case 3 and the average gap for all cases goes greater than 29%. To be more specific, as for the vessel handling-related cost *F1*, M2 generates results that are much better than M3 for all cases. This implies that the joint deployment policy applied in M2, which enables dynamic handling capacity adjustments among blocks, can greatly reduce the time needed by a terminal to handle vessels. Further, although in M2 the costs caused by delayed YC workload for XT services *F2* are higher than those in M4 for half of the cases, the differences are relatively small. In addition, it should be noted that the improvements of *F2* in M4 are achieved by lessening the handling-related workload for blocks in each work shift, which is realized at the expenses of much longer handling times for vessels obtained in M3. To sum up, the joint deployment policy enables container terminals to better utilize the limited resources like QCs and YCs and can greatly improve their efficiency.

As for the computational time, it can be seen from the above table that CPLEX is able to obtain optimal solutions for these instances within very short time (less than 200 seconds). However, additional experimental data show that when the scale of instances grows, the solution time of CPLEX increases drastically (CPLEX fails to solve instances with 5 berths, 12 blocks and 25 vessels in reasonable time). Therefore, it would be favorable to develop some heuristic algorithms to solve instances with large scales.

Sensitivity Analysis

There are several parameters involved in the TJCDP, and some of these parameters, such as number of berths or handling time windows for vessels, are relatively fixed. Here, sensitivity analyses are performed on the two most "manipulable" parameters for a terminal operator, namely the number of QCs (*H*) and the number of YCs (*G*). The sensitivities of these parameters are investigated from two dimensions. When the settings of these parameters change is analyzed, along with how the optimal objectives of the TJCDP change and how the differences between the joint deployment policy and the sequential deployment policy change. In this section, four cases with various demand levels (Cases 1, 4, 7 and 10) are

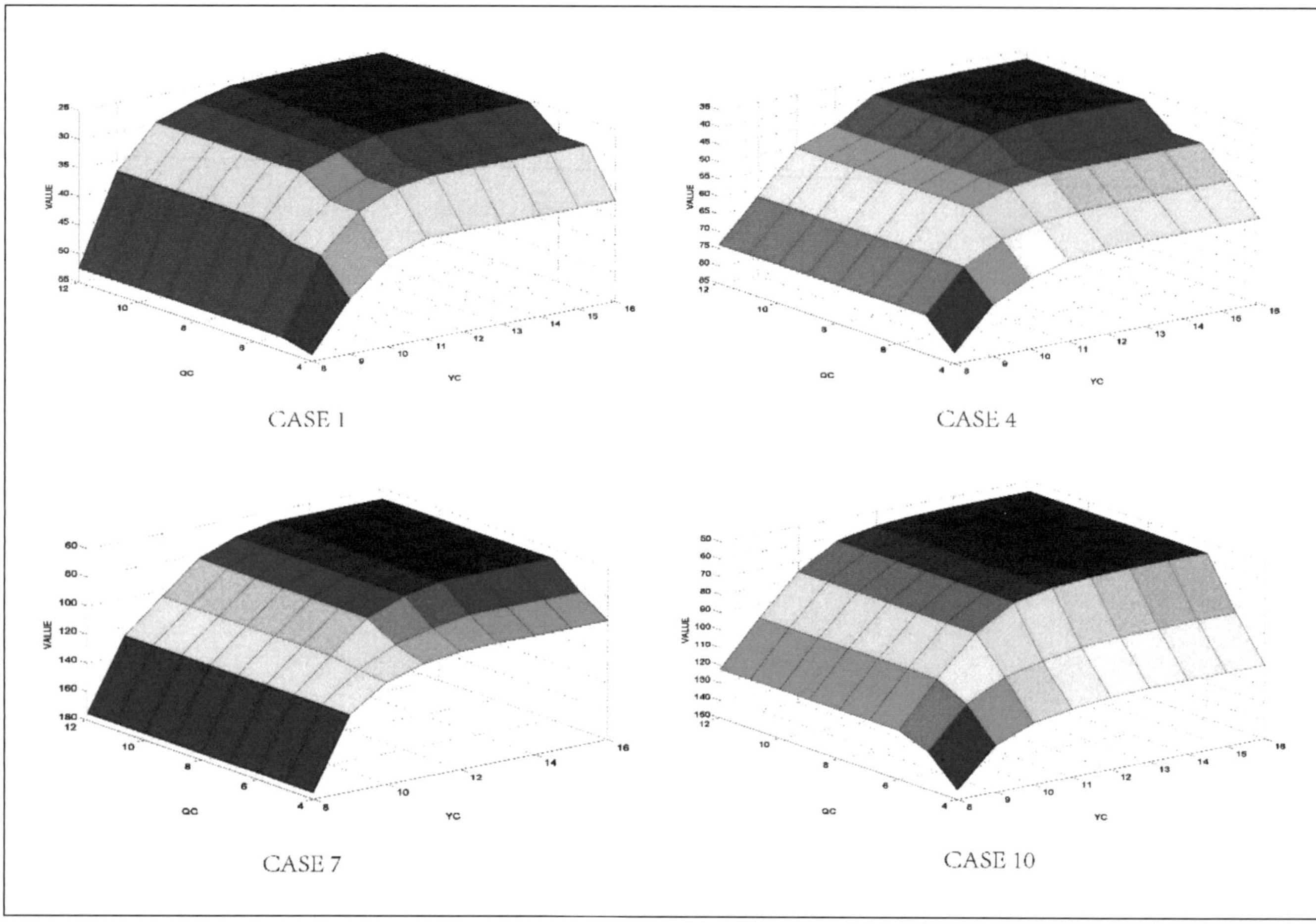

Figure 3. Results of M2 with changing QC and YC numbers.

solved under settings where the numbers of QCs and YCs vary from 4 to 12 and 8 to 16 respectively.

Sensitivity Analysis for the Optimal Value. First the impact of QC and YC numbers on the optimal values of different cases is investigated. Figure 3 shows the optimal solutions of M2 of the four cases with different QC and YC numbers. Note that for a better illustration, the direction of the "VALUE" axes is rotated in this figure.

As shown in this figure, there are diminishing returns to additional QCs or YCs if all other resources are fixed. Further, the most effective combinations of the two resources (QCs and YCs) for different cases are identified. The most effective combination for Case 1 is 7 QCs plus 12 YCs. For Case 4, the best combination is 7 QCs plus 12 YCs. The most effective combinations for Case 7 and 10 are 6 QCs plus 13 YCs and 6 QCs plus 12 YCs, respectively.

Sensitivity Analysis for the Optimality Gap. Also studied is how optimality gaps between the joint deployment policy and the sequential deployment policy in these cases change when the settings of QCs and YCs change. Figure 4 shows optimality gaps of these cases under different settings of resources. Note that the "YC" axes are rotated in this figure to give a better illustration. It is easy to see that the changes of optimality gaps in the four cases share a similar trend with regard to changing numbers of QCs and YCs.

First, when the numbers of YCs are fixed, how the optimality gaps change with changing QCs numbers is examined. The optimality gaps first sharply increase with increasing numbers of QCs before the QC numbers reach certain values (7, 8, 6, and 6 in Case 1, 4, 7, and 10, respectively) and then remain relatively stable. Next, the relationship between the optimality gaps and YC numbers was investigated, by fixing the numbers of QCs. The optimality gaps increase greatly when the numbers of YCs increase from 8 to certain values (11, 10, 12, and 11 in Case 1, 4, 7, and 10, respectively) and then experience sharp drops. It is obvious that the joint deployment policy gains the greatest superiority when the numbers of QCs and YCs are in a moderate range.

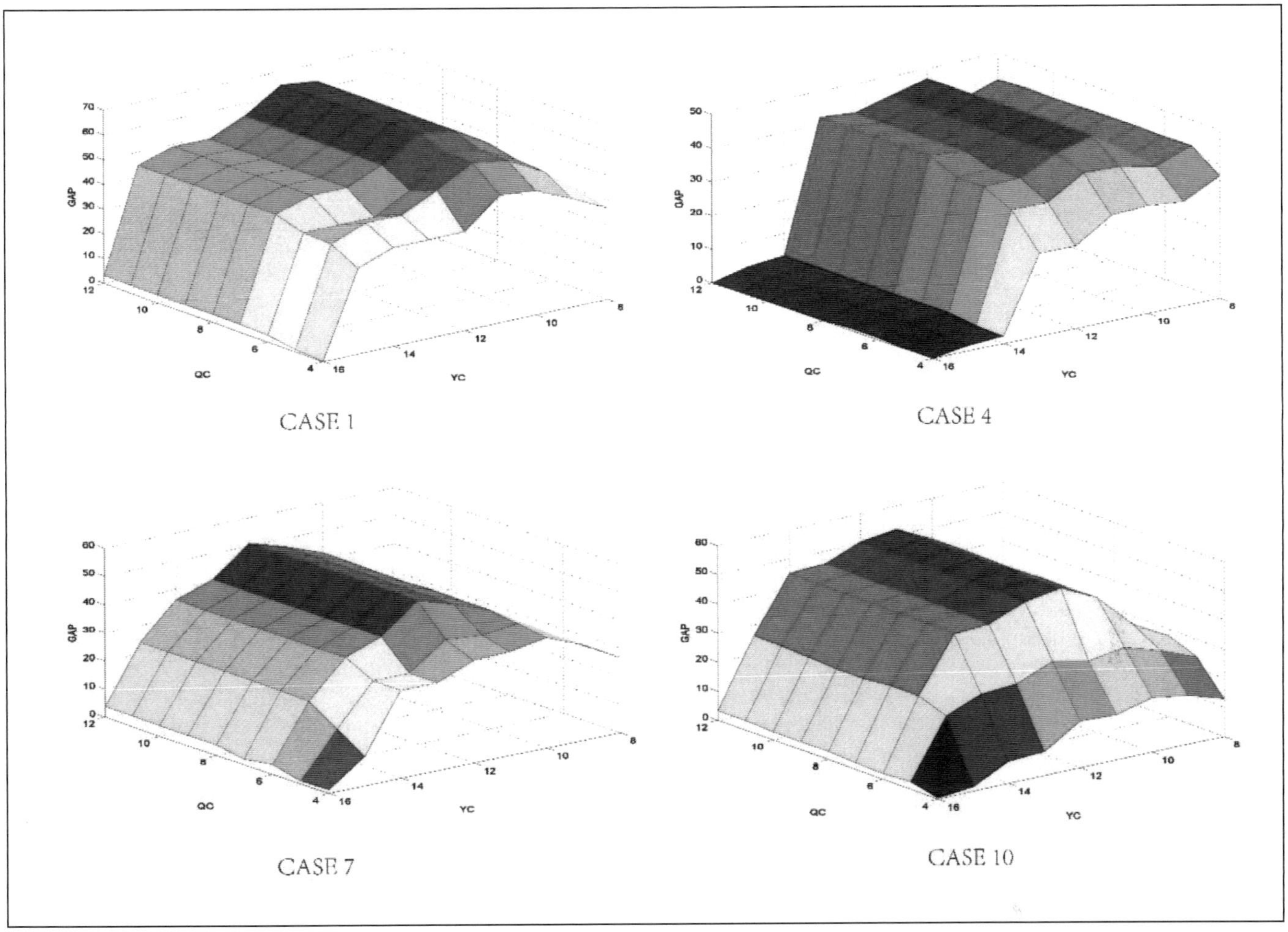

Figure 4. Optimality gaps with changing QC and YC numbers.

Conclusion

The deployments of QCs and YCs are closely intertwined and are of critical importance to the efficiency of container terminals. This paper studied the TJCDP. A nonlinear mixed-integer programming model was proposed for the considered problem and then linearized to make it easy to solve. It was proved that the problem is NP-hard in the strong sense. A series of case applications were solved by the model and discussed in detail. The computational results demonstrated the effectiveness of the model. The impacts of the numbers of QCs and YCs on the performance of the proposed model were analyzed, and the most effective combinations of QCs and YCs for a container terminal at various demand levels were identified. It was also found that the joint deployment policy gained the greatest superiority against its sequential counterpart when the numbers of QCs and YCs are modest.

For future studies, there are two promising directions. First, in view of the complexity of the problem, it would be interesting to develop efficient heuristic algorithms for solving the TJCDP on a large scale. Second, sea transportation faces great uncertainties, and future research can study the problem with uncertain berth allocation or XT transportation plans.

Acknowledgments

This study is supported by the National Natural Science Foundation of China (No. 71701178).

Author Contributions

The authors confirm contribution to the paper as follows: study conception and design: Wu L, Wang S; data collection: Wu L; analysis and interpretation of results: Wu L, Wang S; draft manuscript preparation: Wu L. All authors reviewed the results and approved the final version of the manuscript.

References

1. UNCTAD. Review of Maritime Transportation 2015. Presented at the United Nations Conference on Trade and Development, New York and Geneva, 2017. http://unctad.org/en/PublicationsLibrary/rmt2016_en.pdf. Accessed July 7, 2017.

2. Moorthy, R., and C. P. Teo. Berth Management in Container Terminal: The Template Design Problem. *OR Spectrum*, Vol. 28, No. 4, 2006, pp. 495–518.
3. Giallombardo, G., L. Moccia, M. Salani, and I. Vacca. Modeling and Solving the Tactical Berth Allocation Problem. *Transportation Research Part B: Methodological*, Vol. 44, No.1, 2010, pp. 232–245.
4. Zhen, L., Z. Xu, K. Wang, and Y. Ding. Multi-period Yard Template Planning in Container Terminals. *Transportation Research Part B: Methodological*, Vol. 93, 2016, pp. 700–719.
5. Zhen, L., E. P. Chew, and L. H. Lee. An Integrated Model for Berth Template and Yard Template Planning in Transshipment Hubs. *Transportation Science*, Vol. 45, No. 4, 2011, pp. 483–504.
6. Jin, J. G., D.-H. Lee, and H. Hu. Tactical Berth and Yard Template Design at Container Transshipment Terminals: A Column Generation Based Approach. *Transportation Research Part E: Logistics and Transportation Review*, Vol. 73, 2015, pp. 168–184.
7. Golias, M. M., M. Boile, and S. Theofanis. Discrete Berth-scheduling Problem. *Transportation Research Record: Journal of the Transportation Research Board*, 2010. 2168: 1–8.
8. Theofanis, S., M. Boile, and M. M. Golias. Container Terminal Berth Planning. *Transportation Research Record: Journal of the Transportation Research Board*, 2009. 2100: 22–28.
9. Daganzo, C. F. The Crane Scheduling Problem. *Transportation Research Part B: Methodological*, Vol. 23, No. 3, 1989, pp. 159–175.
10. Kim, K. H., and Y. M. Park. A Crane Scheduling Method for Port Container Terminals. *European Journal of Operational Research*, Vol. 156, No. 3, 2004, pp. 752–768.
11. Zhang, C., Y. Wan, J. Liu, and R. J. Linn. Dynamic Crane Deployment in Container Storage Yards. *Transportation Research Part B: Methodological*, Vol. 36, No. 6, 2002, pp. 537–555.
12. Ng, W. C., and K. L. Mak. Yard Crane Scheduling in Port Container Terminals. *Applied Mathematical Modelling*, Vol. 29, No. 3, 2005, pp. 263–276.
13. Meisel, F., and C. Bierwirth. Integration of Berth Allocation and Crane Assignment to Improve the Resource Utilization at a Seaport Container Terminal. In *Operations Research Proceedings 2005*, Springer, Berlin, Heidelberg, 2006, pp. 105–110.
14. Blazewicz, J., T. C. E. Cheng, M. Machowiak, and C. Oguz. Berth and Quay Crane Allocation: A Moldable Task Scheduling Model. *Journal of the Operational Research Society*, Vol. 62, No. 7, 2011, pp. 1189–1197.
15. Cheung, R. K., C. L. Li, and W. Lin. Interblock Crane Deployment in Container Terminals. *Transportation Science*, Vol. 36, No. 1, 2002, pp. 79–93.
16. Linn, R. J., and C. Q. Zhang. A Heuristic for Dynamic Yard Crane Deployment in a Container Terminal. *IIE Transactions*, Vol. 35, No. 2, 2003, pp. 161–174.
17. Jin, J. G., D.-H. Lee, and J. X. Cao. Storage Yard Management in Maritime Container Terminals. *Transportation Science*, Vol. 50. No. 4, 2014, pp. 1300–1313.

The Standing Committee on Intermodal Freight Terminal Design and Operations (AT050) peer-reviewed this paper (18-02151).

Article

TRR
JOURNAL OF THE TRANSPORTATION RESEARCH BOARD

Transportation Research Record
2018, Vol. 2672(9) 47–58

DOI: 10.1177/0361198118782754
journals.sagepub.com/home/trr
SAGE

Integrated Scheduling of Rail-Mounted Gantry Cranes, Internal Trucks and Reach Stackers in Railway Operation Area of Container Terminal

Baicheng Yan[1], Xiaoning Zhu[1], Li Wang[1], and Yimei Chang[1]

Abstract

In this paper, the integrated scheduling of handling equipment at the railway handling area in container terminals is studied, where rail-mounted gantry cranes, internal trucks, and reach stackers are deployed. In the course of the handling operation, loading and unloading containers are handled simultaneously. The handling process is first studied and some scheduling schemes are put forward. Based on the analysis, the problem is formulated as a mixed-integer programming model, with the objectives of minimizing the makespan and the total waiting time of all equipment. Then, to solve the problem, a genetic algorithm is employed, where the first available machine rule is applied in the selection of trucks and reach stackers. Sets of numerical experiments are conducted to verify the effect of the proposed algorithm. Based on the results of experiments, some key indicators are calculated and the effects of different equipment configuration schemes are studied.

In 2015, the Chinese government proposed the "Belt and Road" as an important development strategy. According to the development plan, railroads will cover around 80% of the major ports in China before 2018 to shorten the transfer time of containers between trains and vessels.

In China's railway container terminals, the storage yards are usually set under rail-mounted gantry cranes (RMGs) and no more than three loading–unloading tracks are laid, which are normally longer than 900 m. Nevertheless, in container terminals, the length of the loading–unloading tracks is limited, and more tracks need to be introduced under RMGs. Consequently, if the storage yards are set under RMGs, their capacity may be insufficient. Moreover, inland and outland containers have to be inspected, but the storage yards under RMGs are unsuitable for customs inspection. Therefore, blocks of storage yards in container terminals are used, and containers are transferred between areas by internal trucks (ITs).

Although rubber-tired gantry cranes (RTGCs) are widely deployed in container terminals, the movement of RTGCs between yards can be time consuming. Because of the smaller volume of sea–rail intermodal containers, reach stackers (RSs), which are cheaper and more flexible than RTGCs, are used in storage yards. In this paper, equipment configuration schemes consisting of RMGs, ITs, and RSs are studied, and the integrated scheduling of equipment is addressed.

The remainder of this paper is organized as follows. Relevant literature is first summarized. Then, the integrated scheduling problem is described and formulated as a mixed-integer programming (MIP) model, and a genetic algorithm (GA)-based optimization is proposed. Numerical experiments are conducted to evaluate the effectiveness of the proposed solution methods, and conclusions and future research are discussed in the last section.

Literature Review

To date, there have been many studies focusing on the equipment scheduling problem of container terminals. The scheduling of quay cranes (QCs), yard cranes (YCs), and yard trucks (YT) has been widely studied (*1*, *2*). Most of these studies focus on improving the efficiency and throughput of the aprons and storage blocks of

[1]School of Traffic and Transportation, Beijing Jiaotong University, Beijing, P. R. China

Corresponding Author:
Address correspondence to Xiaoning Zhu: xnzhu@bjtu.edu.cn

container terminals by selecting the makespan and the vessels' leaving time as objectives.

As the most important equipment in container terminals, the scheduling of QCs has been studied by many researchers. The work of Kim and Park has been widely cited when the scheduling of QCs is discussed (*3*). They formulated the QC scheduling as an m-parallel machine scheduling problem with the goal of minimizing the completion time of all QCs. To solve the problem effectively, a heuristic search algorithm was proposed. Lee et al. studied the handling sequence of QCs (*4*). A MIP model considering the interference between QCs was proposed. The problem of the double cycling of QCs was addressed by Goodchild and Daganzo (*5*). Strategies for adjusting the port's operation to support double cycling were proposed. In addition to the interference between QCs, some other constraints were taken into consideration. The stability of vessels was considered by Wang et al., and a MIP model regarding the stability of vessels and scheduling of QCs was proposed (*6*). Li et al. studied the problem of the disruption recovery of integrated berth allocation and QC assignment (*7*).

For the scheduling of YCs, the scheduling of straddle cranes was studied by Kim and Kim, where the total travel distance was optimized (*8*, *9*). The dynamic deployment of RTGCs among blocks was studied by Zhang et al., with the aim of minimizing the total delayed workload (*10*). To solve the problem, a MIP model was formulated and the Lagrangian relaxation model was developed. The problem of bay allocation and scheduling of YCs for transshipment containers was addressed by Lee et al. (*11*). With the development of automated container terminals (ACT), research related to automated YCs is getting more attention. Hu et al. studied the problem of the sequencing and routing of automated lifting vehicles in ACT (*12*). The interference between stacking cranes is formulated by analyzing the minimal temporal intervals between any two tasks.

For the scheduling of YTs, the deployment of automated guided vehicles (AGVs) was analyzed by Vis et al., where the optimum number of AGVs in a semi-automated terminal and the fleet size of AGVs were calculated (*13*, *14*). The routing of yard trailers was studied by Nishimura et al., and GA was employed to solve the problem (*15*). The sharing of trucks among container terminals has been studied by He et al., with the aim of the minimization of the total overflowed workloads and total transferring costs (*16*). The rolling-horizon approach was employed for the immediate scheduling.

The problem of the integrated scheduling of different types of handling equipment has also been addressed in recent years. A problem including storage allocation, dispatch of vehicles, and scheduling of cranes was studied by Bish et al., and a heuristic algorithm was developed (*17*). Chen et al. studied the integrated scheduling of handling equipment in container terminals, which was formulated as a hybrid flow shop problem, and Tabu Search was employed (*18*). The integrated scheduling of crane handling and truck transportation in container terminals was further studied by Chen et al., where a crane scheduling plan, truck routing plan, and a complete solution were constructed separately by a three-stage algorithm (*19*). The scheduling of ITs, YCs, and QCs in container terminals was addressed by He et al. (*20*). A MIP model considering the minimizing of total departure delay of vessels and energy consumption was developed, and a GA-based algorithm proposed.

The scheduling of RSs, which are widely used in railway container terminals, has not been well studied. Moreover, the handling operation for sea–rail intermodal container transportations is scarcely studied, and existing research related to intermodal transportation has mainly focused on the coordination of different modes (*21*), network planning (*22*), and evaluation (*23*). Therefore, the integrated scheduling of the RMGs, RSs, and ITs in container terminals for sea–rail intermodal transportation is studied in this paper.

Problem Description and Formulation

Railway Handling Area Configuration

The configuration of the railway handling area considered in this paper consists of railway handling lines, connecting area, and container yard (Figure 1). RMGs span several parallel railway tracks and driving lines for trucks. A waiting area is established beside the tracks, for waiting ITs before handling operations are started. Also, a floor storage area is set in the waiting area for the temporary depositing of intermediate containers that cannot be immediately loaded onto ITs.

RSs are used in container yards, and each of them operates in a specific block during the handling process. For the movement and handling operation of RSs within the block, each block is divided into several sub-blocks. Transfer areas are set near the blocks, where the handling operations between RSs and trucks can be conducted. There is a connecting area between the railway handling lines and container yard. In practice, the connecting area can be the road between container yards and railway handling area, whose length can influence the makespan.

Equipment Scheduling Scheme

To organize handling operations efficiently, some scheduling rules are proposed in practice. In this paper, as ITs are shared within the area and each block is served by more than one RS, the scheduling of handling

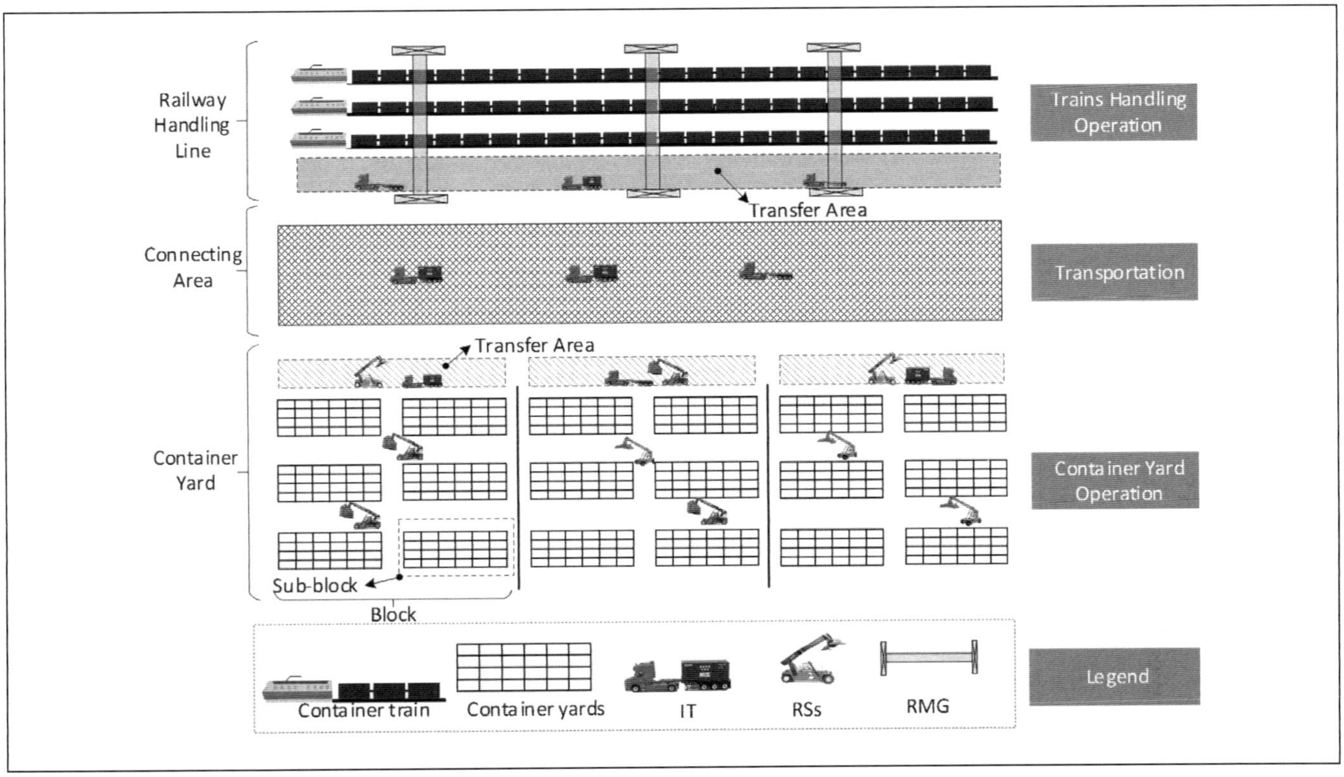

Figure 1. Configuration of railway handling area in container terminal.

equipment can be very complicated. Therefore, a scheduling scheme is given. First of all, several assumptions are proposed.

1. The locations of containers are given, including their locations on the train and corresponding slots in container yards.
2. Only 40ft containers are discussed, which means that only one container can be loaded on each wagon.
3. Import and export containers are stacked in different sub-blocks.
4. The reshuffle and housekeeping operations in the container yard are not considered, which means that during the handling operation, RSs serve the loading and unloading of the containers only.
5. The conflict between RSs and traffic congestion caused by ITs are not considered.

Based on the above assumptions, the scheduling scheme can be elaborated as follows. The operating range for each of the RMGs is fixed and interference between them can be avoided. Specifically, during the loading operation, the RMG moves to the specific location of the task and waits for the arrival of the container. After the IT arrives, the RMG transfers the container to the specified wagon and then moves to the next position. The operations of RMGs are similar in the unloading operation. However, when a loading operation is conducted, as the specified wagon may be occupied, the container on the wagon needs to be stored in the floor storage area first and transferred to the corresponding ITs later in the process.

For the scheduling of RSs and ITs, the first available equipment is selected according to the time that containers arrive at the specific area. In the unloading operation, as the arrival time of the ITs can be calculated, RSs that can be used first are selected. Then, RSs move to the specified positions with containers and stack them in the assigned slots. In the loading process, RSs handle the containers according to the operation sequences of QCs, and the RSs that can arrive at the slots first are selected. The containers will then be transferred to the transfer areas, and the ITs are selected according to the arrival time.

Formulation

Parameters and Decision Variables

Parameters. T = The set of all tasks to be handled (indexed by i), where $i = 0$ denotes a virtual container, $|T| = n$;

Ω = The set of all RMGs that can be scheduled (indexed by r), $|\Omega| = m_1$;

Γ = The set of all ITs that can be scheduled (indexed by k), $|\Gamma| = m_2$;

B = The set of blocks in which containers stacked (indexed by b), $|B| = m_b$;

Y_b = The set of all RSs that can be scheduled in block b (indexed by rs_b), $|Y_b| = m_{3b}$;

W = The set of wagons that be assigned to tasks, denoted by w_i;

$x^r{}_i = \in \{0, 1\}$, 1—if task i is conducted by RMG r, otherwise $x^r{}_i = 0$;

l_i = The slot of task i in container yards and $l_i = \{l_{1i}, l_{2i}, l_{3i}, l_{4i}, l_{5i}\}$, which represent the block number, the sub-block number, the bay number, the stack number and the tier number of container i respectively;

$dr^r{}_{i'i}$ = The traveling distance of RMG r from task i' to task i;

$dr^r{}_{0i}$ = The traveling distance of RMG r from its original position to the wagon of task i;

$ds_{i'i}$ = The traveling distance of reach stacker rs_b from the end position of task i' to the start position of task i in block b;

ds_{0i} = The distance of reach stacker rs_b from its original position to the original position of task i in block b;

ds_i = The traveling distance of reach stacker rs_b for task i in the block;

$dt^k{}_{i'i}$ = The distance of IT k from the destination of task i' to the origin of task i, where the destination and origin mean the wagon number or the block number of the container, respectively;

$dt^k{}_{0i}$ = The distance of IT k from its original position to the origin of task i;

$dt^k{}_i$ = The distance of IT k from the origin to the destination of task i, where the destination and origin mean the wagon number or the block number of the container, respectively;

vr_e = The running speed of RMGs in unloaded condition;

vr_l = The running speed of RMGs in loaded condition;

vs_e = The running speed of RSs in empty condition;

vs_l = The running speed of RSs in loaded condition;

vt_e = The velocity of ITs in unloaded condition;

vt_l = The velocity of ITs in loaded condition;

p_i = The handling time of container i by RMGs;

p_{1i} = The handling time of transfer container i from IT to wagon or in reverse;

p_{2i} = The handling time of transfer container i from wagon to floor storage area;

p_{3i} = The handling time of transfer container i from floor storage area to IT;

τ_{1i} = The handling time of task i by a reach stacker at transfer area;

τ_{2i} = The handling time of task i by a reach stacker in the block;

$\lambda_i = \in \{0, 1\}$, 1 if task i is a loading operation; $\lambda_i = 0$ otherwise;

M = A large positive number;

Decision Variables. $\eta_i = \in \{0, 1\}$, 1 if container i needed to be stored in the floor storage area first, 0 otherwise;

$\alpha^r{}_{i'i} = \in \{0, 1\}$, 1 if RMG r conducts task i right after i', 0 otherwise;

$\beta^{rs_b}{}_{i'i} = \in \{0, 1\}$, 1 if reach stacker rs_b conducts task i right after i', 0 otherwise;

$\gamma^k{}_{i'i} = \in \{0, 1\}$, 1 if IT k travels from the destination of container i' to the origin of container i, 0 otherwise;

$\mu_{i'i} = \in \{0, 1\}$, 1 if task i' is completed before task i, 0 otherwise;

cr_i = The completion time of task i by RMG, where $i \in T$;

cs_i = The completion time of task i by RS;

ct_i = The completion time of task i by IT;

atr_i = The time that IT arrives at the wagon location of task i;

atb_i = The time that IT arrives at the transfer area of task i;

The Integrated Scheduling Model

Objectives. The minimization of makespan and total waiting time are included in objectives. The waiting time for the handling operation includes the waiting time between RMGs and ITs and the waiting time between RSs and ITs. The waiting time is mainly determined by the arrival time of the interfacing equipment. Therefore, the waiting time of ITs can be calculated as

$$wt^k{}_i = \max(0, x^r_i \cdot ar_i - \sum_{i' \in \Gamma, i' \neq i} \gamma^k_{i'i} \cdot atr_i) + \max(0, \sum_{i' \in \Gamma, i' \neq i} \beta^{rs_b}_{i'i} \cdot as_i - \sum_{i' \in \Gamma, i' \neq i} \gamma^k_{i'i} \cdot atb_i) \quad (1)$$
$$\forall i \in T, r \in \Omega, k \in \Gamma, rs_b \in Y_b$$

The waiting time of RMGs can be calculated as

$$wr^r{}_i = \max(0, \sum_{i' \in \Gamma, i' \neq i} \gamma^k_{i'i} \cdot atr_i - x^r_i \cdot ar_i), \quad \forall i \in T, r \in \Omega, k \in \Gamma \quad (2)$$

The waiting time of RSs can be calculated as

$$ws^{rs_b}{}_i = \max(0, \sum_{i' \in \Gamma, i' \neq i} \gamma^k_{i'i} \cdot atb_i - \sum_{i' \in \Gamma, i' \neq i} \beta^{rs_b}_{i'i} \cdot as_i), \quad \forall i \in T, rs_b \in Y_b, k \in \Gamma \quad (3)$$

ar_i and as_i are the time that RMGs and RSs arrive at the corresponding points, which can be calculated as

$$ar_i = \sum_{i' \in \Gamma, i' \neq i} \alpha^r_{i'i} \cdot cr_{i'} + \sum_{i' \in \Gamma, i' \neq i} \alpha^r_{i'i} \cdot dr_{i'i} \Big/ vr_e, \forall i \in T, r \in \Omega \tag{4}$$

$$as_i = \sum_{i' \in \Gamma, i' \neq i} cs_{i'} \cdot \beta^{rs_b}_{i'i} + \sum_{i' \in \Gamma, i' \neq i} (1 - \lambda_{i'}) ds_{i'} \cdot \beta^{rs_b}_{i'i} \Big/ vs_e, \forall i \in T, rs_b \in \Upsilon_b \tag{5}$$

The integrated scheduling problem in this paper can be formulated as a MIP model. Based on the calculation of waiting time, two objective of this problem can be formulated as follow.

$$\min f_1 = \max(\lambda_i \cdot cr_i + (1 - \lambda_i) \cdot cs_i), \forall i \in T \tag{6}$$

$$\min f_2 = \sum_{i \in T} \sum_{k \in \Gamma} wt^k_i + \sum_{i \in T} \sum_{r \in \Omega} wr^r_i + \sum_{i \in T} \sum_{b \in B} \sum_{rs_b \in \Upsilon_b} ws^{rs_b}_i \tag{7}$$

The first objective of the model is the minimization of the makespan of all tasks. The second objective is the minimization of the waiting time of all equipment, which will guarantee a smaller unproductive time.

Constraints. The constraints of the integrated scheduling problem can be divided into several parts: the constraints related to time, the constraints related to operation sequence, and the constraints related to the range of value.

- Constraints related to time

$$\eta_i \leq (1 - \lambda_i) \cdot \mathrm{M} \quad \forall i \in T \tag{8}$$

$$p_i = (1 - \eta_i) \cdot p_{1i} + \eta_i \cdot p_{3i} \quad \forall i \in T \tag{9}$$

$$cr_i - (dr_{i'i}/vr_e + p_i + \eta_i \cdot p_{2i}) + (1 - \alpha^r{}_{i'i})M \geq cr_{i'}, \forall i, i' \in \mathrm{T}, i \neq i', r \in \Omega, i' = 0 \tag{10}$$

$$cr_i - (dr_{i'i}/vr_e + p_i + \eta_i \cdot p_{2i}) + (1 - \alpha^r{}_{i'i})M \geq cr_{i'}, \forall i, i' \in \mathrm{T}, i \neq i', r \in \Omega, i' > 0 \tag{11}$$

$$\tau_i = \tau_{1i} + \tau_{2i}, \quad \forall i \in \mathrm{T} \tag{12}$$

$$cs_i - (ds_{i'i}/vs_e + \tau_i + ds_i/vs_l) + (1 - \beta^{rs_b}{}_{i'i})M \geq cs_{i'}, \forall i, i' \in \mathrm{T}, i \neq i', rs_b \in \Upsilon_b, \Upsilon_b \in \mathrm{B}, i' = 0 \tag{13}$$

$$cs_i - (ds_{i'i}/vs_e + \tau_i + ds_i/vs_l) + (1 - \beta^{rs_b}{}_{i'i})M \geq cs_{i'}, \forall i, i' \in \mathrm{T}, i \neq i', rs_b \in \Upsilon_b, \Upsilon_b \in \mathrm{B}, i' > 0 \tag{14}$$

$$atr_i + (1 - \gamma^k_{i'i})M \geq (ct_{i'} + dt_{i'i}/vt_l) \cdot (1 - \lambda_i), \forall i, i' \in \mathrm{T}, i \neq i', k \in \Gamma, i' = 0 \tag{15}$$

$$atb_i + (1 - \sum_{i' \in \mathrm{T}, i' \neq i} \gamma^k_{i'i})M \geq (atr_i + p_i + dt^k{}_i/vt_l) \cdot (1 - \lambda_i), \quad \forall i \in \mathrm{T}, k \in \Gamma \tag{16}$$

$$atb_i + (1 - \gamma^k_{i'i})M \geq (ct_{i'} + dt_{i'i}/vt_l) \cdot \lambda_i, \forall i, i' \in \mathrm{T}, i \neq i', k \in \Gamma, i' = 0 \tag{17}$$

$$atr_i + (1 - z^k_i)M \geq (atb_i + \tau_{1i} + dt^k{}_i/vt_l) \cdot \lambda_i, \forall i, i' \in \mathrm{T}, i \neq i', k \in \Gamma, i' = 0 \tag{18}$$

$$cr_i \geq atr_i + p_i, \quad \forall i \in \mathrm{T} \tag{19}$$

$$cs_i \geq atb_i + \tau_{1i} + (ds_i/vs_l + \tau_{2i}) \cdot (1 - \lambda_i), \quad \forall i \in \mathrm{T} \tag{20}$$

Constraint 8–20 are related to the time of operation. Constraint 8 defines the relationship between η_i and λ_i, which means only the unloading containers can be temporarily stored in the floor storage area. Constraint 9 defines the handling time of RMGs. Constraints 10 and 11 indicate the relationship between cr_i and $cr_{i'}$ when RMGs conduct task i and task i' continuously. Constraint 10 defines the completion time of the first task of each RMG. Constraints 12–14 are constraints related to the operation time of RSs. Constraint 12 defines the handling time τ_i of RSs. Constraint 13 defines the completion time of the first task. Constraint 14 shows the relationship of the completion time of two successive tasks conducted by the same RS, which indicates that the time interval between cs_i and $cs_{i'}$ satisfies the time requirement for transferring tasks. Constraints 15–18 define the arrival time of ITs and completion time of RMGs and RSs. Constraints 15 and 17 mean that for unloading and loading operations, the time intervals that ITs transfer between tasks satisfies the minimum running time between different locations. Constraints 16 and 18 indicate the relationship between atb_i and atr_i, in which the operation time and running time are considered. Constraint 19 guarantees that the completion time of task i by RMG is later than the arrival time of IT. Constraint 20 ensures that for RSs the completion time of task i is later than the sum of ITs' arrival time, the handling time of RSs, and the running time of RSs inside the block.

- Constraints related to operation sequences

$$\sum_{r \in \Omega} \sum_{i' \neq i, i' \in \mathrm{T}} \alpha^r{}_{i'i} = 1, \quad \forall i \in \mathrm{T}, i \neq 0 \tag{21}$$

$$\sum_{r\in\Omega}\sum_{i'\neq i,i\in T}\alpha^{r}{}_{i'i}=1,\quad \forall i'\in T,i'\neq 0 \tag{22}$$

$$\sum_{r\in\Omega}\alpha^{r}{}_{0i}\leq 1,\quad \forall i\in T,i\neq 0 \tag{23}$$

$$\sum_{i\in T,i\neq 0}\alpha^{r}{}_{0i}=1,\quad \forall r\in\Omega \tag{24}$$

$$\sum_{r\in\Omega}\alpha^{r}{}_{i0}\leq 1,\quad \forall i\in T,i\neq 0 \tag{25}$$

$$\sum_{i\in T,i\neq 0}\alpha^{r}{}_{i0}=1,\quad \forall r\in\Omega \tag{26}$$

$$\alpha^{r}_{i'i}+\alpha^{r}_{ii'}\leq 1,\quad \forall i,i'\in T,r\in\Omega \tag{27}$$

$$\sum_{i'\in T,i'\neq i}\alpha^{r}{}_{i'i}=\sum_{i''\in T,i''\neq i}\alpha^{r}{}_{ii''},\quad \forall i\in T,i\neq 0,r\in\Omega \tag{28}$$

$$\sum_{rs_b\in Y_b}\sum_{i'\neq i,i'\in T}\beta^{rs_b}{}_{i'i}=1,\quad \forall i\in T,Y_b\in B \tag{29}$$

$$\sum_{rs_b\in Y_b}\sum_{i'\neq i,i\in T}\beta^{rs_b}{}_{i'i}=1,\quad \forall i'\in T,Y_b\in B \tag{30}$$

$$\sum_{rs_b\in Y_b}\beta^{rs_b}{}_{0i}\leq 1,\quad \forall i\in T,i\neq 0,Y_b\in B \tag{31}$$

$$\sum_{i\in T,i\neq 0}\beta^{rs_b}{}_{0i}=1,\quad \forall rs_b\in Y_b,Y_b\in B \tag{32}$$

$$\sum_{rs_b\in Y_{l_{1i}}}\beta^{rs_b}{}_{i0}\leq 1,\quad \forall i\in T,i\neq 0,Y_b\in B \tag{33}$$

$$\sum_{i\in T,i\neq 0}\beta^{rs_b}{}_{i0}=1,\quad \forall rs_b\in Y_{l_{1i}},Y_b\in B \tag{34}$$

$$\beta^{rs_b}{}_{i'i}+\beta^{rs_b}{}_{ii'}\leq 1,\quad \forall i,i'\in T,rs_b\in Y_b,Y_b\in B \tag{35}$$

$$\sum_{i'\neq i,i\in T}\beta^{rs_b}{}_{i'i}=\sum_{i''\neq i,i\in T}\beta^{rs_b}{}_{ii''},\quad \forall i\in T,\; i\neq 0,rs_b\in Y_b,Y_b\in B \tag{36}$$

$$\sum_{k\in\Gamma}\sum_{i'\in T,i\neq i'}\gamma^{k}{}_{i'i}=1,\quad \forall i\in T,i\neq 0 \tag{37}$$

$$\sum_{k\in\Gamma}\sum_{i\in T,i\neq i'}\gamma^{k}{}_{i'i}=1,\quad \forall i'\in T,i\neq 0 \tag{38}$$

$$\sum_{k\in\Gamma}\gamma^{k}{}_{0i}\leq 1,\quad \forall i\in T,i\neq 0 \tag{39}$$

$$\sum_{i\in T,i\neq 0}\gamma^{k}{}_{0i}=1,\quad \forall k\in\Gamma \tag{40}$$

$$\sum_{k\in\Gamma}\gamma^{k}{}_{i0}\leq 1,\quad \forall i\in T,i\neq 0 \tag{41}$$

$$\sum_{i\in T,i\neq 0}\gamma^{k}{}_{i0}=1,\quad \forall k\in\Gamma \tag{42}$$

$$\gamma^{k}{}_{ii'}+\gamma^{k}{}_{i'i}\leq 1,\quad \forall i,i'\in T,i\neq i',k\in\Gamma \tag{43}$$

$$\sum_{i'\in T,i\neq i'}\gamma^{k}{}_{i'i}=\sum_{i''\in T,i\neq i''}\gamma^{k}{}_{ii''},\quad \forall i\in T,i\neq 0,k\in\Gamma \tag{44}$$

Constraints 21–44 are related to the operation sequence of different equipment, in which constraints 21–28, constraints 29–36, and constraints 37–44 are related to the operation sequences of RMGs, RSs, and ITs, respectively. As the operation area of each RMGs is given and the safety distance between RMGs can be guaranteed, the constraints related to operation sequence of RMGs, RSs, and ITs are similar in expressions and meaning. Therefore, only constraints related to RMGs are discussed, as the meanings of other constraints are similar. Constraint 21 ensures that for each task i exactly one RMG and one preceding task can be assigned. Similarly, constraint 22 guarantees that for each task i' exactly one RMG and one succeeding task can be assigned. Constraint 24 means that each RMG conducts one task after it leaves its initial position. Constraints 23 and 25 ensure that for each task i there is at most one inflow or outflow arc connecting the virtual start or end position of RMGs. Constraint 26 guarantees that each RMG conducts one task as its last task. Constraint 27 ensures the operation sequence between task i and i'. Constraint 28 ensures that each RMG can conduct tasks continuously.

- Constraints related to ranges of decision variables

$$\eta_i,\alpha^{r}{}_{i'i},\beta^{rs_b}{}_{i'i},\gamma^{k}{}_{i'i}\in\{0,1\},\quad \forall i,i'\in T,i\neq i',\; r\in\Omega,k\in\Gamma,rs_b\in Y_b,Y_b\in B \tag{45}$$

$$cr_i,cs_i,ct_i,atr_i,atb_i\geq 0,\quad \forall i\in T,i\neq 0 \tag{46}$$

Constraint 45 defines the binary variables and constraint 46 defines the integer variables.

Genetic Algorithm

The integrated scheduling of handling equipment at container terminals can be formulated as a hybrid flow shop scheduling problem, where QCs, ITs, and YCs are the machines of different stages and the loading and unloading processes for containers are jobs that need to be completed (*18*); this is well known as a non-deterministic polynomial hard problem (NP-hard). As a block is served by several RSs, the integrated scheduling of RMGs, ITs, and RSs can be very complicated. Therefore heuristic algorithms are widely used to obtain near-optimal scheduling plans (*24*). To solve the problem efficiently, the following GA was designed.

Framework of the GA

The integrated scheduling of handling equipment in container terminals can be divided into two steps: the

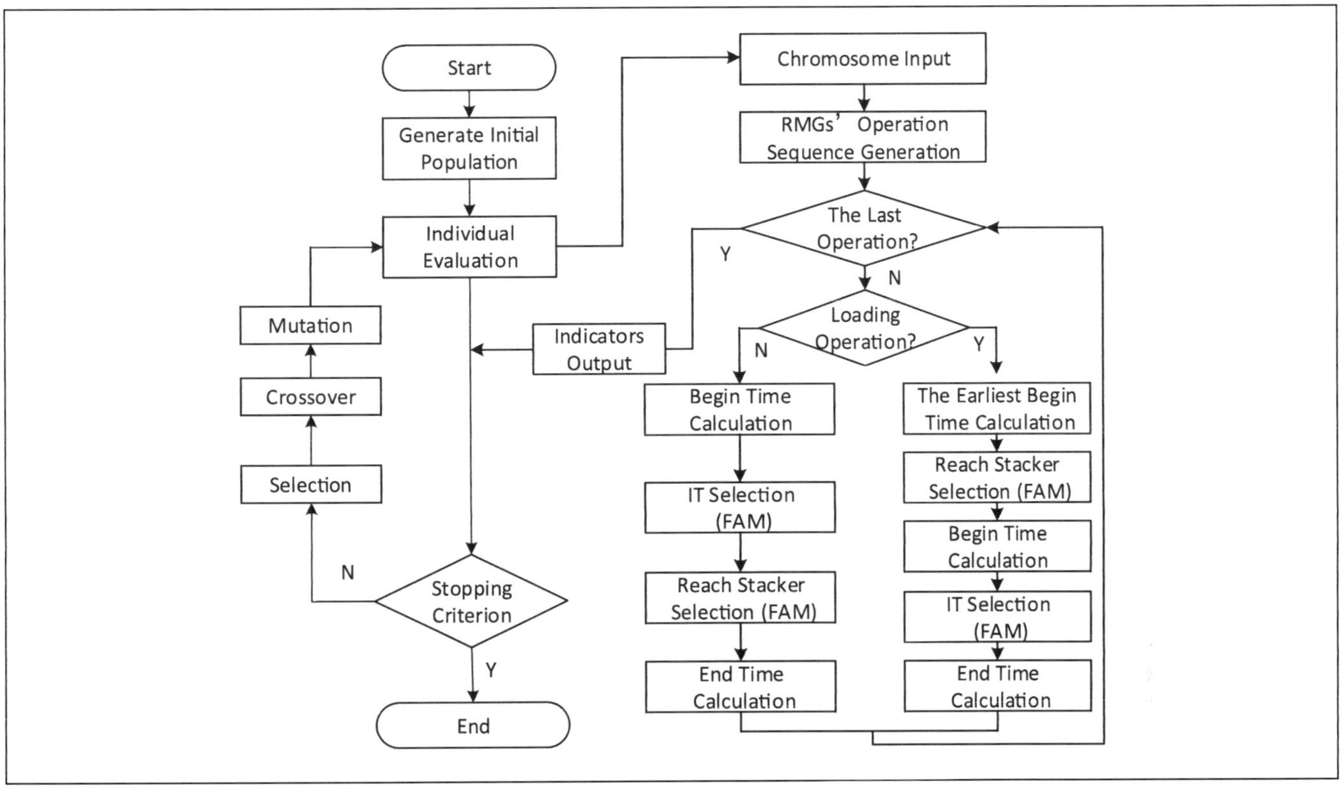

Figure 2. Flowchart of the GA-based optimization algorithm.

allocation of tasks and the calculation of the operation sequence. To solve these problems effectively and efficiently, heuristic rules are adopted. Generally, two rules are adopted in the equipment scheduling problem: first come first service and first available machine (FAM) (*18*). As the minimization of total waiting time of all equipment is considered, the FAM rule is applied. The GA procedure here is illustrated in Figure 2.

Representation

The main purpose of the chromosome is to give the operation sequences of all RMGs. In the proposed GA, the operation sequences of RMGs are represented as a one-dimensional chromosome, whose genes are the number of the tasks. For example, when 8 tasks are handled by 2 RMGs, the chromosome can be denoted as {83415762}. When tasks for RMGs are given, for example, tasks 1–4 are assigned to RMG 1 and tasks 6–8 are assigned to RMG 2, the operation sequences of the two RMGs can be obtained as {3412} and {8576}.

For unloading operations, the start time is mainly influenced by the RMGs' operation sequences. When the operation sequences of RMGs are given, ITs and RSs can be chosen according to FAM. For loading operations, the start time of the loading operations is mainly affected by the completion times of preceding tasks. When the completion times of the preceding tasks are known, the earliest start time of RMGs for loading tasks can be calculated, which can be further used in the assignment of tasks.

Evaluation and Selection

To integrate the makespan and total waiting time, normalization of the objectives is adopted, which is shown as function 47, where $f_{1\max}, f_{1\min}, f_{2\max}$ and $f_{2\min}$ represent the maximum and minimum value of the makespan and total waiting time, respectively. α_1 and α_2 are the weights of makespan and waiting time, where $\alpha_1 + \alpha_2 = 1$. In the selection operation, as the original objective has been transferred to the maximization of the normalization value, f_i can be used as the fitness value and roulette selection is applied.

$$f_i = \max \alpha_1 \cdot \frac{f_{1\max} - f_{1i}}{f_{1\max} - f_{1\min}} + \alpha_2 \cdot \frac{f_{2\max} - f_{2i}}{f_{2\max} - f_{2\min}} \qquad (47)$$

Crossover

In this paper, partially mapping crossover (PMX) is adopted in the crossover operation. An example of the crossover operation is shown in Figure 3. During the operation, the chromosomes are divided into the operation sequences of RMGs. Then, for each RMG, crossover positions are generated randomly and PMX is

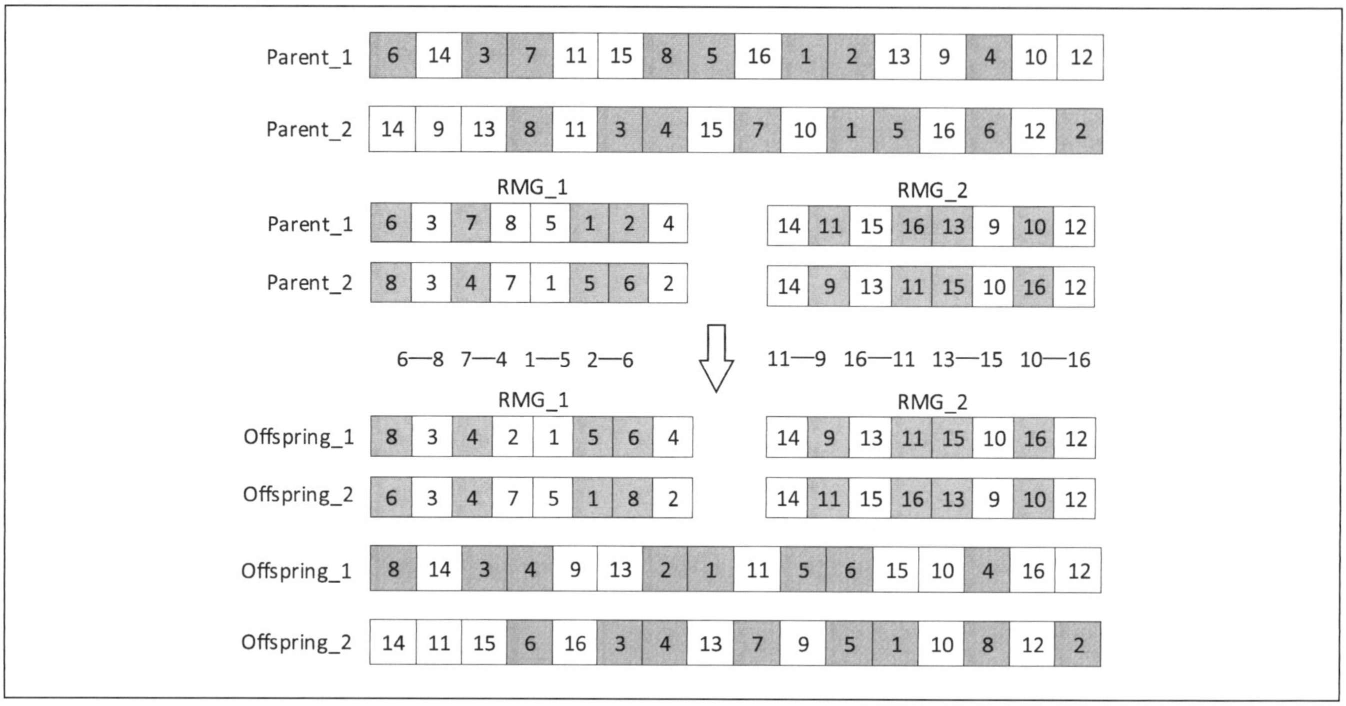

Figure 3. An example of the crossover operation.

conducted to obtain new operation sequences. Lastly, the new operation sequences are used to replace the original sequences to obtain the offspring.

Mutation

The mutation operation in this paper consists of both reverse operator and swap operator. For reverse operator, two positions of the operation sequence are generated randomly and the tasks between them are reversed. For swap operator, several positions are selected randomly and the tasks at these positions are swapped.

Numerical Experiments

To evaluate the effectiveness of the proposed model and algorithm, a series of experiments were conducted. First, the effectiveness of the algorithm was verified with a case of 40 tasks (20 loading and 20 unloading) and the values of α_1 and α_2 were determined. Then, experiments of large size were conducted, and cases with different equipment number compared. All experiments were run on a computer with Intel Core i5-4210M @ 2.3GHz processors, 8 GB RAM, and 64 bit operation system.

Parameters of the Experiment

The key parameters of the experiment are set as shown in Table 1. The parameters related to RMGs and RSs are taken from the parameters of SRMG5507 and SRSC4535GC of SANY. The parameters related to the ITs are set according to the Hova tractors of SINOTRUK.

Small-Size Problems Study

As the minimization of makespan cannot guarantee a reduced waiting time, to evaluate the effect of different weights, several sets of experiments with different values of α_1 are conducted, in which 40 tasks are handled by 3 RMGs, five ITs, and 4 RSs (two in each block). For each pair of weights, the experiment is conducted 10 times. The population size (*pop*), maximum generations (MAXGEN), probability of crossover (p_c), probability of mutation (p_m), and gap (*GAP*) between generations are set to be 100, 500, 0.8, 0.2 and 0.9, respectively.

The performance comparison of different weights is shown in Figure 4, in which the number below represents the value of α_1 and α_2; for example, "0307" means $\alpha_1 = 0.3$ and $\alpha_2 = 0.7$, respectively. It is clear that, when $\alpha_1 = 0.6$ and $\alpha_2 = 0.4$, a better result can be achieved. Therefore, to achieve a better optimization effect, the multi-objective approach is adopted and the weights of makespan and waiting time are set as 0.6 and 0.4, respectively.

Equipment Configuration Schemes Study

To verify the effect of different equipment configuration schemes, sets of large-scale experiments (240 tasks) were

Table 1. Key Parameters of the Experiments

Name of parameter	Range	Value
Gantry traveling speed of RMG (Full Load)	30–100 m /min	50 m/min
Trolley traversing speed of RMG	50–100 m /min	50 m/min
Hoisting speed of RMG (Empty Load)	30–80 m /min	30 m/min
Hoisting speed of RMG (Full Load)	15–40 m /min	20 m/min
Hoisting height of RMG	9.5–21.2 m	10 m
Travel speed of IT (Empty Load)	≤39 km/h	21.6 km/h
Travel speed of IT (Full Load)	≤39 km/h	18 km/h
Travel speed of reach stacker (Empty Load)	≤25 km/h	25 km/h
Travel speed of reach stacker (Full Load)	≤21 km/h	20 km/h
Lifting speed of reach stacker (Empty Load)	≤420 mm/s	250 mm/s
Lifting speed of reach stacker (Full Load)	≤250 mm/s	250 mm/s
Length of the wagon	——	15 m
Length of the container (including the space between containers)	——	12.5 m
Height of the container	——	2.5 m
Maximum number of the layers in sub-block	——	4
Maximum number of containers in one bay	——	16
Distance between railway lines	——	6.5m
Distance between railway handling area and container yard	——	800m

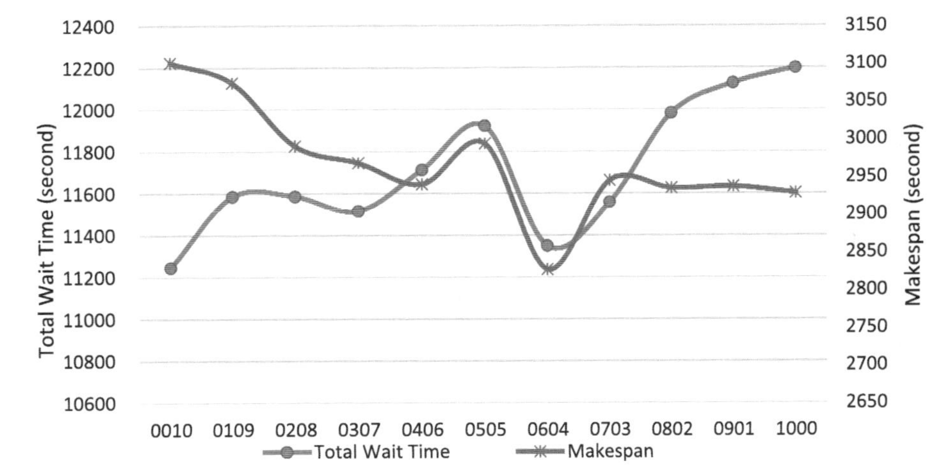

Figure 4. The performance comparison of different weights.

conducted and key indicators, including makespan and types of unproductive time, compared. The *pop*, MAXGEN, p_c, p_m, *GAP*, α_1, and α_2 in the GA are set as 200, 500, 0.8, 0.2, 0.9, 0.6, and 0.4 respectively. Each set of experiment was conducted 10 times and the mean value calculated. To evaluate the general effects of different schemes and the efficiency variances of equipment, both the total and mean unproductive time were calculated and compared.

The comparison of the makespan and the sum of unproductive time for each kind of equipment among different schemes is shown in Figure 5, in which the schemes on the *x*-axis represent the number of RMGs, RSs in each block, and ITs separately. For example, 3-3-9 means that, in this case, three RMGs, three RSs (each block), and nine ITs are deployed. It can be observed that, first, compared with 3-3-9, the increase of the RSs and ITs can decrease the makespan of all tasks and the unproductive time of RMGs. However, when more equipment is deployed, the unproductive time of the handling process, of which waiting time accounts for a great proportion, increases significantly; this can result in higher operation costs and greater fuel consumption.

Specifically, on the one hand, compared with 3-3-9, increasing RS (3-4-9) shows a better effect than increasing IT (3-3-9), because of the decrease in the total empty

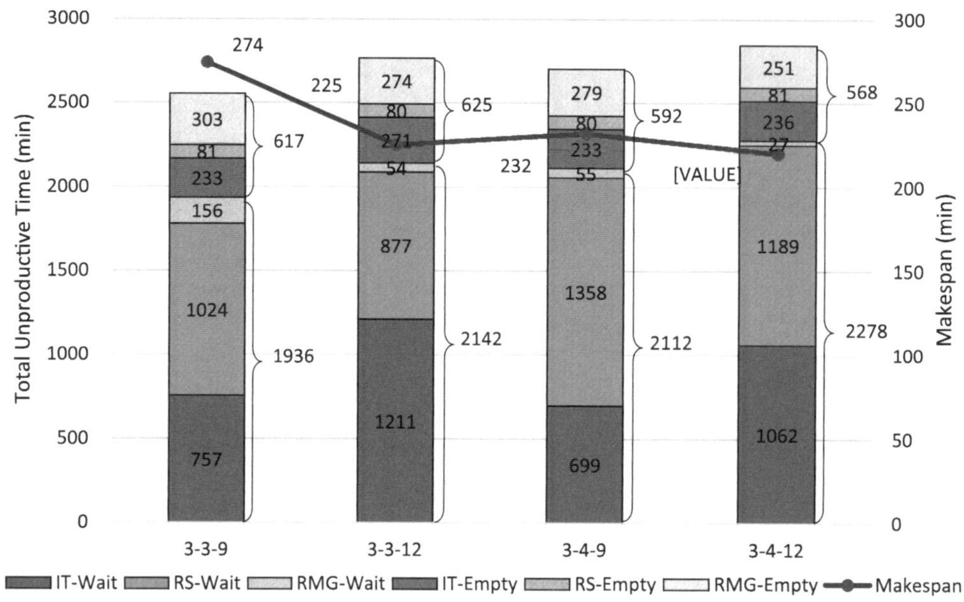

Figure 5. Comparison of the total value of key indicators.

time. Moreover, the increase of RSs is beneficial to the operational efficiency of ITs and RMGs. On the other hand, in this case, increasing ITs is helpful to the further utilization of the capabilities of the handling equipment. At a cost of greater waiting time of ITs, deploying more ITs can effectively reduce the unproductive time of RMGs and RSs; this means that, in schemes 3-3-9 and 3-4-9, the operational capabilities of RSs and RMGs have not been fully utilized.

When more equipment is used, the workloads for each can be different. Therefore, it is necessary to compare their mean unproductive time, as shown in Figure 6. The time variations are the changes of mean unproductive time between schemes. For instance, "No.1—No.2" means the mean unproductive time increases of 3-3-12 compared with 3-3-9.

It can be observed that, first, deploying more ITs and RSs can greatly improve the working efficiency of RMGs because the empty and waiting time of RMGs decrease significantly. Second, deploying more ITs and RSs shows different effects on the other operations. When more RSs are deployed, the empty time of ITs reduces, especially in the comparative group of No.2–No.4. In consideration of the distance differences between block–block and block–tracks, reduced empty time means that ITs are more likely to transfer the loading and unloading containers jointly. However, the increase of ITs does not show an obvious effect on the mean empty time of RSs. Third, the increase of equipment can result in greater equipment waiting time, which means that their operational efficiency is greatly affected. In this case, comparing the effects of increasing RSs with increasing ITs, the former shows a better effect on the handling operation because the mean waiting time of RSs is not significantly increased.

Conclusions and Future Research Directions

As sea–rail intermodal transportation gradually receives more attention (*25*), the significant problem of the integrated scheduling of RMGs, RSs, and ITs in the railway handling area of container terminals was studied here. First, the handling area configuration and equipment scheduling schemes was analyzed. Then, a MIP model with the objective of minimizing the makespan and total waiting time of the handling operations was proposed. To solve the problem efficiently, a GA was developed that takes FAM rules into consideration. Last, sets of experiments were conducted to validate the effect and efficiency of the model and algorithm. In this paper, a set of experiments for 240 tasks was analyzed to differentiate the effects of various equipment configuration schemes on the handling operation. The unproductive time of equipment was adopted as a main indicator for system performance. The results reveal that, although the increase of equipment can shorten the makespan effectively, more RSs and ITs can also result in more unproductive time, especially waiting time, which means a higher handling cost and greater fuel consumption.

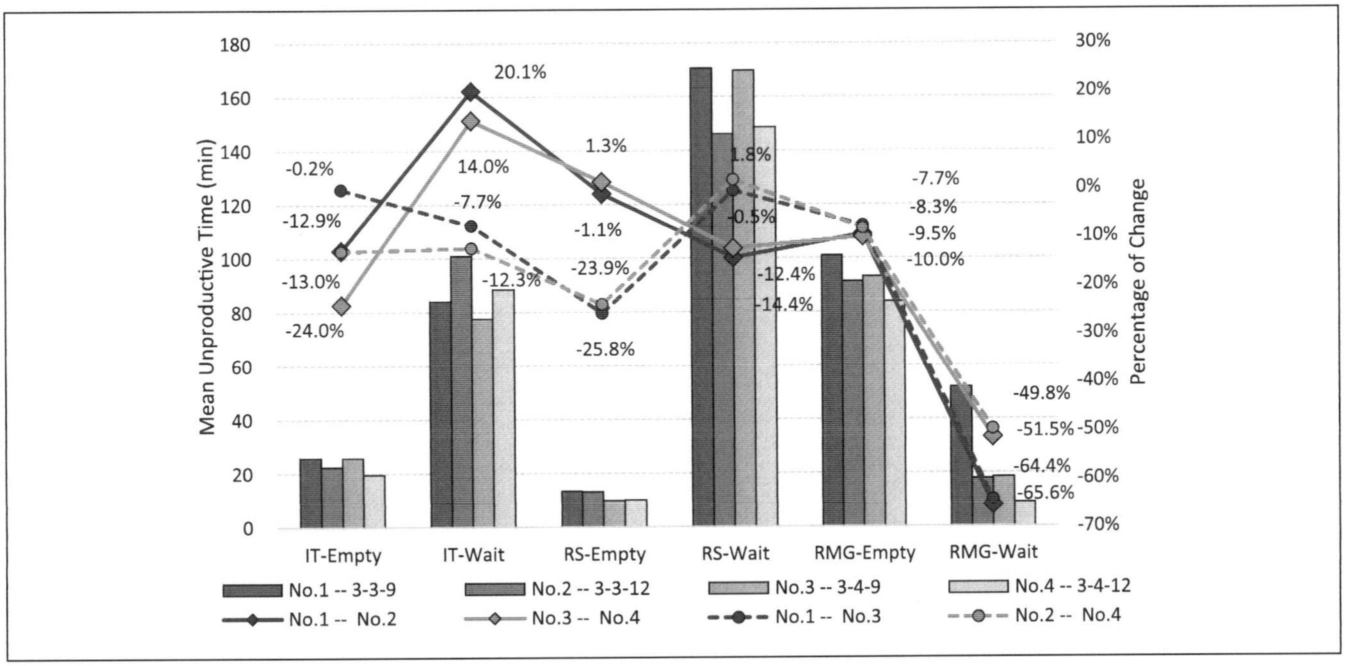

Figure 6. Comparison of the mean value of key indicators.

Moreover, increasing ITs and RSs results in different effects on the overall handling process. In this case, increasing RS produces a better effect. In practice, enterprises can choose the best equipment configuration based on their time, fuel consumption, and cost requirements.

This research can be further extended in several ways, as here only 40ft containers are studied. In practice, the handling of 20ft containers is common, which means the operation sequences can be greatly affected because the loading tasks can only be conducted after all containers in the wagon are unloaded. In addition, the operation time of the equipment is calculated via equipment indicators, which is deterministic. However, the operation time can be affected by factors related to the operators.

Acknowledgments

This work was supported by the Major Research Plan of the National Natural Science Foundation of China [grant number 71390332], the Fundamental Research Funds for the Central Universities [grant number 2017YJS100] and Major Project of Civil Aviation Science and Technology Project [project number 201501].

Author Contributions

The authors confirm contribution to the paper as follows: study conception and design: Baicheng Yan, Xiaoning Zhu, Li Wang; data collection: Baicheng Yan, Yimei Chang; analysis and interpretation of results: Baicheng Yan; draft manuscript preparation: Baicheng Yan. All authors reviewed the results and approved the final version of the manuscript.

References

1. Steenken, D., S. Voß, and R. Stahlbock. Container Terminal Operation and Operations Research - a Classification and Literature Review. *Or Spectrum*, Vol. 26, No. 1, 2004, pp. 3–49.
2. Stahlbock, R., and S. Voß. Operations Research at Container Terminals: A Literature Update. *Or Spectrum*, Vol. 30, No. 1, 2008, pp. 1–52.
3. Kim, K. H., and Y. Park. A Crane Scheduling Method for Port Container Terminals. *European Journal of Operational Research*, Vol. 156, No. 3, 2004, pp. 752–768.
4. Lee, D.-H., H. Q. Wang, and L. X. Miao. Quay Crane Scheduling with Non-interference Constraints in Port Container Terminals. *Transportation Research Part E: Logistics and Transportation Review*, Vol. 44, No.3, 2008, pp.124–135.
5. Goodchild, A. V., and C. F. Daganzo. Crane Double Cycling in Container Ports: Planning Methods and Evaluation. *Transportation Research Part B Methodological*, Vol. 41, No. 8, 2007, pp. 875–891.
6. Wang, J., H. Hu, and Y. Song. Optimization of Quay Crane Scheduling Constrained by Stability of Vessels. *Transportation Research Record: Journal of the Transportation Research Board*, 2013. 2330: 47–54.
7. Li, M. Z., J. G. Jin, and C. X. Lu. Real-Time Disruption Recovery for Integrated Berth Allocation and Crane Assignment in Container Terminals. *Transportation Research Record Journal of the Transportation Research Board*, 2015. 2479: 49–59.
8. Kim, K. H., and K. Y. Kim. Routing Straddle Carriers for the Loading Operation of Containers Using a Beam Search Algorithm. *Computers & Industrial Engineering*, Vol. 36, No. 1, 1999, pp. 109–136.

9. Kim, K. H., and K. Y. Kim. A Routing Algorithm for a Single Straddle Carrier to Load Export Containers Onto a Containership. *International Journal of Production Economics*, Vol. 59, No. 1–3, 1999, pp. 425–433.
10. Zhang, C., Y. W. Wan, J. Liu, and R. J. Linn. Dynamic Crane Deployment in Container Storage Yards. *Transportation Research Part B*, Vol. 36, No. 6, 2002, pp. 537–555.
11. Lee, D.-H., J. G. Jin, and J. H. Chen. Integrated Bay Allocation and Yard Crane Scheduling Problem for Transshipment Containers. *Transportation Research Record: Journal of the Transportation Research Board*, 2011. 2222: 63–71.
12. Hu, Z. H., J. B. Sheu, and J. X. Luo. Sequencing Twin Automated Stacking Cranes in a Block at Automated Container Terminal. *Transportation Research Part C*, Vol. 69, 2016, pp. 208–227.
13. Vis, I. F. A., de R. Koster, K. J. Roodbergen, and L. W. P. Peeters. Determination of the Number of Automated Guided Vehicles Required at a Semi-automated Container Terminal. *Journal of the Operations Research Society*, Vol.52, No. 4, 2001, pp. 409–417.
14. Vis, I. F. A. Minimum Vehicle Fleet Size Under Time-window Constraints at a Container Terminal. *Transportation Science*, Vol. 39, No. 2, 2005, pp. 249–260.
15. Nishimura, E., A. Imai, and S. Papadimitriou. Yard Trailer Routing at a Maritime Container Terminal. *Transportation Research Part E: Logistics and Transportation Review*, Vol. 41, No. 1, 2005, pp. 53–76.
16. He, J. L., W. M. Zhang, Y. F. Huang, and W. A. Yan. A Simulation Optimization Method for Internal Trucks Sharing Assignment Among Multiple Container Terminals. *Advanced Engineering Informatics*, Vol. 27, No. 4, 2013, pp. 598–614.
17. Bish, E. K., T. Y. Leong, C. L. Li, J. W. C. Ng, and D. Simchi-Levi. Analysis of a New Vehicle Scheduling and Location Problem. *Naval Research Logistics*, Vol. 48, No. 5, 2001, pp. 363–385.
18. Chen, L., N. Bostel, P. Dejax, L. Xi, and J. Cai. A Tabu Search Algorithm for the Integrated Scheduling Problem of Container Handling Systems in a Maritime Terminal. *European Journal of Operational Research*, Vol. 181, No. 1, 2007, pp. 40–58.
19. Chen, L., A. Langevin, and Z. Q. Lu. Integrated Scheduling of Crane Handling and Truck Transportation in a Maritime Container Terminal. *European Journal of Operational Research*, Vol. 225, No. 1, 2013, pp. 142–152.
20. He, J., Y. Huang, W. Yan, and S. Wang. Integrated Internal Truck, Yard Crane and Quay Crane Scheduling in a Container Terminal Considering Energy Consumption. *Expert Systems with Applications*, Vol. 42, No. 5, 2015, pp. 2464–2487.
21. Xie, Y., X. Liang, L. Ma, and H. Yan. Empty Container Management and Coordination in Intermodal Transport. *European Journal of Operational Research*, Vol. 257, No.1, 2016, pp. 223–232.
22. Caballini, C., S. Fioribello, S. Sacone, and S. Siri. An MILP Optimization Problem for Sizing Port Rail Networks and Planning Shunting Operations in Container Terminals. *IEEE Transactions on Automation Science & Engineering*, Vol. 13, No. 4, 2016, pp. 1492–1503.
23. Lämmel, G., M. Chraibi, A. U. K. Wagoum, and B. Steffen. Hybrid Multimodal and Intermodal Transport Simulation. *Transportation Research Record Journal of the Transportation Research Board*, 2016. 2561: 55–64.
24. Kang, L., J. Wu, H. Sun, X. Zhu, and B. Wang. A Practical Model for Last Train Rescheduling with Train Delay in Urban Railway Transit Networks. *Omega*, Vol. 50, 2015, pp. 29–42.
25. Kang, L., Q. Meng, and Q. Liu. Fundamental Diagram of Ship Traffic in the Singapoe Strait. *Ocean Engineering*, Vol. 13, No. 147, 2018, pp. 340–354.

The Standing Committee on Intermodal Freight Terminal Design and Operations (AT050) peer-reviewed this paper (18-03371).

Article

TRR
JOURNAL OF THE TRANSPORTATION RESEARCH BOARD

Transportation Research Record
2018, Vol. 2672(9) 59–69

DOI: 10.1177/0361198118786620
journals.sagepub.com/home/trr
SAGE

Fuel Consumption Optimization Model for the Multi-Period Inventory Routing Problem

Chun Cheng[1], Mingyao Qi[2], and Louis-Martin Rousseau[1]

Abstract

In the traditional multi-period inventory routing problem (MIRP), traveling distance is considered as the only measurement of vehicles' variable transportation cost; however, it is in fact the fuel consumption cost, not the distance, which is the greater concern. This paper evaluates vehicles' variable transportation cost by fuel consumption, which is influenced by distance, load, and fuel price. It presents an integer program to formally characterize the fuel consumption considered MIRP (FCMIRP), which can help enterprises obtain a more accurate tradeoff between transportation and inventory costs. It also benefits the environment, because reducing fuel consumption will curb carbon dioxide (CO_2) emissions. Valid inequalities are added to strengthen the model and use a branch-and-cut algorithm. Computational tests indicate that the FCMIRP can decrease fuel consumption and total cost over the traditional model. Factors that influence the results of FCMIRP are also discussed.

With growing energy shortages and rising prices, governments and enterprises have begun to pay close attention to sustainable development, to save energy and protect the environment. For example, the U.S. government has made energy policies to reduce fossil fuel consumption (*1*). The United Nations and the European Union have enacted legislation to control energy consumption. Companies like Wal-Mart and IKEA have started to invest in renewable energy, hoping to decrease traditional energy (petroleum or coal) consumption.

Supply chain activities, such as production, transportation, and inventory, all consume energy. Therefore, in recent years, researchers have begun to consider energy consumption in the supply chain (*2*). Among all the logistics activities, transportation is regarded as one of the largest consumers of petroleum (*3*). Thus, reducing fuel consumption in transportation will contribute to green logistics and cost reduction.

Generally, transportation costs are divided into two parts: (i) fixed cost, including driver wages, vehicle maintenance and depreciation, and so on; and (ii) variable cost, which mainly refers to fuel cost (*4*). Therefore, accurate computation of transportation cost can help plan more practical vehicle routes. Because fuel cost is related to many factors (distance, load, road condition, speed, etc.) (*2, 4*), it is not accurate to consider travel distance alone. For example, the fuel consumption of a fully loaded vehicle is always greater than an identical but empty vehicle when they are traveling along a specific route at the same speed.

Up to now, many studies on inventory routing problems (IRPs) have used distance as the sole measurement of variable transportation cost. Because in IRPs there is a tradeoff to be made between transportation and inventory costs, if the computation of transportation cost is inaccurate, it will not only lead to suboptimal traveling schedules, but will also fail to achieve exact inventory strategies. Therefore, in this paper fuel cost is used as vehicles' variable transportation cost. It is worth mentioning that considering fuel consumption in MIRP also benefits the environment, because transportation is logistics' largest source of CO_2 emissions (*5*), which are caused by fuel consumption.

To conclude, this study makes the following contributions:

- As far as is known, this paper is among the first to examine the influence of fuel cost on MIRP in detail.

[1]Department of Mathematics and Industrial Engineering and CIRRELT, Polytechnique Montréal and CIRRELT, Montreal, Quebec, Canada
[2]Research Center on Modern Logistics, Graduate School at Shenzhen, Tsinghua University, Shenzhen, China

Corresponding Author:
Address correspondence to Mingyao Qi: qimy@sz.tsinghua.edu.cn

- Numerical experiments demonstrate that FCMIRP can reduce overall cost and achieve a better environmental benefit. This could provide managerial insights for enterprises and governments in green logistics.

The remainder of the paper is organized as follows. Section 2 reviews related literature. Section 3 presents mathematical models for the FCMIRP and the traditional MIRP. Section 4 describes the solution method, and Section 5 presents numerical results. Conclusions follow in Section 6.

Literature Review

The IRP is a classic combinatorial optimization problem which determines simultaneously the optimal inventory strategy and vehicle schedules to minimize the supply chain's total cost. IRP has been extensively studied since its introduction in (*6*). By considering different industrial applications and constraints, various types of IRPs have been formulated. For industrial applications, IRPs are classified as road-based IRPs or maritime IRPs. As to constraints, IRPs can be defined according to the following criteria: time horizon, supply chain topology, routing component, inventory policy, and fleet composition and size (*7*). For more details on IRPs see (*7, 8*).

In a MIRP, the total cost is minimized by determining retailers' optimal replenishment time and vehicle routes. (*9*) addresses a single period model which schedules multi-item replenishment with uncertain demand. (*10*) considers a many-to-one distribution network with multi-supplier and multi-product. (*11*) studies a multi-product multi-vehicle MIRP. There are also extensive studies on the algorithm for solving MIRP. The first type is the exact algorithm, including the branch-and-cut (B&C) algorithm and the branch-price-and-cut (BPC) algorithm. (*12*) develops the first B&C algorithm: in that paper, only one vehicle is available for each period. (*13*) uses the B&C algorithm for several classes of MIRPs: multi-vehicle MIRPs with homogeneous and heterogeneous vehicles, MIRPs with transshipment options, and MIRPs with added consistency features. (*14*) uses a BPC algorithm to solve a one-to-many MIRP with maximum-level (ML) inventory replenishment policy. Under an ML policy, any quantity of products can be delivered to a retailer as long as the retailer's maximal inventory capacity is respected. The MIRP subsumes a vehicle routing problem (VRP) which is NP-hard, so (meta)heuristics are developed, including Variable Neighborhood Search (*15*), Tabu Search (*16*), Genetic Algorithm (*10*), and so on.

With increasing environmental pressures, researchers have started to consider fuel consumption and CO_2 emissions in the supply chain. (*17, 18*) summarize recent models for sustainable supply chain management (SSCM). In SSCM, one of the most extensively studied areas is green VRP which explicitly includes fuel or emission cost in VRP models. (*2*) addresses a capacitated VRP that minimizes fuel consumption. (*19*) explores cumulative VRP, in which both distance and vehicle load are considered for calculating fuel consumption. (*20*) compares different objective functions for VRP to quantify their influence on cost and emissions, including minimizing comprehensive cost, minimizing travel distance, minimizing weighted load, and minimizing energy consumption. (*21*) extends the traditional heterogeneous VRP objective to include fuel and emission costs, and analyzes their influences. (*22*) proposes to consider CO_2 emissions in the facility location problem and suggests the difference between cost-minimization solutions and emission-minimization solutions. For detailed reviews of green logistics see (*1, 23*).

Problem Formulation

In this section, the mathematical models for the FCMIRP and the traditional MIRP are presented.

Problem Description

Consider an outbound product distribution system consisting of a supplier and N geographically dispersed retailers. In each period, vehicles depart from the supplier to distribute products to the retailers. Assumptions include:

- The supplier only offers one type of product. At the beginning of each period, products are ready for distribution.
- Vehicle routes start and end at the supplier, and can be completed in one time period.
- Waiting, loading, and unloading times at each vertex are not considered.
- Vehicles are homogeneous and capacitated.
- Split delivery is not allowed: a retailer is visited by at most one vehicle in each period.
- Retailers' demands are deterministic but variable over periods. Their demands must be fulfilled; therefore, backorders are not allowed.
- Retailers' inventory capacities are limited, and the ML replenishment policy is applied.
- Inventory holding costs are considered for both the supplier and the retailers. At the beginning of the time horizon, the initial inventory levels at each vertex are known.

The proposed FCMIRP uses the following notations:

Sets

$V = \{0, 1, \ldots, N\}$ A set of all vertices, $i,j \in V$; vertex 0 is the supplier

$V' = \{1, 2, \ldots N\}$ A set of retailers

$T = \{1, 2, \ldots p\}$ Time horizon, $t \in T$

Parameters

d_i^t Retailer i's demand in period t (kg)

c_{ij} Distance between vertex i and vertex j (km)

Q The weight capacity of each vehicle (kg)

h_i Inventory holding cost at vertex i per unit weight per period (\$/kg/period)

I_i^0 Initial inventory level of vertex $i$at the beginning of the time horizon (kg)

C_i Maximal inventory holding capacity at retailer i (kg)

r^t The product weight supplier receives (or produces) in period t (kg)

u Unit fuel price (\$/L)

u' Variable transportation cost per unit distance per vehicle (\$/km)

F Fixed vehicle cost per trip (\$/trip)

f_d Driver wage per hour (\$/h)

v Vehicle speed (km/h)

Decision variables

x_{ij}^t $x_{ij}^t = 1$ if arc (i,j) is traversed by a vehicle in period t, otherwise 0

a_{ij}^t The product weight carried by a vehicle through arc (i,j) in period t (kg)

q_i^t The product weight delivered to retailer i in period t (kg)

I_i^t The inventory level at vertex i at the end of period t (kg)

ρ_{ij}^t Fuel consumption rate of each vehicle from vertex i to vertex j in period t (L/km)

FC_{ij}^t Fuel cost of each vehicle from vertex i to vertex j in period t (\$)

Fuel Cost

The model proposed by *(4, 5)* is used to compute fuel consumption. These papers state that fuel consumption mainly results from two elements: distance and fuel consumption rate (FCR)—fuel consumption per unit of distance, where FCR is linearly associated with a vehicle's load according to statistical data. Some other complicated models for fuel calculation are also available. For example, it is possible to consider the effect of vehicle speed, road angle, acceleration, weather, and traffic condition. However, here a choice has been made to neglect some factors, based on the following reasons: (i) it is found that in many pieces of research the road angle and acceleration are set to 0 and the speed is assumed to be constant (*21, 24, 25*); (ii) it is impractical and unreasonable to attempt to quantify factors such as weather and traffic, because they vary significantly in different regions (*5*); (iii) IRP is a problem that is even more difficult than VRP, and it is necessary to keep the model tractable.

FCR is first formulated as

$$\rho_{ij}^t = \rho^0 + \frac{\rho^* - \rho^0}{Q} a_{ij}^t \tag{1}$$

where ρ^0, ρ^* are the empty-load and full-load FCR respectively. As ρ^0, ρ^* and Q are fixed for a specific type of vehicle, Equation 1 is simplified as

$$\rho_{ij}^t = \rho^0 + \beta a_{ij}^t, \text{in which } \beta = \frac{\rho^* - \rho^0}{Q} \tag{2}$$

Therefore, the fuel cost from node i to node j in period t is

$$FC_{ij}^t = u\rho_{ij}^t c_{ij} = u\left(\rho^0 + \beta a_{ij}^t\right) c_{ij} \tag{3}$$

Mathematical Formulation

The completed formulation for FCMIRP is

$$\text{Min} f_1 = \sum_{t \in T} \sum_{i \in V} h_i I_i^t + \sum_{t \in T} \sum_{j \in V'} F x_{0j}^t + \sum_{t \in T} \sum_{i \in V} \sum_{j \in V} f_d c_{ij} x_{ij}^t / v + \sum_{t \in T} \sum_{i \in V} \sum_{j \in V} u\left(\rho^0 x_{ij}^t + \beta a_{ij}^t\right) c_{ij} \tag{4}$$

subject to

$$I_0^t = I_0^{t-1} + r^t - \sum_{i \in V'} q_i^t \quad \forall t \in T \tag{5}$$

$$I_i^t = I_i^{t-1} + q_i^t - d_i^t \quad \forall i \in V', t \in T \tag{6}$$

$$I_i^t \geq 0 \quad \forall i \in V, t \in T \tag{7}$$

$$I_i^t \leq C_i \quad \forall i \in V', t \in T \tag{8}$$

$$q_i^t \leq C_i - I_i^{t-1} \quad \forall i \in V', t \in T \tag{9}$$

$$a_{ij}^t \leq Q x_{ij}^t \forall i \in V, \quad \forall j \in V, t \in T \tag{10}$$

$$\sum_{j \in V} x_{ij}^t = \sum_{j \in V} x_{ji}^t \quad \forall i \in V, t \in T \tag{11}$$

$$\sum_{j \in V} a_{ij}^t = \sum_{j \in V} a_{ji}^t - q_i^t \quad \forall i \in V', i \neq j, t \in T \tag{12}$$

$$\sum_{j \in V} x_{ji}^t \leq 1 \quad \forall i \in V', t \in T \tag{13}$$

$$x_{ii}^t = 0 \quad \forall i \in V, t \in T \tag{14}$$

$$a_{ij}^t \geq 0 \quad \forall i \in V, j \in V, t \in T \tag{15}$$

$$q_i^t \geq 0 \quad \forall i \in V', t \in T \tag{16}$$

$$x_{ij}^t \in \{0,1\} \quad \forall i \in V, j \in V, t \in T \tag{17}$$

The objective function (4) minimizes the total cost, including inventory cost, fleet cost, driver wage, and fuel cost. Constraints (5) and (6) are the inventory balance equations at the supplier and the retailers respectively. Constraint (7) means that inventories cannot be negative at each vertex. Constraint (8) imposes maximal inventory level at retailers. Constraint (9) limits the product weight delivered to each retailer, to satisfy demand and to respect maximal inventory capacity. Constraint (10) guarantees that vehicles' capacities are not violated. Constraint (11) indicates that the number of vehicles leaving a node is equal to that arriving at a node. Constraint (12) is the product flow balance equation at retailers and eliminates all subtours. Constraint (13) ensures that split delivery is not allowed. Constraints (14)–(17) define variable types.

For contrast, the mathematical formulation for traditional MIRPs is

$$\text{Min } f_2 = \sum_{i\in V}\sum_{t\in T} h_i I_i^t + \sum_{t\in T}\sum_{j\in V'} F x_{0j}^t + \sum_{t\in T}\sum_{i\in V}\sum_{j\in V} f_d c_{ij} x_{ij}^t / v + \sum_{i\in V}\sum_{j\in V}\sum_{t\in T} u' c_{ij} x_{ij}^t \tag{18}$$

subject to (5)–(17) and

$$\sum_{i\in V'} q_i^t = \sum_{j\in V'} a_{0j}^t \quad \forall t \in T \tag{19}$$

Constraint (19) ensures that in each period the product weight delivered to all retailers is equal to that provided by the supplier. If this constraint is absented from the traditional model, sometimes a vehicle may carry more product (more than the weight delivered to all retailers) from the supplier, and then at the end of the trip the vehicle carries surplus product back to the supplier. However, these constraints are not necessary in FCMIRP, as vehicles will consume more fuel in such a scenario, which is not a wise choice for FCMIRP.

Valid Inequality and Solution Method

Valid Inequalities

(*11*, *12*, *26*) have introduced several classes of valid inequalities for the IRPs. Some of them are extended to strengthen the models.

$$\sum_{j\in V}\sum_{l=1}^{t} x_{ji}^l \geq \left(\sum_{l=1}^{t-1} d_i^l - I_i^0\right) / C_i \quad \forall i \in V', \forall t \in T \tag{20}$$

Inequality (20) means that for retailer i, if its total demand over $[1, t-1]$ is greater than its initial inventory level, then it has to be visited at least the number of times that corresponds to the right-hand side of the inequality in the interval $[1, t]$.

Constraint (20) can be extended to any time interval $[t_1, t_2]$; consider that the maximal delivery weight takes the minimal value between a vehicle's capacity and retailers' inventory capacities. This yields inequality (21):

$$\sum_{j\in V}\sum_{t'=t_1}^{t_2} x_{ji}^{t'} \geq (\sum_{t'=t_1}^{t_2} d_i^{t'} - I_i^{t_1-1}) / \min\{Q, C_i\} \quad \forall i \in V', \forall t_1, t_2 \in T, t_2 \geq t_1 \tag{21}$$

For retailer i, if its inventory at the end of period $t_1 - 1$ is sufficient to meet its demands over $[t_1, t_2]$, then it is not mandatory for vehicles to visit it. Otherwise, there has to be at least one visit to retailer i during $[t_1, t_2]$. This yields inequality (22):

$$\sum_{j\in V}\sum_{t'=t_1}^{t_2} x_{ji}^{t'} \geq (\sum_{t'=t_1}^{t_2} d_i^{t'} - I_i^{t_1-1}) / \sum_{t'=t_1}^{t_2} d_i^{t'} \quad \forall i \in V', \forall t_1, t_2 \in T, t_2 \geq t_1 \tag{22}$$

Note that the right-hand side of Equation 20 is rounded up, because it takes a constant value. However, this is not allowed in Equations 21 and 22, because their numerators contain variables, which will make them nonlinear otherwise.

Solution Method

For exact display of the different solutions obtained from the two models, exact algorithms are preferred in this study. Therefore, a B&C algorithm is used here. First, valid inequalities (20)–(22) are added into the model at the root node of the search tree. Then a commercial solver is used to solve the linear relaxation problems at each node. The best bound strategy is employed to select nodes.

Computational Experiments and Analyses

The aim of this section includes:

- To display the solution differences between the two models in detail;
- To discuss the factors that influence FCMIRP's decisions.

Instance Data

As to MIRP, this study uses data sets generated by (*11*) for a single product. The names of instances indicate the numbers of retailers, product types, vehicles, periods, and the instance number. For example, label "20-1-3-5-4" is the fourth instance for 20 customers, one type of product, three vehicles, and five periods. Note that the number of vehicles is unlimited in this study. For a specific combination of customer and period, there are 15 instances by changing the number of vehicles. The distance between nodes is computed as $c_{ij} = \left\lfloor \sqrt{(x_i - x_j)^2 + (y_i - y_j)^2} + 0.5/a \right\rfloor$, where (x_i, y_i) are the coordinates of vertex i, and $\alpha = 6$. The vehicle speed $v = 80$. As to fuel consumption, according to (*27*), the values set are $\rho^0 = 0.296$ and $\rho^* = 0.390$. Additionally, based on (*21, 24, 25*), $F = 35.19$, $f_d = 10.35, u = 0.96, u' = 0.75$.

The B&C algorithm was implemented in Python language using Gurobi 7.5.1 as the solver. One thread was used. All computations were executed on a PC with an Intel Core i5 Processor (2.3 GHz) and 8GB RAM. A time limit of 1 h was imposed on each instance.

Comparisons between Models

Cost Configuration. To study the differences between models, four aspects are compared: fuel consumption, inventory cost, transportation cost, and total cost. Only instances with optimal solutions under the time limit are reported. Note that in traditional MIRP, the amount of fuel consumption might be different for a given route, because the distance matrix is symmetric and a vehicle may travel from two opposite directions. In the following analyses, the worst-case fuel consumption is reported for the traditional model to indicate the maximal economic and environmental benefits which can be achieved from FCMIRP. Detailed comparison results are given in Table 1 and graphical comparison is presented in Figures 1 and 2. For ease of exposition, each instance is given a new number, for which refer to Table 1.

From Table 1 and Figure 1, in most instances (38 out of 45, about 84.4%), the fuel consumption of the FCMIRP is less, and it can achieve a saving of 2.21% on average. For instances 1, 5, 12, 16, 42, and 43, the saving is over 6%. Because fuel consumption is the direct cause of CO_2 emissions, the conclusion is that in most cases, FCMIRP can help enterprises achieve a better environmental benefit. Moreover, in all instances, FCMIRP generates solutions with lower cost, leading to an average saving of 1.09%. Figure 2 shows that in some cases, even though the inventory cost of the two models is the same, the transportation cost of the FCMIRP is lower, because less fuel is consumed. For instances 6, 15, 18, 26, 28, and 33, the transportation cost has a slight increase, to lower inventory cost. This suggests that considering fuel consumption in MIRP will lead to a new balance between inventory and transportation costs.

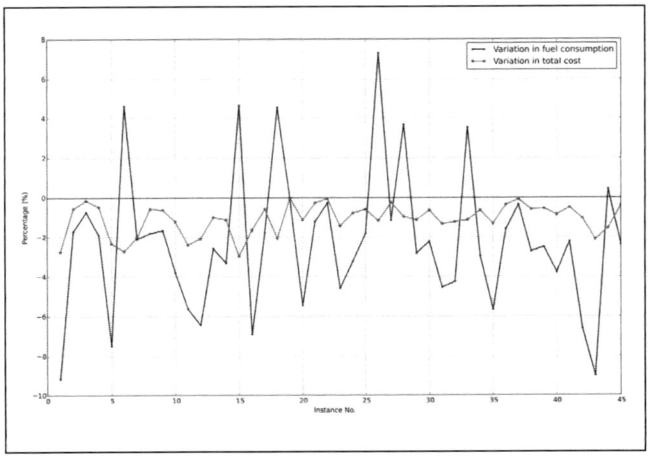

Figure 1. Gap in fuel consumption and total cost compared with traditional MIRP.

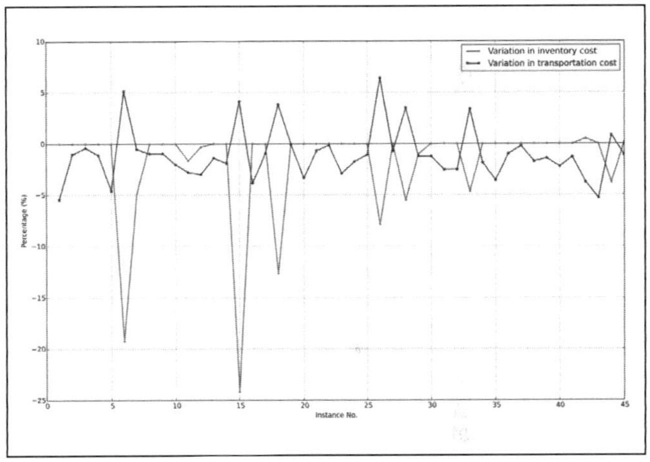

Figure 2. Gap in inventory and transportation cost compared with traditional MIRP.

In seven instances, vehicles consume more fuel in the FCMIRP. This counter-intuitive phenomenon can be explained by the FCMIRP's cost-oriented nature; sometimes a small increase in fuel consumption can lead to a large decrease in inventory cost, resulting in a lower total cost. In the following sections, four instances will be used to exemplify the specific differences in the solutions of these two models.

Route Direction and Order. FCMIRP will lead to a difference in route direction, because decision makers will only choose the direction in which vehicles consume less energy. However, in traditional MIRP, a vehicle may travel in either direction. Taking instance "10-1-1-3-1" as

Table 1. Detailed Comparison between the FCMIRP and the Traditional MIRP

No.	Instance	FCMIRP				Traditional MIRP				Gap (%)			
		Fuel (L)	Inventory cost ($)	Transport cost ($)	Total cost ($)	Fuel (L)	Inventory cost ($)	Transport cost ($)	Total cost ($)	Fuel	Inventory cost	Transport cost	Total cost
1	10-1-1-3-1	142.94	225.88	231.66	457.54	156.03	225.88	244.23	470.11	−9.16	0.00	−5.43	−2.75
2	10-1-1-3-2	146.91	208.85	238.71	447.56	149.41	208.85	241.11	449.96	−1.70	0.00	−1.01	−0.54
3	10-1-1-3-3	92.64	239.77	162.80	402.57	93.31	239.77	163.45	403.22	−0.73	0.00	−0.40	−0.16
4	10-1-1-3-4	127.75	281.30	211.13	492.43	130.17	281.30	213.45	494.75	−1.89	0.00	−1.10	−0.47
5	10-1-1-3-5	179.55	273.79	281.04	554.83	192.98	273.79	293.93	567.72	−7.48	0.00	−4.59	−2.32
6	10-1-3-3-1	228.82	180.61	382.04	562.64	218.27	215.47	362.46	577.93	4.61	−19.30	5.13	−2.72
7	10-1-3-3-2	199.64	174.04	340.3	514.34	203.82	182.56	342.12	524.68	−2.10	−4.90	−0.53	−2.01
8	10-1-3-3-3	178.05	220.08	311.3	531.38	181.26	220.08	314.25	534.33	−1.80	0.00	−0.95	−0.56
9	10-1-3-3-4	200.75	168.97	341.5	510.47	204.06	168.97	344.68	513.65	−1.65	0.00	−0.93	−0.62
10	10-1-3-3-5	159.30	193.41	286.57	479.99	165.36	193.41	292.39	485.80	−3.81	0.00	−2.03	−1.21
11	10-1-5-3-1	214.20	203.92	396.33	600.24	226.23	207.28	407.36	614.64	−5.62	−1.65	−2.78	−2.40
12	10-1-5-3-2	194.80	190.79	370.98	561.77	207.34	191.41	381.99	573.40	−6.44	−0.32	−2.97	−2.07
13	10-1-5-3-3	253.05	189.79	446.57	636.35	259.58	189.79	452.84	642.63	−2.58	0.00	−1.40	−0.99
14	10-1-5-3-4	207.69	236.50	350.62	587.12	214.56	236.50	357.22	593.72	−3.31	0.00	−1.88	−1.12
15	10-1-5-3-5	290.00	165.49	495.75	661.24	276.56	205.51	475.35	680.86	4.63	−24.18	4.11	−2.97
16	10-1-1-5-3	206.90	479.58	353.74	833.32	221.17	479.58	367.31	846.89	−6.90	0.00	−3.84	−1.63
17	10-1-3-5-2	350.40	380.92	611.56	992.48	356.59	380.92	617.25	998.17	−1.77	0.00	−0.93	−0.57
18	10-1-3-5-5	351.23	360.42	647.94	1008.35	335.20	406.07	623.1	1029.17	4.56	−12.67	3.83	−2.06
19	20-1-1-3-1	194.32	522.88	301.82	824.71	194.44	522.88	301.93	824.81	−0.06	0.00	−0.04	−0.01
20	20-1-1-3-2	179.67	552.59	282.32	834.91	189.48	552.59	291.74	844.33	−5.46	0.00	−3.34	−1.13
21	20-1-1-3-3	160.55	386.60	254.4	641.00	162.49	386.60	256.13	642.73	−1.21	0.00	−0.68	−0.27
22	20-1-1-3-4	171.38	447.59	270.09	717.68	171.83	447.59	270.53	718.12	−0.27	0.00	−0.16	−0.06
23	20-1-1-3-5	230.76	360.69	351.16	711.86	241.35	360.69	361.33	722.02	−4.59	0.00	−2.90	−1.43
24	20-1-3-3-1	206.56	440.33	350.96	791.29	213.24	440.33	357.11	797.44	−3.23	0.00	−1.75	−0.78
25	20-1-3-3-2	271.27	361.31	438.05	799.37	276.20	361.31	442.78	804.09	−1.82	0.00	−1.08	−0.59
26	20-1-3-3-3	225.03	418.64	372.96	791.60	208.63	451.69	349.06	800.75	7.29	−7.89	6.41	−1.16
27	20-1-3-3-4	173.44	481.09	268.32	749.41	175.41	481.09	270.21	751.30	−1.14	0.00	−0.70	−0.25
28	20-1-3-3-5	221.94	360.41	370.38	730.79	213.76	380.35	357.48	737.83	3.68	−5.53	3.48	−0.96
29	20-1-5-3-1	276.93	411.56	478.67	890.23	284.75	415.70	484.5	900.20	−2.82	−1.01	−1.22	−1.12
30	20-1-5-3-2	277.29	448.62	479.67	928.28	283.45	448.62	485.58	934.20	−2.22	0.00	−1.23	−0.64
31	20-1-5-3-3	275.73	423.72	477.52	901.25	288.26	423.66	489.55	913.21	−4.54	0.01	−2.52	−1.33
32	20-1-5-3-4	248.58	428.00	406.31	834.31	259.15	428.00	416.45	844.45	−4.25	0.00	−2.50	−1.22
33	20-1-1-5-2	344.39	684.85	537.48	1222.33	332.19	716.87	519.3	1236.17	3.54	−4.68	3.38	−1.13
34	30-1-1-3-1	206.07	598.85	319.31	918.16	212.20	598.85	325.19	924.04	−2.97	0.00	−1.84	−0.64
35	30-1-1-3-2	240.21	633.86	363.34	997.20	253.80	633.86	376.39	1010.25	−5.66	0.00	−3.59	−1.31
36	30-1-1-3-3	224.93	571.96	342.46	914.42	228.52	571.96	345.78	917.74	−1.60	0.00	−0.97	−0.36
37	30-1-1-3-4	247.80	587.98	375.16	963.14	248.66	587.98	375.98	963.96	−0.34	0.00	−0.22	−0.09
38	30-1-1-3-5	218.64	626.66	333.31	959.98	224.57	626.66	339.01	965.67	−2.72	0.00	−1.71	−0.59
39	30-1-3-3-1	218.19	587.88	367.94	955.83	223.64	587.88	373.17	961.05	−2.50	0.00	−1.42	−0.55
40	30-1-3-3-2	247.86	651.09	404.84	1055.93	257.20	651.09	413.8	1064.89	−3.77	0.00	−2.21	−0.85
41	30-1-3-3-3	238.06	616.95	392.85	1009.8	243.35	616.95	397.93	1014.88	−2.22	0.00	−1.29	−0.50
42	30-1-3-3-4	222.05	658.25	371.91	1030.17	236.74	654.99	385.88	1040.87	−6.61	0.50	−3.76	−1.04
43	30-1-3-3-5	264.42	649.00	429.53	1078.53	288.22	649.00	452.38	1101.38	−9.00	0.00	−5.32	−2.12
44	30-1-5-3-1	323.47	580.91	540.69	1121.59	322.03	602.86	536.07	1138.93	0.44	−3.78	0.85	−1.55
45	30-1-5-3-2	271.55	562.45	436.51	998.96	277.99	561.55	441.27	1002.82	−2.37	0.16	−1.09	−0.39
	Average	224.57	409.53	372.86	782.39	228.97	414.86	375.80	790.66	−2.21	−1.89	−1.00	−1.09

Note: No. = number.

Table 2. Instance 10-1-1-3-1 to Show the Difference in Route Direction

FCMIRP	Traditional MIRP
Period 1: No route Period 2: 0→9→4→5→6→2→3→1→10→0 (fuel consumption: 142.94) Period 3: No route	Period 1: No route Period 2: 0→10→1→3→2→6→5→4→9→0 (fuel consumption: 156.03) Period 3: No route

Table 3. Instance 10-1-3-5-2 to Show the Difference in Route Order

FCMIRP	Traditional MIRP
Periods 1, 2: No route Period 3: 0→2→3→4→0 0→7→10→8→1→0 Period 4: 0→6→**2**→**9**→0 0→1→5→8→0 Period 5: No route	Periods 1, 2: No route Period 3: 0→4→3→2→0 0→1→8→10→7→0 Period 4: 0→**2**→**9**→6→0 0→8→5→1→0 Period 5: No route

Table 4. Instance 10-1-5-3-1 to Show the Difference in Inventory Strategy

FCMIRP	Traditional MIRP
Periods 1: No route Period 2: 0→10(45)[a]→4(116)→9(37)→5(28)→1(1)→0 0→2(61)→8(119)→0 Period 3: 0→2(75)→**7(24)**→0	Period 1: No route Period 2: 0→10(45)→4(116)→9(37)→5(28)→1(1)→0 0→2(61)→8(119)→**7(24)**→0 Period 3: 0→2(75)→0

Note: 10(45)[a] means retailer 10 is visited and the product weight delivered is 45 kg.

an example, in the two models the route length and inventory cost are the same whereas the route direction is converse, as shown in Table 2.

From Table 2 one may conclude that it is possible to directly transform traditional MIRP to FCMIRP by simply reversing the route direction, but this method does not always work. Take instance "10-1-3-5-2," presented in Table 3, for example. For traditional MIRP, if one route in period 2 is reversed, becoming "0→6→9→2→0," then the fuel consumption will be 47.64. However, if the visiting order of retailers 2 and 9 is further exchanged, just as the route in the FCMIRP, the fuel consumption will be 45.27. Therefore, a simple reverse operation cannot guarantee a maximal saving of energy.

Inventory Strategy. Sometimes even though the routes of the two models are similar, their inventory strategy may be different. Take instance "10-1-5-3-1," for example.

Table 4 shows that the routes of the two models are very similar except that retailer 7 is visited in different periods. For traditional MIRP, visiting retailer 7 in period 2 will reduce route length, result in lower driver cost and variable transportation cost. However, for FCMIRP, it is better to visit retailer 7 later, because its inventory holding cost coefficient is much higher than that of the supplier, and the saving of inventory cost will exceed the increase of driver cost.

Route Schedule and Inventory Strategy. FCMIRP may result in a totally different route schedule and inventory strategy. Take instance "10-1-5-3-2" as an example, shown in Table 5. It demonstrates that the number of trips in each period is different and that the total product weight delivered to a retailer (such as retailer 10) varies. From instance data, the inventory holding cost coefficient of retailer 10 is 0; in traditional MIRP, then, more products are carried to the retailer, to control inventory cost, which on the other hand causes more fuel consumption.

To conclude, the solutions of the two models differ from each other in route direction, route order, and inventory strategy, which leads to a new tradeoff between inventory and transportation cost, and also generates different environmental implications.

Parameter Analyses

The study now examines the factors that may influence the decision of FCMIRP. The analyses are performed based on two randomly selected small instances (10-1-5-

Table 5. Instance 10-1-5-3-2 to Show the Difference in Route Schedule and Inventory Strategy

FCMIRP	Traditional MIRP
Periods 1: No route	Period 1: No route
Period 2: 0→**1(43)**→3(71)→4(88)→9(17)→0	Period 2: 0→**10(44)**→9(17)→4(88)→3(71)→0
Period 3: 0→**1(86)**→6(52)→7(64)→0	0→**1(53)**→0
0→2(28)→**10(10)**→0	Period 3: 0→2(28)→7(64)→6(52)→**1(76)**→0

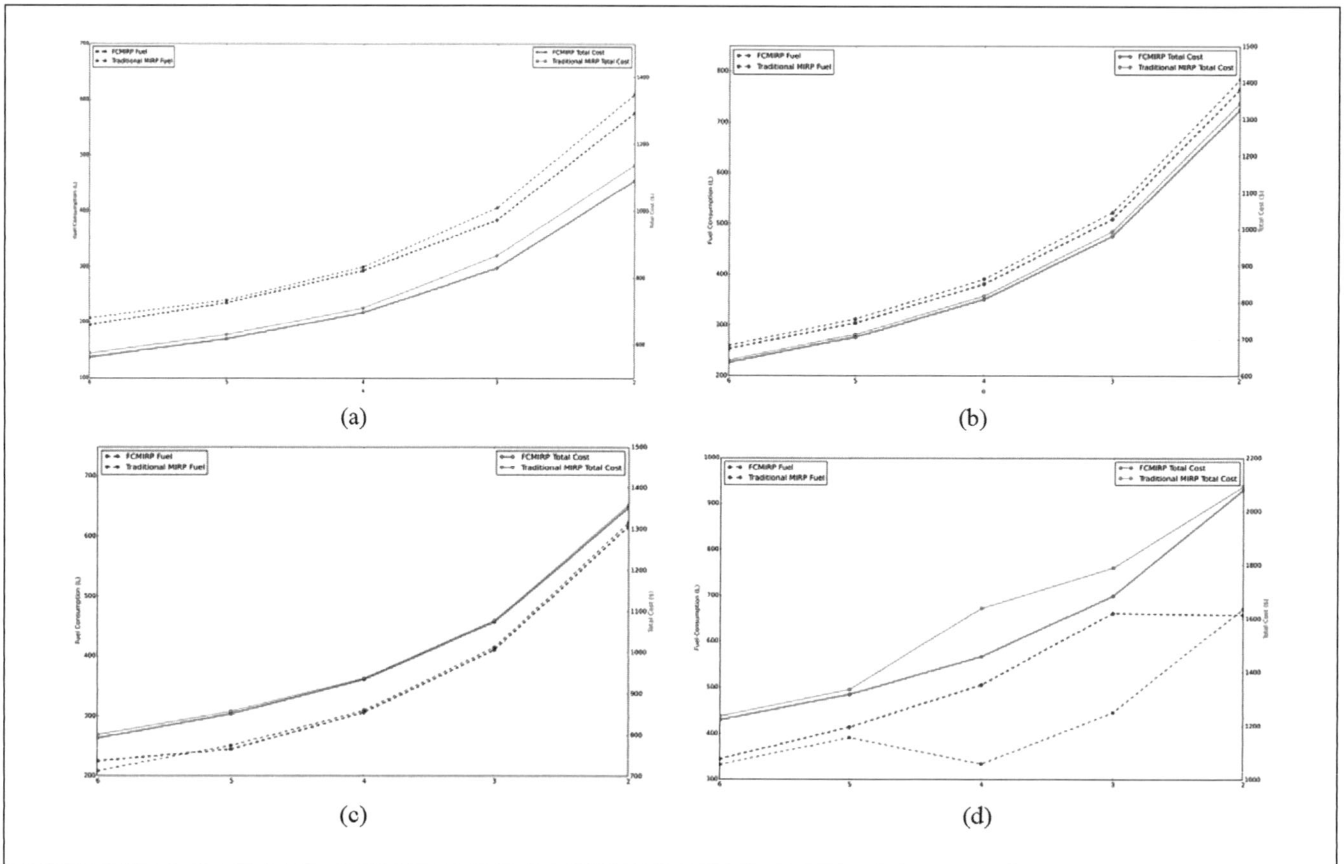

Figure 3. The influence of the distance factor for (*a*) instance 10-1-5-3-2, (*b*) instance 10-1-5-3-3, (*c*) instance 20-1-3-3-3, and (*d*) instance 20-1-1-5-2.

3-2, 10-1-5-3-3) and two medium instances (20-1-3-3-3, 20-1-1-5-2), because they are more time-efficient compared with large ones.

Distance Factor. The distance between nodes in the previous test has been scaled down, to consider the situation that during each trip the driver's work time is shorter than 8 h (actually the results show that the average work time is 5.12 h per trip). Here this constraint is relaxed to model the case of long distance transport, in order to figure out FCMIRP's potential for this application. The process in this case is to gradually increase the distance between nodes by changing α from 6 to 2 and observe its effect (Figure 3); other parameters' values are the same as already given above as instance data.

Figure 3 shows that as distance increases the total cost will increase as expected in these two models. However, the total cost of the FCMIRP is lower under any case. The average saving for each instance is 2.99%, 1.18%, 0.60%, and 4.34%, respectively. The maximal saving is 12.41% for instance 20-1-1-5-2 with $\alpha = 4$. Thus, the conclusion is that the FCMIRP is also capable of saving cost for long distance transport, and sometimes it has a great potential to reduce cost.

The general fuel consumption trend is increasing for both models. For the first two instances, the FCMIRP can save 4.45% and 2.57% fuel on average. For the last

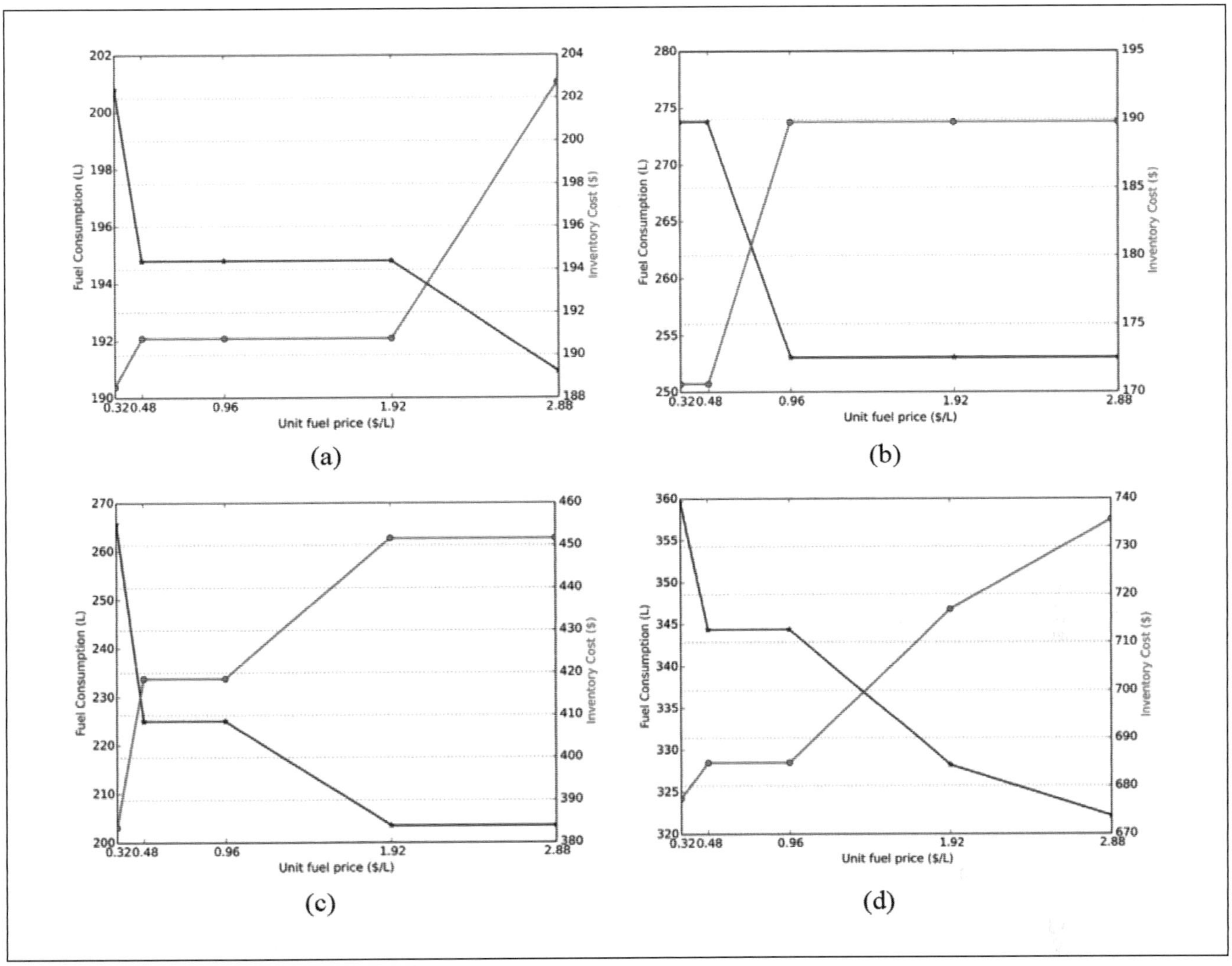

Figure 4. Trend of inventory cost and fuel consumption as *u* changes for (*a*) instance 10-1-5-3-2, (*b*) instance 10-1-5-3-3, (*c*) instance 20-1-3-3-3, and (*d*) instance 20-1-1-5-2.

instance, FCMIRP consumes more fuel in order to lower inventory cost and total cost.

Unit Fuel Price. For FCMIRP, u is changed step by step and its influence on fuel consumption and inventory cost is observed (Figure 4).

As expected, when u increases the general trend of inventory cost is increasing and that of fuel consumption is decreasing. However, the process reveals that sometimes when the fuel price varies, both the inventory cost and fuel consumption remain constant. In these situations, if decision makers choose to further increase inventory expenses, it will cost more money, more than the decrease of transportation cost. Therefore, it is better to keep the current decisions. Because CO_2 emissions and fuel consumption are linearly dependent, it is also possible to recognize the fuel price here as an emission price (or carbon tax). Then Figure 4 can also represent the trend of CO_2 emissions with varying carbon taxing. Thus, the conclusion is that a higher fuel or emission price does not always mean a lower emission level.

Fleet Size. Instances 10-1-5-3-2 and 20-1-3-3-3 are used to analyze the influence of fleet size, as shown in Table 6. Constraint (23) is added to both models, where s is the maximal allowable trip number in each period. Meanwhile, the vehicle capacity is halved in each instance.

$$\sum_{j \in V'} x_{0j}^t \leq s \quad t \in T \tag{23}$$

Table 6 shows that FCMIRP can still save fuel and total cost when this new constraint is added. As expected,

Table 6. The Influence of Fleet Size

	10-5-3-2							20-3-3-3					
	FCMIRP		Traditional MIRP		Gap (%)			FCMIRP		Traditional MIRP		Gap (%)	
s	Fuel	Cost	Fuel	Cost	Fuel	Cost	s	Fuel	Cost	Fuel	Cost	Fuel	Cost
2	282.13	770.68	291.51	780.43	−3.32	−1.27	1	270.55	928.63	283.33	942.81	−4.72	−1.53
3	282.13	767.32	291.51	777.07	−3.32	−1.27	2	270.32	897.66	281.19	908.53	−4.02	−1.21

when the allowable trip number is small, the total cost will increase, which mainly results from increasing inventory cost. This is because if fewer vehicles are available, it is necessary to move some trips to the next a few periods to be executed, and the inventory holding cost coefficient of the supplier is higher than that of some retailers.

Conclusion

This paper addresses a MIRP with environmental consideration. It is different from previous studies in that a new model considering fuel consumption is proposed, which can help enterprises make a more accurate trade-off between transportation and inventory costs, and assist them to lower fuel consumption and CO_2 emissions. Numerical tests demonstrate that FCMIRP indeed generates different vehicle schedules and inventory strategies. For all instances, the FCMIRP is capable of saving overall cost, and in most cases it can decrease fuel consumption and CO_2 emissions.

Based on the numerical tests, it is possible to derive managerial insights from two points of view: (1) enterprises should reconsider their inventory replenishment (tactic) plan and vehicle routing (operational) plan by taking fuel consumption into account, as these have been proved to be quite different from the traditional distance-based plans; and (2) for government regulators, higher fuel prices or carbon taxing do not always lead to a better environmental benefit, because to pursue overall profits, enterprises may sacrifice transportation costs (therefore more emissions) for savings in inventory costs in cases in which inventory costs are a major component.

Future studies could be conducted on two aspects of the problem. First, it would be possible to develop powerful heuristic algorithms to solve large FCMIRP instances. Second, more factors, such as vehicle speed, road angles, and different vehicle types, could be incorporated when calculating fuel consumption.

Acknowledgments

This work is supported by the National Natural Science Foundation of China (71772100), Shenzhen Science and Technology Project (JCYJ20170412171044606), and National Key Technologies Research and Development Program (2016YFB0502601, 2016YFC0803107). The authors would like to thank anonymous referees for their helpful comments.

Author Contributions

The authors confirm contribution to the paper as follows: study conception and design: Cheng, C., Qi, M.; data collection: Cheng, C.; analysis and interpretation of results: Cheng, C., Qi, M., Rousseau, L. M; draft manuscript preparation: Cheng, C, Rousseau, L. M. All authors reviewed the results and approved the final version of the manuscript.

References

1. Lin, C., K. L. Choy, G. T. S. Ho, S. H. Chung, and H. Y. Lam. Survey of Green Vehicle Routing Problem: Past and Future Trends. *Expert Systems with Applications*, Vol. 41, No. 4 PART 1, 2014, pp. 1118–1138. https://doi.org/10.1016/j.eswa.2013.07.107.
2. Kara, I., B. Y. Kara, and M. K. Yetis. Energy Minimizing Vehicle Routing Problem. In *Combinatorial Optimization and Applications: First International Conference, COCOA 2007, Xi'an, China, August 14-16, 2007. Proceedings* (A. Dress, Y. Xu, and B. Zhu, eds.), Springer Berlin Heidelberg, Berlin, Heidelberg, pp. 62–71.
3. Salimifard, K., and R. Shahbandarzadeh, H and R. Raeesi. Green Transportation and the Role of Operation Research. *Proc., 2012 International Conference on Traffic and Transportation Engineering*, Vol. 26, Singapore, IACSIT Press 2012, pp. 74–79.
4. Xiao, Y., Q. Zhao, I. Kaku, and Y. Xu. Development of a Fuel Consumption Optimization Model for the Capacitated Vehicle Routing Problem. *Computers and Operations Research*, Vol. 39, No. 7, 2012, pp. 1419–1431. https://doi.org/10.1016/j.cor.2011.08.013.
5. Zhang, S., C. K. M. Lee, K. L. Choy, W. Ho, and W. H. Ip. Design and Development of a Hybrid Artificial Bee Colony Algorithm for the Environmental Vehicle Routing Problem. *Transportation Research Part D: Transport and Environment*, Vol. 31, 2014, pp. 85–99. https://doi.org/10.1016/j.trd.2014.05.015.
6. Bell, W. J., L. M. Dalberto, M. L. Fisher, A. J. Greenfield, R. Jaikumar, P. Kedia, R. G. Mack, and P. J. Prutzman. Improving the Distribution of Industrial Gases with an On-Line Computerized Routing and Scheduling Optimizer. *Interfaces*, Vol. 13, No. 6, 1983, pp. 4–23.
7. Andersson, H., A. Hoff, M. Christiansen, G. Hasle, and A. Løkketangen. Industrial Aspects and Literature Survey:

Combined Inventory Management and Routing. *Computers and Operations Research*, Vol. 37, No. 9, 2010, pp. 1515–1536. https://doi.org/10.1016/j.cor.2009.11.009.

8. Coelho, L. C., J. F. Cordeau, and G. Laporte. Thirty Years of Inventory Routing. *Transportation Science*, Vol. 48, No. 1, 2014, pp. 1–19. https://doi.org/10.1287/trsc.2013.0472.
9. Huang, S. H., and P. C. Lin. A Modified Ant Colony Optimization Algorithm for Multi-Item Inventory Routing Problems with Demand Uncertainty. *Transportation Research Part E: Logistics and Transportation Review*, Vol. 46, No. 5, 2010, pp. 598–611. https://doi.org/10.1016/j.tre.2010.01.006.
10. Moin, N. H., S. Salhi, and N. A. B. Aziz. An Efficient Hybrid Genetic Algorithm for the Multi-Product Multi-Period Inventory Routing Problem. Vol. 7595, No. 225, 2011, pp. 1–21. https://doi.org/10.1016/j.ijpe.2010.06.012.
11. Coelho, L. C., and G. Laporte. A Branch-and-Cut Algorithm for the Multi-Product Multi-Vehicle Inventory-Routing Problem. *International Journal of Production Research*, Vol. 51, No. 23–24, 2013, pp. 7156–7169. https://doi.org/10.1080/00207543.2012.757668.
12. Archetti, C., L. Bertazzi, G. Laporte, and M. G. Speranza. A Branch-and-Cut Algorithm for a Vendor-Managed Inventory-Routing Problem. *Transportation Science*, Vol. 41, No. 3, 2007, pp. 382–391. https://doi.org/10.1287/trsc.1060.0188.
13. Coelho, L. C., and G. Laporte. The Exact Solution of Several Classes of Inventory-Routing Problems. *Computers and Operations Research*, Vol. 40, No. 2, 2013, pp. 558–565. https://doi.org/10.1016/j.cor.2012.08.012.
14. Desaulniers, G., J. G. Rakke, and L. C. Coelho. A Branch-Price-and-Cut Algorithm for the Inventory-Routing Problem. *Transportation Science*, Vol. 50, No. 3, 2015, pp. 1060–1076. https://doi.org/10.1287/trsc.2015.0635
15. Mjirda, A., B. Jarboui, R. Macedo, S. Hanafi, and N. Mladenović. A Two Phase Variable Neighborhood Search for the Multi-Product Inventory Routing Problem. *Computers and Operation Research*, Vol. 52, 2014, pp. 291–299. https://doi.org/10.1016/j.cor.2013.06.006.
16. Zachariadis, E. E., C. D. Tarantilis, and C. T. Kiranoudis. An Integrated Local Search Method for Inventory and Routing Decisions. *Expert Systems with Applications*, Vol. 36, No. 7, 2009, pp. 10239–10248. https://doi.org/10.1016/j.eswa.2009.01.069.
17. Seuring, S. A Review of Modeling Approaches for Sustainable Supply Chain Management. *Decision Support Systems*, Vol. 54, No. 4, 2013, pp. 1513–1520. https://doi.org/10.1016/j.dss.2012.05.053.
18. Brandenburg, M., K. Govindan, J. Sarkis, and S. Seuring. Quantitative Models for Sustainable Supply Chain Management: Developments and Directions. *European Journal of Operational Research*, Vol. 233, No. 2, 2014, pp. 299–312.
19. Cinar, D., K. Gakis, and P. M. Pardalos. A 2-Phase Constructive Algorithm for Cumulative Vehicle Routing Problems with Limited Duration. *Expert Systems with Applications*, Vol. 56, 2016, pp. 48–58.
20. Bektaş, T., and G. Laporte. The Pollution-Routing Problem. *Transportation Research Part B: Methodological*, Vol. 45, No. 8, 2014, pp. 1232–1250. https://doi.org/10.1016/j.trb.2011.02.004.
21. Koç, Ç., Ç. Koç, T. Bekta, O. Jabali, and G. Laporte. The Fleet Size and Mix Pollution-Routing Problem. *Transportation Research Part B: Methodological*, Vol. 70, 2014, pp. 239–254. https://doi.org/10.1016/j.trb.2014.09.008.
22. Martínez, J. C. V., and J. C. Fransoo. Green Facility Location. In *Sustainable Supply Chains* (Y. Bouchery, C. J. Corbett, J. C. Fransoo, and T. Tan, eds.), Springer International Publishing, New York, 2017, pp. 219–234.
23. Dekker, R., J. Bloemhof, and I. Mallidis. Operations Research for Green Logistics: An Overview of Aspects, Issues, Contributions and Challenges. *European Journal of Operational Research*, Vol. 219, No. 3, 2012, pp. 671–679. https://doi.org/10.1016/j.ejor.2011.11.010.
24. Treitl, S., P. C. Nolz, and W. Jammernegg. Incorporating Environmental Aspects in an Inventory Routing Problem: A Case Study from the Petrochemical Industry. *Flexible Services and Manufacturing Journal*, Vol. 26, No. 1–2, 2014, pp. 143–169. https://doi.org/10.1007/s10696-012-9158-z.
25. Soysal, M., J. M. Bloemhof-Ruwaard, R. Haijema, and J. G. A. J. van der Vorst. Modeling a Green Inventory Routing Problem for Perishable Products with Horizontal Collaboration. *Computers and Operations Research*, Vol. 89, 2018, pp.168–182. https://doi.org/10.1016/j.cor.2016.02.003.
26. Coelho, L. C., and G. Laporte. Improved Solutions for Inventory-Routing Problems through Valid Inequalities and Input Ordering. *International Journal of Production Economics*, Vol. 155, 2014, pp. 391–397. https://doi.org/10.1016/j.ijpe.2013.11.019.
27. Ubeda, S., F. J. Arcelus, and J. Faulin. Green Logistics at Eroski: A Case Study. *International Journal of Production Economics*, Vol. 131, No. 1, 2011, pp. 44–51. https://doi.org/10.1016/j.ijpe.2010.04.041.

The Standing Committee on Freight Transportation Planning and Logistics (AT015) peer-reviewed this paper (18-00143).

Article

TRR
JOURNAL OF THE TRANSPORTATION RESEARCH BOARD

Transportation Research Record
2018, Vol. 2672(9) 70–80

DOI: 10.1177/0361198118777084
journals.sagepub.com/home/trr
SAGE

Supplier Evaluation Model in Freight Activity Microsimulation Estimator

Seyed Mehdi Mahmoudifard[1], Ramin Shabanpour[2], Nima Golshani[2], Kiana Mohammadian[3], and Abolfazl Mohammadian[2]

Abstract

The supplier selection process is one of the main components of the Freight Activity Microsimulation Estimator (FAME), which is a disaggregated and comprehensive framework that simulates the freight movements for all industries and all commodities in the U.S. However, the supplier selection and supplier evaluation models in the FAME face computational issues. Using the result of a nationwide establishment survey, this study analyzes the supplier selection problem by evaluating the potential suppliers. The buyer's behavior on selecting the distance range in which the trade forms is analyzed using both machine-learning and statistical approaches. A decision-tree model and an ordered probit model are estimated and compared to better comprehend the supplier evaluation process. The results indicate that several factors such as the type of the business, commodity type, number of orders, and the value of orders are significant factors. In addition, the decision-tree model is reliable in forecasting the consumer's behavior.

Supplier selection is one of the most important stages in the supplier management process. It directly affects the profit of a company and the final quality of products. Therefore, poor decision-making in the supplier selection process could widely affect a company's failure (*1*).

For a company, selecting the right supplier comprises much more than skimming a series of price lists. The decision-maker's choices depend on a wide range of features such as quality, reliability, and service attributes of the supplier. How the business weighs up the importance of these features will be based on its priorities and strategies. For instance, if a company wants to cut down the time it takes to serve its customers, suppliers which offer faster delivery will rank higher than those who compete only on price.

A wide variety of studies have been focused on how companies weigh the importance of these factors. Deboer et al. (*2*) categorized these models into five types: 1—linear weighting models; 2—total cost of ownership models; 3—mathematical programming models; 4—statistical models; and 5—Artificial Intelligence (AI) models.

In linear weighting models, certain weights are assigned to different criteria. A larger amount of weight indicates the greater importance of a criterion. The weights are multiplied by the rates on the criteria and then summed up to create a single score for each supplier. The supplier with the highest score will then be chosen as the proper option to trade with (*3–7*). A common technique in linear weighting models is called Analytic Hierarchy Process (AHP). Several studies have proposed using AHP to manage imprecision in the supplier selection process, including (*8–11*). AHP avoids the difficulties of assigning exact numbers to weigh the criteria; instead, it uses verbal statements to compare different criteria (*2*).

Total Cost of Ownership (TCO) models attempt to consider all lifecycle costs of an item including: pre-transaction costs, transaction costs, and post-transaction costs (*12*). Furthermore, the cost-ratio is considered a TCO method (*4*). In this method, all costs corresponding to delivery, quality, and service are collected and then expressed as a penalty or benefit percentage on the unit price. Several studies have been performed to increase the accuracy of cost indication for service and delivery criteria (*13–15*).

[1]Data Scientist and Transportation Planner at CDM Smith, Chicago, IL
[2]Department of Civil and Materials Engineering, University of Illinois at Chicago, Chicago, IL
[3]Neuqua Valley High School, Chicago, IL

Corresponding Author:
Address correspondence to Seyed Mehdi Mahmoudifard: mahmoudifardsm@cdmsmith.com

In Mathematical Programming (MP) models, the decision problem is formulated in terms of an objective function that needs to be maximized or minimized by changing the values of the variables in the objective function (e.g., amount ordered from a supplier). Weber, Current, and Desai used Data Envelopment Analysis (DEA) and MP to create a tool for negotiations with suppliers (*16*, *17*). DEA is a linear programming (LP) technique for determining the relative performance of structural units (*18*, *19*). A DEA-based methodology was proposed by Saen (*20*) that considered both imprecise data and undesirable outputs simultaneously. In another study, Zeydan et al. (*21*) considered both quantitative and qualitative variables in assessment of a supplier's efficiency. In that study, qualitative variables were transformed to quantitative variables to be operational in the DEA method.

To solve some of the problems regarding the stochastic uncertainty of supplier choice, statistical models have been proposed. Although most purchasing situations have some sort of uncertainty, the studies that have considered this stochastic uncertainty are rare (*2*, *22–24*).

AI models incorporate historical data and computer-aided systems to train the models in such a way that in similar situations, a non-expert will be able to consult the model (*2*, *24*). Neural networks (NN), expert systems, and case-based-reasoning systems are the best examples of AI models (*24*). Numerous researchers that have used AI models in their studies (*25–34*). In general, AI models can manage the complexity and uncertainty more efficiently than traditional models, the reason being that AI models are designed to operate in an approach similar to human judgment.

Supplier evaluation and supplier selection are very important from a freight transportation perspective (*35*, *36*). Since the results of such models finally resulted in the recognition of trade between businesses and, consequently, the stream of commodities between them, the flows between firms are the inputs to freight transportation models.

The Freight Activity Microsimulation Estimator (FAME) is a framework, which simulates freight movement in the United States. One of the key elements of this framework is the supplier evaluation and the supplier selection models (*37–39*, *41*, *42*). The supplier evaluation and selection structure of FAME is a two-step process in which, first, a rank-ordered probit model is used to determine the weight of each criterion. An optimization model is then used to determine the supplier-buyer chain (*34–37*).

Although FAME is a suitable framework for a freight transportation simulation, it suffers from inevitable computational problems. The supplier-buyer chain formation confronts massive data loss that consequently results in inaccurate traffic assignment. To address these issues, we have proposed a procedure that reduces the computational restriction of the model.

This paper employs both data mining and statistical techniques to evaluate potential suppliers for a firm. A decision-tree model and an ordered probit model are estimated to analyze the behavior of the buyers and determine the distance range in which the trade between supplier and buyer forms. The utilized data is drawn from a nationwide freight survey conducted at the University of Illinois at Chicago (UIC) in 2012 (*42–44*).

The general notion is that local businesses would perform their trade locally, meaning that small businesses would choose their suppliers within a limited range of distance. However, by performing the analysis on the data, we found that several other factors can affect the supplier distance which will be discussed in the result section.

The rules that are found in the decision tree are applied to the supplier selection structure to decrease the number of potential suppliers. The aforementioned method consequently addresses the computational issues of the FAME framework by solving the supplier evaluation part.

The rest of this study is organized as followed: after the introduction, a descriptive analysis of the survey results is presented. Following that, the methodological approach that is used in this study will be elaborated. Then, the decision tree results are presented and discussed. Finally, in the conclusion, the findings of this study are presented.

Data

To develop a freight activity microsimulation framework, Samimi et al. (*42*) conducted an establishment survey. The first wave of this survey was completed back in 2009 at UIC and was focused primarily on collecting the essential data for developing the basic components of the FAME. The collected data was mainly used to develop the mode choice model. Initially, three methods of data collection were evaluated, including telephone interview, mail questioner, and web-based methods (which were selected because of lower cost, a higher response rate, and more convection factors) (*42*, *45*, *46*).

Two new waves of the survey were run in 2010–2011 by another team at UIC to fill in the gaps of the previous research and to collect detailed information about logistic choices and supply chain formation. Around 219,000 contacts were attempted, using all different contact techniques, including: web crawling, telephone introductions, and e-mail blast campaigns. One introductory e-mail and two reminders were sent to the representatives of freight handling in the companies. In total, 657 surveys were completed which resulted in 970 useable shipment forms (*45*).

The survey comprised three major parts. In the first part, questions about the characteristics of the business establishment (e.g., firm location, value of total annual shipments, number of employees, number of weekly inbound/outbound shipments, supply chains, and major suppliers etc.) were asked. The second part of the survey collected the information about the five most recent shipments and their corresponding attributes, such as origin, destination, commodity type, mode of transportation, weight and value of the shipment, and so forth (*45*). In the third part, several new questions were added to collect the information on the features of decision-makers and the other logistics modules of supply chain.

One of the main purposes of the survey was to collect data to understand the process of decision-making in a supply chain. Respondents were asked to rate eight potential criteria—cost, credit and finance, distance and convenience, delivery, loyalty, manufacturing capacity and reliability, management and service, and technology and quality—on a scale from one to five. As presented in Figure 1, the results show a considerably different picture of how suppliers were selected. Quality, delivery, manufacturing capacity, and cost have been given values higher than four. In addition, Figure 2 shows how respondents assigned scores for each criterion (*43*).

Using the origin and destination zip code, the distance range is calculated as the Great Circle Distance (GCD). The distance at which the major supplier is located is categorized in one of six ranges (i.e., "200 miles or less," "200–500 miles," "500–1,000 miles," "1,000–1,500 miles," "1,500–2,000 miles," and "2,000 miles and more"). The survey results were used to determine the distance range in which the trades are forming by developing a decision-tree model and an ordered probit model.

Methodology

Decision Tree

Data Mining is defined as the procedure that uses mathematical, statistical, machine-learning techniques, and artificial intelligence to identify and extract useful information and consequently gain knowledge out of large databases (*47–51*). Decision-tree learning is one of the most practical procedures for Inductive Learning. Rule induction denotes the rules derived from the decision-tree techniques. The dataset is divided into many segments in a way which maximizes the purity, which is defined as the degree to which the dependent variable fits to a certain class. The rules that are used for separating the data are called the inducted rules. Decision tree is considered as a non-parametric procedure and it is suitable for finding the interaction effect or nonlinearity (*52*). Also, decision trees are very convenient for the sake of interpretation of the analysis results.

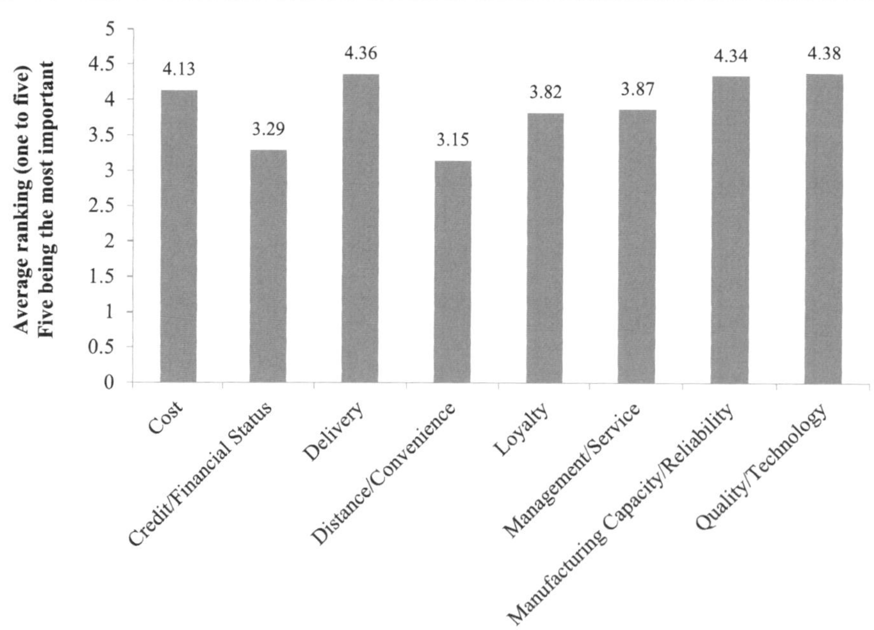

Figure 1. Average scores of supplier selection criteria (*43*).

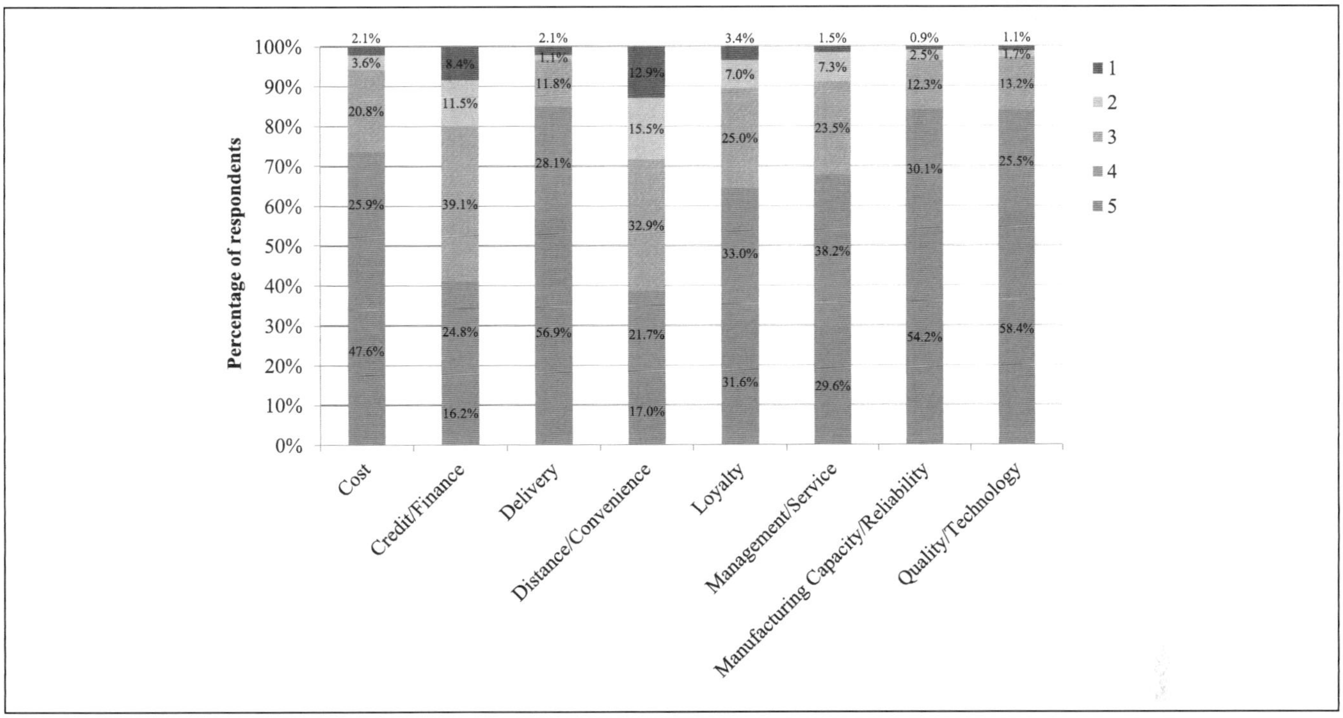

Figure 2. Rankings of supplier selection criteria (*43*).

A decision tree has four types of procedure such as chi-squared automatic interaction detection (CHAID) (*53*), classification and regression tree (CART) (*54*), quick, unbiased, efficient statistical tree (QUEST) (*55*), and commercial version 5.0 (C5.0) (*56*).

- The CHAID method is built on the chi-square test of association. A CHAID tree is a decision tree that is built by continually separating subsets of the data into two or more child nodes, starting with the entire dataset (*57*). To specify the best split at any node, any permissible pair of categories of the predictor variables is combined till there is no statistically major difference within the pair according to the target variable. The final nodes identify subcategories built by different sets of independent variables (*58*).
- CART is a recursive partitioning procedure to be used both for classification and regression. CART is built on splitting subgroups of the dataset using all predictor variables to generate two child nodes repeatedly. The best predictor is selected using a variety of diversity measures or impurity. CART tries to produce subgroups of the data that are as homogeneous as possible according to the target variable (*54*).
- QUEST is a binary-split decision tree method for categorization and data mining. It can be used with linear or univariate combination splits. A unique feature of QUEST is that its attribute selection procedure has negligible bias. If all the attributes are uninformative, according to the class attribute, then almost each one has the same chance of being selected to split a node (*55*).
- C5.0 is a supervised learning classification process that is used to build the decision trees from the data (*56*). It uses a divide-and-conquer procedure to make the decision trees. The main difference between C5.0 and other decision-tree building methods is in the test selection and evaluation procedure.

Association rules based models: association rule learning is a well-researched method for revealing interesting relations among variables in a large database. Extraction of regular patterns leads to data patterns with some predetermined level of frequency. The quality of extracted patterns is evaluated using measures such as confidence and support. For instance, for the association rule $X => Y$, confidence is the fraction of times Y is presented in the data when X also exists, and support denotes the number of times X exists as a fraction of the total number of samples. An association is called strong when there is high confidence and support and it could be potentially useful (*59*).

In this study, we employed the decision-tree procedure to analyze the results of the UIC freight establishment survey. We applied the classification and regression tree

(CRT) method to categorize the data into homogeneous categories which have similar patterns, considering their supplier selection and transportation pattern (*54*). CRT resulted in a series of decision rules to define the probability of choosing a supplier in a specific distance range. Applying this method decreases the computational burden of the supplier selection problem in FAME. The results of the estimated decision tree model are presented in the results section.

Ordered Probit Model

In addition to the decision tree model, we have estimated an ordered probit model to evaluate the performance of statistical models in the context of the supplier evaluation decision. As indicated in the previous section, six possible ranges of distance have been considered for a company's supplier selection. Since these categories can be considered in an ordinal scale, ordered probability models seem to be appropriate for solving the supplier evaluation problem. Therefore, we estimated an ordered probit model which does not rely on the Independence of Irrelevant Alternatives (IIA) assumption, as follows (*60*):

$$z_i = \beta' X_i + \varepsilon_i, y_i = j \ if \ \mu_{j-1} < z_i < \mu_j, j = 0, \ldots, J \quad (1)$$

Where z_i is an unobserved variable that is used as the basis for modeling, X_i is the vector of independent variables that determines company supplier selection decision-making, β' is the vector of estimable parameters, μ_j is a threshold that defines y_i and it is estimated jointly with β', y_i corresponds to integer ordering, j is the integer ordered choice for the dependent variable, and ε_i is the random error term assumed to be normally distributed with mean zero and variance of one. Furthermore, the ordered selection probability of each choice can be written as (*61*):

$$P(y = j) = \Phi\left(\mu_j - \beta' X\right) - \Phi\left(\mu_{j+1} - \beta' X\right) \quad (2)$$

Where, $P(y = j)$ is the probability of outcome j, and $\Phi(.)$ is the cumulative normal distribution, and μ_j and μ_{j+1} are the lower and upper thresholds of outcome j, respectively. The likelihood function can then be written as in (*61*):

$$L = \prod_{n=1}^{N} \prod_{j=1}^{J} \left(\Phi\left(\mu_j - \beta' X_n\right) - \Phi\left(\mu_{j+1} - \beta' X_n\right)\right)^{\delta_{jn}} \quad (3)$$

Where, N is the total number of observations, J is the total number of choices, and δ_{in} is a binary variable indicating if observation n belongs to choice j.

Results

Ordered Probit Model

The dataset is divided into a training set for model estimation and a test set for model validation with 80% and 20% of the observations, respectively. Table 1 presents the estimation results of the ordered probit model for the supplier selection decision. All possible variables and variable interactions are tested, and only the statistically significant variables at 90%, 95%, and 99% levels of confidence are presented in the table. For model interpretation, a positive value of a coefficient implies that increasing the explanatory parameter increases the

Table 1. Ordered Probit Model Estimation Results

Variable	Coefficient	*t*-stat	*p*-value
Model			
Constant	−0.13*	−1.65	.10
Average weekly number of outbound shipping (in thousands)	0.01*	1.82	.07
Number of employees	0.04*	1.95	.05
Primary business indicator (1 if agriculture, forestry, fishing and hunting, 0 otherwise)	−0.95**	−2.35	.02
Primary business indicator (1 if transportation and warehousing, 0 otherwise)	0.58***	4.62	.00
Primary business indicator (1 if wholesale trade, 0 otherwise)	0.59***	4.42	.00
Commodity indicator (1 if other, 0 otherwise)	0.21**	1.96	.05
Commodity indicator (1 if prepared foodstuffs, 0 otherwise)	−0.42**	−2.11	.04
Model summary			
μ_1	0.39***	11.36	.00
μ_2	0.77***	17.01	.00
μ_3	1.08***	20.17	.00
μ_4	1.22***	21.09	.00
Log-likelihood at convergence	−980.25		
Number of observations	661		

Note: *significant at 90%. **significant at 95%. ***significant at 99%.

Table 2. Partial Derivatives and Pseudo-Elasticities

Variable	Choices					
	0–200	200–500	500–1000	1000–1500	1500–2000	>2000
Average weekly number of outbound shipping (in thousands)	−0.009	0.001	0.006	0.011	0.015	0.001
Number of employees	−0.077	−0.005	0.048	0.092	0.123	0.208
Primary business indicator (1 if agriculture, forestry, fishing and hunting, 0 otherwise)	0.674	−0.395	−0.637	−0.820	−0.928	−1.133
Primary business indicator (1 if transportation and warehousing, 0 otherwise)	−0.415	−0.088	0.152	0.429	0.654	1.455
Primary business indicator (1 if wholesale trade, 0 otherwise)	−0.418	−0.095	0.146	0.428	0.657	1.486
Commodity indicator (1 if other, 0 otherwise)	−0.160	−0.003	0.089	0.185	0.256	0.467
Commodity indicator (1 if prepared foodstuffs, 0 otherwise)	0.321	−0.098	−0.257	−0.396	−0.488	−0.699

likelihood of the last response (i.e., 2000 miles and greater) and decreases the probability of the first response (i.e., less than or equal to 200 miles).

To better interpret the results, mean partial derivatives (for continuous variables) and pseudo-elasticities (for dummy variables) are calculated for each exploratory variable and each category as follows (*60*):

$$\frac{\partial P(y=I)}{\partial X} = -\phi(\mu_{I-2} - \beta X)\beta' \quad (4)$$

$$E = \phi\left(\frac{\beta_j X_{j,1}}{\sigma}|X_i = 1\right) - \phi\left(\frac{\beta_j X_j}{\sigma}|X_i = 0\right) \quad (5)$$

Where, $P(y = I)$ is the probability of outcome I, and $\emptyset(.)$ is the probability density function of the standard normal distribution. Table 2 presents the direct effects of the independent variables.

The results indicate that several variables representing characteristics of the company affect its supplier selection. For example, both "Wholesale Trade", and "Transportation and Warehousing" companies are less likely to acquire their required goods from 500 miles or less and are, therefore, more likely to select suppliers located further than 500 miles away. On the other hand, "Agriculture, Forestry, Fishing and Hunting" businesses tend to source their goods from local suppliers (as shown in Table 2, these businesses increase the probability of selecting a supplier from 200 miles or closer by 0.674). These variables along with the food commodity indicators are possibly capturing the effect of the type of supplies required by the food industries, which may go bad in long trips.

The results also indicate that an increase in the number of employees in a company increases the probability of selecting a supplier from "500 to 1,000 miles", "1,000 to 1,500 miles", "1,500 to 2,000 miles", and "2,000 miles and more" by 0.048, 0.092, 0.123, and 0.208, respectively. On a similar note, a higher number of weekly outbound shipments by a company increases the chance of sourcing its goods from a more distant supplier. These variables may be capturing the effect of companies' size on their supplier selection where larger companies can afford to pay higher transportation costs if the initial costs are lower.

Decision Tree Model

Figure 3 shows the decision tree in which the depth of the tree was assumed to be three levels. The CRT basics for the development of the tree includes a minimum number of 100 observations in parent nodes and 50 in child nodes. The Gini procedure was used as the impurity measurement in which splits are found that maximize the homogeneity of child nodes with respect to the value of target variable, and a cross-validation process performed on the results.

We found that a firm's primary business, commodity type, average number of weekly outbound shipments, major supplier value (Cost), and number of orders from the major supplier are the significant variables in this decision. Primarily, two variables including the size of a business and the warehouse situation of that business are considered to be important. The size of a business was represented by its floor area or number of employees. However, according to the results of the survey and the decision-tree model, it is shown that the distance to the supplier is not affected significantly by these two variables.

Commodity type shows a significant influence on the decision-tree splits. According to Figure 3, chemical-pharmaceutical products, gravel-natural sands-cement, prepared foodstuffs, and mixed freight-miscellaneous show similar patterns and other commodities show different patterns considering the distance to the main supplier.

Finally, having the decision tree, the split rules can be extracted and then used in the first step of supplier

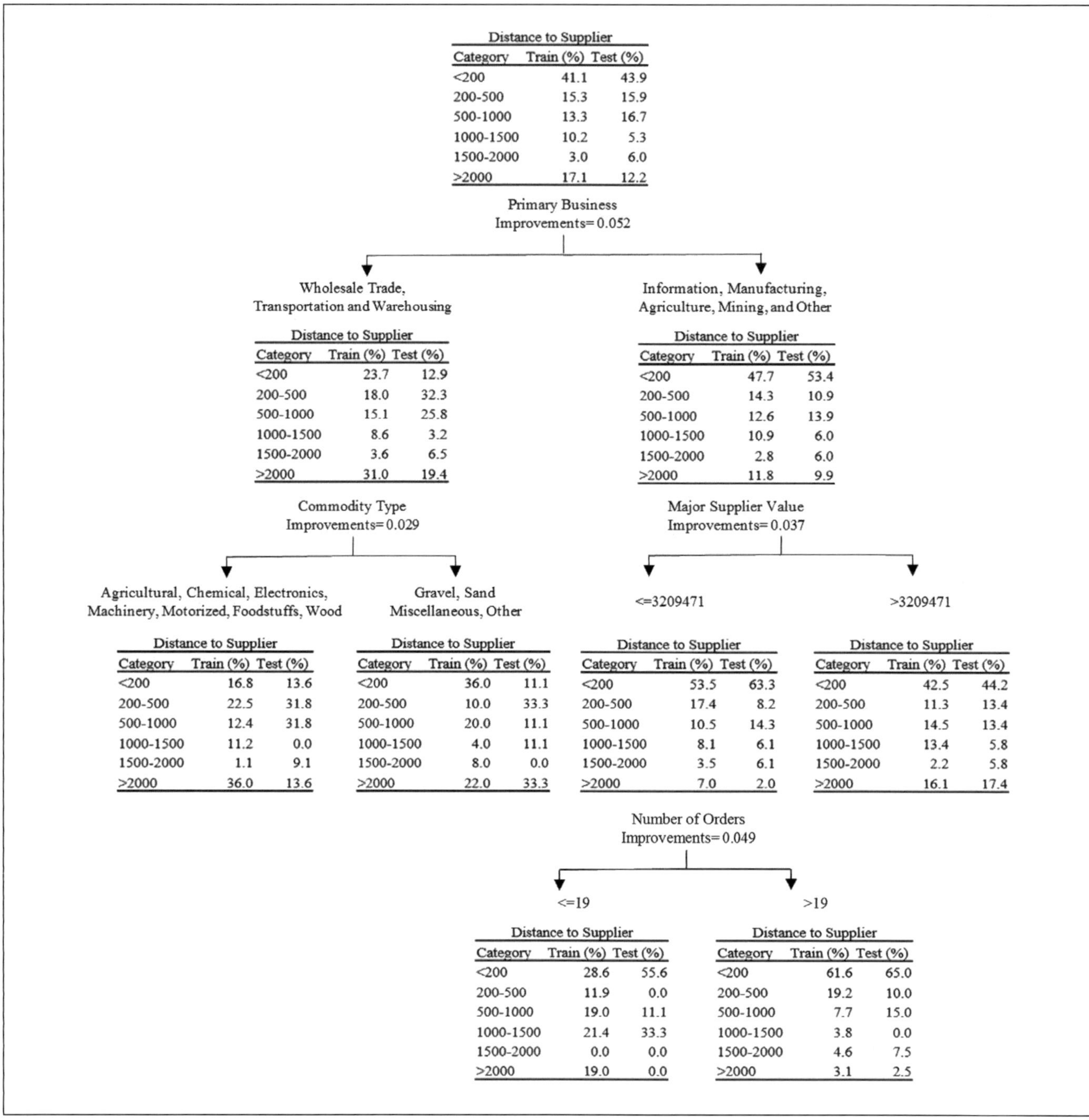

Figure 3. Decision tree cluster for distance to supplier category.

evaluation of the FAME framework. The FAME model then uses an optimization process to allocate the freight flows to the firm types in the U.S. (*38*).

To visualize the differences between observed and prediction models, a group of companies located in Illinois were selected, all having similar attributes, such as "primary business" and similar range of "value of orders." The results were plotted in Figure 4, which shows that decision trees are more successful in predicting the behavior of those specific companies. The outputs of the models for this group of firms are relatively similar, showing that in the distance range of 200 miles or less, the probability of trade formation is higher than other distance ranges.

Figure 5 compares the prediction accuracy of both decision-tree and ordered probit models with respect to supplier distance ranges. The results indicate that decision-tree models can correctly predict the outcome in

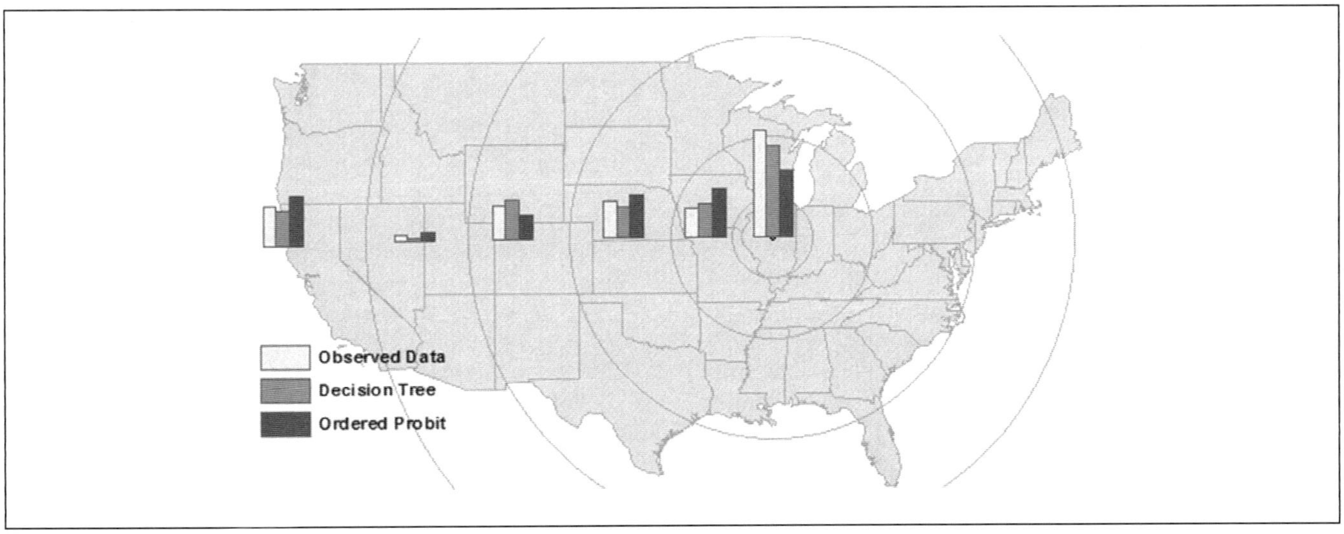

Figure 4. Visualization of the probability distribution of firm's supplier selection distance, located in Illinois.

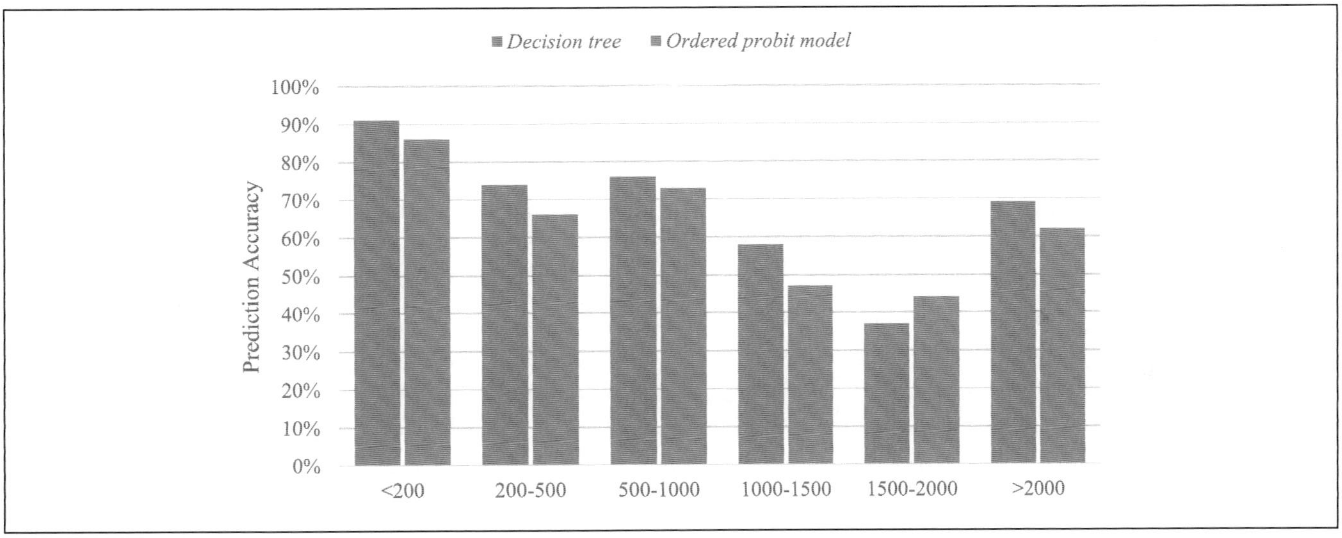

Figure 5. Supplier distance prediction accuracy in decision-tree and ordered probit models.

77.62% of cases, while the prediction accuracy rate in ordered probit models is 71.84%. It should be noted that, in general, machine learning methods can in fact be superior to statistical models since they can effectively capture nonlinearities in behavioral factors to a change in a variable (*50*, *51*). The prediction for distance ranges of 1,500–2,000 miles is relatively poor mainly because the number of observations for this distance range is not sufficient which prevents models from finding the correct pattern for this range. The best prediction accuracy belongs to "under 200" miles in which both models can correctly classify the observations in 91% and 86% of cases in the dataset, respectively.

Conclusion

To analyze the result of an establishment survey, several methods were considered, such as: neural networks (NN), decision trees, and ordered probit models. Among machine-learning approaches, decision trees were selected for several reasons. Firstly, NN do not present an easily understandable model. In contrast, in a decision tree, it is easy to observe that some initial variables separate the data into two or more categories, and then other variables further divide the resulting child groups. These facts can be very useful to the researchers who are trying to recognize the underlying perspectives of the data being analyzed.

Moreover, if a challenge is made to a decision according to an NN model, it is very difficult to justify and explain the causes of the decision to non-technical people. In contrast, a decision tree is straightforwardly explainable, allowing the process of decision-making to be readily shown. Furthermore, it is difficult to implement an NN model in a computer system without assigning an interpreter. However, it is very easy to convert a decision tree into "if…then…else" statements once a decision-tree model has been built.

After building a machine-learning-based decision tree and an econometric-based ordered probit model and analyzing the results of the two approaches, it was revealed that a firm's primary business type could affect the distance to the main supplier. Other significant variables that were found to be important in identifying the distance to the main supplier were a firm's commodity type, number of orders from the major supplier, and average number of weekly outbound shipments. The variables such as floor space, number of employees, and warehouse condition did not show significant influence on the final results of the decision-tree model.

References

1. Gonzalez, M. E., G. Quesada, and C. A. Mora Monge. Determining the Importance of the Supplier Selection Process in Manufacturing: A Case Study. *International Journal of Physical Distribution & Logistics Management*, Vol. 34, No. 6, 2004, pp. 492–504.
2. De Boer, L., E. Labro, and P. Morlacchi. A Review of Methods Supporting Supplier Selection. *European Journal of Purchasing & Supply Management*, Vol. 7, No. 2, 2001, pp. 75–89.
3. Zenz, G. J. *Purchasing and the Management of Materials*. John Wiley & Sons, New York, 1987.
4. Timmerman, E. An Approach to Vendor Performance Evaluation. *IEEE Engineering Management Review*, Vol. 3, No. 15, 1987, pp. 14–20.
5. De Boer, L., L. van der Wegen, and J. Telgen. Outranking Methods in Support of Supplier Selection. *European Journal of Purchasing & Supply Management*, Vol. 4, No. 2, 1998, pp. 109–118.
6. Grando, A., and A. Sianesi. Supply Management: A Vendor Rating Assessment. *CEMS Business Review*, Vol. 1, No. 1, 1996, pp. 199–212.
7. Gregory, R. E. Source Selection: A Matrix Approach. *Journal of Purchasing and Materials Management*, Vol. 22, No. 2, 1986, pp. 24–29.
8. Nydick, R. L., and R. P. Hill. Using the Analytic Hierarchy Process to Structure the Supplier Selection Procedure. *Journal of Supply Chain Management*, Vol. 28, No. 2, 1992, pp. 31–36.
9. Barbarosoglu, G., and T. Yazgac. An Application of the Analytic Hierarchy Process to the Supplier Selection Problem. *Production and Inventory Management Journal*, Vol. 38, No. 1, 1997, p. 14.
10. Narasimhan, R. An Analytical Approach to Supplier Selection. *Journal of Purchasing and Materials Management*, Vol. 19, No. 4, 1983, pp. 27–32.
11. Masella, C., and A. Rangone. A Contingent Approach to the Design of Vendor Selection Systems for Different Types of Co-Operative Customer/Supplier Relationships. *International Journal of Operations & Production Management*, Vol. 20, No. 1, 2000, pp. 70–84.
12. Ellram, L. M., and A. Carr. Strategic Purchasing: A History and Review of the Literature. *International Journal of Purchasing and Materials Management*, Vol. 30, No. 1, 1994, pp. 9–19.
13. Monczka, R. M., and S. J. Trecha. Cost-Based Supplier Performance Evaluation. *Journal of Supply Chain Management*, Vol. 24, No. 1, 1988, pp. 2–7.
14. Smytka, D. L., and M. W. Clemens. Total Cost Supplier Selection Model: A Case Study. *International Journal of Purchasing and Materials Management*, Vol. 29, No. 4, 1993, pp. 42–49.
15. Sweeney, E., and W. Bahr. *Practitioner Perspectives on Supply Chain Management and Logistics: A Study from the United Kingdom*. Proceedings of the 20th Annual Logistics Research Network Conference, Derby, UK, 2015.
16. Weber, C. A., and A. Desai. Determination of Paths to Vendor Market Efficiency Using Parallel Coordinates Representation: A Negotiation Tool for Buyers. *European Journal of Operational Research*, Vol. 90, No. 1, 1996, pp. 142–155.
17. Weber, C. A., J. R. Current, and A. Desai. Non-Cooperative Negotiation Strategies for Vendor Selection. *European Journal of Operational Research*, Vol. 108, No. 1, 1998, pp. 208–223.
18. Wu, D. D., and D. Olson. Enterprise Risk Management: A DEA VaR Approach in Vendor Selection. *International Journal of Production Research*, Vol. 48, No. 16, 2010, pp. 4919–4932.
19. Wu, D. Supplier Selection: A Hybrid Model Using DEA, Decision Tree and Neural Network. *Expert Systems with Applications*, Vol. 36, No. 5, 2009, pp. 9105–9112.
20. Saen, R. F. Developing a New Data Envelopment Analysis Methodology for Supplier Selection in the Presence of Both Undesirable Outputs and Imprecise Data. *The International Journal of Advanced Manufacturing Technology*, Vol. 51, No. 9–12, 2010, pp. 1243–1250.
21. Zeydan, M., C. Çolpan, and C. Çobanoğlu. A Combined Methodology for Supplier Selection and Performance Evaluation. *Expert Systems with Applications*, Vol. 38, No. 3, 2011, pp. 2741–2751.
22. Soukup, W. R. Supplier Selection Strategies. *Journal of Supply Chain Management*, Vol. 23, No. 2, 1987, pp. 7–12.
23. Ronen, B., and D. Trietsch. A Decision Support System for Purchasing Management of Large Projects. *Operations Research*, Vol. 36, No. 6, 1988, pp. 882–890.
24. Wu, C., and D. Barnes. A Literature Review of Decision-Making Models and Approaches for Partner Selection in Agile Supply Chains. *Journal of Purchasing and Supply Management*, Vol. 17, No. 4, 2011, pp. 256–274.
25. Zhao, K., and X. Yu. A Case Based Reasoning Approach on Supplier Selection in Petroleum Enterprises. *Expert*

Systems with Applications, Vol. 38, No. 6, 2011, pp. 6839–6847.
26. Faez, F., S. Ghodsypour, and C. O'brien. Vendor Selection and Order Allocation Using an Integrated Fuzzy Case-Based Reasoning and Mathematical Programming Model. *International Journal of Production Economics*, Vol. 121, No. 2, 2009, pp. 395–408.
27. Humphreys, P., Y. Wong, and F. Chan. Integrating Environmental Criteria into the Supplier Selection Process. *Journal of Materials Processing Technology*, Vol. 138, No. 1, 2003, pp. 349–356.
28. Choy, K., W. Lee, H. Lau, D. Lu, and V. Lo. Design of an Intelligent Supplier Relationship Management System for New Product Development. *International Journal of Computer Integrated Manufacturing*, Vol. 17, No. 8, 2004, pp. 692–715.
29. Choy, K. L., W. Lee, and V. Lo. An Intelligent Supplier Management Tool for Benchmarking Suppliers in Outsource Manufacturing. *Expert Systems with Applications*, Vol. 22, No. 3, 2002, pp. 213–224.
30. Yigin, I., H. Taşkin, I. Cedimoglu, and B. Topal. Supplier Selection: An Expert System Approach. *Production Planning and Control*, Vol. 18, No. 1, 2007, pp. 16–24.
31. Guo, X., Z. Yuan, and B. Tian. Supplier Selection Based on Hierarchical Potential Support Vector Machine. *Expert Systems with Applications*, Vol. 36, No. 3, 2009, pp. 6978–6985.
32. Lee, C. C., and C. Ou-Yang. A Neural Networks Approach for Forecasting the Supplier's Bid Prices in Supplier Selection Negotiation Process. *Expert Systems with Applications*, Vol. 36, No. 2, 2009, pp. 2961–2970.
33. Montazer, G. A., H. Q. Saremi, and M. Ramezani. Design a New Mixed Expert Decision Aiding System Using Fuzzy ELECTRE III Method for Vendor Selection. *Expert Systems with Applications*, Vol. 36, No. 8, 2009, pp. 10837–10847.
34. Aksoy, A., and N. Öztürk. Supplier Selection and Performance Evaluation in Just-in-Time Production Environments. *Expert Systems with Applications*, Vol. 38, No. 5, 2011, pp. 6351–6359.
35. Roorda, M., S. McCabe, and H. Kwan. Design of a Shipper-Based Survey of Freight Movements in the Greater Golden Horseshoe. *Proc., Annual Canadian Transportation Research Forum (CTRF) Conference, Quebec*, Quebec, 2006.
36. Roorda, M. J., R. Cavalcante, S. McCabe, and H. Kwan. A Conceptual Framework for Agent-Based Modelling of Logistics Services. *Transportation Research Part E: Logistics and Transportation Review*, Vol. 46, No. 1, 2010, pp. 18–31.
37. Pourabdollahi, Z., B. Karimi, A. K. Mohammadian, and K. Kawamura. A Multi-Criteria Supplier Evaluation Model for Supply Chains. Presented at 93rd Annual Meeting of the Transportation Research Board, Washington, D.C., 2014.
38. Mahmoudifard, S. M., R. Shabanpour Anbarani, N. Golshani, and A. K. Mohammadian. An Activity Microsimulation Framework for the U.S. Grain Transportation: An Expansion to the Freight Activity Microsimulation Estimator (FAME) Model. Presented at 96th Annual Meeting of the Transportation Research Board, Washington, D.C., 2017.
39. Mahmoudifard, S. M. *Travel Behavior Analysis of Agricultural Commodities in a Freight Activity Microsimulation Framework*. PhD dissertation. University of Illinois, Chicago, 2016.
40. Pourabdollahi, Z., M. Javanmardi, B. Karimi, A. Mohammadian, and K. Kawamura. Mode and Shipment Size Choice Models in the FAME Simulation Framework. Presented at 92nd Annual Meeting of the Transportation Research Board, Washington, D.C., 2013.
41. Samimi, A., A. Mohammadian, and K. Kawamura. Behavioral Freight Movement Modeling. *Proc., The 12th International Conference on Travel Behaviour Research, Resource Paper, Workshop*, No. 2, 2009.
42. Samimi, A., A. Mohammadian, and K. Kawamura. A Behavioral Freight Movement Microsimulation Model: Method and Data. *Transportation Letters*, Vol. 2, No. 1, 2010, pp. 53–62.
43. Pourabdollahi, Z., B. Karimi, A. Mohammadian, and K. Kawamura. An Agent-Based Supply Chain and Freight Transportation Model: Case Study for Chicago Metropolitan Area. Presented at 95th Annual Meeting of the Transportation Research Board, Washington, D.C., 2016.
44. Mahmoudifard, S. M., S. Ko, and K. Mohammadian. *Assessing Sustainable Freight Policies*, 2014.
45. Pourabdollahi, Z. *An Agent-Based Freight Transportation Modeling Framework*. PhD dissertation. University of Illinois, Chicago, 2015.
46. Mahmoudifard, S. M., S. Ko, and K. Mohammadian. *CFIRE*, 2014.
47. Turban, E., J. Aronson, T. Liang, and R. Sharda. *Decision Support and Business Intelligence Systems*, 8th ed. Prentice Hall, Upper Saddle River, NJ, 2007.
48. Ngai, E. W., L. Xiu, and D. C. Chau. Application of Data Mining Techniques in Customer Relationship Management: A Literature Review and Classification. *Expert Systems with Applications*, Vol. 36, No. 2, 2009, pp. 2592–2602.
49. Berson, A., S. Smith, and K. Thearling. *Building Data Mining Applications for CRM*. McGraw-Hill, New York, 2000.
50. Golshani, N., R. Shabanpour, S. M. Mahmoudifard, S. Derrible, and A. K. Mohammadian. Comparison of Artificial Neural Networks and Statistical Copula-Based Joint Models. Presented at the Transportation Research Board, Washington, D.C., 2017.
51. Golshani, N., R. Shabanpour, S. M. Mahmoudifard, S. Derrible, and A. Mohammadian. Modeling Travel Mode and Timing Decisions: Comparison of Artificial Neural Networks and Copula-Based Joint Model. *Travel Behaviour and Society*, Vol. 10, 2017, pp. 21–32.
52. Bae, J. K., and J. Kim. Product Development with Data Mining Techniques: A Case on Design of Digital Camera. *Expert Systems with Applications*, Vol. 38, No. 8, 2011, pp. 9274–9280.
53. Kass, G. V. An Exploratory Technique for Investigating Large Quantities of Categorical Data. *Applied Statistics*, 1980, pp. 119–127.

54. Breiman, L., J. Friedman, C. J. Stone, and R. A. Olshen. *Classification and Regression Trees*. CRC Press, Boca Raton, 1984.
55. Loh, W. Y., and Y. S. Shih. Split Selection Methods for Classification Trees. *Statistica Sinica*, 1997, pp. 815–840.
56. Quinlan, J. R. *C4. 5: Programming for Machine Learning*. Morgan Kaufmann, San Mateo, 1993, p. 38.
57. Michael, J., and S. L. Gordon. *Data Mining Technique: For Marketing, Sales and Customer Support*. John Wiley & Sons, New York, Vol. 445, 1997.
58. Magidson, J. M., M. J. M. J. Norusis, L. Hernández, L. Olariaga, L. Lizasoain, and L. Joaristi. *SPSS: SPSS for Windows CHAID Release 6.0*. SPSS Inc., 1993.
59. Agrawal, R., T. Imieliński, and A. Swami. Mining Association Rules Between Sets of Items in Large Databases. *Acm Sigmod Record, No*. 22, 1993, pp. 207–216.
60. Washington, S. P., M. G. Karlaftis, and F. Mannering. *Statistical and Econometric Methods for Transportation Data Analysis*. CRC Press, Boca Raton, 2010.
61. Greene, W., and D. Hensher. *Modeling Ordered Choices: A Primer*. Cambridge University Press, Cambridge, 2010.

The Standing Committee on Freight Transportation Planning and Logistics (AT015) peer-reviewed this paper (18-02772).

Article

TRR
JOURNAL OF THE TRANSPORTATION RESEARCH BOARD

Transportation Research Record
2018, Vol. 2672(9) 81–92

DOI: 10.1177/0361198118787105
journals.sagepub.com/home/trr
SAGE

Dimensionality Reduction to Reveal Urban Truck Driver Activity Patterns

Fangping Lu[1], Fang Zhao[2], and Lynette Cheah[1]

Abstract

This paper studies the activity profiles of truck drivers in urban areas. Finding repeating dynamical patterns is important in understanding freight behaviors, and aids freight-friendly planning. In the digital age, data on truck drivers is becoming more available with heterogeneous demographic and work profiles. By synthesizing such pervasive data and applying machine learning concepts, this paper proposes to identify signature travel activity patterns via dimensionality reduction. Based on driver survey data, truck drivers' behaviors are represented as longitudinal activity sequences. Dimensionality reduction and activity reconstruction via principal components analysis (PCA), logistic PCA, and autoencoder were conducted to reveal fundamental activity features and approximate the underlying data-generating function. In the driver survey dataset, 243 truck drivers in Singapore reported their daily activities for 1,099 weekdays. This study found that PCA produced the most faithful reconstruction of drivers' activities. When projecting the input data down from 2,592 to 82 dimensions, PCA explained 77% of variances with a reconstruction error of 0.99%. Logistic PCA is a useful extension of PCA to study the pattern of a single activity. It captures the variation of infrequent activities such as truck queuing, which PCA fails to reconstruct. Autoencoder was found to be more powerful than PCA in reconstructing activities – with 1% of original dimensions, it reconstructed the activities with an error rate of 1.24%. Moreover, when implemented as a variational autoencoder, autoencoder generated realistic-looking samples of driver activities. The top three most distinctive activity patterns of Singapore truck drivers are reported using PCA.

Passenger and freight transportation systems vary fundamentally, therefore, different types of models have been developed to study their respective travel behaviors. For passenger mobility, different modeling frameworks, such as the traditional four-step travel demand model, are readily available. Household travel surveys are used to collect data for these models. In recent years, pervasive sensing has provided a novel data source to study passenger travel behaviors. At the same time, advances in data mining have also inspired new analytical frameworks for studying the dynamics of passenger activities (*1–4*). Freight travel, however, is less well researched due to scarcity of freight data, the involvement of multiple decision makers and supply chain actors, and complexity in modeling. In modeling travel behaviors, freight agents are more heterogeneous and their behaviors vary more than passengers. For example, a typical passenger's day may consist of a few key locations such as home, school, or work, whereas a truck driver can visit dozens of locations and perform different activities in a day. Consequently, characterizing activity patterns of freight travel is less intuitive and straightforward than doing so for passenger travel.

This study set out to explore the activity profiles of truck drivers in cities. To address the first issue of the lack of freight data, GPS tracking and a stop-based driver survey were conducted to capture the activities that truck drivers performed at each stop over several days. Because a truck driver typically visits many locations and engages in multiple activities daily, such as pickups or deliveries, the corresponding activity profile requires a high-dimensional representation. Therefore aggregation of information is needed to discover meaningful activity patterns while preserving critical details of the activity profiles. Tour-based models are important approaches in studying freight behaviors (*5, 6*). To complement these models, this paper proposes a framework in which patterns are extracted automatically without a priori knowledge on how truck drivers should behave.

[1]Engineering Systems & Design, Singapore University of Technology and Design, Singapore
[2]Singapore-MIT Alliance for Research and Technology, Singapore

Corresponding Author:
Address correspondence to Lynette Cheah: lynette@sutd.edu.sg

Inspired by Eagle and Pentland's pioneering work in which principal components analysis (PCA) was first used to analyze human mobility patterns (*2*), this paper proposes a methodology framework where dimensionality reduction is used to identify activity patterns. The central idea of this approach is to represent the high-dimensional activity sequence in lower dimensions while retaining the essential information. By forcing a model to prioritize which aspects of the input activity data to keep in lower dimension, it often learns key features of the original activities (*7*). From the lower-dimension representation, travel activities can be reconstructed. The discrepancy between the reconstruction and the original activity sequences is used as the main metric for evaluating how well the lower-dimension representation preserves the inherent structure of the observed behavior. In the proposed framework, dimensionality reduction techniques of PCA, logistic PCA, and autoencoder are implemented independently. Using high-resolution real-life freight activity data, their performances on discovering underlying patterns in truck driver activities are evaluated.

In line with this objective, the contribution of this paper is twofold. First, in terms of methodology, it provides novel implementations of logistic PCA, autoencoder, and variational autoencoder (VAE) in uncovering the inherent travel activity patterns. Although PCA has been applied in human mobility research (*2*, *4*, *8*), other dimensionality reduction techniques such as autoencoder, which are widely used in image processing, natural language processing and representation learning are almost never implemented in detecting mobility patterns. This paper develops practical implementations of these four techniques in analyzing freight activity patterns, and compares their performance. Second, from an empirical perspective, this study exposes the fundamental heterogeneity in urban truck drivers' activity sequences by applying dimensionality reduction to a large dataset collected from 243 truck drivers in Singapore. Due to scarcity of data, few studies have comprehensively described freight activities at an aggregate level. The patterns identified from this dataset provide evidence on the existence of distinct and repeating activity chains in urban freight transport.

The remainder of this paper is organized as follows. Section 2 describes the dataset used, and the data transformation methodology. In section 3, PCA, logistic PCA, autoencoder, and VAE are introduced. Section 4 presents the best model from each of the dimensionality reduction methods described in section 3 when applied to the input dataset. The validity of these four methods is discussed. Section 5 describes the Singapore truck drivers' activity patterns discovered from the dataset. Finally, this paper concludes with comparisons on methods developed, their significance, and application for future work.

Data

Data Description

The sample data used in this study orginates from an ongoing freight vehicle survey in Singapore. Using a fully digital data collection platform – Future Mobility Sensing-Freight (FMS-Freight) – truck and driver activity data of high granularity was acquired (*9*). For each truck driver participating in the survey, a GPS logger was installed in the vehicle that continuously collected the vehicle's location data. Based on the raw logger data, as well as other contextual information (e.g., land use, points of interest etc.), FMS-Freight inferred drivers' stop times, locations, and activities. The driver is presented with the inferred timeline in a web interface, in which they can verify the details and answer additional questions such as stop location functions, commodity type, shipment volume, and so forth. Each driver was required to verify the activities performed for five consecutive weekdays. The data collection project started in February 2017 and aimed to track over 6,000 truck drivers.

At the time this paper was produced, only data on heavy goods vehicles were available, which are trucks with a maximum laden weight of more than 3,500 metric tons. After removing activity data that was either incomplete or with missing GPS information, a dataset with a total of 1,099 weekdays of activities reported by 243 heavy goods vehicle drivers was used as the input. Among the 243 drivers, 62% worked in the construction industry, followed by 13% in manufacturing and 9% in transportation and storage. Among the 243 vehicles operated by drivers, 213 were heavy goods vehicles. In the remaining 30 non-goods vehicles, there were 24 prime movers, three cranes, two concrete mixers and a recovery vehicle. Activity options posed in the survey questionnaire are presented in Table 1.

Data Transformation

In studying a truck driver's activities, a suitable representation of an activity sequence from the raw survey data is needed. Two possible approaches have been reported in the literature: 1) represent an activity sequence as a vector $\boldsymbol{a} \in \boldsymbol{R}^D$ with 0 and 1 as elements (*2*, *4*); 2) denote the activity sequence as a matrix, in which rows represent the time of the day and the columns represent the days in a week (*8*). The main difference between these approaches is that the second separates the longitudinal variation into two orthogonal time scales – time of day and days in a week. For urban freight, it is generally assumed that activity variation within a day is larger than on a day-to-day basis. This study hence adopts the first approach. Nevertheless, to study the behaviors of long-haul drivers

Table 1. Aggregated Truck Driver Activity Types

Activity type ID	Activity type	Activities in survey
1	ChangeShift	1) Start/end my shift, change driver
2	Pickup	2) Pick up cargo; 3) Pick up trailer
3	Delivery	4) Deliver cargo; 5) Drop off trailer
4	Pickup/Delivery	When both Pickup and Delivery are declared
5	Work (other)	6) Fueling; 7) Maintenance; 8) Other work-related business; 9) Provide services
6	Rest	10) Meal; 11) Resting
7	Anomaly	12) Fail to deliver/pick up; 13) Queuing; 14) Personal errands
8	Travel	15) Travel as indicated by GPS and verified by user
9	Gap	16) Other please specify; 17) GPS data gap

whose behavior may be more regular on a daily basis, the orthogonal design of the second approach could be more suitable.

Using the same notation as in Jiang et al. (*4*), each day was divided into a total of 288 time intervals, each with a 5 min interval duration. For each individual driver, the activity occurring at the first second of a time interval would be noted as the activity for the entire time interval.

Let G be the total number of activity types. Jiang et al. propose the G activities satisfy the "compatibility condition" before applying PCA (*4*). This condition requires that at any given time interval, a driver should engage in one – and only one – activity. Jiang et al.'s rationale is mainly driven by the need for the multivariate Gaussian assumption of applying PCA. Their reconstruction method is also based on their formulation of an activity sequence in which each time interval corresponds to only one activity type (*4*). For logistic PCA, the compatibility condition can be relaxed due to its Bernoulli probability assumption; therefore, it allows multiple activities to happen within the same time interval. For autoencoder and VAE, although the compatibility condition is not required, the same input as with PCA is used for comparison.

Formally, an activity sequence $\boldsymbol{a}$ in this study is defined as follows:

$$\boldsymbol{a} \equiv \begin{pmatrix} a_1 \\ a_2 \\ \vdots \\ a_D \end{pmatrix},$$

$$\text{where } a_{t+288\times(l-1)} = \begin{cases} 1 & \text{if activity } l \text{ is performed at } t\text{-th time interval} \\ 0 & \text{otherwise} \end{cases}$$
$$\forall i \in \{0, 1, \ldots, D\},\ D = 288 \times G,\ t \in \{0, 1, \ldots, 288\},$$
$$l \in \{0, 1, \ldots, D\} \quad (1)$$

The space of individuals' daily behaviors can thus be defined as the matrix space $\boldsymbol{A}$ of activity sequences in Equation 2:

$$\boldsymbol{A} \equiv \begin{pmatrix} \boldsymbol{a}_1^T \\ \boldsymbol{a}_2^T \\ \vdots \\ \boldsymbol{a}_N^T \end{pmatrix}, \text{ where } \boldsymbol{a}_i \in \{0, 1\}^D \quad (2)$$

In the dataset, each driver could select multiple activities at the same stop. However, when the compatibility condition is binding, activities are aggregated into $G = 9$ types in Table 1. A main activity is assigned to each interval. The activity with a smaller ID would be used as the main one. Without the compatibility condition, each interval can have multiple activities. Activities are aggregated into $G = 8$ types, excluding "Pickup/Delivery" in Table 1.

By counting the percentage of drivers performing each of the G activities in Table 1, the temporal rhythm of truck drivers' activities can be revealed. Figure 1*a* depicts the variation in truck drivers' activities on an average weekday. It describes the average percentage of drivers engaging in a specific activity at any given time in a weekday. Drivers begin their work shift around 6:00 a.m., and the number increases rapidly until 10:00 a.m. What accompanies this trend in the start of the shifts is traveling (driving). As expected, drivers start traveling around the same time when they start their shift. After 3:00 p.m., drivers begin to end their shift and this continues until 12:00 a.m., suggesting that the end-shift time varies greatly across drivers.

Approximately 30 min after traveling, "Pickup" activity increases, followed by "Work(other)" and "Delivery." As service trips are categorized under "Work(other)," it could be the case that service trucks start working later than cargo-carrying ones. Around 12:00 p.m., drivers reduce their traveling and the number of "Rest" entries start to peak. This implies that, although not many drivers rest during a shift, if they do it tends to be around noon. In the afternoon, "Pickup" declines before

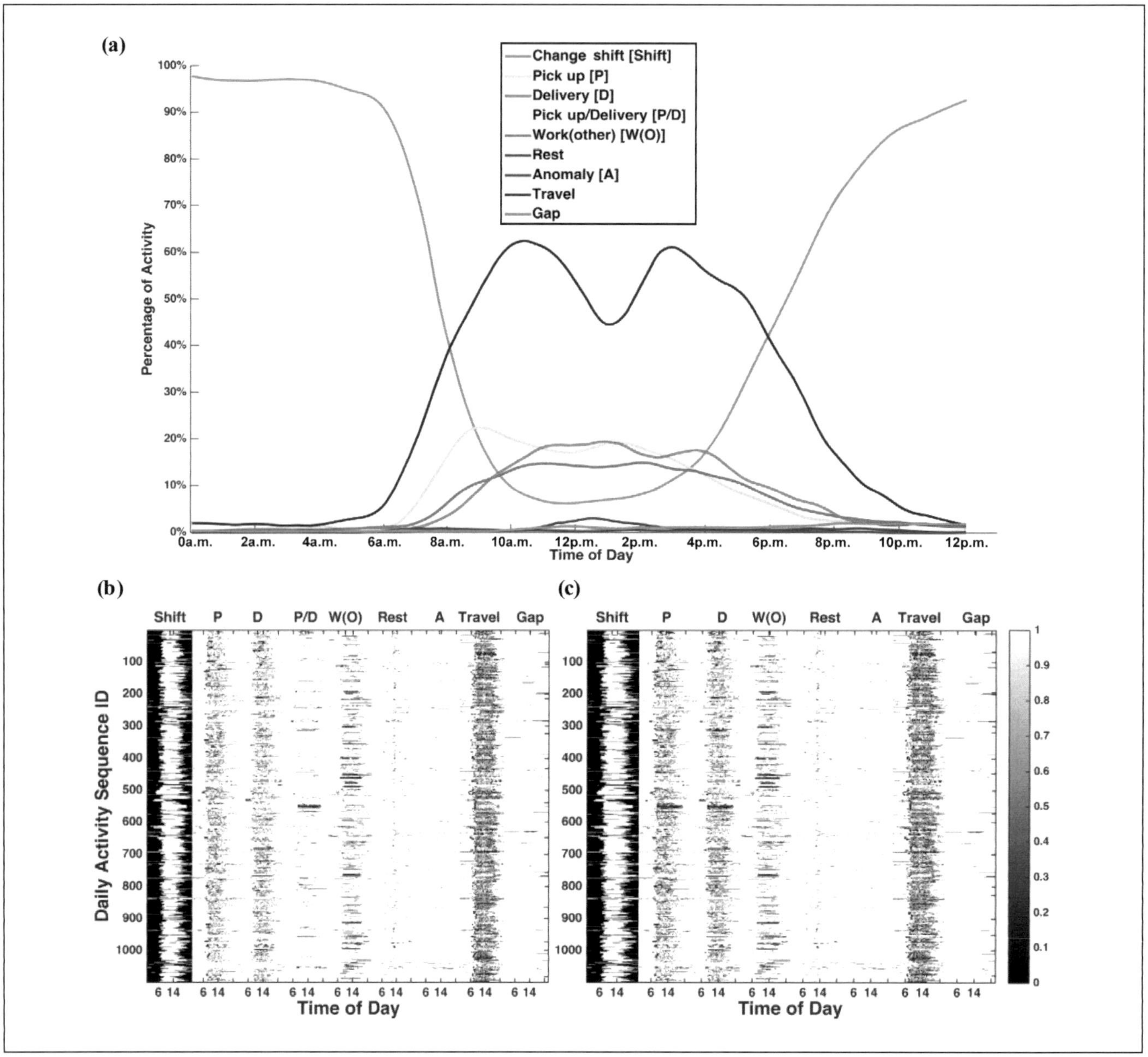

Figure 1. (*a*) aggregated temporal variation of 243 truck drivers' activities on an average weekday; (*b*) activity matrix when $G = 9$; (*c*) activity matrix when $G = 8$.

"Work(other)" and "Delivery," which is consistent with the morning trend. Overall, "Pickup" activities have two peaks during the day, one around 9:00 a.m. and another at 1:00 p.m. Similar trends held for "Delivery," but happened at later time intervals. "Anomaly" and "Gap" had a small presence in the dataset, and their occurrences were observed to be independent of the time of day.

Following the notation in Equation 1 and Equation 2, the input dataset for dimensionality reduction is visualized in Figure 1, *b* and *c* as black-and-white images. For A_{ij} in the input activity matrix A, if $A_{ij} = 1$, then it is colored black. If $A_{ij} = 0$, then it is colored white. Figure 1*b* plots the input data when aggregated into nine activity types; whereas Figure 1*c* plots the case with eight activity types. The activity sequences formulated as shown in Figure 1*b* were used as inputs in PCA, autoencoder, and VAE; whereas the activity sequences shown in Figure 1*c* were used in logistic PCA. The key difference between the two input formulations arises when considering whether the compatibility condition is binding.

Methodology

Related Work

PCA was first used to extract mobility patterns from human activities (Eagle and Pentland [*2*]). Over 100 subjects' daily activities were reconstructed using six primary characteristic vectors, termed "eigenbehaviors," with an accuracy of 90%. Their method can be generalized into the framework described below:

1. Define the activity space of all sample subjects as a matrix $A \in R^{N \times D}$, which are represented with notations in equations 1 and 2.
2. Represent the activity matrix in a lower-dimension matrix $\boldsymbol{A_1} \in R^{N \times K}$, where $K < D$.
3. Reconstruct the activity data as $A_2 \in R^{N \times D}$.

Step 1 transforms the dataset into activity sequences as denoted in Equation 1 and Equation 2; whereas Step 2 performs dimensionality reduction on the transformed data obtained in Step 1. Depending on which dimensionality reduction method is used in Step 2, activity reconstruction in Step 3 will vary. Currently, PCA is the main technique used to analyze human mobility patterns (*2*, *4*, *8*). However, because of the fundamental differences between freight and passenger activities, PCA may not have been the most suitable for the truck driver activity dataset used in this paper. Logistic PCA, autoencoder, and VAE were therefore explored because they had more relaxed probabilistic assumptions on the input variables.

The idea behind the work of Eagle and Pentland is widely applied in fields such as image processing, bioinformatics, and natural language processing (*2*, *8–10*). Pattern, or feature discovery via dimensionality reduction is an important problem in representation learning. Intuitively, what happens during dimensionality reduction is that more "noisy" data is thrown out so that what remains is more relevant in representing the key features. By assuming there exists some underlying structure in an activity sequence or behavior space, dimensionality reduction can uncover this structure by also removing the noise. Jiang et al. document a step-by-step implementation of PCA in revealing passenger activity patterns (*4*). This paper applied the same PCA method to examine truck driver activity patterns, and expanded the framework to incorporate other dimensionality reduction methods: logistic PCA, autoencoder, and VAE. As all three methods are well developed, the contribution of this paper in implementing them in activity pattern discovery mainly lies in activity reconstruction. The formulation of PCA follows the work of Pearson (*11*); whereas the formulation of logistic PCA used is from Landgraf and Lee (*12*). As Tensorflow is used to implement autoencoder and VAE, their formulation follows those incorporated in Tensorflow (*13*).

Logistic PCA and autoencoder were selected as comparisons to PCA mainly because these three methods were compared in performing similar tasks. Hinton and Salakhutdinov compared the performance of PCA, logistic PCA, and autoencoder in learning the lower-dimensional representation of high-dimensional data (*10*). The data used in their work was randomly chosen from the MNIST dataset (*12*) of handwritten digits. Using Hinton and Salakhutdinov's method of training deep autoencoder networks, autoencoder produced a much clearer reduced image than PCA and logistic PCA (*10*). The standard autoencoder can be implemented as a variational autoencoder, which is a popular approach not only in dimensionality reduction, but also in generating new information (*14*).

In evaluating the performance of pattern discovery via different dimensionality reduction methods, one important diagnostic metric is reconstruction error. This is defined as the percentage of incorrectly reconstructed entries out of the total number of entries. Let $\boldsymbol{A}_2$ be the reconstructed matrix. Since all elements in $\boldsymbol{A}$ and $\boldsymbol{A}_2$ are binary, the reconstruction error can be computed as follows:

$$e(\boldsymbol{A}) = \frac{\|\boldsymbol{A} - \boldsymbol{A}_2\|_1}{N \times D} \tag{3}$$

A low reconstruction error means the reconstruction is a truthful representation of the original activity data. However, a reconstruction error of zero is not desirable, because a model learns nothing useful about an input by copying it. To address this issue, another diagnostic metric in selecting the best method is the intermediate dimensions, that is, K (see next section). The goal is to maintain a relatively faithful reconstruction of the input data while minimizing the number of intermediate dimensions. Furthermore, in the context of freight behavior research, interpretability of the output is equally important.

PCA

PCA has been widely used as a dimensionality reduction technique (*15*). From its mathematical formulation, PCA finds the optimal representation of multivariate data in lower dimensions by minimizing the mean squared error (MSE). PCA thus possesses an appealing geometric property: it is a linear projection in which high-dimensional data is projected into a lower-dimensional space. Following Jiang et al. (*4*), applying PCA to reconstruct travel activity was formally formulated as below.

Let $\boldsymbol{A} \in \boldsymbol{R}^{N \times D}$; $\boldsymbol{a}_i \in \boldsymbol{R}^D$ is the i-th row of $\boldsymbol{A}$; and $\boldsymbol{a}_i$ denotes one activity sequence. Let $\bar{\boldsymbol{a}} \in \boldsymbol{R}^D$ denote the sample mean vector, $\boldsymbol{V} = (\boldsymbol{v}_1, .., \boldsymbol{v}_K)$ in which $\boldsymbol{V} \in \boldsymbol{R}^{D \times K}$ denote the K eigenvectors computed from PCA, and $\boldsymbol{x}$

$= (x_1, \ldots, x_K) \in \boldsymbol{R}^K$ the coordinates of projecting $(\boldsymbol{a}_i - \bar{\boldsymbol{a}})$ onto the column space of $\boldsymbol{V}$. $\boldsymbol{V}\,\boldsymbol{x}$ is the projection of original activity space into the space spanned by K eigenvectors.

Let $\hat{\boldsymbol{a}} \in \boldsymbol{R}^D$ be the reconstructed activity sequence:

1. Compute $\boldsymbol{r} = \bar{\boldsymbol{a}} + \boldsymbol{V}\,\boldsymbol{x}$, $\boldsymbol{r} \in \boldsymbol{R}^D$.
2. For any $t \in \{1, 2, \ldots, 288\}$, $l \in \{1, 2, \ldots, G\}$.

$$\widehat{\boldsymbol{a}}_{t+288\times(l-1)} = \begin{cases} 1, & \text{if } r_{t+288\times(l-1)} = \max\{r_{t+288\times(l-1)} \in R \mid \forall l \in \{0, 1, \ldots, D\}\} \\ 0, & \text{otherwise} \end{cases} \quad (4)$$

Logistic PCA

Similar to PCA, the geometric interpretation of logistic PCA is also a projection, but nonlinear.

Dissimilar from PCA, logistic PCA assumes that each element $\boldsymbol{A}_{ij}$ in the input matrix $\boldsymbol{A}$: 1) is binary; and 2) follows a Bernoulli distribution with parameter p_{ij}. Because of these two assumptions, it allows the study of the pattern of a single activity and is more suitable for binary input. While PCA minimizes MSE, logistic PCA finds the principal components by minimizing the Bernoulli deviance. This paper follows the formulation proposed by Landgraf and Lee in computing principal components (*16*). In computing principal components, a parameter m, which approximates a Bernoulli distribution, needs to be selected. In Landgraf and Lee, m satisfies Equation 5 below; section 6 in their work describes selecting m in detail (*16*):

$$p_{ij} = \frac{e^{m(2A_{ij}-1)}}{1 + e^{m(2A_{ij}-1)}} \quad (5)$$

After obtaining principal components using logistic PCA, activities can be reconstructed as follows.

Let $\boldsymbol{A} \in \boldsymbol{R}^{N\times 288}$ denote the activity matrix of one activity type: $\boldsymbol{a}_i \in \boldsymbol{R}^{288}$ is the i-th row of $\boldsymbol{A}$; $\boldsymbol{a}_i$ denotes one activity sequence. $\bar{\boldsymbol{a}} \in \boldsymbol{R}^{288}$ denote the sample mean vector $\boldsymbol{Q} = (\boldsymbol{q}_1, \ldots, \boldsymbol{q}_K)$ where $\boldsymbol{Q} \in \boldsymbol{R}^{D\times K}$ denote the K principal components calculated by logistic PCA and $\boldsymbol{y} = (y_1, \ldots, y_K) \in \boldsymbol{R}^K$, the principal components scores.

Let $\hat{\boldsymbol{a}} \in \boldsymbol{R}^D$ be the reconstructed activity sequence:

1. Compute $\boldsymbol{r} = \bar{\boldsymbol{a}} + \boldsymbol{Q}\,\boldsymbol{y}$, $\boldsymbol{r} \in \boldsymbol{R}^D$.
2. Select a probability threshold $h = 0.9$.

$$\widehat{p} = \frac{e^{ri}}{1 + e^{ri}}, \ \forall i \in \{1, \ldots, 288\}, \qquad \widehat{\boldsymbol{a}} = \begin{cases} 1, & \text{if } \widehat{p}_i > 0.9 \\ 0, & \text{otherwise} \end{cases} \quad (6)$$

Autoencoder and Variational Autoencoder (VAE)

Autoencoder reduces dimensionality via stacked layers of feedforward networks. Conceptually, an autoencoder is an unsupervised representation of original data. It generally learns the identity function $F(X) = X$ under the constraint on dimensionality (*7*). The advantage of autoencoder is that it can result in powerful outcomes because neural networks are universal function approximators (*19–21*). However, it also suffers from the typical problem of feedforward networks: results are not readily interpretable. Tensorflow was used to develop autoencoder models in this paper; Python codes, used to train the model, is available on GitHub at https://github.com/urbanfreight/Dimensionality-Reduction. The activity reconstruction was performed as below.

Let $\boldsymbol{A} \in \boldsymbol{R}^{N\times D}$ denote the original input, and $\boldsymbol{D} \in \boldsymbol{R}^{N\times D}$ denote the output from autoencoder. For the i-th row in D denoted by $\hat{\boldsymbol{d}}$, let $\hat{\boldsymbol{a}} \in \boldsymbol{R}^D$ be its reconstruction:

$$\widehat{\boldsymbol{a}}_{t+288\times(l-1)} = \begin{cases} 1, & \text{if } \boldsymbol{r}_{t+288\times(l-1)} = \max\{\boldsymbol{r}_{t+288\times(l-1)} \in R \mid \forall l \in \{0, 1, \ldots, D\}\} \\ 0, & \text{otherwise} \end{cases} \quad (7)$$

VAE is one type of autoencoder that learns the latent variable space of its input data. Whereas autoencoder learns to approximate the input, VAE learns the parameters of the probability distribution that generates the input. VAE was defined in 2013 by Kingma and Welling (*14*) and has been widely applied in the top-performing models in image generation and reinforcement learning (*18*).

Implementation and Results

This section presents findings from employing the four machine learning methods introduced. For each method, the parameter selection process is described and the reconstruction yielded by the best model is presented. The validity of applying each method to analyze truck driver activities is discussed.

PCA

Due to the compatibility condition, the input dataset selected for PCA consisted of nine activity types, as shown in Figure 1*a*. At any given time in a day, a driver can engage in one of the nine activity types. PCA requires one parameter to reconstruct the input dataset: K, the number of eigenvectors. Geometrically, K also represents the number of dimensions the original input is projected to.

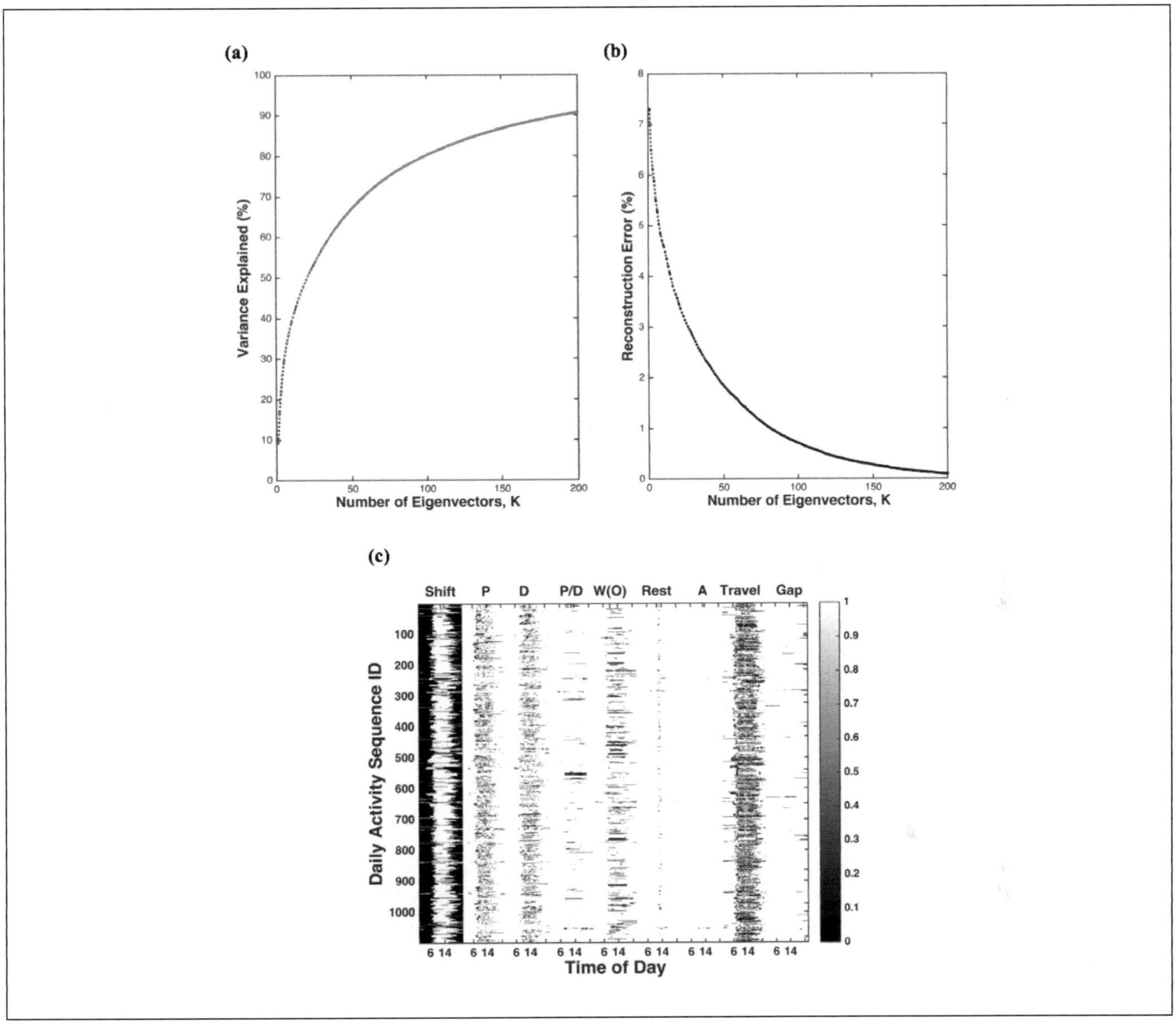

Figure 2. PCA: (*a*) percentage of variance explained; (*b*) reconstruction error; (*c*) the reconstruction matrix with $K = 82$.

In selecting K, two measures were used: 1) the percentage of variance explained; 2) the reconstruction error. In Figure 2*a*, the left panel shows that the percentage of variance explained increases with K; the right panel indicates that the reconstruction error reduces when K increases.

At $K = 82$, the reconstruction error drops to 0.99%. The percentage of variance explained is 77%. Figure 2*b* shows the reconstruction matrix, which provides a faithful reconstruction of the input except for "Anomaly" activity type. When applying the same method to the Chicago Household Travel Survey, Jiang et al. used only 21 eigenvectors to reconstruct 99% of the original weekday activities (*4*). The different results the same method yields when applied on passenger and freight activity datasets suggests that freight activities are inherently less regular and thus less predictable than passenger activities.

Logistic PCA

Logistic PCA assumes that the input data is binary, and each element in the input activity matrix comes from a Bernoulli distribution. If all activities were used as input at the same time, then it assumes that these activities are independent events, which clearly is not the case. For example, if a driver declared "Pickup," "Deliver" is expected at some later time interval. Therefore, when applying logistic PCA to reconstruct the input activities, the reconstruction is performed activity by activity. Specifically, the input activity matrix shown in Figure 1*b*

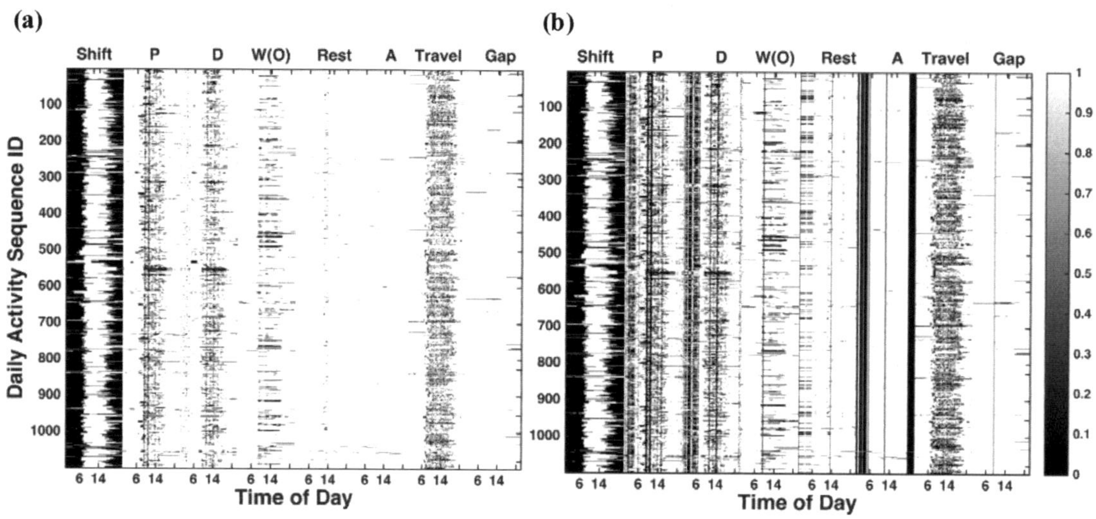

Figure 3. Logistic PCA: (*a*) the reconstructed activity matrix with 93 eigenvectors, $h = 0.9$; (*b*) the reconstructed activity matrix with 93 eigenvectors, $h = 0.5$.

is split into eight smaller activity matrices, with each representing one of the eight activity types. Logistic PCA is employed to reconstruct each of the smaller activity matrices.

Two parameters are required before reconstructing with logistic PCA: m and K. The first parameter m, whose true value is defined in Equation 5, is selected to maximize the log likelihood of generating the input matrix. The R-package "logisticPCA" developed by Landgraf and Lee is used to select m (*16*). After m is determined, K is selected to explain 90% of variances. With m and K, principal components and scores can be computed and reconstruction can then be performed. Combining the best models for each activity type, the total number of eigenvectors used in the reconstruction was 93. By setting the probability threshold, $h = 0.9$, the reconstruction error was 3.21%. When $h = 0.5$, the reconstruction error rose to 7.31%. Figure 3, *a* and *b* show the two reconstructed matrices with different probability threshold values. A threshold h of 0.5 means random reconstruction, or the generation of an input activity matrix based on the Bernoulli parameters estimated with logistic PCA.

Autoencoder and VAE

Tensorflow was used to implement autoencoder and VAE. An autoencoder consists of one encoder and one decoder. Each encoder and decoder can consist of multiple feedforward networks. The encoding dimension refers to the reduced dimension from the original one. When a one-layer autoencoder has MSE loss function, linear activation function, K encoding dimensions, the output from its encoder spans the same subspace as the eigenspace spanned by K eigenvectors computed using PCA (*7*). Conceptually, PCA is nearly a special instance of autoencoder. Nonetheless, although almost any function can be approximated by feedforward networks, it is still worthwhile exploring PCA because it provides interpretability. And interpretability is critical in enabling researchers' understanding of the travel behavior regularities behind the input data.

In training an autoencoder, the input data, as displayed in Figure 1*a*, is randomly split into 75% training and 25% testing. The training dataset has 824 activity sequences whereas the testing dataset has 275. There are three measures in selecting the best model. First, no overfitting. Overfitting describes the scenario where training loss is reducing but testing loss is increasing. The dropout technique and the L1 regularization term are adopted to prevent overfitting (*7*, *20*). Second, a low reconstruction error. Third, a small encoding dimension.

The best autoencoder model implemented achieved a reconstruction error of 1.15%, with an encoding dimension $K = 25$, which is 1% of the input dimensionality of 2,592. The loss function selected was cross-entropy (CE) loss, as the input data had binary values. Evidence from a parameter fine-tuning process also showed that by using CE loss, the reconstruction error dropped to below 10% compared with MSE. The model had LeakyRelu as the activation function for the encoding layer, and Sigmoid as the activation function for the decoding layer. The L1 regularization term of 10e-6 was added to the encoding layer. The dropout rate of 0.1 was also used. The optimization algorithm used was adam, with a learning rate of 0.001. The slow learning rate was selected

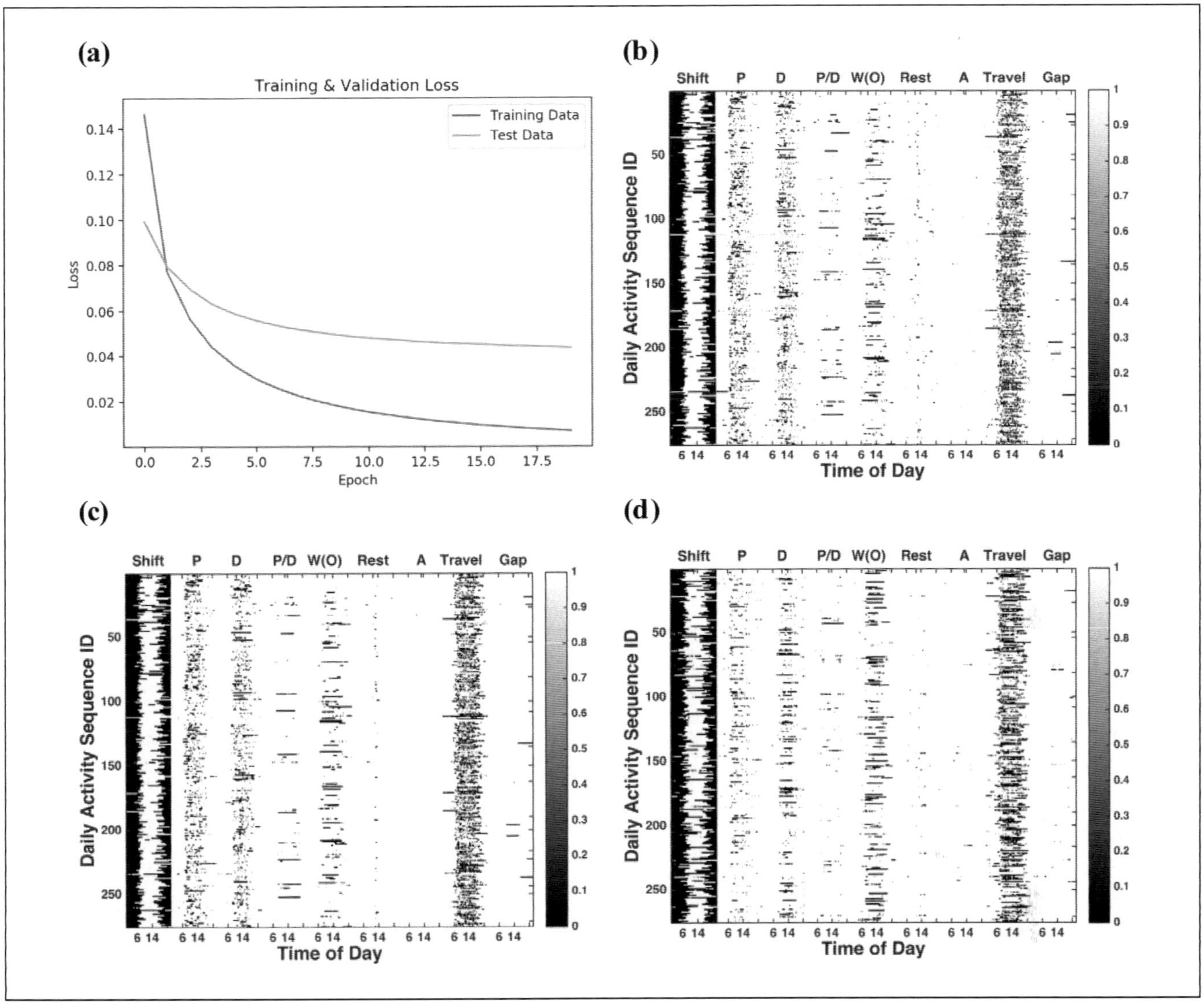

Figure 4. Autoencoder: (*a*) training and testing loss over epochs; (*b*) the test dataset; (*c*) the reconstruction with 25 encoding dimensions; (*d*) the simulated data with VAE.

deliberately because the training process suggested that a higher rate overfits very easily. The batch size was five and number of epochs equaled 20. Figure 4*a* shows how loss changes over different iterations. Figure 4*b* plots the original testing dataset, whereas Figure 4*c* indicates the reconstruction by the best performing autoencoder model.

This paper also experimented with implementing VAE to generate realistic-looking activity sequences. Kingma and Welling provide a reference on how VAE is implemented in Tensorflow (*14*). The major difference from autoencoder is that VAE learns the latent variable space instead of the original input and thus can generate outputs. Logistic PCA can be cast as a VAE with a latent dimension of one, as each Bernoulli distribution has only one parameter.

The best VAE model was formulated as follows. Encoder had two layers of LeakyRelu of dimensions 300 and 25; decoder had one layer of LeakyRelu and one layer of Sigmoid with dimensions of 300 and 2,592. The latent dimension was selected to be 8. Other settings follow from the autoencoder model described previously. Figure 4*d* presents one simulated activity matrix, based on the latent parameter space learned by the best VAE model. The error from the test dataset, 10.43%, was high, Nonetheless, it resembles Figure 4*b*, especially in capturing less major activities such as "Anomaly." Further research is required to evaluate whether insights on freight behaviors derived based on the simulated activity sequences are the same as the real ones.

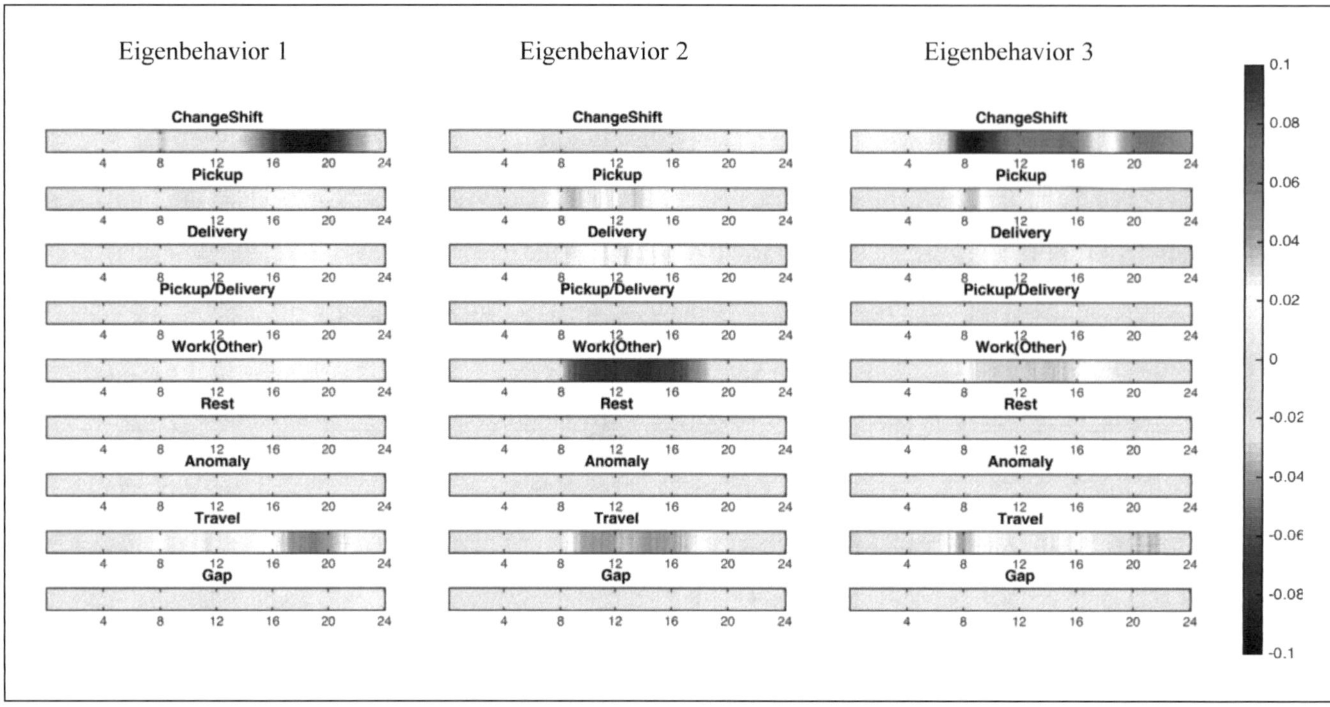

Figure 5. Top three activity patterns of urban truck drivers identified via PCA.

Validity of Applying PCA, Logistic PCA, Autoencoder and VAE

For PCA to apply, it is generally assumed that the input data follows a multivariate Gaussian distribution. When the input data is an activity matrix, as defined in Equation 2, it is non-Gaussian. Consequently, the derived principal components, that is, eigenbehaviors, are not strictly orthogonal, though, overlaps in the behaviors different principal components measure can exist. However, as commented by Jiang et al. PCA is still employed in a wide range of similar studies involving non-Gaussian data because orthogonality of principal components is not needed if the study purpose is not related to regression, or intended to reveal the probability distribution of input data (*4*). Logistic PCA only assumes each element in the input data to be a Bernoulli random variable. The binary nature of the activity matrix is more applicable to logistic PCA. However, logistic PCA requires estimating the natural parameters of Bernoulli distributions before computing principal components, which is a non-trivial task and usually only an approximated value is available. Autoencoder and VAE are built on feedforward networks, which do not have probabilistic assumptions, and thus remain valid for the task of reconstructing travel activity data.

Findings: Patterns of Truck Drivers' Daily Activities

By employing the best PCA model to the dataset, the top three eigenbehaviors, which are eigenvectors computed by PCA, revealed interesting facts about truck drivers' behaviors. In Figure 5, the color in each eigenbehavior represents the likelihood of one activity occurring across different times in a day: red at 0.1 denotes the highest probability of engaging in one activity; blue at −0.1 denotes the highest probability of the activity not happening; green at 0 denotes no inference.

Eigenbehavior 1 revealed that the most distinguishing feature of a driver's activity sequence was whether the driver was on shift from 4:00 to 8:00 p.m., and traveled extensively during this period. This may be because 4:00 to 8:00 p.m. is not a common work shift, so it captures the most variance in truck driver behaviors.

Eigenbehavior 2 explained the second largest behavior variance. It revealed a pattern in which drivers engage in "Pickup" and "Delivery" activities from 8:00 a.m. to 7:00 p.m., and were not likely to perform "Work (other)" during this period. As providing services is also categorized as "Work (other)," this pattern mainly describes the drivers who do not provide services but travel frequently during work hours to pick up and deliver goods. Delivery activity happened slightly after Pickup, but both had similar durations.

Eigenbehavior 3 described a pattern as follows: on shift from 7:00 a.m. to 4:00 p.m., traveled extensively around 8:00 a.m., performed the first "Pickup" around 8:00 a.m. and engaged in other work-related activities until going off shift. This might depict the daily routine of a road maintenance worker who picks up tools first and provides services afterwards.

Conclusion

This study explored different dimensionality reduction methods to reveal the activity patterns of truck drivers. Although PCA has been widely applied in uncovering patterns from passenger mobility (*2*, *4*, *8*), other dimensionality methods have not been developed in identifying activity patterns. As freight activity varies fundamentally from passenger activity, this paper set out to find a suitable method. Logistic PCA, autoencoder, and VAE techniques were investigated to reconstruct travel activity, to reveal the underlying behavior structure. In terms of reconstruction performance, all methods yielded less than 5% error. PCA gave the most truthful reconstruction, with an error of 0.99%, followed by autoencoder and logistic PCA. Nonetheless, autoencoder brought down the dimensions needed to 1% of the original dimensionality while sacrificing 0.5% in reconstruction error. When the dataset becomes too large for PCA to be applied efficiently, autoencoder is an ideal alternative. When assessing how interpretable the patterns discovered by each method were, PCA and logistic PCA were readily interpretable, whereas autoencoder was not. Finally, one critical advantage that autoencoder and VAE have is that they can easily incorporate the "space" dimension. PCA and logistic PCA take 2D matrices as input, whereas autoencoder and VAE can process multi-dimensional tensors. In studying freight behaviors, activity location has a direct impact on traffic modeling. Revealing patterns with time, location, and activity is an important aspect that autoencoder and VAE are able to address. In future work, identifying activity patterns with a location element will be explored with autoencoders.

Overall, PCA appeared to be most suitable for intuitively understanding the existing patterns in the dataset. Logistic PCA, as a variation of PCA, provided the flexibility to study a single activity such as "Anomaly" in the truck driver dataset used, which occurs much less frequently. VAE was most suitable for extrapolation, that is, synthesizing new realistic-looking datasets.

Dimensionality reduction is an important technique that reveals the underlying structure of data from vast amount of inputs. By imposing constraints on dimensionality, the most salient aspects of the input are preserved (*7*). The methods developed in this paper are not restricted to freight activity datasets, they could be generalized to any travel dataset for unsupervised pattern discovery. With dimensionality reduction, researchers are able to comprehend the "signature" of an individual's mobility and even compare how similar one individual is to another based on this signature.

One acknowledged limitation in the Singapore case study was the sampling bias. A large portion of drivers came from the construction industry, so the patterns identified may not be representative of the population. As the survey dataset expands, the plan is to continue exploring activities.

Acknowledgments

This research was supported by the Singapore Ministry of National Development and the National Research Foundation under the Land and Liveability National Innovation Challenge (Award no. L2NICTDF1-2016-1). We thank the Urban Redevelopment Authority of Singapore, JTC Corporation, and Land Transport Authority of Singapore for their support.

Author Contributions

The authors confirm contribution to the paper as follows: study conception and design: Lu, Cheah; data collection: Lu, Zhao, Cheah; analysis and interpretation of results: Lu, Zhao, Cheah; draft manuscript preparation: Lu, Zhao, Cheah. All authors reviewed the results and approved the final version of the manuscript.

References

1. Gonzalez, M. C., A. H. Cesar, and B. Albert-Laszlo. Understanding Individual Human Mobility Patterns. *Nature*, Vol. 453, No. 7196, 2008, pp. 779–782.
2. Eagle, N., and A. Pentland. Eigenbehaviors: Identifying Structure in Routine. *Behavioral Ecology and Sociobiology*, Vol. 63, No. 7, 2009, pp. 1057–1066.
3. Song, C., Z. Qu, N. Blumm, and A. L. Barabasi. Limits of Predictability in Human Mobility. *Science*, Vol. 327, No. 5968, 2010, pp. 1018–1021.
4. Jiang, S., J. Ferreira, and M. C. Gonzalez. Clustering Daily Patterns of Human Activities in the City. *Data Mining and Knowledge Discovery*, Vol. 25, No. 3, 2012, pp. 478–510.
5. You, S. I., J. Y. Chow, and S. G. Ritchie. Inverse Vehicle Routing for Activity-Based Urban Freight Forecast Modeling and City Logistics. *Transportmetrica A: Transport Science*, Vol. 12, No. 7, 2016, pp. 650–673.
6. Sakai, T., K. Kawamura, and T. Hyodo. Logistics Chain Modelling for Urban Freight. *Transportation Research Record: Journal of the Transportation Research Board*, 2017. 2609: 55–66.
7. Goodfellow, I., Y. Bengio, and A. Courville. *Deep Learning*. MIT Press, Cambridge, MA, 2016. http://www.deeplearningbook.org/.
8. Goulet-Langlois, G., H. N. Koutsopoulos, and J. Zhao. Inferring Patterns in the Multi-Week Activity Sequences of

Public Transport Users. *Transportation Research Part C: Emerging Technologies*, Vol. 64, 2016, pp. 1–16.
9. Zhao, F., F. Pereira, R. Ball, Y. Kim, Y. Han, C. Zegras, and M. Ben-Akiva. Exploratory Analysis of a Smartphone-Based Travel Survey in Singapore. *Transportation Research Record: Journal of the Transportation Research Board*, 2015. 2494: 45–56.
10. Hinton, G. E., and R. R. Salakhutdinov. Reducing the Dimensionality of Data with Neural Networks. *Science*, Vol. 313, No. 5786, 2006, pp. 504–508.
11. Pearson, K. On Lines and Planes of Closest Fit to Systems of Points in Space. *The London, Edinburgh, and Dublin Philosophical Magazine and Journal of Science*, Vol. 2, No. 11, 1901, pp. 559-572.
12. Lecun, Y., C. Cortes, C. Burges. The MNIST Database of Handwritten Digits (Images), 1999. http://yann.lecun.com/exdb/mnist/.
13. Abadi, M., A. Agarwal, P. Barham, E. Brevdo, Z. Chen, C. Citro, G. S. Corrado, A. Davis, J. Dean, M. Devin, and S. Ghemawat. Tensorflow: Large-Scale Machine Learning on Heterogeneous Distributed Systems. *arXiv preprint arXiv:1603.04467*, 2016.
14. Kingma, D. P., and M. Welling. Auto-Encoding Variational Bayes. *arXiv preprint arXiv:1312.6114*, 2013.
15. Jolliffe, I. T. Principal Component Analysis, (2nd ed.). Springer, New York, NY, 2002, pp 1–6. https://doi.org/10.1007/b98835.
16. Landgraf, A. J., and Y. Lee. Dimensionality Reduction for Binary Data through the Projection of Natural Parameters. *arXiv preprint arXiv:1510.06112*, 2015.
17. Hornik, K. Approximation Capabilities of Multilayer Feedforward Networks. *Neural Networks*, Vol. 4, No. 2, 1991, pp. 251–257.
18. Zhang, C., S. Bengio, M. Hardt, B. Recht, and O. Vinyals. Understanding Deep Learning Requires Rethinking Generalization. *Proc., International Conference on Learning Representations*, Toulon, France, 2017, pp. 1–15.
19. Wang, H., and B. Raj. On the Origin of Deep Learning. *arXiv preprint arXiv:1702.07800*, 2017. http://dx.doi.org/10.1139/f56-020.
20. Srivastava, N., G. Hinton, A. Krizhevsky, I. Sutskever, and R. Salakhutdinov. Dropout: A Simple Way to Prevent Neural Networks from Overfitting. *Journal of Machine Learning Research*, Vol. 15, 2014, pp. 1929–1958.

The Standing Committee on Freight Transportation Planning and Logistics (AT015) peer-reviewed this paper (18-03813).

The opinions, findings, and conclusions or recommendations expressed in this material are those of the authors only.

Article

TRR
JOURNAL OF THE TRANSPORTATION RESEARCH BOARD

Transportation Research Record
2018, Vol. 2672(9) 93–106

DOI: 10.1177/0361198118797806
journals.sagepub.com/home/trr
SAGE

Advanced Woody Biomass Logistics for Co-Firing in Existing Coal Power Plant: Case Study of the Great Lakes States

Sangpil Ko[1] and Pasi Lautala[2]

Abstract

One of the core decisions on securing the supply of biomass to co-firing power plants is the design of a proper logistics system. Though technologies have been developed to improve the characteristics of feedstock, their effects in the logistics systems have received less attention. This paper compares the conventional woody biomass logistics system with the advanced woody biomass logistics system that includes the torrefaction process to upgrade the feedstock. More specifically, this study uses a mixed integer linear program to integrate conventional and advanced biomass logistics and applies them to a case study of 26 coal power plants in the Great Lakes states to determine the most feasible logistics system for each plant. The results revealed that the amount of cost savings from the advanced woody biomass logistics depends on the capacity and location of the power plant. To categorize the cost savings of individual plants, it was found that plants can be divided into seven groups based on three parameters: 1) annual coal tonnage shipped (total capacity), 2) availability of biomass in the vicinity, and 3) average distance from biomass collecting sites. Overall savings from advanced logistics were small, but plants within each of the groups demonstrated similar cost reductions. The sensitivity analysis showed that trucking costs and feedstock purchasing costs were most sensitive to variation. From torrefaction effects, energy content after torrefaction had more significance than mass reduction.

The electricity generation sector has been the largest source of U.S. greenhouse gas emissions, accounting for approximately 30% of the United States' total (*1*). Burning coal produces many gases, including CO_2, SO_2, and NOx, as well as heavy metals that affect the environment and human health (*2*). In 2015, coal accounted for approximately 71% of all CO_2 emissions from electricity generation (*3*).

One of the most implementable technologies to reduce the use of fossil fuels in the electricity generation sector is co-firing with biomass in existing boilers. There are several attractive aspects of biomass co-firing (*4–6*), but also challenges. Biomass logistics is one of the challenges due to the intrinsic characteristics of biomass, such as scattered harvesting areas, lower energy density than fossil fuels, and seasonality. Technologies, such as wood torrefaction are being investigated to overcome the challenges and to contribute to the cost savings in logistics systems. Torrefaction shows promise with regard to increasing the energy density per unit of mass by approximately 30%, as torrefied biomass at optimum condition contains 70% to 80% of the original mass while retaining 80% to 90% of the original energy (*7*, *8*).

Conventional woody biomass logistics use the existing infrastructure as described in Figure 1, *a* and *b*. Wood is harvested and dried prior to collection. Dried wood is chipped at the landing and loaded into a truck for transport, either directly to the power plant, or to an intermediate terminal for trans-loading. When received at the power plant, the chips are cleaned, dried, and used for power generation.

An alternative, called advanced woody biomass logistics, was developed in Idaho National Laboratory (INL) (*9*). A key feature of the advanced woody biomass logistics is the integration of a local biomass processing depot where the biomass is torrefied (densified) prior to transport to the plant (Figure 1*c*).

[1]Department of Civil and Environmental Engineering, Michigan Technological University, Houghton, MI
[2]Department of Civil and Environmental Engineering, Michigan Tech Transportation Institute, Michigan Technological University, Houghton, MI

Corresponding Author:
Address correspondence to Sangpil Ko: sangpilk@mtu.edu

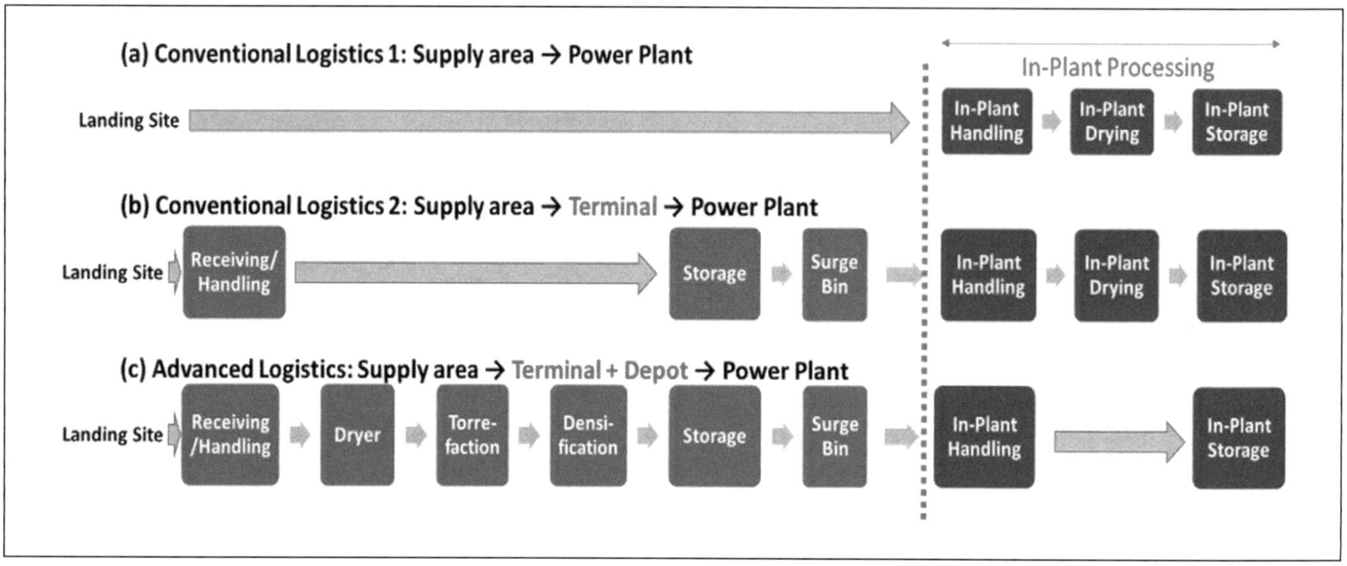

Figure 1. Conventional ([*a*] and [*b*]) and advanced (*c*) woody biomass logistics.

Harvesting and in-field processing use same methods in both the conventional and advanced logistics systems, but there is no intermediate depot in conventional feedstock logistics. In the advanced logistics system, torrefaction is conducted in those depots, increasing both conversion performance and transportation efficiency, whether transported by truck or rail. In this research, depot and terminal are considered to be distinct: the former is a place for the feedstock upgrading process (i.e., torrefaction) and only applies to the advanced logistics system, whereas the latter is a place for the collection/trans-loading of large quantities of feedstock in either the conventional or advanced system. Note that a depot can also fulfill the function of a terminal if it is constructed within an existing terminal.

Although it is recognized that various other logistics systems can be considered for the woody biomass supply chain, this paper concentrates on a comparative study solely between the conventional and advanced woody biomass logistics systems described above. The main objective is to determine the type of the logistics system that would be most appropriate for coal-fired power plants, based on plant capacity and the proximity of biomass feedstock sourcing locations.

Literature Review

Most of the literature on biomass co-firing focuses on analyzing its technological potential (*10–12*) and economic feasibility (*13–15*). A small number of papers have dealt directly with the logistics costs of delivering biomass to co-firing power plants. Roni et al. suggested a hub and spoke supply chain network for biomass co-firing in coal-fired power plants that considered unit trains for long-haul transportation (*5*). However, in many cases, even long-haul transportation is undertaken by trucks due to low accessibility/availability of rail terminals/sidings, non-competitive rates, or the difficulty generating sufficient volumes for unit train quantities.

Some research has integrated the logistics model with co-firing analysis. For example, Eksioglu et al. proposed a model that offers a holistic view of the processes including production, storage, and transportation (*16*). Their mathematical models integrate plant operations with transportation and other logistics-related decisions and thus assist in co-firing decisions by determining how much biomass should be used to replace coal at a plant level. More recently, Poudel et al. suggested a two-stage stochastic programming model for the design and management of a biomass co-firing supply chain network under feedstock supply uncertainty (*17*). They extended prior optimization models by incorporating seasonality and uncertainty in the modeling framework under an assumption that a fixed ratio (6%) of coal would be replaced with biomass.

The limitations of previous researches include the exclusion of various logistics options for biomass procurement. Furthermore, there does not seem to be literature that considers the additional logistics benefits from the increased efficiency of torrefaction, as most literature related to the technology mainly addresses its technical aspects (*18–20*). The general objective of this research is to provide enhanced understanding of well-organized woody biomass logistics for co-firing in coal power plants. Specifically, this study focuses on a case study using local transportation parameters and data to explore the applicability of advanced woody biomass logistics, including torrefaction, as a feedstock upgrading option.

Mathematical Model

The main objective of our model is to determine an optimal logistics system of woody biomass co-firing with a coal-fired power plant including torrefaction option. The second objective is to compare the cost difference between conventional and advanced logistics system. The following cost factors are used for calculating total logistics cost:

$$\begin{aligned} \text{Cost}^{\text{Total Logistics}} &= \text{Cost}^{\text{Purchasing}} + \text{Cost}^{\text{Transport}} \\ &+ \text{Cost}^{\text{Loading}} + \text{Cost}^{\text{Capital}} + \text{Cost}^{\text{Operation}} \end{aligned}$$

The cost factors depend on the type of logistics system and include feedstock purchasing cost, transportation cost by truck, rail, or both truck and rail, as well as the loading cost in the intermediate facility. Capital and operation costs are accounted for at the power plant and at the intermediate facilities (depots or terminals), when applicable.

A mixed integer linear programming (MILP) was developed to integrate the following six conventional/advanced biomass logistics options and to determine the most cost-efficient logistics system for each power plant:

- Conventional logistics
  - Direct trucking to plant only (Type A)
  - Direct trucking to plant, or indirect trucking through terminal (Type B)
  - Direct trucking to plant, or indirect trucking to terminal, followed by rail transportation to plant (Type C)
- Advanced logistics (with torrefaction)
  - Direct trucking to plant, or indirect trucking through depot (Type D)
  - Direct trucking to plant, or indirect trucking to depot, followed by rail transportation to plant (Type E)
  - Direct trucking, indirect trucking through truck depot (torrefaction), or indirect trucking to depot, followed by rail transportation of torrefied feedstock to plant (Type F).

If no savings are secured with advanced logistics options, direct trucking (without torrefaction) is used for the plant. The biomass that flows through the truck or rail depot includes a torrefied woody biomass shipment from the depot to the power plant with improved energy density per unit of mass.

All notation for sets, decision variables, and parameters used in our model are presented in Table 1 for both non-torrefied (N) and torrefied (T) biomass. The model includes two groups of biomass transportation flows as decision variables: X and Y. X indicates the inbound truck flows of biomass transported from supplier to intermediate facility (depot or terminal) or to the power plant directly. Y means the outbound truck/rail flows shipped from an intermediate facility to the power plant.

Our model is applied to each power plant (i.e., $n[P] = 1$) independently, as it addresses a user (or company) optimization rather than system optimization. However, the model can also be applied for multiple power plants at the same time, so parameters and variables are written to allow such application.

The mathematical formulation of this problem is as follows:

$$Min \sum_{l} \sum_{p} \left[\sum_{k} \left(f_{lp}^{VK} \cdot V_{klp}^{K} + f_{lp}^{WK} \cdot W_{klp}^{K} \right) + \sum_{j} \left(f_{lp}^{VJ} \cdot V_{jlp}^{J} + f_{lp}^{WJ} \cdot W_{jlp}^{J} \right) \right] \quad (1)$$

$$\begin{aligned} &\sum_{b} \sum_{p} \Bigg[\sum_{i} \sum_{k} (C_b^{TF} + C_b^{TV} \cdot D_{ik}^{IK}) \left(X_{ibkp}^{KN} + X_{ibkp}^{KT} \right) + \sum_{i} \sum_{j} \left(C_b^{TF} + C_b^{TV} \cdot D_{ij}^{IJ} \right) \left(X_{ibjp}^{JN} + X_{ibjp}^{JT} \right) \\ &+ \sum_{k} \left(C_b^{TF} + C_b^{TV} \cdot D_{kp}^{KP} + C_b^{L} \right) \left(Y_{bkp}^{KN} + Y_{bkp}^{KT} \right) + \sum_{j} \left(C_b^{RF} + C_b^{RV} \cdot D_{jp}^{JP} + C_b^{L} \right) \left(Y_{bjp}^{JN} + Y_{bjp}^{JT} \right) \quad (2) \\ &+ \sum_{i} \left(C_b^{TF} + C_b^{TV} \cdot D_{ip}^{IP} \right) X_{ibp}^{P} \Bigg] \end{aligned}$$

$$+ \sum_{i} \sum_{b} \sum_{p} \sum_{k} \sum_{j} C_b^{PU} \left(X_{ibkp}^{KN} + X_{ibkp}^{KT} + X_{ibjp}^{JN} + X_{ibjp}^{JT} + X_{ibp}^{P} \right) \quad (3)$$

Table 1. Notation for Set, Decision Variables, and Input Parameters

Sets

- B = Set of biomass feedstock, $b \in B$
- I = Set of biomass feedstock suppliers, $i \in I$
- J = Set of rail intermediate facilities (terminals or depots), $j \subset J$
- K = Set of truck intermediate facilities (terminals or depots), $k \in K$
- P = Set of co-firing power plants, $p \in P$
- L = Set of capacity level of intermediate facilities (terminals or depots), $l \in L$

Decision variables

Truck and rail decision variables (X, Y)

- X_{ibkp}^{KN}, X_{ibkp}^{KT} = Flow of biomass b shipped from supplier i to truck terminal k and truck depot k, respectively
- X_{ibjp}^{JN}, X_{ibjp}^{JT} = Flow of biomass b shipped from supplier i to rail terminal j and rail depot j, respectively
- X_{ibp}^{P} = Flow of biomass b transported from supplier i to plant p directly
- Y_{bkp}^{KN}, Y_{bjp}^{JN} = Flow of untorrefied biomass b transported from truck terminal k to plant p, and from rail terminal j to plant p, respectively
- Y_{bkp}^{KT}, Y_{bjp}^{JT} = Flow of torrefied biomass b transported from truck depot k to plant p, and from rail depot j to plant p, respectively

Terminal and depot binary variables (V, W)

- V_{klp}^{K} = 1, if truck terminal k with capacity level l is used for plant p, 0 otherwise
- W_{klp}^{K} = 1, if truck depot with capacity level l is built at truck terminal k, 0 otherwise
- V_{jlp}^{J} = 1, if rail terminal j with capacity level l is used for plant p, 0 otherwise
- W_{jlp}^{J} = 1, if rail depot with capacity level l is built at rail terminal j, 0 otherwise

Input parameters

Cost parameters (C, f)

- C_b^{TF}, C_b^{RF} = Fixed Transport cost of biomass feedstock b by truck and rail, respectively (US$/ton)
- C_b^{TV}, C_b^{RV} = Variable Transport cost of biomass feedstock b by truck and rail, respectively ($/ton-mile)
- C_b^{L} = Trans-loading cost of biomass feedstock b (US$/ton)
- C_b^{PU} = Purchasing cost of biomass feedstock b (US$/ton)
- f_l^{VK}, f_l^{VJ} = Capital cost of truck and rail terminal with capacity l, respectively (US$/ton)
- f_l^{WK}, f_l^{WJ} = Capital cost of truck and rail depot with capacity l, respectively (US$/ton)
- C_b^{OIN}, C_b^{OIT} = Operation cost in terminal and depot for biomass feedstock b, respectively (US$/ton)
- C_b^{CPN}, C_b^{CPT} = Capital cost in plant for untorrefied and torrefied biomass b, respectively (US$/ton)
- C_b^{OPN}, C_b^{OPT} = Operation cost in plant for untorrefied and torrefied biomass b, respectively (US$/ton)

Distance parameters (D)

- D_{ik}^{IK}, D_{ij}^{IJ} = Distance between supplier i and truck intermediate facility j, and between supplier i and rail intermediate facility k, respectively (miles)
- D_{ip}^{IP} = Distance between supplier i and plant p (miles)
- D_{kp}^{KP}, D_{jp}^{JP} = Distance between plant p and truck intermediate facility k, and between plant p and rail intermediate facility j (miles)

Supply and demand quantity parameters (Q, R, E)

- Q_{lp} = Capacity of intermediate facility (terminal or depot) with capacity level l for plant p (ton)
- R_{ib} = Maximum quantity of biomass b usable from supplier i (ton)
- E_{pt} = Demand quantity of coal shipped to plant p (ton)

Conversion value parameters (H, Θ)

- H^{C} = Heating value of coal (gigajoules/ton)
- H_b^{N}, H_b^{T} = Heating value of biomass feedstock b, untorrefied and torrefied, respectively (gigajoules/ton)
- θ_b^{M} = Mass conversion factor of biomass feedstock b after torrefaction process (%)

$$+\sum_{i}\sum_{b}\sum_{p}\sum_{k}\sum_{j}\left\{C_{b}^{OIN}\left(X_{ibkp}^{KN}+X_{ibjp}^{JN}\right)+C_{b}^{OIT}\left(X_{ibkp}^{KT}+X_{ibjp}^{JT}\right)\right\} \quad (4)$$

$$+\sum_{b}\sum_{p}\sum_{k}\sum_{j}\left\{C_{b}^{CPN}\left(Y_{bkp}^{KN}+Y_{bjp}^{JN}+X_{ibp}^{P}\right)+C_{b}^{CPT}\left(Y_{bkp}^{KT}+Y_{bjp}^{JT}\right)\right\} \quad (5)$$

$$+\sum_{b}\sum_{p}\sum_{k}\sum_{j}\left\{C_{b}^{OPN}\left(Y_{bkp}^{KN}+Y_{bjp}^{JN}+X_{ibp}^{P}\right)+C_{b}^{OPT}\left(Y_{bkp}^{KT}+Y_{bjp}^{JT}\right)\right\} \quad (6)$$

subject to

$$\sum_{i}\sum_{b}X_{ibkp}^{KN}\leqslant\sum_{l}Q_{lp}\left(V_{klp}^{K}-W_{klp}^{K}\right)\forall k\in K,p\in P \quad (7)$$

$$\sum_{i}\sum_{b}X_{ibjp}^{JN}\leqslant\sum_{l}Q_{lp}\left(V_{jlp}^{J}-W_{jlp}^{J}\right)\forall j\in J,p\in P \quad (8)$$

$$\sum_{i}\sum_{b}X_{ibkp}^{KT}\leqslant\sum_{l}Q_{lp}\cdot W_{klp}^{K}\forall k\in K,p\in P \quad (9)$$

$$\sum_{i}\sum_{b}X_{ibjp}^{JT}\leqslant\sum_{l}Q_{lp}\cdot W_{jlp}^{J}\forall j\in J,p\in P \quad (10)$$

$$\sum_{i}X_{ibkp}^{KN}=Y_{bkp}^{KN}\forall k\in K,b\in B,p\in P \quad (11)$$

$$\theta_{b}^{M}\cdot\sum_{i}X_{ibkp}^{KT}=Y_{bkp}^{KT}\forall k\in K,b\in B,p\in P \quad (12)$$

$$\sum_{i}X_{ibjp}^{JN}=Y_{bjp}^{JN}\forall j\in J,b\in B,p\in P \quad (13)$$

$$\theta_{b}^{M}\cdot\sum_{i}X_{ibjp}^{JT}=Y_{bjp}^{JT}\forall j\in J,b\in B,p\in P \quad (14)$$

$$W_{klp}^{K}\leqslant V_{klp}^{K}\forall k\in K,l\in L,p\in P \quad (15)$$

$$W_{jlp}^{J}\leqslant V_{jlp}^{J}\forall j\in J,l\in L,p\in P \quad (16)$$

$$\sum_{l}V_{klp}^{K}\leqslant 1\forall k\in K,p\in P \quad (17)$$

$$\sum_{l}V_{jlp}^{J}\leqslant 1\forall j\in J,p\in P \quad (18)$$

$$\sum_{k}\sum_{j}\sum_{p}\left(X_{ibkp}^{KN}+X_{ibkp}^{KT}+X_{ibjp}^{JN}+X_{ibjp}^{JT}+X_{ibp}^{P}\right)\leqslant R_{ib}\forall i\in I,b\in B \quad (19)$$

$$\sum_{k}\sum_{j}\sum_{b}\left\{H_{b}^{N}\left(Y_{bkp}^{KN}+Y_{bjp}^{JN}+X_{ibp}^{P}\right)+H_{b}^{T}\left(Y_{bkp}^{KT}+Y_{bjp}^{JT}\right)\right\}\geqslant H^{C}\cdot E_{p}\forall p\in P \quad (20)$$

$$V_{klp}^{K},W_{klp}^{K},V_{jlp}^{J},W_{jlp}^{J}\in\{0,1\}, \textit{All other decision variables}\geqslant 0 \quad (21)$$

Term 1 of the objective function indicates the capital costs for a new intermediate facility (terminal, depot, or both terminal and depot) using truck or rail. Each terminal/depot has a maximum annual capacity in terms of the tonnage limit of biomass processing. Term 2 indicates the sum of shipping costs for single- and multi-modal transportation between supply areas, terminals/depots, and power plants. Transportation cost is divided into fixed costs, distance variable costs, and trans-loading costs for shipments that go through terminals/depots. Term 3 expresses the purchasing costs for each type of biomass feedstock and Term 4 describes the operation costs of each terminal/depot. In depots, the operation costs include feedstock upgrading with torrefaction. Term 5 covers capital costs for in-plant processing such as handling, drying, and storage; and Term 6 indicates operation costs in the power plant. As seen in Figure 1, untorrefied and torrefied biomass require different in-plant processing and as such incur different in-plant operation costs.

Constraints 7 to 10 ensure that biomass can only be transported to an intermediate facility after the terminal is built and dictate the capacity limitation for the facility. These constraints also ensure that all biomass feedstock shipped to a depot are upgraded by torrefaction process when a depot is built. Constraints 11 to 14 balance the flow for terminals/depots, ensuring equal inbound and outbound flows. Constraints 12 and 14 provide mass conversion factors to account for the mass reduction from torrefaction.

Constraints 15 to 16 ensure that a depot can be built only within a terminal selected as an intermediate location. Constraints 17 to 18 make sure that no more than one facility (for terminal and depot, respectively) can open at a given location for a power plant. Constraint 19 ensures that the biomass outflow from a supply area is less than its supply availability. Demand satisfaction in terms of energy value is addressed in Constraint 20, and

Constraint 21 defines the binary and non-negativity requirements for the decision variables.

The costs of a conventional logistics system are included in the study to allow for comparison with the advanced logistics system. The mathematical model for the conventional logistics system is developed by removing the biomass upgrading options included in the advanced logistics model. In other words, the conventional logistics costs can be calculated by restricting the values of decision variables relating to building depots to "zero" (i.e., $W_{klp}^{K} = W_{jlp}^{J} = 0$).

The Solution Approach

Our MILP model addresses the multiproduct/multitype capacitated facility location problem, which belongs to the class of NP-hard problems (*21*). We used the MILP solver of CPLEX 12.7, which is based on a branching and cut (BC) algorithm. CPLEX 12.7 provides two solving approaches to the MILP problem: the regular BC algorithm and the benders-based BC algorithm. In this study, the performance of these two algorithms was compared, based on the elapsed time. In a regular BC algorithm, the parameters of CPLEX solver were set at their defaults that include presolve, cuts generation, and the periodic heuristic algorithm. We solved the equivalent MILP formulations by the built-in benders-based BC algorithm of CPLEX. The benders-based BC algorithm used the CPLEX strategy that automatically decomposes MILP into two problems, that is, CPLEX puts all integer variables into the master problem and continuous variables into subproblems. Although the master problem suggests tentative facility location sites, the subproblem is the simple transportation problem used to calculate the actual transportation costs for a set of facility sites suggested by the master problem. The details of two built-in CPLEX algorithms are explained in IBM's *CPLEX User's Manual* (*22*). All experiments were run on a 3.6 GHz Core i7 processor with 16 GB of RAM.

Case Study: Advanced Woody Biomass Logistics for Co-Firing in Coal Power Plant in the Great Lakes States

This section describes the case study used to test the model. The study evaluated the effect of advanced biomass logistics for existing coal power plants in the Great Lakes states: Michigan (MI), Wisconsin (WI), and Minnesota (MN). The section describes data collection and pre-processing for transportation costs, followed by the experimental results and sensitivity analysis to investigate the uncertainty of critical parameters.

Data Collection and Pre-Processing

Feedstock Supply Network: Origin and Destination. Though various types of biomass can fuel co-firing power plants, forest and mill residue were used for this research, as they produce low sulfur and chemical levels (*16*) and are widely available in the study region. Furthermore, a potential recovery in the U.S. housing and lumber markets could increase the availability of sawmill residues (*23*).

In general, the supply network consists of the origins and destinations (nodes) and arcs that connect the nodes. In this research, the origins are sites where the two types of woody biomass are produced and the final destinations are the existing coal power plants (or the intermediate terminals/depots). Actual road and rail networks provided by the U.S. Department of Transportation were used as arcs (*24*). The procurement areas for woody biomass feedstock in this study cover softwood sawmills (for mill residues) and the collecting points (for forest residues) that are located in the centroid of each county within 16 states that cover the study region (Figure 2). The actual locations and capacities of the softwood sawmills were used for the mill residue data (*25*).

Figure 2 also illustrates the locations of 26 coal power plants used for the study that operate in the three Great Lakes states region, according to the 2016 EIA-923 database (*26*). The annual coal tonnage shipped and heat input rates were obtained from the U.S. Energy Information Administration (EIA) (*26*).

Feedstock Availability. The availability of forest residues within the 845 counties in the study area is based on the U.S. Department of Agriculture (USDA) report (*27*). Counties with annual forest residue generation under a thousand dry tons were omitted from the research.

The available quantity of mill residue for 74 softwood sawmills included is based on a 2009 USDA Forest Service report (*28*). It is well known that most mill residues are used to generate on-site heating energy, or sold for other purposes. In this study, approximately 40% of total mill residue were considered available as a feedstock for biomass co-firing (*23*, *29*).

Alternative Location For Truck and Rail Terminals. To narrow down the candidate sites for truck terminals, only centroids of counties with a minimum of 100 dry tons of forest residues per day were selected. The rail terminals were selected through a three-step process that first considered all the existing rail yards operated in the study area, then screened for duplicate yards in any location, and finally checked that the candidate rail yards were located in counties in which at least 100 dry tons of forest residues were produced each day. After the pre-

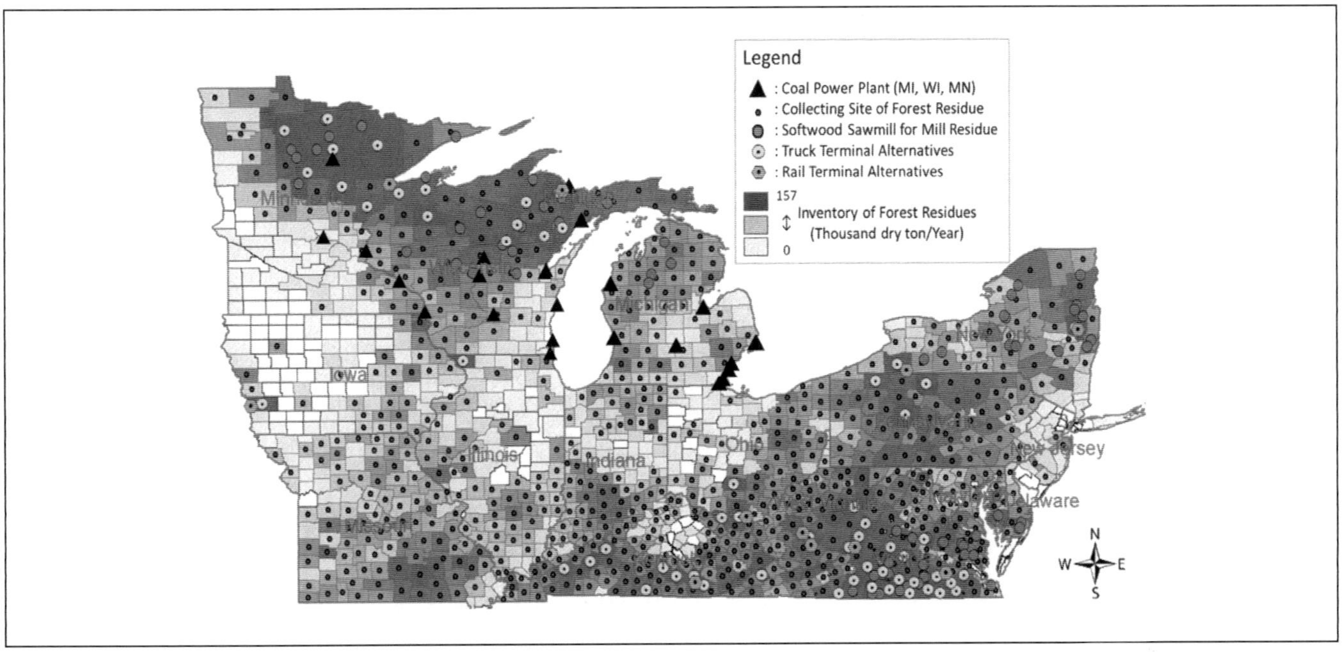

Figure 2. Feedstock locations, target coal power plants, and potential terminals.

screening, 58 truck and 11 rail locations were identified as potential terminal/depot sites for the case study. ArcGIS Desktop 10.3.1 was used to identify and screen alternative locations.

Purchase Cost of Woody Biomass Feedstock. The purchase cost for forest residue (US$14.99 per green ton) and mill residue (US$22.12 per green ton), were generated from the EIA *2016 Billion-Ton Report* (*23*). All costs were adjusted to 2016 U.S. dollars. Although the purchase price of forest residues is lower than that of mill residues, the amount of energy per unit mass is also lower. It is assumed that the energy content of forest residue, mill residue, and coal are 6.6, 9.9, and 27.6 GJ per green ton, respectively (*30*).

Transportation Costs. Truck rates are for full truckloads based on 1,300 origin–destination (O-D) pairs of actual truck price data obtained from local forest products companies in the study region. The trucking dataset was divided into two parts based on intermediate facility: 1) shipments from collecting sites to the truck terminals, and 2) shipments from terminals to the plant. Generally, truck terminals serve as truck-to-truck transfer locations, as well as a buffer for inventory management. Two regression equations developed for truck rates to differentiate the movements before and after truck terminal are as follows:

$$Y_{\text{Rate of Truck Shipped to Terminal (\$/Ton)}} = 0.0852x + 5.3066$$

$$Y_{\text{Rate of Transferred Truck in Terminal (\$/Ton)}} = 0.0582x + 7.1467$$

where x indicates a one-way road distance between O-D pair. The R^2 values for these regressions are 0.847 and 0.699, respectively. In this paper, all distances between O-D pairs were calculated by ArcGIS Network Analyst from actual road and rail networks.

There have been attempts to establish rail rates for biomass transportation (*5*, *31*), but those cases are based on publicly available tariffs, which exclude the contract rates between carriers and shippers. In this research, rail rates were identified through a three-step process that first derived public rates from official tariffs of the main freight rail operators within our study region, then compared them with a real sample of rail rate data provided by local forest companies, and finally calculated the possible rate discount due to a contract between forest shipper and rail carrier. The majority of freight rail transportation in the United States is handled by Class I railroad carriers, and of these, our study area is served by Norfolk Southern Railway (NS) and CSX Transportation (CSXT) for eastern regions and by Canadian National Railway (CN), Union Pacific Railroad (UP), and Burlington Northern Santa Fe Railway (BNSF) for western regions. In this research, rates for two main rail operators, CN and NS, were derived from official tariffs for transporting woody biomass within the study region. It is assumed that rates for UP and BNSF are equal to CN, as their service regions overlap. The same assumption is used for CSXT and NS rates.

One hundred fifty-three CN O-D pairs and 163 NS O-D pairs of rail movements used to ship forest products were obtained from the 2016 tariff data publicly available on each company's website (*32*, *33*). Regression analyses of tariffs were conducted to determine the rail rate. To reduce complexity/uncertainty of rates, all movements used in the study were expected to originate and terminate within the same company's network, excluding movements requiring interchanges with other railroads. The regression equations for CN and NS carload tariff rates of wood products are as follows:

$$y_{cn(\$/carload)} = 4.8698x + 1512$$

$$y_{ns(\$/carload)} = 4.1774x + 2136$$

where x indicates a one-way rail distance between O-D pair. The R^2 values for these regressions are 0.916 and 0.890, respectively.

After these public rail tariffs were compared with real price samples provided by forest companies in the study area, it was found that the actual rail price was 55% to 83% of the price estimated by using tariff regression curves. It appears price discounts of 17% to 45% over public tariffs are used in the actual contract rail rates for our study area. In our study, the final rail rate is discounted by 30% from public tariffs.

Capital and Operation Cost. Capital costs can occur in power plants and in terminals/depots. In the conventional woody biomass logistics, capital costs include equipment and building costs for receiving/handling and storage within the terminal or plant. Advanced logistics systems require additional equipment and facilities for the torrefaction process in depots. Operation costs include ownership costs, such as annual depreciation and insurance, and operating costs, such as labor and maintenance costs (*9*). Operation costs occur at both terminal/depot and power plant locations.

In this paper, INL cost data was used to estimate capital and operation costs (*9*). INL used a fixed logistics scenario designed to co-fire 10% biomass for a 400 MW capacity power plant, but it is assumed that unit costs per ton could be linearly adjusted to other plant capacities.

The capital and operation unit costs for each process in conventional and advanced logistics systems, as obtained from INL are presented in Table 2. Three capacity levels of terminal/depot were assumed: 50,000, 100,000, and 150,000 dry tons per year. Thus, for example, the installed capital cost of a 50,000 dry ton truck terminal would be $748,000 (US$14.96 * 50,000 tons). For the advanced logistics system, the cost would be increased to US$2,607,833 (US$128.1 * 50,000 tons) due to additional costs of upgrading equipment. The capital cost of a rail terminal depends on the length of track and number of turnouts. The track cost for holding 25 rail cars and one turnout were used for rail terminals in this analysis, regardless of terminal capacity. It was assumed that capacity differences could be handled by varying the frequency of rail service. The cost for building rail track is US$727.72 per yard, and a turnout cost is US$111,520 (*5*). As we use annual costs in our calculations, the annual equivalent for these investments over 30 years is used with a discount factor of 10%.

Experimental Results

The mathematical model was applied to five different co-firing ratios (from 10% to 50%) and a total of 10 different scenarios, as presented in Table 3. There were a total of six different logistics alternatives available, three based on conventional systems (Type A through C) and three based on advanced system (Type D through F). Table 3 illustrates the logistics alternatives and presents the preferred logistics type for each plant in each co-firing scenario. The plants were sorted by the tonnage of coal shipped annually. The results suggest that the plants needing a high tonnage of coal generally performed better with the advanced logistics system. Direct trucking (Type A) prevailed as the modal choice in all conventional logistics scenarios and in several cases when an advanced logistics option was also available. Most plants with an advanced logistics system used rail transportation to deliver torrefied feedstock, especially for higher levels of the co-firing ratio (Type E and F). Medium capacity power plants used rail transportation from the depot (Type E), rather than truck (Type D).

A comparison of the time performance of a regular BC and a benders-based BC algorithm in CPLEX for solving the conventional and advanced logistics system were conducted. The model for our case study contains over 3 million decision variables, including 10,764 integer variables per plant. The benders-based BC algorithm showed slightly better performance of elapsed time (average 4.3 s) than a regular BC algorithm (average 5.1 s) under the conventional logistics system for both the small (10%) and large (50%) co-firing scenarios. The solving time for advanced logistics showed more fluctuations by both BC algorithm approaches. We obtained optimal solutions for all plants, so there were no significant differences between the two approaches in the solution quality.

Cost Savings from the Advanced Biomass Logistics System. The cost savings from the advanced logistics system were determined by calculating the cost difference between conventional and advanced logistics scenarios for each co-firing ratio. When comparing cost differences between individual plants, three parameters were used to divide them into seven groups (Figure 3, left-hand side): 1)

Table 2. Capital and Operation Costs of Conventional and Advanced Logistics for Woody Biomass Co-Firing

Capital cost/ operation cost (2016 US$ per dry ton)		Processes in terminal or depot							Processes in power plant			
		Receiving/ handling	Dryer	Torrefaction	Densification	Storage	Surge bin	Sum	In-plant handling	In-plant drying	In-plant storage	Sum
Conventional logistics	1) Supply area → Power plant	na	na	na	na	na	na	0.0 / 0.0	17.92 / 4.28	35.51 / 11.5	3.29 / 1.06	56.72 / 16.84
	2) Supply area → Terminal → Power plant	10.92 / 2.46	na	na	na	3.26 / 1.03	0.78 / 0.07	14.96 / 3.56	17.92 / 4.28	35.51 / 11.5	3.29 / 1.06	56.72 / 16.84
Advanced Logistics	3) Supply area → Terminal + Depot → Power plant	10.92 / 2.46	41.58 / 15.73	68.7 / 10.14	2.87 / 5.59	3.26 / 1.03	0.78 / 0.07	128.1 / 35.02	0.61 / 0.18	na	3.29 / 1.06	3.9 / 1.24

Note: na = not applicable.
Source: Adapted from Boardman et al. (9).

annual coal tonnage shipped (plant capacity), 2) biomass inventory within 50 mi of power plant, and 3) average distance from biomass collecting sites. The average distance was calculated based on distances from county-based centroids (forest residues) and sawmills (mill residues) to each power plant.

The ranking percentile was calculated for each parameter and then used to evaluate the relative standing of a value (quartile) within the parameter. If a ranking percentile of a power plant for a parameter is over 75%, it is classified into the "Very High (= 4)" category. Likewise, "High (= 3)", "Low (= 2)", and "Very Low (= 1)" indicate the ranking percentile over 50%, over 25%, and under 25%, respectively. Figure 3 (right-hand side) takes the same plants and identifies the change in logistics costs with different co-firing ratios. Note that a negative value indicates cost reduction.

The results show that overall savings from advanced logistics are small, but plants within each of the groups demonstrated similar cost reductions. Power plants with low capacity, high quantities of biomass in close vicinity, and a high average distance from collecting sites (Group 1) witnessed no cost reduction from advanced logistics, regardless of the level of co-firing. Similarly, low capacity plants with low biomass availability in the vicinity, but high average distances from collections sites (Group 2) only showed a cost reduction at the highest co-firing levels. All other groups showed higher, although varying levels of cost savings. As expected, most cost savings went to high capacity plants with little biomass within 50 mi (groups 5 and 6), but the trend was different with increasing co-firing ratios (benefits decreased for Group 5, but increased for Group 6). Furthermore, somewhat surprisingly, fairly low capacity plants with limited biomass in the vicinity, but low average distance from collecting sites (Group 3) showed more savings than high capacity plants with high biomass levels in the vicinity (Group 4). It was also noted that the plants in Group 7 that are a low average distance from collecting sites showed few cost savings, despite their very high capacity.

Sensitivity Analysis

This research used fixed torrefaction costs and parameters, such as mass reduction (70% of original mass) and energy contents reserved (90% of original energy content) after torrefaction. Since there is still some uncertainty in the actual impacts of torrefaction, we conducted a sensitivity analysis for the 10% co-firing scenario by independently decreasing and increasing each cost factor by 10%, and recalculating the advanced logistics costs for all 26 power plants. As shown in Figure 4, trucking costs and feedstock purchasing costs are most sensitive to variation. For example, if truck costs decreased by 10%,

Table 3. Results on the Logistics Type Selected by Coal Power Plant in Each Scenario

Plant	Co-firing ratio	10%		20%		30%		40%		50%	
	Annual coal shipped	CL	AL	CL	AL	CL	AL	CL	AL	CL	AL
		Scenario 1	Scenario 2	Scenario 3	Scenario 4	Scenario 5	Scenario 6	Scenario 7	Scenario 8	Scenario 9	Scenario 10
2	13,114	A	A	A	A	A	A	A	A	A	A
25	81,005	A	A	A	A	A	A	A	A	A	A
17	120,213	A	A	A	A	A	A	A	A	A	A
5	142,731	A	A	A	A	A	A	A	A	A	D
4	170,635	A	A	A	A	A	A	A	A	A	E
24	186,495	A	A	A	A	A	A	A	A	A	A
8	194,419	A	A	A	A	A	E	A	E	A	F
12	247,459	A	A	A	A	A	E	A	E	A	E
7	293,749	A	A	A	A	A	E	A	E	A	E
10	398,117	A	A	A	A	A	E	A	E	A	E
19	831,562	A	A	A	E	A	E	A	E	A	E
9	981,161	A	D	A	E	A	E	A	E	A	F
20	1,160,302	A	A	A	E	A	E	A	E	A	E
1	1,404,069	A	A	A	E	A	E	A	E	A	E
14	1,416,005	A	A	A	E	A	E	A	E	A	F
16	1,872,569	A	E	A	E	A	E	A	F	A	F
18	1,985,783	A	E	A	E	A	E	A	E	A	F
15	2,204,624	A	E	A	F	A	F	A	F	A	F
23	2,698,901	A	E	A	E	A	F	A	F	A	F
26	3,586,307	A	F	A	F	A	F	A	F	A	F
3	3,733,052	A	E	A	F	A	F	A	F	A	F
13	3,908,727	A	A	A	F	A	F	C	F	C	F
22	4,037,218	A	E	A	F	A	F	A	F	A	F
11	5,454,623	A	F	A	F	A	F	A	F	A	F
6	6,646,027	A	F	A	F	A	F	A	F	A	F
21	6,958,505	A	F	A	F	C	F	C	F	C	F

Note: CL = conventional logistics; AL = advanced logistics; Type A = conventional logistics with direct trucking to plant only; Type B = conventional logistics with direct trucking to plant, or indirect trucking through terminal; Type C = conventional logistics with direct trucking to plant, or indirect trucking with rail through terminal; Type D = advanced logistics with direct trucking to plant, or indirect trucking through depot; Type E = advanced logistics with direct trucking to plant, or indirect trucking with rail through depot; Type F = Advanced logistics with direct trucking, indirect trucking through truck depot, or indirect trucking with rail through rail depot.

Plant Group	Logistics Condition	Annual coal tons	Biomass within 50mi	Supply distance
Group 1	Plant 2	1	4	4
	Plant 17	1	4	3
	Plant 24	1	4	3
	Plant 25	1	4	3
	Average	1.0	4.0	3.3
	Rank	VL	VH	VH
Group 2	Plant 4	1	2	2
	Plant 5	1	2	1
	Average	1.0	2.0	1.5
	Rank	VL	L	L
Group 3	Plant 7	2	3	1
	Plant 8	1	1	2
	Plant 12	2	3	3
	Plant 10	2	2	1
	Average	1.9	2.0	1.8
	Rank	L	L	L
Group 4	Plant 1	3	4	4
	Plant 13	4	4	4
	Plant 14	3	3	4
	Plant 18	3	4	3
	Plant 19	2	3	4
	Plant 20	2	3	4
	Average	2.5	3.3	3.8
	Rank	H	VH	VH
Group 5	Plant 9	2	3	1
	Plant 15	3	1	2
	Plant 16	3	1	2
	Plant 23	3	2	3
	Average	2.8	1.8	2.0
	Rank	H	L	L
Group 6	Plant 6	4	2	1
	Plant 11	4	1	2
	Plant 21	4	1	4
	Average	4.0	1.3	2.3
	Rank	VH	L	H
Group 7	Plant 3	4	2	1
	Plant 22	4	1	1
	Plant 26	4	1	2
	Average	4.0	1.3	1.3
	Rank	VH	L	L

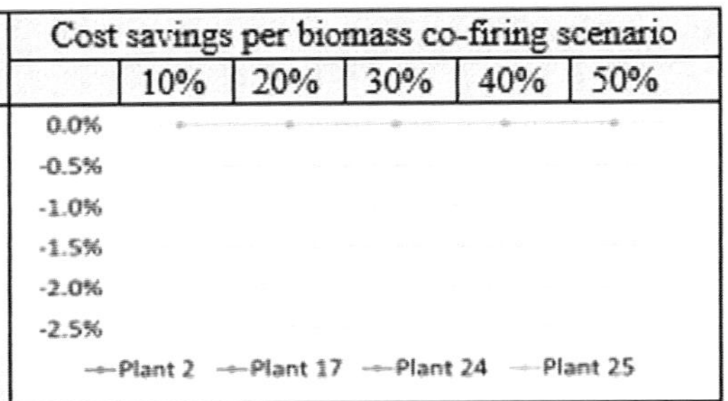

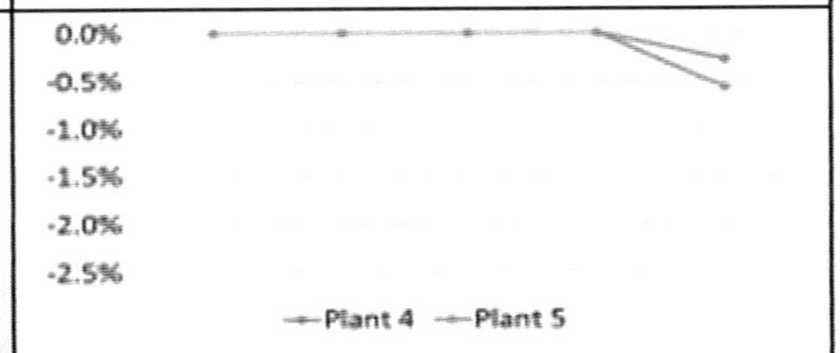

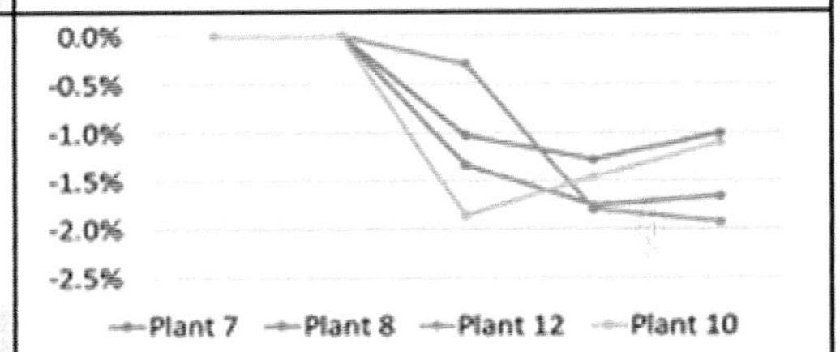

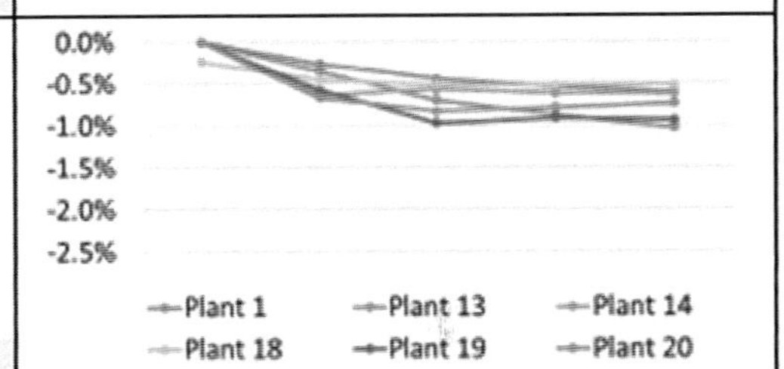

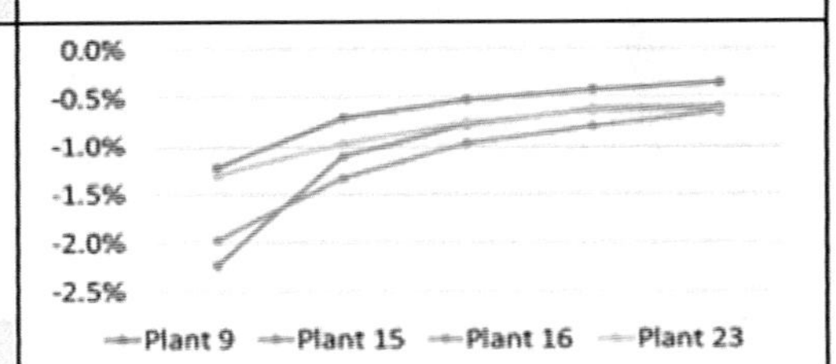

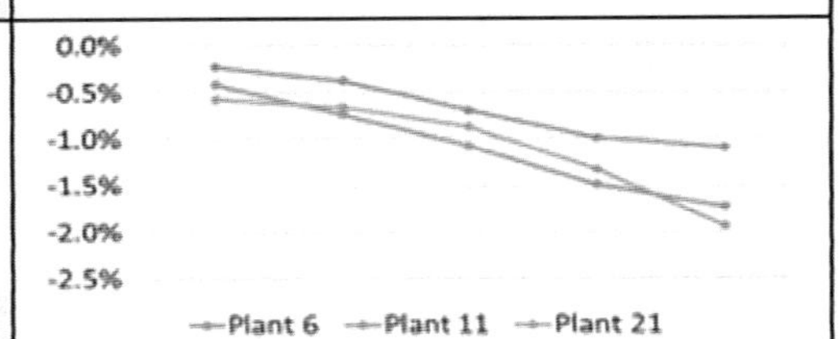

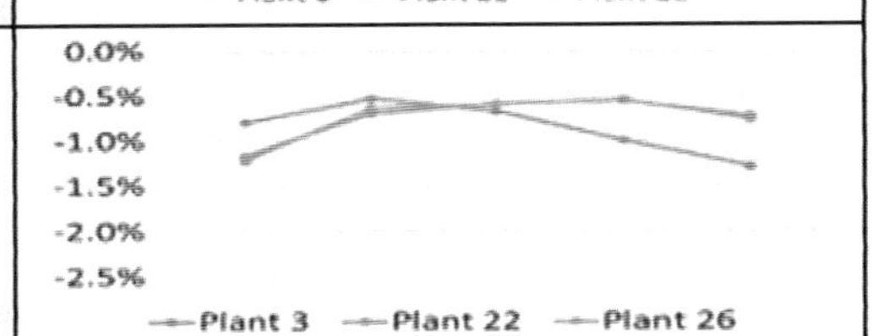

Figure 3. Quartile ranks of plants by logistics conditions, and change in cost trends.
Note: VH = very high (>3); H = high (<=3); L = low (<=2); VL = very low (<=1).

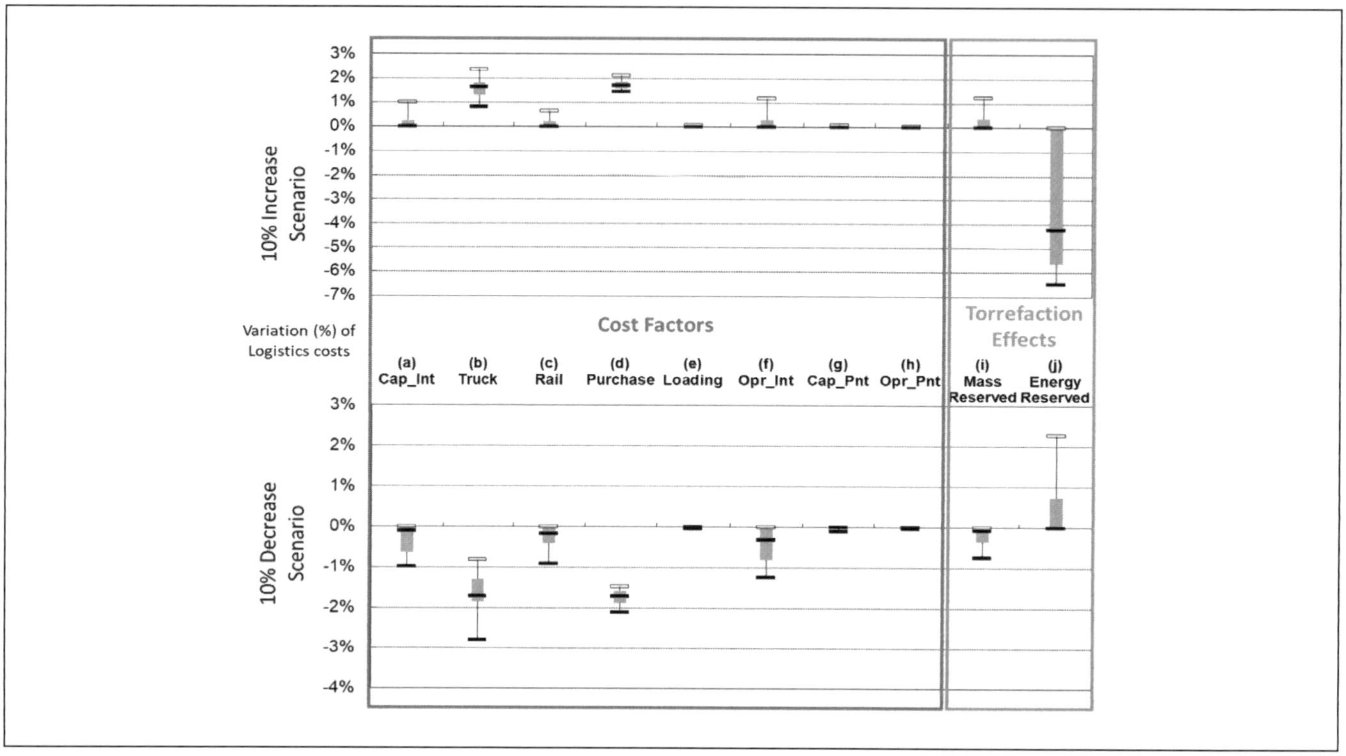

Figure 4. Cost variation of advanced logistics system for all coal power plants (10% co-firing scenario).
Note: Cap_Int = capital cost for intermediate facility; Opr_Int = operation cost for intermediate facility; Cap_Pnt = capital cost for in-plant process; Opr_Pnt = operation cost for in-plant process.

the average advanced logistics costs for all 26 power plants would decrease by 1.7% (Figure 4*b*). However, a 10% decrease of rail rates would lead to only 0.3% cost decrease (Figure 4*c*). For torrefaction effects, energy content after torrefaction has more significance than mass reduction. If energy content was increased by 10%, the average advanced logistics costs would decrease by 4.2% (Figure 4*j*). Note that this percentage indicates the average change in advanced logistics costs for all 26 coal power plants. The percent change on an individual power plant may be different. For example, the logistics costs of plant 16 decreased by almost 6.5% when the energy content after torrefaction was increased by 10%.

Conclusions

This study investigates logistics costs when using biomass feedstock for co-firing in existing coal power plants. More specifically, it attempts to identify and minimize the logistics costs, using either conventional or advanced woody biomass logistics. The analysis was conducted through the MILP model that minimizes logistics costs including feedstock purchasing, transportation, transloading, capital, and operation costs. The results provide insight for existing coal power plants interested in woody biomass co-firing, based on three geographical parameters: plant size, availability of biomass in the vicinity, and the average distance from feedstock collecting sites.

The MILP model developed in the study was applied to coal power plants in three Great Lakes states region. The results revealed that the amount of cost savings from the advanced woody biomass logistics depends on the capacity and location of the power plant. To categorize the cost savings of individual plants, it was found that plants can be divided to seven groups based on three parameters: 1) annual coal tonnage shipped (total capacity), 2) availability of biomass in the vicinity, and 3) average distance from biomass collecting sites. Overall savings from advanced logistics are small, but plants within each of the groups demonstrated similar cost reductions. Power plants with low capacity, high quantities of biomass in close vicinity, and a high average distance from collecting sites witnessed no cost reduction from advanced logistics, regardless of the level of co-firing. Similarly, low capacity plants with low biomass availability in the vicinity, but high average distances from collections sites only showed cost reductions at the highest co-firing levels. All other groups showed higher, although varying levels of cost savings. The sensitivity analysis revealed that trucking costs and feedstock purchasing costs were most sensitive to variation. For torrefaction effects, energy content after torrefaction had more significance than mass reduction.

Though this study tried to use as much actual local data as possible, it is recognized that several assumptions were made, such as the terminal capacity, exclusion of feedstock seasonality, and the flexibility of train service frequency in the study area. In future studies, access to actual capital and operation costs would facilitate the development of a more accurate tool for advanced logistics analysis.

Acknowledgments

This research was supported by the National Science Foundation's Partnerships in International Research and Education Program and National University Rail Center, a US DOT-OST Tier 1 University Transportation Center.

Author Contributions

The authors confirm contribution to the paper as follows: study conception and design: KS, PL; data collection: KS, PL; analysis and interpretation of results: KS, PL; draft manuscript preparation: SK. All authors reviewed the results and approved the final version of the manuscript.

References

1. U.S. Environmental Protection Agency. *Sources of Greenhouse Gas Emissions*. U.S. Environmental Protection Agency, Washington, D.C., 2016.
2. U.S. Energy Information Administration. *Annual Energy Outlook*. U.S. Energy Information Administration, Washington, D.C., 2014.
3. U.S. Energy Information Administration. *CO_2 Emissions from U.S. Electric Power Sector by Source, 2015*, U.S. Energy Information Administration, Washington, D.C., 2016.
4. Tillman, D. A. Biomass Cofiring: The Technology, The Experience, The Combustion Consequences. *Biomass & Bioenergy*, Vol. 19, No. 6, 2000, pp. 365–384.
5. Roni, M. S., S. D. Eksioglu, E. Searcy, and K. Jha. A Supply Chain Network Design Model for Biomass Co-Firing in Coal-Fired Power Plants. *Transportation Research Part E-Logistics and Transportation Review*, Vol. 61, 2014, pp. 115–134.
6. Basu, P., J. Butler, and M. A. Leon. Biomass Co-Firing Options on the Emission Reduction and Electricity Generation Costs in Coal-Fired Power Plants. *Renewable Energy*, Vol. 36, No. 1, 2011, pp. 282–288.
7. Van der Stelt, M. J. C., H. Gerhauser, J. H. A. Kiel, and K. J. Ptasinski. Biomass Upgrading by Torrefaction for the Production of Biofuels: A Review. *Biomass & Bioenergy*, Vol. 35, No. 9, 2011, pp. 3748–3762.
8. Dutta, A., and M. A. Leon. *Pros and Cons of Torrefaction of Woody Biomass*. Joint eco ETI & CEF Projects Workshop Turfgrass Institute, University of Guelph, Guelph, ON, Canada, 2011.
9. Boardman, R. D., K. G. Cafferty, C. Nichol, E. M. Searcy, T. Westover, R. Wood, M. D. Bearden, J. E. Cabe, C. Drennan, S. B. Jones, and J. L. Male. *Logistics, Costs, and GHG Impacts of Utility Scale Cofiring with 20% Biomass*. Idaho National Laboratory (INL), Idaho Falls, ID, Pacific Northwest National Laboratory (PNNL), Richland, WA, 2014.
10. Li, J., A. Brzdekiewicz, W. Yang, and W. Blasiak. Co-Firing Based on Biomass Torrefaction in a Pulverized Coal Boiler with Aim of 100% Fuel Switching. *Applied Energy*, Vol. 99, 2012, pp. 344–354.
11. Steer, J., R. Marsh, A. Griffiths, A. Malmgren, and G. Riley. Biomass Co-Firing Trials on a Down-Fired Utility Boiler. *Energy Conversion and Management*, Vol. 66, 2013, pp. 285–294.
12. Baxter, L. Biomass-Coal Co-Combustion: Opportunity for Affordable Renewable Energy. *Fuel*, Vol. 84, No. 10, 2005, pp. 1295–1302.
13. De, S., and M. Assadi. Impact of Cofiring Biomass with Coal in Power Plants–A Techno-Economic Assessment. *Biomass and Bioenergy*, Vol. 33, No. 2, 2009, pp. 283–293.
14. O'Mahoney, A., F. Thorne, and E. Denny. A Cost-Benefit Analysis of Generating Electricity from Biomass. *Energy Policy*, Vol. 57, 2013, pp. 347–354.
15. Griffin, W. M., J. Michalek, H. S. Matthews, and M. N. A. Hassan. Availability of Biomass Residues for Co-Firing in Peninsular Malaysia: Implications for Cost and GHG Emissions in the Electricity Sector. *Energies*, Vol. 7, No. 2, 2014, pp. 804–823.
16. Eksioglu, S. D., H. Karimi, and B. Eksioglu. Optimization Models to Integrate Production and Transportation Planning for Biomass Co-Firing in Coal-Fired Power Plants. *IIE Transactions*, Vol. 48, No. 10, 2016, pp. 901–920.
17. Poudel, S. R., M. Marufuzzaman, and L. Bian. A Hybrid Decomposition Algorithm for Designing a Multi-Modal Transportation Network Under Biomass Supply Uncertainty. *Transportation Research Part E-Logistics and Transportation Review*, Vol. 94, 2016, pp. 1–25.
18. Tumuluru, J. S., R. D. Boardman, and C. T. Wright. Response Surface Analysis of Elemental Composition and Energy Properties of Corn Stover During Torrefaction. *Journal of Biobased Materials and Bioenergy*, Vol. 6, No. 1, 2012, pp. 25–35.
19. Couhert, C., S. Salvador, and J. -M. Commandre. Impact of Torrefaction on Syngas Production from Wood. *Fuel*, Vol. 88, No. 11, 2009, pp. 2286–2290.
20. Chen, Y.-C., W.-H. Chen, B.-J. Lin, J.-S. Chang, and H. C. Ong. Impact of Torrefaction on the Composition, Structure and Reactivity of a Microalga Residue. *Applied Energy*, Vol. 181, 2016, pp. 110–119.
21. Mirchandani, P. B., and R. L. Francis. *Discrete Location Theory*. John Wiley & Sons, New York, NY, 1990.
22. IBM. *CPLEX User's Manual, Version 12 Release 7. IBM Corp*. Studio, IBM ILOG CPLEX Optimization, 2016.
23. U.S. Department of Energy. *2016 Billion-Ton Report: Advancing Domestic Resources for a Thriving Bioeconomy, Volume 1: Economic Availability of Feedstocks* (Langholtz, M. H., B. J. Stokes, and L. M. Eaton, eds.), Oak Ridge, TN, Oak Ridge National Laboratory, 2016, p. 448.
24. U.S. Department of Transportation. *Transportation Networks, National Transportation Atlas Database 2015.*

Bureau of Transportation Statistics, Washington, D.C., 2016.
25. Diffen. *Hardwood vs Softwood.* 2017, http://www.diffen.com/difference/Hardwood_vs_Softwood. Accessed February 5, 2017.
26. U.S. Energy Information Administration. *Form 906/920/923: Utility, Non-Utility, and Combined Heat And Power Plant Database. Monthly Time Series.* U.S. Energy Information Administration, Washington, D.C., 2017.
27. U.S. Department of Agriculture. *Timber Product Output (TPO) Reports.* Knoxville, TN, USDA Forest Service, Southern Research Station, 2012.
28. Spelter, H., D. McKeever, and D. Toth. *Profile 2009: Softwood Sawmills in the United States and Canada.* U.S. Department of Agriculture Forest Service, Forest Products Laboratory, Madison, WI, 2009, p. 55.
29. Fehrs, J. E., and V. Williston, *Secondary Mill Residues and Urban Wood Waste Quantities in the United States.* Northeast Regional Biomass Program, Washington, D.C., 1999.
30. Krajnc, N. *Wood Fuels Handbook.* Food and Agriculture Organization of the United Nations, Pristina, Kosovo, 2015.
31. Gonzales, D., E. M. Searcy, and S. D. Eksioglu. Cost Analysis for High-Volume and Long-Haul Transportation of Densified Biomass Feedstock. *Transportation Research Part A-Policy and Practice*, Vol. 49, 2013, pp. 48–61.
32. Canadian National Railway. *Canadian National Railway: Prices, Tariffs & Transit Times.* 2016, https://www.cn.ca/en/customer-centre/prices-tariffs-transit-times. Accessed June 25, 2016.
33. Norfolk Southern Railway. *Norfolk Southern Railway: Public Price Publications.* 2016, http://www.nscorp.com/mktgpublic/publicprices/. Accessed June 25, 2016.

The Standing Committee on Freight Transportation Planning and Logistics (AT015) peer-reviewed this paper (18-05365).

Article

TRR
JOURNAL OF THE TRANSPORTATION RESEARCH BOARD

Transportation Research Record
2018, Vol. 2672(9) 107–119

DOI: 10.1177/0361198118782256
journals.sagepub.com/home/trr
SAGE

Freight Mode Choice: A Regret Minimization and Utility Maximization Based Hybrid Model

Nowreen Keya[1], Sabreena Anowar[1], and Naveen Eluru[1]

Abstract

With the introduction of automated vehicles, the performance of the trucking industry is expected to be improved. In fact, this may impact the entire freight transportation system as trucks possess the highest mode share in freight transportation. To investigate this impact, a hybrid utility–regret-based mode choice model accommodating for shipper level unobserved heterogeneity is proposed in this study. It recognizes that not all attributes influencing shipment mode are evaluated following a homogenous decision rule (solely random utility maximization/solely random regret minimization). The proposed model system is developed using 2012 Commodity Flow Survey data. To demonstrate the applicability of the proposed model system, a detailed policy analysis is conducted considering several futuristic scenarios such as implementation of automation and controlled access of truck traffic to an urban region. The results indicate that introduction of automation in the freight industry would be more beneficial for long-haul hire truck mode than short-haul private truck mode. An increase in truck shipping time due to re-routing of truck traffic away from urban regions causes a modal shift from truck to parcel and "other" mode (rail, water or multiple modes).

An efficient and cost-effective freight transportation system is the prerequisite for a region's economic growth and prosperity. In the U.S., about 122.5 million households, 7.5 million businesses and 90 thousand government units, daily depend on the efficient movement of about 55 million tons of freight valued at around $49 billion (*1*). The demand for goods has grown steadily over the past half century and is expected to increase with the growth in population. The percentage share of freight transported in 2013 by weight and value by mode are as follows: truck (70 and 64), rail (9 and 3), water (4 and 1.5), air (0.1 and 6.5), and pipeline (7.7 and 6.0) (*1*). The remainder of the freight is transported by multiple modes, mail and unknown modes. This percentage clearly indicates that road-based freight transportation is an important component of the supply chain in the U.S. and trucks are the preferred mode of shipping for most manufacturers and distributors in the country. Higher percentage of truck mode share is associated with negative externalities including, air pollution, traffic congestion, increase in accident severity, and expeditious deterioration of road and bridge infrastructure. Heavy trucks comprise only 3 percent of the total registered vehicles in the U.S. and comprise 7% of the total vehicle miles driven, yet they are involved in 11% of the total road fatalities (*2*). Usually multiple axle trucks produce rutting damage, and single and tandem axles cause cracking on road surfaces (*3*).

There is growing recognition among transportation researchers that addressing the freight industry associated challenges needs us to examine several dimensions including freight mode choice, freight infrastructure, pricing strategies across modes, and wages. In our research, we focus our attention on identifying and quantifying the influence of factors affecting mode choice for freight shipments. With the emerging advances in vehicle technology—connected and autonomous vehicles—there is likely to be a seismic shift in the freight industry in the near future. While level 4 adoption, which is a fully self-driving vehicle in all conditions, (as defined by NHTSA, (*4*)) is likely to take time, several intermediate levels of vehicle technologies are already being introduced by private and public companies. These vehicular advances offer significant advantages to the trucking

[1]Department of Civil, Environmental & Construction Engineering, University of Central Florida, Orlando, FL

Corresponding Author:
Address correspondence to Sabreena Anowar: sabreena.anowar@ucf.edu

industry in terms of fuel, time, and labor cost savings. For instance, a platoon of connected trucks in a formation can reduce the impact of wind resistance by maintaining a shorter distance between them (15 m instead of 50 m) thus saving fuel and reducing CO_2 emission by around 7% for a platoon of three trucks (*5*). Further, adoption of fully autonomous vehicles will allow the trucking industry to circumvent the need for federally mandated driver breaks for long-haul trips. These are instances of how vehicle technology can offer environmental and financial benefits. While these changes are likely to improve the performance of the trucking industry, their impact on the overall shipment mode choice is less straightforward.

The proposed research effort contributes to our understanding of the impact of these technological adoptions, by developing advanced discrete choice models for freight mode choice analysis. Toward that end, we adopt a three-pronged research approach. First, we contribute to the existing literature by examining freight mode choice from the perspectives of alternative behavioral paradigms including classical random utility (RU) framework, newly emerging random regret (RR) framework, and hybrid framework (that builds on both utility and regret). Two kinds of hybrid models are considered: (1) hybrid framework with single utility equation accommodating regret and utility terms, and (2) latent class model with one segment following RU structure and another following RR structure. Second, a national level dataset drawn from the Commodity Flow Survey (CFS) 2012 is augmented with a host of exogenous variables generated at origin and destination CFS areas and used for model building exercise. Finally, based on these variable effects, a host of policy scenarios are identified and evaluated employing the best-specified model structure. Based on the policy scenario outcomes, recommendations for freight planning process are given.

Earlier Work and Current Study Context

A detailed review of literature on freight mode choice models is available in our previous study (*6*). From our review, we found that the following categories of exogenous variables influence shipment mode choice: (1) level of service (LOS) measures (such as shipping time, shipping cost, speed, delay, fuel cost); (2) freight characteristics (such as commodity group, commodity size, commodity density, commodity value, commodity weight, product state, temperature controlled or not, perishability, trade type, quantity); (3) transportation network and origin–destination (O-D) attributes (such as shipment O-D, distance, ratio of highway and railway miles in origin and in destination); and (4) others (such as service reliability, service frequency, loss and damage, shipper's characteristics).

On the methodological front, the majority of earlier studies have employed traditional RU based multinomial logit (MNL) model (*7–11*) and its several extensions such as nested logit model (*10*, *12* and *13*), mixed logit model (*6*, *8*), heteroscedastic extreme value model (*14*, *15*), latent class multinomial logit model (*8* and *16*), and copula based joint model embedded with an MNL model (*17*). Alternative approaches such as artificial neural network (*18*, *19*) and neuro-fuzzy models (*19*) have also been developed. RU framework is mainly a compensatory behavioral framework that might not be optimal in determining choice behavior with alternative specific attributes. An alternative, RR framework that allows for pairwise attribute comparison among alternatives has been successfully applied in several fields including transportation (for travel mode choice (*20*) or route choice (*21*), road pricing (*22*), departure time (*23*), automobile fuel choice (*24*), online dating (*25*), healthcare (*26*), and recreational site choice (*27*)). Recently, Boeri and Masiero (*28*) used RR based multinomial logit (RRMNL) model to study mode choice based on a stated preference survey conducted on some Swiss medium to large industries. In their study, the authors found that the RRMNL model performed slightly better than its utility counterpart.

Although comparison between RU maximization (RUM) and RR minimization (RRM) based approaches is beneficial, it is also possible that attribute impact on choice behavior could follow either approach. Toward accommodating such flexibility, a hybrid approach that allows attribute impacts to follow both RU and RR is employed in our analysis. While behavioral paradigm is quite important, the presence of unobserved heterogeneity is also likely to affect choice behavior. To accommodate for alternative behavioral paradigms and potential presence of unobserved heterogeneity, we develop the following models: (1) RU based mixed MNL, (2) RR based mixed MNL, (3) a hybrid utility–regret mixed MNL model combining both RU and RR based attribute processing, and (4) latent class models with hybrid segments—one segment following RU decision rule with the other following RR decision rule. These models are estimated using data from the 2012 U.S. CFS.

Empirical Data

Data Source

The main data source for this study is the 2012 CFS data. The survey is conducted every 5 years since 1993 and is the only publicly available source of commodity flow information at a national level. The public use microdata (PUM) file of CFS 2012 contains a total of 4,547,661 shipment records from approximately 60,000 responding industries. A sample of 5,565 records is drawn from the original dataset to manage the burden of generating LOS variables (shipping cost and shipping time) ensuring that the weighted mode share in the random sample is the same as the weighted mode share in the original dataset. Of this, 4,000 records were randomly chosen for estimation purpose and 1,565 records were set aside for validation exercise.

Dependent Variable Generation. A total of twenty-one shipping modes are reported in CFS 2012. In our study, based on sample share, the reported modes were categorized into five classes: (1) hire truck (including truck and hire truck), (2) private truck, (3) air, (4) parcel or courier service, and (5) other mode (predominantly rail mode and the rest of the modes). Hire truck refers to those trucks operated by non-governmental business units to provide transport services to customers for a payment. On the other hand, private truck is not available to the public and is owned and used by an individual business unit for shipping its own freight. The air mode consists of combination of air and truck, as truck is needed to pick up or deliver the commodity from or to a particular place, which cannot be accessed by air mode. Parcel or courier service mainly refers to multiple modes. The "other" mode refers to rail, water, pipeline or a combination of non-parcel multiple modes. The distribution of the weighted mode share in the sample is as follows: hire truck (16.57%), private truck (25.97%), air (1.42%), parcel (55.73%), and other (0.31%). We also created alternative availability following a heuristic approach based on shipment weight and routed distance (*6*).

Independent Variable Generation. Several data compilation steps were undertaken in our study. First, we generated LOS variables employing information from several sources for all available modes. For instance, shipping cost by hire truck and private truck was estimated using the 2007 revenue per ton-mile from National Transportation Statistics (NTS) with appropriate regional and temporal correction factors. For parcel mode, using FedEx, pricing functions were generated with distance and weight as variables for the seven zones in the U.S. The pricing functions also accommodated for shipping speed—express overnight (1 day), express deferred (3 days), and ground service (5 days)—based on observed shares of these shipping options from FedEx 2015 annual report. For shipping time by hire and private truck, three different speed bands were considered based on trip distance while considering the required break times according to the service regulations provided by the Federal Motor Carrier Safety Administration (FMCSA; see Keya et al. [*6*] for a detailed discussion on how mode shipping time and cost variables were generated for each mode). Second, using GIS layers from different sources, we generated several O-D attributes. For example, from the National Transportation Atlas Database 2012 (NTAD 2012) and the Highway Performance Monitoring System (HPMS), we collected roadway and railway network files and generated the roadway (including length of tolled road and truck route) and railway lengths. Other information collected from the same sources include number of airports, number of seaports, and number of intermodal facilities. Truck annual average daily traffic was collected from National Highway Freight Transportation (NHFN). Third, from census, the following data were collected: population count, counts of employees and establishments by NAICS industry types (manufacturing, mining, retail trade, warehouse and storage, company and enterprise, wholesale and information), mean household income, number of warehouses, number of freight transportation establishments, and percentage of population below poverty level in 2012. The origin and destination area type (urban or rural) was classified based on the percentage of population residing in each area. If more than 50% population lived in the area it was classified as urban; rural otherwise. The CFS area was categorized into low, medium, and high-income category groups based on annual average household income (<\$50,000, \$50,000–\$80,000, and >\$80,000, respectively). A state was recognized as a cold state if the average annual temperature was below or equal to 60°F; warm otherwise. The statewise temperature data was collected from Current Results: Weather and Science Facts (*29*). Based on the highest number of industries located in an area, the area was classified as a manufacturing, mining, wholesale, information, retail trade, warehouse and storage, or company and enterprise major area.

Descriptive Statistics

Figure 1 illustrates the shipment weight distribution by mode. It shows that private trucks carry increased tonnage in the California, Piedmont Atlantic, and Gulf Coast regions. Air and parcel modes mainly carry loads less than or equal to 30 lbs in the majority of the CFS areas. In Figure 2, the shipping cost by different modes across the CFS areas is presented. It can be observed from the figure that shipping cost is comparatively higher in California and Great Lake mega regions for hire and private truck (more than \$370 and \$100, respectively). The shipping cost by air mode is relatively higher in northern states (>\$450). The reason might be the cold weather in these states. Shipping cost by parcel mode is lower than other modes across the whole U.S. with very few CFS areas with shipping cost more than \$80. The

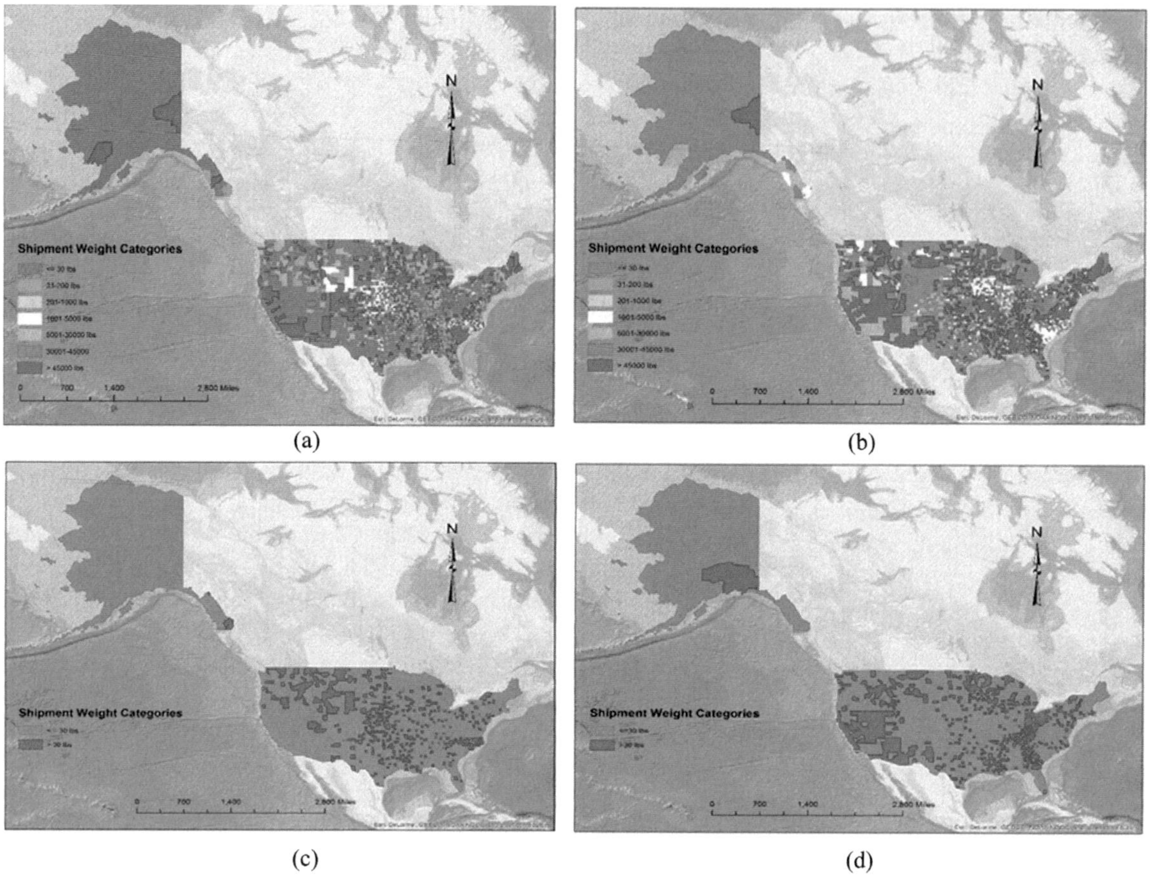

Figure 1. Shipment weight distribution in CFS areas: (*a*) hire truck; (*b*) private truck; (*c*) air; (*d*) parcel.

shipping cost by parcel mode in most of the areas is less than \$80. Figure 3 demonstrates the shipping time distribution by mode across the entire U.S. In most of the regions, shipping time varies between 12 to 63 hours for hire truck and 1 to 3 hours for private truck. Very few regions have shipping time as high as 100 hours by hire truck. Shipping time by private truck is more than 6 hours in very few areas, because private truck usually travels shorter distance compared with hire truck. The shipping time by air mode in most CFS areas is less than 3 hours. For parcel mode, shipping time is greater than 94 hours in the majority of the CFS areas, as typically parcel mode takes 3 to 5 days to deliver a product (except express delivery option which usually takes 1 or 2 days). Barely any areas can be found in the figure where shipping time is 1 to 3 days.

Econometric Framework

In this section, we discuss the econometric frameworks employed in the study.

Mixed Hybrid Model (Combination of RUM and RRM)

Let $s(s = 1, 2, \ldots\ldots, S)$ be the index for shippers, and $i(n = 1, 2, \ldots\ldots, I)$ be the index for mode alternatives characterized by $m(m = 1, 2, \ldots N, \ldots, M)$ attributes. Let us also consider that N attributes are evaluated following RU maximization principle while the rest $(M - N)$ are evaluated following RR minimization principle. With these notations, the systematic part of the hybrid (or modified) utility–regret equation would take the form

$$\mathrm{HU}_i = \sum_{m=1}^{N} \beta'_m x_i - \sum_{j \neq i} \sum_{m=N+1}^{M} \ln\left[1 + \exp\left\{\beta'_m \left(x_{jm} - x_{im}\right)\right\}\right] \tag{1}$$

In the above formula, the linear in parameter portion represents RU maximization and the non-linear part represents RR minimization based attribute processing. Considering, the error term to be standard type-1 extreme value distributed, the mathematical expression for the unconditional probability of the hybrid utility–regret

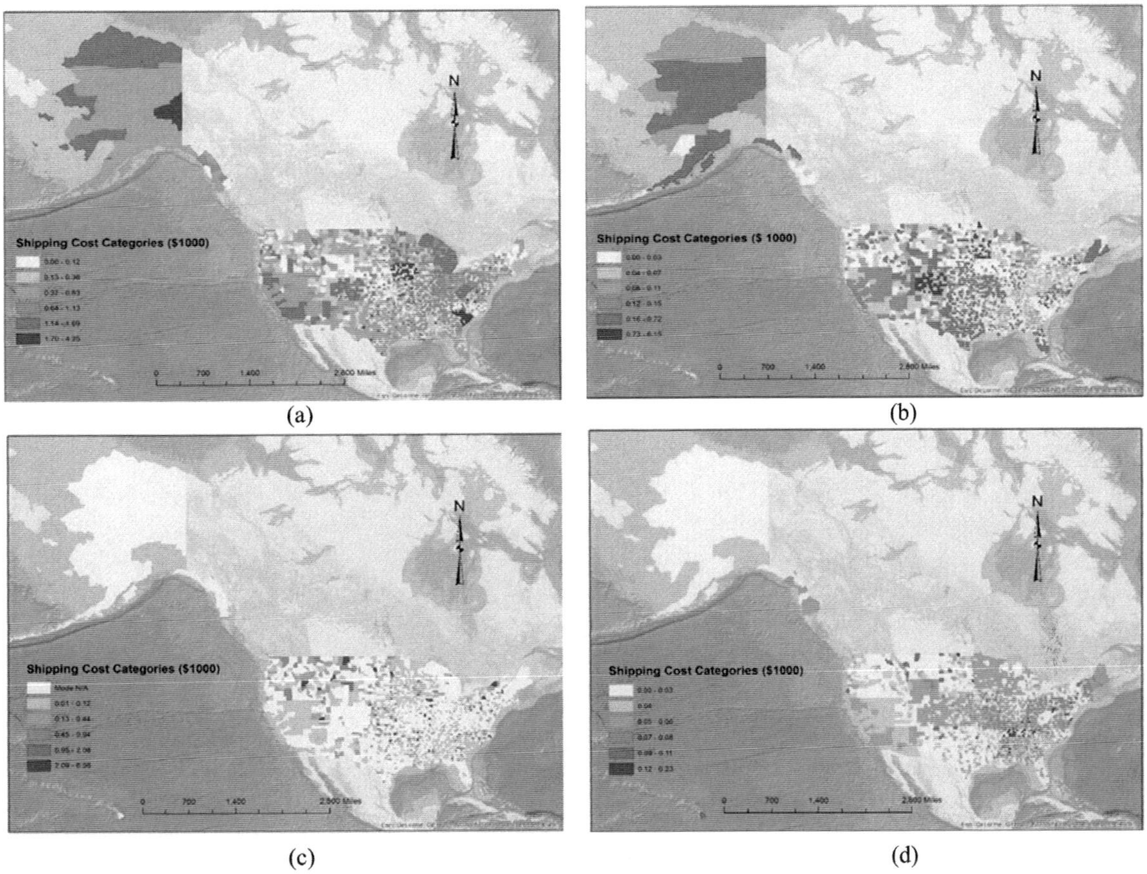

Figure 2. Shipping cost ($1,000) distribution in CFS areas: (*a*) hire truck; (*b*) private truck; (*c*) air; (*d*) parcel.

model could be written (accommodating for unobserved heterogeneity) as

$$P_i^{\mathrm{HU}} = \int\left(\left[\frac{\exp(\mathrm{HU}_i)}{\sum_{i=1}^{I}\exp(\mathrm{HU}_i)}\right]^{d_i}\right) f(\beta)d\beta \quad (2)$$

where $f(\beta)$ is a density function specified to be normally distributed with mean 0 and variance σ^2 and d_i is a binary variable which is equal to 1 if shipper s choose mode i or 0 otherwise. There is no *a priori* expectation regarding which attributes are likely to be processed in utility theoretic fashion and which are likely to be processed by RR approach. If all parameters are evaluated based on RU maximization principle, then the hybrid model collapses to traditional RU based mixed MNL model and if all parameters are evaluated based on RR minimization principle, then it collapses to regret based mixed MNL model. To estimate parameters, maximum simulated likelihood (MSL) estimation technique is employed. For this study, we use a quasi-Monte Carlo approach (Scrambled Halton draws) with 200 draws for the MSL estimation (see Bhat (*30*) for more details).

Latent Class Two Segment Model with RUM and RRM

In the two-class latent segment model, Segment 1 follows RU principle and segment 2 follows an RR based decision rule. The latent-segmentation-based models assign shipments probabilistically into k ($k = 1, 2$) segments based on a host of explanatory variables (for example, freight characteristics). The mathematical expression for the probability of a shipment s belonging to segment k can be expressed as

$$P_{sk} = \frac{\exp(\gamma_k' z_s)}{\sum_{k=1}^{2}\exp(\gamma_k' z_s)} \quad (3)$$

where, z_s is a vector of shipment attributes that influences the propensity of belonging to segment k, and γ_k' is a vector of estimable coefficients. Within the latent class approach, the unconditional probability of a shipment s being shipped by mode i is given as

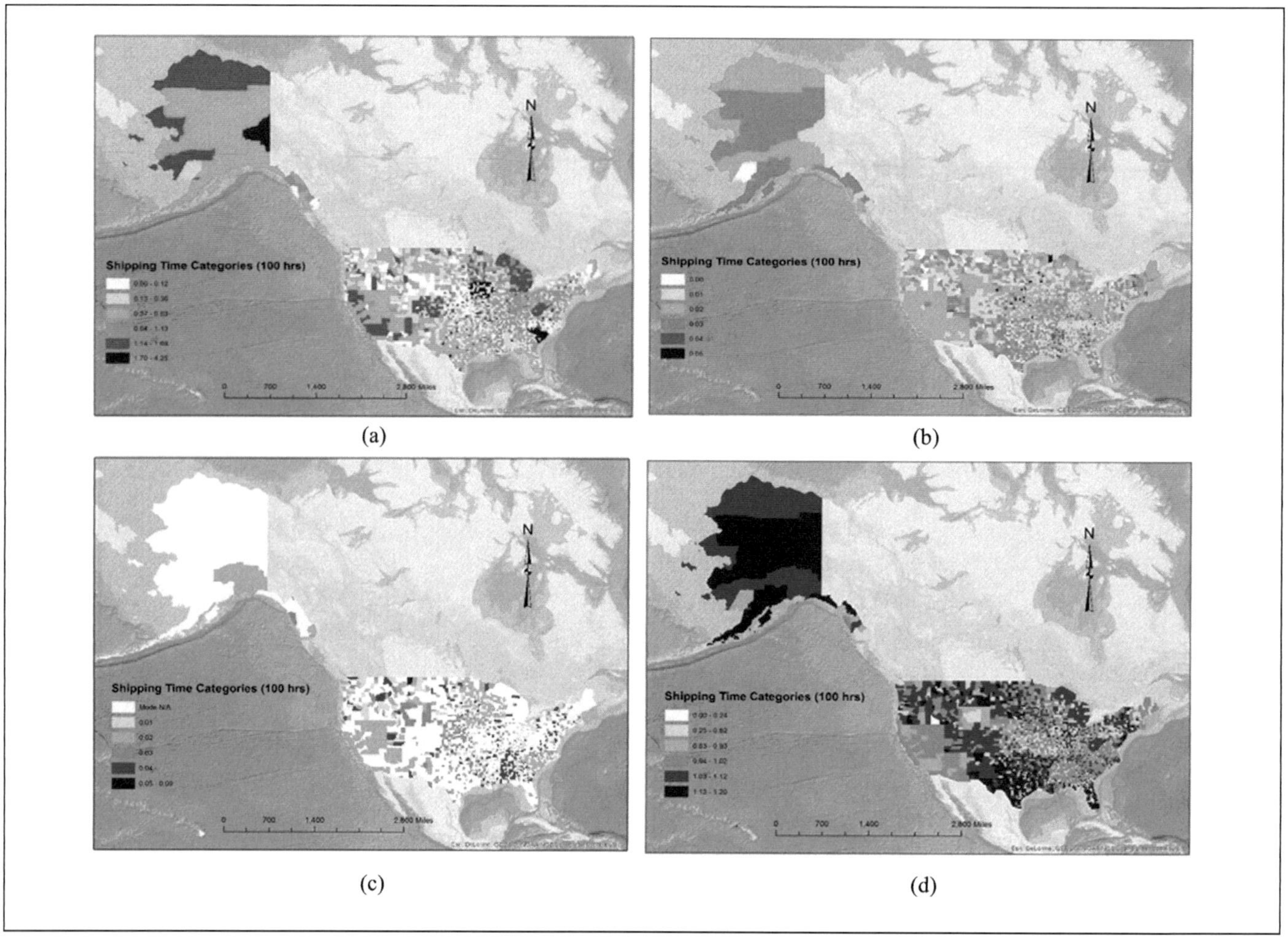

Figure 3. Shipping time (100 hrs) in CFS areas: (*a*) hire truck; (*b*) private truck; (*c*) air; (*d*) parcel.

$$P_s(i) = \sum_{k=1}^{2} (P_s(i)|k)(P_{sk}) \tag{4}$$

where $P_s(i)|k$ represents the conditional probability of shipment s being shipped by mode i within the segment k. Using the notations mentioned above, the conditional probability for segment 1 (considering RU maximization principle) would be

$$P_s(i)|1 = \frac{\exp(\alpha_k' x_{si})}{\sum_{i=1}^{I} \exp(\alpha_k' x_{si})} \tag{5}$$

Here, α_s' represents a vector of coefficients, and x_{si} is a vector of attributes influencing mode choice. On the other hand, for segment 2 (considering RR based decision), the conditional probability would be given as

$$P_s(i)|2 = \frac{\exp(-R_{si})}{\sum_{i=1}^{I} \exp(-R_{si})} \tag{6}$$

Here, $R_{si} = \sum_{j \neq i} \sum_{m=1}^{M} \ln\left[1 + \exp\left\{\delta_m\left(x_{sjm} - x_{sim}\right)\right\}\right]$; δ_m is a vector of estimable coefficients associated with attribute x_m; x_{im} and x_{jm} are (Lx1) column vector of mode attributes for the considered alternative i and another alternative j, respectively. The log-likelihood function for the entire dataset with appropriate $P_s(i)|k$ is

$$\text{LL} = \sum_{s=1}^{S} \log(P_s(i)) \tag{7}$$

Empirical Analysis

Model Fit

In this study, a series of models was estimated including traditional RU maximization based MNL (RUMNL), RR minimization based MNL (RRMNL), RU based mixed MNL (RUMMNL), RR based mixed MNL

Table 1. Comparison of Different Models

Model	Log-likelihood at convergence	No. of parameters	No. of observation	BIC values
RUMNL	−1782.95	41	4000	3905.96
RRMNL	−1769.30	40	4000	3870.36
HUMNL	−1769.69	38	4000	3854.55
RUMMNL	−1772.06	42	4000	3892.75
RRMMNL	−1759.83	41	4000	3859.72
HUMMNL	−1758.52	39	4000	3840.52
LSRURR	−1857.98	36	4000	4014.55

(RRMMNL), hybrid utility–regret-based MNL (HUMNL), hybrid utility–regret-based mixed MNL (HUMMNL), and latent class two segment (RU and RR) model (LSRURR). To compare these models, Bayesian Information Criterion (BIC) values were computed (presented in Table 1). The BIC value for a given empirical model can be calculated using (– 2 [LL] + K ln [Q]), where (LL) is the log-likelihood value at convergence, K is the number of parameters, and Q is the number of observations. The lowest BIC value was found for HUMMNL (3840.49). Therefore, we present and discuss the results obtained from this model only (Table 2). Please note that we considered a 90 percent significance level. The last column of Table 2 identifies whether the variable was evaluated following RU structure or RR structure. We discuss the results for RUM variables followed by RRM variables.

Exogenous Variable Effects (RUM)

The LOS variables (shipping cost and shipping time) negatively influence mode share. This is expected, as shippers naturally would prefer modes offering faster shipping time and lower carrying cost. We also allowed for the presence of unobserved heterogeneity across shipping cost and time. It was found that shipping cost has a statistically significant standard deviation. The coefficient of cost follows a normal distribution with a mean value of -0.8097 and a standard deviation of 0.4639. The distribution infers that shipping cost impacts most of the observation negatively with a very small proportion (4.09%) of cases having a positive impact of cost. In addition to an overall shipping time coefficient, shipping time interactions with different commodity types were examined (observed and unobserved). Of the various commodity types, only the raw food and prepared products presented a statistically significant result for observed effects. The estimated parameters imply that raw and prepared foods are more sensitive to shipping time compared with other commodity types. The result is reasonable because these products are usually perishable and require timely delivery. For export freight, air is more likely to be the preferred alternative compared to hire truck (see Wang et al. (*31*) for similar result). Private truck is more likely to be chosen when the shipment value is less than $5000.

The transportation network and demographic attributes offer intuitive results as well. With increasing highway density at origin, the propensity to choose parcel mode increases. The result indicates that increasing roadway connectivity increases the accessibility for parcel mode. Densely populated areas attract more freight flows; hence, the probability of choosing private truck, air, and parcel mode also increases with increasing population density at destination. The utility for using private trucks decreases with increase in inter-modal facilities in the area. The result also shows that probability of choosing private truck decreases when density of warehouse and super center increases at origin. Air mode is less likely to be chosen for destinations with population below poverty level, presumably since shipping through air mode is expensive. Moreover, the impoverished destinations may not have necessary provisions for air mode as well (airports or freight airstrips). With increasing number of employee density in manufacturing industries at origin, the probability of choosing private truck decreases.

Exogenous Variable Effects (RRM)

The constants do not possess any substantive interpretation after introducing other exogenous variables. The coefficients of freight characteristics treated with RRM approach bear intuitive results. The probability of choosing parcel mode decreases when the commodity is non-flammable liquid or other hazardous material. It is expected because this type of commodity needs special care for handling and advanced safety precautions. Probability of choosing private truck increases when the commodity to be shipped needs temperature control as desired temperature control facilities can be provided by private truck providers. Hence, regret would be lesser compared with any other mode when private truck is chosen for temperature-controlled products. In addition,

Table 2. Estimation Result of Mixed Hybrid Model Combination of RUM and RRM Based Approaches

	Hire truck		Private truck		Air		Parcel/courier		Other		
Explanatory variables	Parameter	*t*-stat	Parameter	*t*-stat	Parameter	*t*-stat	Parameter	*t*-stat	Parameter	*t*-stat	Type
Constant	0	–	0.2222	2.680	−0.3997	−1.021	1.3049	7.959	−1.7770	−3.532	RRM
LOS variables											
Shipping cost (1000 $)	−0.8097	−2.239	−0.8097	−2.239	−0.8097	−2.239	−0.8097	−2.239	−0.8097	−2.239	RUM
Std. dev.	0.4639	1.751	0.4639	1.751	0.4639	1.751	0.4639	1.751	0.4639	1.751	RUM
Shipping time (hrs)	−0.0059	−3.648	−0.0059	−3.648	−0.0059	−3.648	−0.0059	−3.648	−0.0059	−3.648	RUM
Interaction variables											
Shipping time * Raw food (hrs)	−0.0169	−2.625	−0.0169	−2.625	−0.0169	−2.625	−0.0169	−2.625	−0.0169	−2.625	RUM
Shipping time * prepared products (hrs)	−0.0086	−2.129	−0.0086	−2.129	−0.0086	−2.129	−0.0086	−2.129	−0.0086	−2.129	RUM
Freight characteristics											
Hazardous material (base: not hazardous)											
Non-flammable liquid and other hazardous material	–	–	–	–	–	–	−0.6022	−3.557	–	–	RRM
Temperature controlled (base: No)											
Yes	–	–	0.2743	2.366	–	–	–	–	–	–	RRM
Export (base: No)											
Yes	–	–	–	–	2.4275	5.664	–	–	–	–	RUM
SCTG commodity type (base: wood, papers and textile)											
Prepared products	–	–	0.5488	4.064	–	–	–	–	–	–	RRM
Stone and non-metallic minerals	–	–	−0.3178	−3.381	–	–	–	–	–	–	RRM
Petroleum and coals	–	–	0.5279	3.220	–	–	–	–	–	–	RRM
Chemicals	–	–	−0.1538	−2.300	–	–	–	–	–	–	RRM
Electronics	–	–	−0.1552	−2.354	0.6292	3.146	–	–	–	–	RRM
Furniture and others	–	–	0.1544	2.394	–	–	–	–	–	–	RRM
Shipment value ($) (base: value >5000)	–	–			–	–	–	–	–	–	
Value ≤ 1000	–	–	1.6217	10.484	–	–	–	–	–	–	RUM
1000 < value ≤ 5000	–	–	0.9355	5.254	–	–	–	–	–	–	RUM
Value > 5000	–	–	–	–	–	–	−0.3176	−2.787	–	–	RRM
Transportation network and demographic variables											
Origin mega region (base: non mega region)											
Florida	–	–	0.2998	2.198	–	–	–	–	–	–	RRM
Destination mega region (base: non mega region)											
North-East	–	–	–	–	–	–	−0.1356	−1.653	–	–	RRM
Origin area type (base: rural)											
Urban	–	–	0.2787	2.593	–	–	–	–	–	–	RRM
Avg. temperature at origin (base: warm (>60°F))											
Cold (≤60°F)	–	–	–	–	–	–	0.1850	2.826	–	–	RRM
Major industry at origin (base: manufacturing)											
Wholesale	–	–	0.1209	1.850	–	–	–	–	–	–	RRM
Major industry at destination (base: manufacturing)											
Wholesale	–	–	−0.1093	−1.788	–	–	–	–	–	–	RRM
Origin highway density (mi/mi^2)	–	–	–	–	–	–	2.2970	1.974	–	–	RUM
Density of interstate highways and freeways at destination (mi/mi^2)	–	–	–	–	−0.0283	−1.785	–	–	–	–	RRM

(continued)

Table 2. (continued)

Explanatory variables	Hire truck		Private truck		Air		Parcel/courier		Other		Type
	Parameter	*t*-stat	Parameter	*t*-stat	Parameter	*t*-stat	Parameter	*t*-stat	Parameter	*t*-stat	
Density population at destination (pop/mi^2)	–	–	0.0011	3.500	0.0011	3.500	0.0007	3.733	–	–	RUM
No. of intermodal facility at destination	–	–	–0.0067	–2.869	–	–	–	–	–	–	RUM
Density of warehouse and super center at origin (per mi^2)	–	–	–0.4361	–2.356	–	–	–	–	–	–	RUM
	–	–	–	–	–	–	–0.1903	–2.210	–	–	RRM
Density of wholesale industry at destination (per mi^2)	–	–	–0.2117	–2.978	–	–	–	–	–	–	RRM
Percentage of population below poverty level at destination	–	–	–	–	–10.7827	–1.744	–	–	–	–	RUM
Density of employees in manufacturing industry at origin (per mi^2)	–	–	–0.4453	–7.936	–	–	–	–	–	–	RUM
No. of seaports at destination	–	–	–0.0003	–2.924	–	–	–	–	–	–	RRM
Number of cases						4000					
Log likelihood for constant only model						–2063.51					
Log likelihood at convergence						–1758.52					
No. of parameter						39					
Adjusted rho-square						0.1313					

Note: – = variable insignificant at 90 percent confidence level; RRM = random regret minimization; RUM = random utility maximization.

the probability of choosing private truck increases when the commodity is prepared products, petroleum and coals, or furniture and other commodities. On the other hand, private truck is not preferred when the commodity is stone and non-metallic minerals, chemicals, or electronics. Our findings are in line with the results reported in previous studies (*17* and *31*). Electronic products are comparatively light weight, expensive and need special care while transporting (see Pourabdollahi (*17*) for the same finding) and hence, there would be lesser regret associated with choosing air mode for transporting these commodity types. Parcel mode is less likely to be chosen when the shipment is expensive in terms of its value (more than $5000) (see *16*, *19* and *32* for similar results).

When the origin mega region is Florida, private truck is more likely to be chosen. Again, when destination is northeast region, parcel mode is less likely to be chosen. The probability of choosing private truck increases when the origin is urban area. In cold areas with average temperature below or equal to 60^0 F, parcel mode is more likely to be chosen. The reason may be in colder areas people are more dependent on purchasing products online than going out by themselves to purchase that commodity. Hence, regret would be lesser for this case. The probability of choosing private truck increases when the major industry type at origin is wholesale, but probability of choosing private truck decreases when the major industry type at destination is wholesale. One plausible explanation might be that wholesale dominating origins produce bulk amounts of products, which require shipping by truck rather than air or parcel mode. When the density of interstate highways and freeways at destination increases, the probability of choosing air mode decreases which is expected. With increasing density of warehouse and super centers at destination, the probability of choosing parcel mode decreases. If there is a greater number of seaports at destination, private truck is less likely to be chosen as the shipment mode.

Validation Results

We performed a validation exercise using the 1,565 records to examine the performance of the estimated model. We computed the mean absolute error (MAE) and root mean square error (RMSE) metrics for the predicted mode shares at the aggregate level. The MAE and RMSE values obtained are 0.34 and 0.44, respectively. The results highlight the reasonable performance of the proposed model.

Policy Analysis

To illustrate the applicability of the proposed model, we conducted a detailed policy analysis. The scenarios considered include:

1. A carbon tax on truck mode increasing the shipping cost by 25%, 35%, and 50%
2. A reduction in truck shipping time due to introduction of automated fleets (by eliminating the resting time of heavy vehicle drivers)
3. An increase in truck shipping time by 15%, 25%, and 50% (re-routing of trucks away from the urban region)
4. A combination of 50% increase in truck shipping cost (carbon tax) and reduction in shipping time from scenario 2
5. A carbon tax on air mode increasing the shipping cost by 25% and 50%

Table 3 illustrates the changes in predicted mode shares from base shares for different policy scenarios. In the table, a positive or negative sign specifies an increase or decrease, respectively, from the base mode share. When the shipping cost increases due to carbon tax measure, as expected, the mode share of hire truck and private truck decreases. This reduction ranges from 1.93% to 2.96% for hire truck and 1.08% to 1.77% for private truck. Moreover, percentage share of "other" mode increases significantly under this policy scenario. This is not surprising, because trucks usually carry larger loads which can only be substituted by rail. In the second scenario, shipping times by hire and private trucks are reduced by eliminating the mandatory rest and break times for long haul drivers. As expected, the results illustrate a potential increase in hire truck share (by 6.91%). However, there is a slight increment in private truck share because these trucks usually run shorter distance compared to hire truck and hence, rest or break time is not usually needed for the drivers. This essentially signifies that vehicle automation might be more beneficial for long-haul modes. On the other hand, reduction in truck shipping time decreases the share of air and parcel mode substantially. To reduce congestion, to reduce conflicts between heavy vehicle and automobiles and pedestrians/cyclists on the roadways within cities, and to reduce air pollution, city officials might decide to reroute truck flows to by-pass roadways located at the periphery of the cities. This will apparently benefit passenger traffic but will lead to increased shipping time for trucks. We capture the effect of such rerouting in the third scenario. As expected, increase in shipping time leads to a substantial decrease in truck share. More specifically, hire truck share decreases between the ranges of 2.35% to 7.85%. In contrast, share of private trucks does not decrease remarkably. Under this scenario, shippers are more likely to opt for parcel and other modes if truck flows are rerouted. More interestingly, a simultaneous increase in truck shipping cost (carbon tax) and reduction in shipping time leads to an increase in the share of hire truck

Table 3. Percentage Changes of Mode Share from Base Prediction under Different Policy Scenarios

Mode	Truck shipping cost 25% increase	Truck shipping cost 35% increase	Truck shipping cost 50% increase	Truck shipping time under automated vehicles	Truck shipping time 15% increase	Truck shipping time 25% increase	Truck shipping time 50% increase	Truck shipping cost 50% increase and truck shipping time reduction	Air shipping cost 25% increase	Air shipping cost 50% increase
Hire truck	−1.93	−2.41	−2.96	6.91	−2.35	−3.68	−7.85	4.83	0.42	0.48
Private truck	−1.08	−1.54	−1.77	0.27	−1.09	−1.13	−1.21	0.08	−1.16	−1.14
Air	−4.39	−4.29	−4.15	−7.16	−2.70	−2.04	−0.33	−6.22	−7.71	−11.92
Parcel	1.01	1.29	1.42	−2.20	1.22	1.60	2.82	−1.69	0.72	0.75
Other	35.75	51.55	76.23	0.68	12.74	13.82	16.63	72.12	3.45	3.45

indicating that shippers are usually more sensitive to shipping time than shipping cost. At the same time, share of "other" mode increases by almost 72% under this policy scenario. Finally, a carbon tax measure of 25% and 50% on air mode reduces the air mode share by 7.71% and 11.92%, respectively, while increasing parcel and "other" mode share.

Conclusion

An efficient and cost-effective freight transportation system is a prerequisite for a region's economic growth and prosperity. The advanced technology adoption and implementation will benefit the trucking industry both financially and environmentally. However, this change may influence overall freight industry in a complex way. The proposed research effort contributes to our understanding of the impact of these technological adoptions, by developing advanced discrete choice models for freight mode choice analysis.

We contribute to the existing literature by examining shipment mode choice from alternative behavioral paradigms – RU maximization and RR minimization. To capture unobserved heterogeneity of LOS variables, a utility–regret hybrid mixed model was estimated. The corresponding changes predicted to freight mode share under future vehicle technology adoption scenarios are also evaluated. In our empirical analysis, the hybrid utility–regret mixed MNL model performed better compared with all other models. Our finding lends credence to the growing recognition that attributes affecting choice behavior could be treated heterogeneously, using either utility theoretic manner or regret minimization orientation. Overall, the estimated results offer plausible interpretation of the choice behavior. The evaluation of policy scenarios offers reasonable and intuitive results in terms of modal shifts. We found that introduction of automation in the freight industry would be more beneficial for long-haul hire truck mode than short-haul private truck mode. Re-routing of truck flows away from urban region clearly indicates a modal shift from truck to parcel or "other" mode (rail, water or multiple modes). In addition, implementation of carbon tax should be accompanied by shipping time penalty, if modal shift from road-based transportation to rail- or water-vessel-based transportation is to be achieved. These policy insights can be helpful for transportation planners and urban policy makers to provide adequate physical facilities and services for truck transportation. Designated truck routes, controlled access to urban areas and selected parking and loading–unloading infrastructural facilities can improve truck transportation significantly. Also adopting automated truck fleets can cut off the economic and environmental impacts associated with trucking industry to a greater extent.

The study is not without limitations. CFS data does not provide exact geo-coded O-D locations. Several approaches that randomize geo-coded locations to protect privacy are available. CFS data could implement these approaches and provide the geo-coded locations for modeling analysis. The availability of such geo-coded data will improve shipping time computation as well as alternative availability matrices. While our model structures accommodate for the impact of unobserved factors, additional information on shipment frequency, shipper reliability, vehicle fleet ownership of the shipping firm, and shipping time delays would enhance the model developed. Additional work on improving the approaches for LOS computation is beneficial. In future work, analysis of mode choice decisions at regional or state level will enhance the model findings as well as provide policy makers with more customized insights.

Author Contributions

The authors confirm contribution to the paper as follows: Study conception and design: Naveen Eluru, Sabreena Anowar; data collection: Nowreen Keya; analysis and interpretation of results: Nowreen Keya, Sabreena Anowar, Naveen Eluru; draft manuscript preparation: Nowreen Keya, Sabreena Anowar, and Naveen Eluru. All authors reviewed the results and approved the final version of the manuscript.

References

1. Bureau of Transportation Statistics. *Freight Facts and Figures 2015*. Bureau of Transportation Statistics, 2015. http://www.rita.dot.gov/bts/programs/freight_transportation. Accessed July 29, 2017.
2. Bezwada, N. N. K. Characteristics and Contributory Causes Associated with Fatal Large Truck Crashes. PhD dissertation, Kansas State University, 2010.
3. Salama, H. K., K. Chatti, and R. W. Lyles. Effect of Heavy Multiple Axle Trucks on Flexible Pavement Damage Using In-Service Pavement Performance Data. *Journal of Transportation Engineering*, Vol. 132, No. 10, 2006, pp. 763–770.
4. National Highway Traffic Safety Administration. *Preliminary Statement of Policy Concerning Automated Vehicles*. National Highway Traffic Safety Administration, Washington, D.C., 2013, pp. 1–14.
5. Daimler Blog. *Connected Trucks Freight Transport of the Future by Using the Internet*. Daimler Blog. https://www.daimler.com/innovation/digitalization/connectivity/connected-trucks.html. Accessed July 29, 2017.
6. Keya, N., S. Anowar, and N. Eluru. Estimating a Freight Mode Choice Model: A Case Study of Commodity Flow Survey 2012. Presented at 96th Annual Meeting of the Transportation Research Board, Washington, D.C., 2017.
7. Arencibia, A. I., M. Feo-Valero, L. García-Menéndez, and C. Román. Modelling Mode Choice for Freight Transport Using Advanced Choice Experiments. *Transportation Research Part A: Policy and Practice*, Vol. 75, 2015, pp. 252–267.
8. Brooks, M. R., S. M. Puckett, D. A. Hensher, and A. Sammons. Understanding Mode Choice Decisions: A Study of Australian Freight Shippers. *Maritime Economics & Logistics*, Vol. 14, No. 3, 2012, pp. 274–299.
9. Mitra, S., and S. M. Leon. Discrete Choice Model for Air-Cargo Mode Selection. *The International Journal of Logistics Management*, Vol. 25, No. 3, 2014, pp. 656–672.
10. Nugroho, M. T., A. Whiteing, and G. de Jong. Port and Inland Mode Choice from the Exporters' and Forwarders' Perspectives: Case Study—Java, Indonesia. *Research in Transportation Business & Management*, Vol. 19, 2016, pp. 73–82.
11. Yang, D., G. P. Ong, and A. T. H. Chin. An Exploratory Study on the Effect of Trade Data Aggregation on International Freight Mode Choice. *Maritime Policy & Management*, Vol. 41, No. 3, 2014, pp. 212–223.
12. Jiang, F., P. Johnson, and C. Calzada. Freight Demand Characteristics and Mode Choice: An Analysis of the Results of Modeling with Disaggregate Revealed Preference Data. *Journal of Transportation and Statistics*, Vol. 2, No. 2, 1999, pp. 149–158.
13. Rich, J., P. M. Holmblad, and C. Hansen. A Weighted Logit Freight Mode-Choice Model. *Transportation Research Part E: Logistics and Transportation Review*, Vol. 45, No. 6, 2009, pp. 1006–1019.
14. Holguín-Veras, J. Revealed Preference Analysis of Commercial Vehicle Choice Process. *Journal of Transportation Engineering*, Vol. 128, No. 4, 2002, pp. 336–346.
15. Norojono, O., and W. Young. A Stated Preference Freight Mode Choice Model. *Transportation Planning and Technology*, Vol. 26, No. 2, 2003, pp. 1–1.
16. Arunotayanun, K., and J. W. Polak. Taste Heterogeneity and Market Segmentation in Freight Shippers' Mode Choice Behaviour. *Transportation Research Part E: Logistics and Transportation Review*, Vol. 47, No. 2, 2011, pp. 138–148.
17. Pourabdollahi, Z., B. Karimi, and A. Mohammadian. Joint Model of Freight Mode and Shipment Size Choice. *Transportation Research Record: Journal of the Transportation Research Board*, 2013. 2378: 84–91.
18. Abdel Wahab, W., and T. Sayed. Freight Mode Choice Models Using Artificial Neural Networks. *Civil Engineering Systems*, Vol. 16, No. 4, 1999, pp. 267–286.
19. Sayed, T., and A. Razavi. Comparison of Neural and Conventional Approaches to Mode Choice Analysis. *Journal of Computing in Civil Engineering*, Vol. 14, No. 1, 2000, pp. 23–30.
20. Chorus, C. G. A New Model of Random Regret Minimization. *Ejtir*, Vol. 2, No. 10, 2010, pp. 181–196.
21. Chorus, C. G. A Generalized Random Regret Minimization Model. *Transportation Research Part B: Methodological*, Vol. 68, 2014, pp. 224–238.
22. Chorus, C. G., J. A. Annema, N. Mouter, and B. van Wee. Modeling Politicians' Preferences for Road Pricing Policies: A Regret-Based and Utilitarian Perspective. *Transport Policy*, Vol. 18, No. 6, 2011, pp. 856–861.

23. Chorus, C. G., and G. C. De Jong. Modeling Experienced Accessibility for Utility-Maximizers and Regret-Minimizers. *Journal of Transport Geography*, Vol. 19, No. 6, 2011, pp. 1155–1162.
24. Hensher, D. A., W. H. Greene, and C. G. Chorus. Random Regret Minimization or Random Utility Maximization: An Exploratory Analysis in the Context of Automobile Fuel Choice. *Journal of Advanced Transportation*, Vol. 47, No. 7, 2013, pp. 667–678.
25. Chorus, C. G., and J. M. Rose. Selecting a Date: A Matter of Regret and Compromises. In *Choice Modelling: The State of the Art and the State of Practice*, edited by Stephane Hess and Andrew Daly, 2013, pp. 229–242.
26. de Bekker-Grob, E. W., and C. G. Chorus. Random Regret-Based Discrete-Choice Modelling: An Application to Healthcare. *PharmacoEconomics*, Vol. 31, No. 7, 2013, pp. 623–634.
27. Boeri, M., A. Longo, E. Doherty, and S. Hynes. Site Choices in Recreational Demand: A Matter of Utility Maximization or Regret Minimization? *Journal of Environmental Economics and Policy*, Vol. 1, No. 1, 2012, pp. 32–47.
28. Boeri, M., and L. Masiero. Regret Minimisation and Utility Maximisation in a Freight Transport Context. *Transportmetrica A: Transport Science*, Vol. 10, No. 6, 2014, pp. 548–560.
29. Current Results. Weather and Science Facts. *Average Annual Temperature for Each US State*. https://www.currentresults.com/Weather/US/average-annual-state-temperatures.php. Accessed July 29, 2017.
30. Bhat, C. R. Quasi-Random Maximum Simulated Likelihood Estimation of the Mixed Multinomial Logit Model. *Transportation Research Part B: Methodological*, Vol. 35, No. 7, 2001, pp. 677–693.
31. Wang, Y., C. Ding, C. Liu, and B. Xie. An Analysis of Interstate Freight Mode Choice between Truck and Rail: A case Study of Maryland, United States. *Procedia-Social and Behavioral Sciences*, Vol. 96, 2013, pp. 1239–1249.
32. Moschovou, T., and G. Giannopoulos. Modeling Freight Mode Choice in Greece. *Procedia-Social and Behavioral Sciences*, Vol. 48, 2012, pp. 597–611.

The Standing Committee on Freight Transportation Planning and Logistics (AT015) peer-reviewed this paper (18-06118).

Article

TRR
JOURNAL OF THE TRANSPORTATION RESEARCH BOARD

Transportation Research Record
2018, Vol. 2672(9) 120–129

DOI: 10.1177/0361198118791667
journals.sagepub.com/home/trr
SAGE

Weigh Station Impact on Truck Travel Time Reliability: Results and Findings from a Field Study and a Simulation Experiment

M. Shoaib Samandar[1], Billy M. Williams[2], and Ishtiak Ahmed[1]

Abstract

Weigh stations are necessary for safeguarding highway infrastructure by enforcing truck weight limits. However, mandating all trucks to stop at all weigh stations decreases travel time reliability. This decrease in travel time reliability adversely impacts the productivity of the trucking industry and to a lesser degree impacts personal travel reliability as well. This study, conducted at the Lumberton weigh station on Interstate 95 in North Carolina, quantifies the impact of weigh stations on truck travel time reliability. Truck travel times were observed over periods of weigh station operation and weigh station closure. Comparison of these two states sheds light on the variability in travel time caused by weigh station operation. Results show that when the weigh station is operational, truck travel time reliability degrades significantly. VISSIM microsimulation software was used to quantify the expected impact of weigh in motion (WIM) on truck travel time reliability assuming different scenarios of WIM truck bypass. The model results indicate that WIM technology does increase travel time reliability and provides benefits to both trucking companies and enforcement agencies. It was also found that increases in the proportion of WIM bypass result in increases in route travel time reliability. However, this simulation model improvement was not uniform, with the highest marginal improvement occurring in the 30%–40% WIM bypass range.

Travel time reliability deals with measuring the unforeseen deviation from the expected duration of travel that travelers build through their travel experiences or from external sources. Unreliable travel times in general have an adverse effect on all modes of freeway traffic. However, the effect is much more pronounced and complicated in freight transportation because logistic decisions involve more than one party (shippers, receivers, and carriers) (*1*). The magnitude of this travel time unreliability is itself a variable of freeway segment among other factors (incident, work zone, adverse weather, etc.). Potentially, freeway segments with weigh stations could be listed as the most unreliable segments of rural and interstate highways for truck traffic. The unreliability of these segments can be directly associated with the weigh station operational design and capacity and with fluctuations in freight traffic demand.

On the other hand, weigh stations are necessary tools for safeguarding the freeway infrastructure and improving traffic safety by enforcing truck weight limits. Out of eight weigh stations in the state of North Carolina, seven are equipped with static scales. Under the current weight enforcement guidelines, all large trucks are routed to the weigh station's static scales whenever the weigh station is open. However, trucks loaded with legal weights and appropriate credentials will experience unnecessary increase in travel time. Re-routing all trucks to the weigh station and mandating them to stop for weighing reduces the travel time reliability of the trucking industry and results in high enforcement costs for the regulating agency.

For weight enforcement purposes, static scales are very accurate and reliable. However, when demand levels grow, trucks can back up onto the freeway causing significant increase in travel times and safety hazards. To remedy this situation, weigh stations are often equipped with systems that detect potentially dangerous truck queues

[1]Institute for Transportation Research and Education (ITRE), North Carolina State University, Raleigh, NC
[2]North Carolina State University, Raleigh, NC

Corresponding Author:
Address correspondence to M. Shoaib Samandar: smsamand@ncsu.edu

and temporarily close the weigh station. Some trucks that bypass the weigh station during these temporary closures may be overweight or may be operating without proper credentials (*2*).

Weigh in motion (WIM) technology presents a potential solution. When implemented to supplement static scale weighing, these systems can reduce travel time variability at weigh stations. The WIM technology can be set up on the mainline and/or ramp of a weigh station. Al-qadi et al. did a comprehensive review of literature on WIM systems (*3*). WIM systems classify arriving trucks based on a set weight threshold. Trucks equipped with transponders and carrying legal loads are directed to bypass the weigh station's static scales. Trucks classified as potentially overweight and/or with potential credential/safety deficiencies are routed to the weigh station's static scale and/or inspection area for further inspection.

Several previous studies have delved into the operation of weigh stations. Benekohal et al. quantified the delay and traffic conflicts around a weigh station in Springfield, Illinois with the goal to evaluate the effectiveness of WIM systems and potential benefits of Intelligent Transportation System (ITS) technologies (*2, 4*). They found that 30% of trucks were not weighed because of weigh station temporary closure to prevent spillback. Furthermore, they found that trucks with legal weights and appropriate credentials experienced unnecessary delays at the weigh station. Kamyab et al. explored the effects of mainline electronic pre-screening on truck delay and weigh station productivity by developing a new simulation model. Their study found that as participation of transponder-equipped trucks increases, the delay at the simulated weigh station decreases (*5*).

Glassco used the Westa simulation model, a detailed simulation of traffic around inspection stations, to study three weigh stations and found that as percentages of transponder-equipped trucks increase, the benefits of WIM technology increase by reducing not only the travel time for the transponder-equipped trucks, but also for the non-participating trucks. Another finding of this study was that "a WIM scale is very effective in reducing the number of trucks that must be weighed at the static scale" (*6*).

Review of the literature related to weigh station operations reveals a gap in the study of truck travel time reliability along freeway routes where weigh stations are present. The work in this paper attempts to fill the gap through a field study of travel time reliability in the vicinity of a static-scale-equipped weigh station in North Carolina. In order to explore the potential impact of WIM technology, the weigh station is also modeled in VISSIM microsimulation platform. Varying levels of WIM bypass percentage are simulated, and the simulation outputs are compared to the static scale scenario.

Data Collection

Description of Study Area

The study area is located on the northbound direction of I-95, about six miles north of Lumberton in Robeson County, North Carolina. The total length of the study section is approximately 4.16 mi. It includes the northbound direction of Lumberton weigh station and surrounding area. Traffic volume is moderate with an annual average daily traffic of 47,000 in 2015. The study site was the subject of both a field study and a simulation experiment. The next two subsections provide detailed descriptions of the data collection efforts.

Data Collection for Travel Time Reliability Analysis

A field data collection effort was designed to study the impact of weigh station operations on the travel time reliability of trucks and other impacted vehicles. Travel time and speed data were collected for the route described above using two Bluetooth devices located upstream and downstream of the weigh station as shown in Figure 1*a*. Data collection started on June 10, 2015 and ended on September 2, 2015. The study period included 57 weekdays, one of which was a holiday (excluded from the analysis dataset): Independence Day, observed on July 3rd. The study period also included 23 weekend days. Bluetooth observations were unavailable for nine weekdays and four weekend days during the study period, leaving 47 non-holiday weekdays and 18 non-holiday weekend days with Bluetooth observations. During the 47 non-holiday weekdays, the weigh station was operational forty days and closed seven days. Having data on weigh station both when it is open and closed enabled direct analysis of the impact the weigh station has on travel time reliability.

Data Collection for Calibration and Validation of Microsimulation Model

The second set of data collected was for calibration and validation of the microsimulation model. This data collection effort deployed three video cameras and four Bluetooth units. Video was captured at three locations along the highway: one upstream, one downstream, and one at the weigh station scale, as shown in Figure 1*b*. The video recorders were in place and recording for the duration of the study. The cameras were installed in an inconspicuous manner to ensure that truck speeds were not influenced by the data collection efforts. The images of trucks captured by the video camera at Location 1 were matched with the images at the downstream locations, 2 and 3, to determine truck travel times. Location 2 was located at the weigh station. The camera was

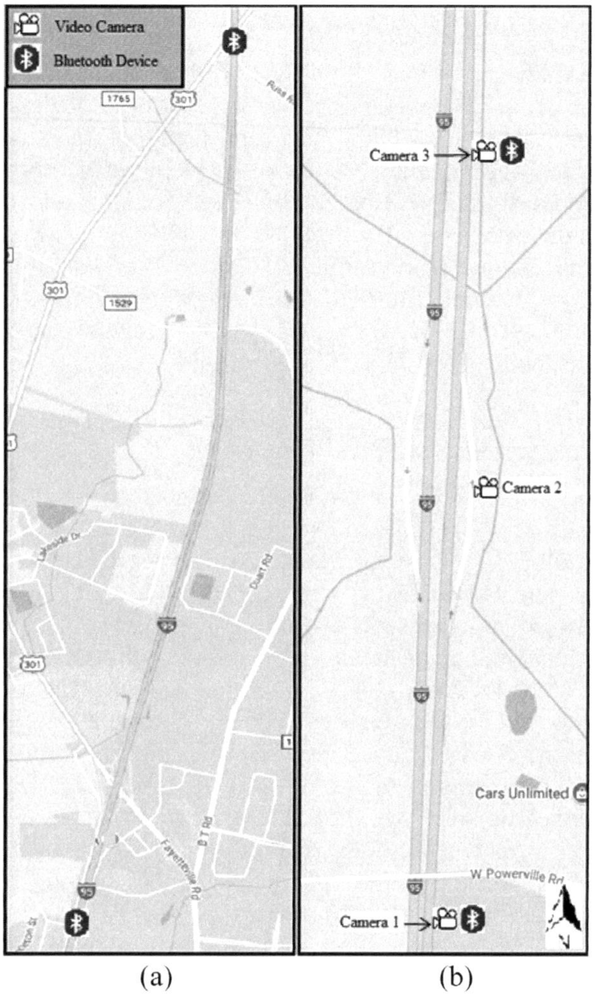

Figure 1. Location of data collection areas around the Lumberton weigh station: (*a*) location of route travel time reliability data collection; (*b*) locations of data collection for microsimulation model calibration and validation.

positioned such that it provided truck travel time from freeway to the weigh station scale and scale service time for trucks could be measured. Location 3 was selected downstream from the weigh station with enough distance for merging trucks to accelerate to normal highway operating speed.

Two of the Bluetooth units were installed at the same location of upstream and downstream cameras. The total length of the route is 0.91 mi. Video data collection started on June 24, 2017 and ended on June 30, 2017. Data was only collected during the operation hours of the weigh station (5:00 a.m. to 6:00 p.m.). Even though data was collected during the weekend, it was not used for the calibration and validation of the simulation model. Data collected during these seven days consisted of: truck service time at the scale, freeway traffic volumes, truck travel times at the weigh station, truck percentage bypassing the weigh station, and lane utilization

Data Processing

The two data sources used in this paper required separate data processing methods, namely Bluetooth data processing and video data processing.

Bluetooth Data Processing

Each Bluetooth sensor unit records one line/record of data for each detection of a Bluetooth device in its vicinity. This record includes a timestamp of the detection, a truncated media access control (MAC) address identifying the device, and the received signal strength indicator (RSSI). As a device passes the sensor unit, it is typically detected many times; for a 100 m Bluetooth detection range, a vehicle traveling at a speed of 65 mph takes about 7 s to pass through a sensor's detection zone. All individual detections are written to a file as separate records. There are two main steps in distilling the raw Bluetooth device detections into travel time records: converting the individual Bluetooth detections into single-time records and matching those records based on their MAC address. To accomplish both these steps, the analysis software BluStats (*7*) was used.

While most of the travel times obtained were representative of the traffic stream, some trips were outliers induced by errors in reading by the sensors or intermediate stops (overweight trucks that had to stop for extended periods of time at weigh station). These travel times needed to be identified and removed from further analysis. A simple non-parametric statistical filter known as IQR4 was used to screen for outlier travel times by the BlueStats software. This statistical filter flags as an outlier any travel time record that is three or more standard deviations away from the mean of the thirty most adjacent travel time records. To approximate the standard deviation, the IQR4 filter uses the inter-quartile range (the difference between the 25th and 75th percentile readings) as an estimate of 0.75 times the standard deviation. This inter-quartile range is then multiplied by four to arrive at the screening buffer value approximately equal to three standard deviations. These travel times were then used for the analysis and generation of travel time distributions and validation of microsimulation model.

Video Data Processing

Data collected by video cameras were exclusively used for the microsimulation model calibration and validation. To determine the travel time for a truck, the images of trucks captured by the video camera at Location 1 were matched with the images at the downstream. Markers were placed on screens used for watching the videos. The time at which a truck passed marker A at Location 1 was read from the video. The time this truck passed Markers B and C at Locations 2 and 3 were also read from the other videos. The travel time for a truck was computed as the difference between the times the truck passed Location 1 and Location 3. If the truck went to the weigh station, the travel time included that stop as well. The setup of camera 2 provided a view of the static scale markers at the weigh station, which helped with collecting static scale service time for trucks.

Analysis and Results

Impact of Static Scale on Travel Time Reliability

The weigh station was operational during 40 days of the study period, operating from 5 a.m. to 6 p.m. daily. During this period, Bluetooth sensors produced a number of Bluetooth device detections. Observed travel times were generated from these detections; while some of these trips were outliers not representative of the traffic stream, the majority of the observations provided useful information. Others, while not outliers, occurred on holidays, weekends, or outside weigh station operational hours.

Plotting the generated travel times against hour of day clearly shows the impact of weigh station on travel time as shown in Figure 2. Figure 2 demonstrates travel times for the route for two different scenarios. The first scenario which is shown in part (*a*) is related to route travel times when the weigh station is open (40 weekdays), and the second scenario illustrated in part (*b*) shows route travel times during periods when the weigh station is closed (7 weekdays). When the weigh station is open two patterns of travel times can be observed. The lower compact band with the majority of travel times within 3–4 min (representing vehicles with minimal impact from the weigh station) and the upper scattered cloud on top of the lower band with travel times greater than 5 min (representing the heavily-impacted vehicles due to operation of the weigh station). However, Figure 2*b* demonstrates only the presence of the lower band when weigh station is closed. This figure provides insight into variability of travel time induced by the operation of weigh station.

The upper scattered cloud of travel times in Figure 2*a* can be intuitively considered to consist primarily of trucks going to the weigh station. Whereas the lower band represents travel times of vehicles that do not go to the weigh station. The majority of these vehicles consist

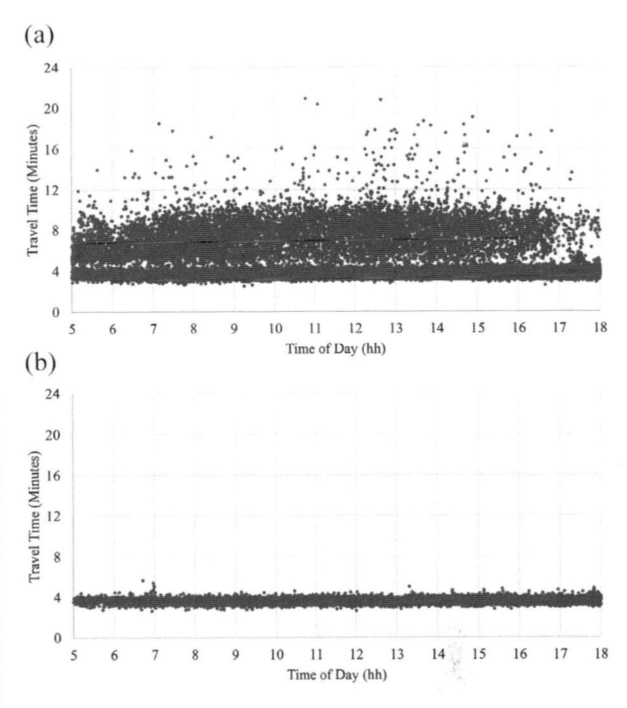

Figure 2. Plot of travel times vs. time of day for periods when (*a*) weigh station is open and (*b*) weigh station is closed.

of passenger cars and trucks that bypass the weigh station due to either temporary closures of weigh station or trucks equipped with NCPass and cleared to bypass the station. NCPass is a transponder system that allows trucks with good compliance and safety records to bypass the weigh station (*8*). Currently, about 1.5% of trucks using the weigh stations in North Carolina are equipped with NCPass.

Furthermore, Bluetooth travel times at the weigh station area were compared to the travel times obtained from I-95's Corridor Coalition Vehicle Probe Project (INRIX) and the Federal Highway Administration's National Performance Management Research Data Set (NPMRDS). The Bluetooth overall travel times were found to be consistent with those obtained from INRIX and NPMRDS.

To quantify the impact of weigh station on truck travel time reliability, truck trips need to be segregated from the rest of the traffic on the freeway. However, there is no way to definitively determine which trips were made by trucks and which were made by passenger cars. Nonetheless, a reasonable approximation was made by fitting a theoretical mixture distribution of two Burr distributions to the empirical travel rate distribution. Figure 3 shows the fit of empirical cumulative distribution function (CDF) and probability density

Table 1. Parameters of Burr Mixture Distributions

	κ (shape)	α (shape)	β (scale)	γ (location)	Mixture percentage
Cars	131.448	4.104	6.100	0.002	78.63%
Trucks	171.279	1.905	1.977	0.749	21.37%

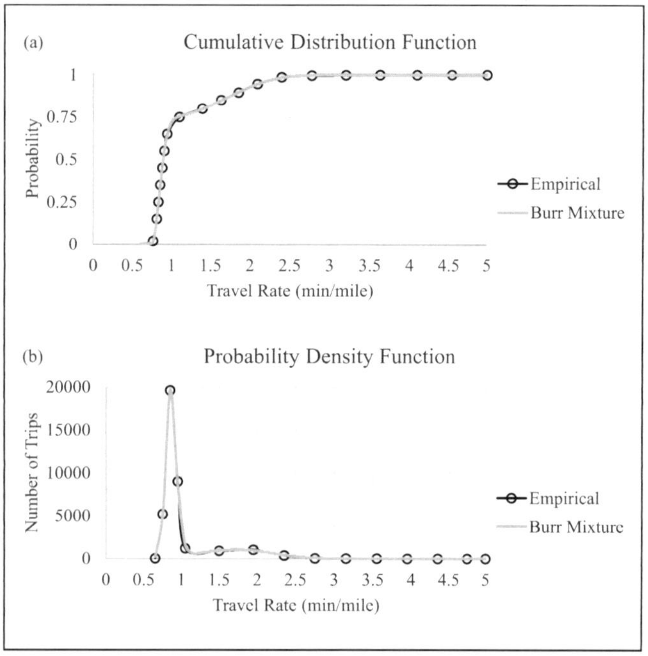

Figure 3. Fit of empirical to theoretical distributions: (*a*) fit to CDF; (*b*) fit to PDF.

function (PDF) to the theoretical distributions. Visual observations show that the Burr mixture fits the data very well.

The theoretical Burr Mixture is composed of two Burr distributions corresponding to population of vehicles significantly impacted by the weigh station (mostly trucks) and vehicles minimally impacted by the weigh station (mostly cars). The lower Burr distribution which corresponds to vehicles minimally impacted (mostly cars) by operation of the weigh station fits the majority of the data shown in the lower band of Figure 2*a*, whereas the upper Burr distribution is related to vehicles heavily impacted (mostly trucks) by the operation of the weigh station and fits the majority of the data points shown on the upper scattered cloud of Figure 2*a*. The two theoretical Burr distributions have the parameters and mixture percentages shown in Table 1.

Table 1 shows that 21.37% of trips detected by Bluetooth units are heavily impacted by the operation of weigh station, the majority of which are trucks. This number may seem high at first but considering the fact that the route is a freight route and the results of the field study (20% trucks on the freeway), this percentage is considered reasonably representative of trucks on the freeway.

While the majority of trucks have travel times that are easily distinguished from cars, there are a small percentage of trucks that have travel times located in the lower band. These trucks are either equipped with NCPass transponders that enables them to bypass the weigh station or they travel the route at times when the demand on the weigh station is high and it is temporarily closed to prevent spillback. During weigh station closures, trucks bypass the weigh station and maintain speeds similar to those of cars. On the other hand, although the majority of car travel times can be distinguished from trucks easily, there are cars that have travel times that can be within the boundaries of trucks entering the weigh station. As a result, the upper tail of minimal impact distribution overlaps with the lower tail of significant impact distribution. A more accurate approximate segregation between cars and trucks in the overlapping region was conducted by creating half-second data bins within the overlapping region. Using area under the curve (probability), each bin's data points were assigned proportionally to either truck or car distributions. Using this approximate segregation of trucks from cars, the truck travel time distributions and travel time performance metrics can be estimated.

To quantify the impact of weigh station, segregated truck travel time distribution and performance measures are compared to the travel time distribution and performance measures obtained during the period when weigh station was not operational (08/23/2015–09/01/2015). It is presumed that trucks and cars maintain travel times very close to each other when the weigh station is closed. This assumption can be validated by referring to Figure 2*b*, where route travel times are similar both for trucks and cars when weigh station is closed. At speed limits of 70 mph, trucks can travel at speeds as high as 85 mph (*9*). However, during the period when weigh station is closed a limited number of trips took place at speeds greater than 85 mph (0.28% of observations). These travel times were considered as outliers and excluded from travel times distribution when the weigh station is closed.

Having the travel times for both scenarios, weigh station open and closed, travel time distributions can be plotted enabling the analyst to compare the travel time

Table 2. Comparison of Reliability Performance Measures for Periods When Weigh Station was Open and Closed

Performance measures	Weigh station open	Weigh station closed	Improvement
Average travel time (minutes)	7.65	3.61	4.04
Buffer time (minutes)	2.82	0.43	2.39
50th percentile TTI	2.01	1.01	1.00
80th percentile TTI	2.45	1.08	1.37
Planning time index	2.82	1.13	1.69
Reliability rating (%)	6.91	99.92	93.00
Standard deviation (minutes)	1.89	0.27	1.62
Misery index	3.27	1.17	2.10

reliability performance measures and get insight into the impact of weigh station on truck travel times. The reliability performance measures compared are those recommended by the Highway Capacity Manual, namely average travel time, buffer time, 50th and 80th percentile travel time indices, planning time index, reliability rating, standard deviation, and misery index (*10*). Table 2 provides a comparison of truck travel time reliability performance measures.

Table 2 shows that when weigh station is open, trucks on average experience more than 100% increase in their travel times compared to their travel time when weigh station is closed (7.65 min vs. 3.61 min). Similarly, buffer time which is an indicator of how much extra travel time a truck would need to budget to ensure arrival on time 95% of the time shows that when weigh station is open an extra 2.39 min should be added to the 0.43min buffer needed when the weigh station is closed.

Index-based reliability measures (50th, 80th, and 95th travel time indices) demonstrate similar trends. When weigh station is open their values are at least twice that of when weigh station is closed. The impact of weigh station is even more pronounced on measures located in the tails of the distribution, such as planning time index and misery index. Furthermore, the values of the standard deviation for the two scenarios are different by more than three times, highlighting the variability of travel times when the weigh station is open. Last, but not least, the reliability rating, which shows the percentage of vehicle miles traveled experiencing a travel time index (TTI) of less than 1.33 for freeways, indicates that when weigh station is open less than 7% of truck miles traveled are reliable, whereas this value is 99.9% when the weigh station is closed.

Impact of WIM on Travel Time Reliability

This section of the paper explores the impact of WIM on travel time reliability of trucks using the VISSIM microsimulation software. VISSIM has built-in vehicle actuated programming (VAP) that enables the user to interact with the software and make necessary decision strategies outside basic simulation model. A weigh station module was developed for the northbound side of the Lumberton weigh station. The system modeled is a mainline WIM with one static scale. Initially, the simulation model consisted of only the static scale for purpose of calibration and validation. Once the model was calibrated and validated, WIM features were added to the model.

The WIM technology was located 0.5 mi upstream of the weigh station. At this location, all trucks need to be in the right-hand lane to be weighed and assigned either clear or report flags. Trucks with clear flags continue at mainline speed and bypass the weigh station. The trucks flagged to report must enter the weigh station unless the queue is full. The weigh station status (open/close) sign is located 600 ft upstream of the weigh station. A presence detector on the weigh station ramp was modeled to emulate the system currently in place. When this detector senses the presence of a truck for more than 8 consecutive seconds, the weigh station closes temporarily to prevent spillback onto the freeway. When the detector is unoccupied, the weigh station is automatically reopened.

Travel time reliability of a weigh station equipped with WIM technology depends on truck demand, scale service time, and WIM bypass percentage. These three factors are considered variables in this study. Truck demand, which depends on the total volume of traffic on the freeway is obtained from field data and varies for each day and hour of the reliability reporting period. Scale service time also is a variable, and its distribution was constructed using field data. A truck going to weigh station is stochastically assigned a scale service time from this distribution.

WIM bypass percentage depends on the combination of WIM accuracy, the percentage of overweight trucks, and the percentage of transponder-equipped trucks. Figure 4 shows how these factors impact the WIM bypass percentage. This study only evaluates the combined impact of these factors on travel time reliability and does not consider their individual impacts separately. As shown in Figure 4, the decision for a truck to either bypass or report is the result of a combination of

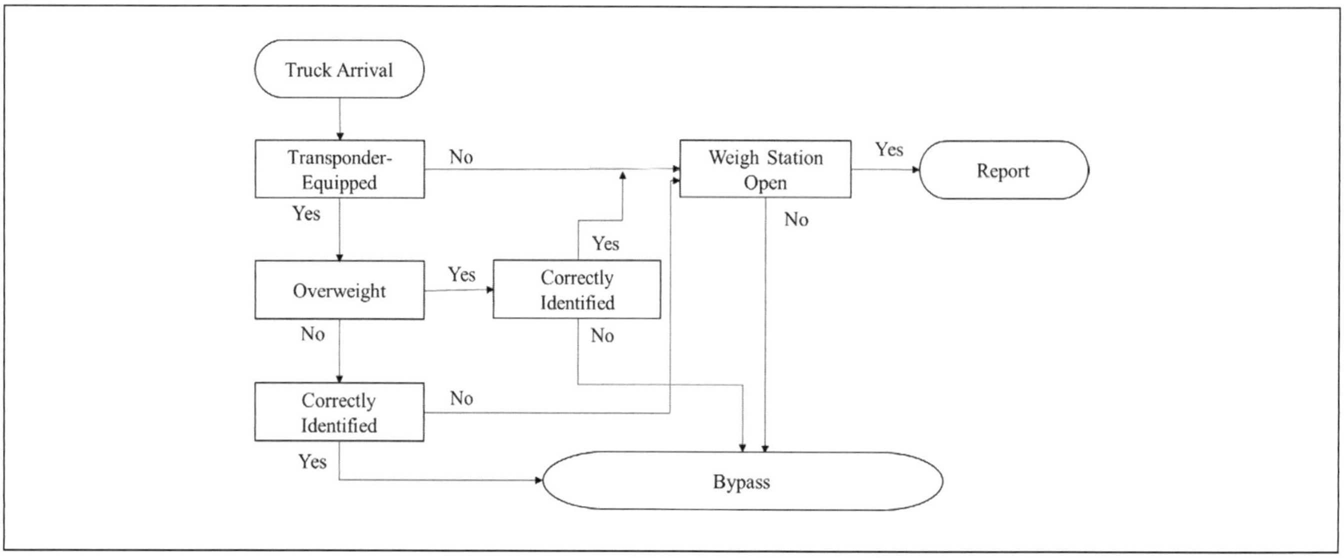

Figure 4. WIM bypass/report logic.

the factors mentioned above. The ten WIM bypass scenarios evaluated in this study span from 10% to 100% in increments of 10%.

Model Inputs and Calibration Factors. The weigh station simulation model was based on actual traffic patterns and geometry data collected at Lumberton weigh station. Following are the input(s) and calibration parameter(s) used in the simulation model:

1. Hourly traffic volume
2. Percentage of trucks in the traffic stream
3. Percentage of trucks equipped with NCPass transponders
4. Look-back distance from route decision points to control lane positioning of trucks (referred to as the lane change distance in VISSIM)
5. Distribution of static scale service time for trucks weighed at the weigh station
6. Freeway field-measured speed distributions for trucks and cars

Model Validation. Mean travel time and speed were used as the key validation parameters to make sure the microsimulation model provides results that are sufficiently close to the ground truth observations. The route was divided into two segments. The first segment consisted of the entire route (4.16 mi) and the second segment, named "weigh station segment", is a partial segment of the entire route with a length of 0.91 mi as shown in Figure 1*b*. The main reason for splitting the route into mentioned segments was the fact that the weigh station partial segment introduces most of the variability in travel time of trucks. Thus, to properly validate the model, it is deemed necessary to both validate the parameters for the entire route and the partial segment where weigh station is located. Data gathered for June 28, 2017 using video cameras and Bluetooth devices were processed and used to validate the model. Table 3 compares the field data with the simulation results, which are obtained from ten 13 h (weigh station operation hours) simulation runs. A 90% confidence level was used in generation of point estimates of means.

Table 3 shows a maximum difference of 8.45% in mean travel time and 7.98% in mean speed between field data and simulation model outputs. What indicates a perfectly calibrated model is still the topic of debate in the simulation modeling community. However, a microsimulation model can be claimed to be successfully calibrated if the mean speed difference between the field measurement and simulation model output is less than 3 mph with at least 90% probability (*11*). Comparison of the field's mean speed with the simulation model's output shows that the maximum difference between the two is 2.09 mph. Thus, it can be claimed that the model is capable of simulating the existing conditions of the freeway.

Travel Time Reliability Results. To obtain the travel time distribution and reliability performance measures for each WIM bypass scenario, a reliability reporting period (RRP) of one month with study period of 13 h (weigh station operation hours) was simulated. Each day of the RRP was simulated 10 times and the output for each day was taken as the average of the ten simulation runs. Thus, with 22 weekdays in a month and 10 WIM bypass levels, a total of 2,200 simulation runs were conducted.

Table 3. Field and Simulation Results

Parameters	Route	Field	VISSIM	Absolute difference
Mean travel time (minutes)	Entire route	7.65	8.17	6.80%
	Weigh station segment	5.01	5.43	8.45%
Mean speed (mph)	Entire route	32.63	30.54	6.41%
	Weigh station segment	10.92	10.05	7.98%

Based on the traffic counts for a week on the freeway, the traffic volumes and volume profile for the rest of the RRP was obtained using Bluetooth detection rates. Bluetooth detection rate is directly related to the volume on the freeway (i.e., as the traffic volume increases, Bluetooth detection rate increases) (*7*). Obtained hourly and daily volumes were used as the input for each day of the RRP in the simulation model. The simulation model introduced these vehicle inputs with stochastic distribution to mimic variability of the traffic stream. However, the percentage of trucks in the traffic stream was held constant at 20% at all times (obtained from field study). Figure 5 shows the simulation output in graphic form and provides a comparison of travel time reliability measures for each scenario of WIM bypass percentage.

Figure 5*a* indicates a reduction in average travel time as the percentage of WIM bypass increases. The decrease is very pronounced until WIM bypass reaches 70%. However, once WIM bypass is above 70%, travel time savings on average get insignificant. With WIM bypass at 100% the average travel time gets close to travel time of freeway when the weigh station was closed.

Interestingly, the buffer time (95th–50th TTI) increases as the WIM bypass percentage increases from 0 to 20 and then starts its descend (Figure 5*b*). This is due to the fact that with WIM bypass less than 20%, the 95th percentile travel time barely changes, while 50th percentile travel time decreases significantly resulting in buffer time going up. However, once WIM bypass is greater than 20%, the value of 95th percentile travel time starts to drop significantly which in turn results in lower buffer times.

The TTI percentile measures (50th, 80th, and 95th) illustrated in Figure 5*c*, *d* and *e*, show significant drops at different levels of WIM bypass percentage. The 50th percentile TTI continuously drops as the percentage of WIM bypass increases from 0 to 50 with most of the drop at 10 and 20 percentage of bypass. However, in the 80th percentile TTI the significant decrease spans from bypass percentage 20 to 50. The 95th percentile TTI begins its significant drop even later at bypass percentage 30 and continues until bypass percentage 90. In addition, misery index shown in part *h* of the figure behaves similar to the 95th percentile TTI due to the fact that both are located at the right tail of the travel time distribution.

Reliability rating demonstrated in Figure 5*f* shows the opposite of what we have been witnessing so far. The value of reliability rating increases as the percentage of WIM bypass trucks increases in the traffic stream, indicating decrease of variability in travel times. In contrast to other performance measures, reliability rating keeps its upward slope even past the 70 percentage. However, its value barely changes when transponder percentage is increased from 0% to 10%.

All reliability performance measures show continuous significant improvement until WIM bypass percentage reaches 70%, beyond which the improvement rates slow down significantly except for the reliability rating. These improvements are indicative that WIM is a promising technology for enhancement of truck travel time reliability at freeway routes with weigh station presence.

Conclusion

This study quantified the impacts of weigh stations on truck travel time reliability. An empirical study comparing periods during which weigh station was operational to periods when weigh station was closed revealed that a static scale weigh station induces a significant degradation in travel time reliability across all travel time reliability performance measures. For the 4.16 mi field study route, truck travel time was observed to be 3.61 min on average during periods when weigh station was not operational (all trucks bypassed the weigh station). This average travel time increased to 7.65 min during periods when the weigh station was open. Similarly, other travel time reliability performance measures saw an increase of at least 100%. The metrics of standard deviation, misery index and buffer time index increased by more than 300%. These degradations in travel time reliability measures underscore the negative impact that weigh station operations have on route travel time reliability. Given the number of weigh stations that a long-haul truck may encounter, these impacts compound and become significantly larger for long-haul service.

Microsimulation results indicate that as WIM bypass percentages grow, travel time reliability performance measures improve. However, at low percentages of WIM bypass where the majority of trucks must report to the weigh station, the improvement is minimal. When this

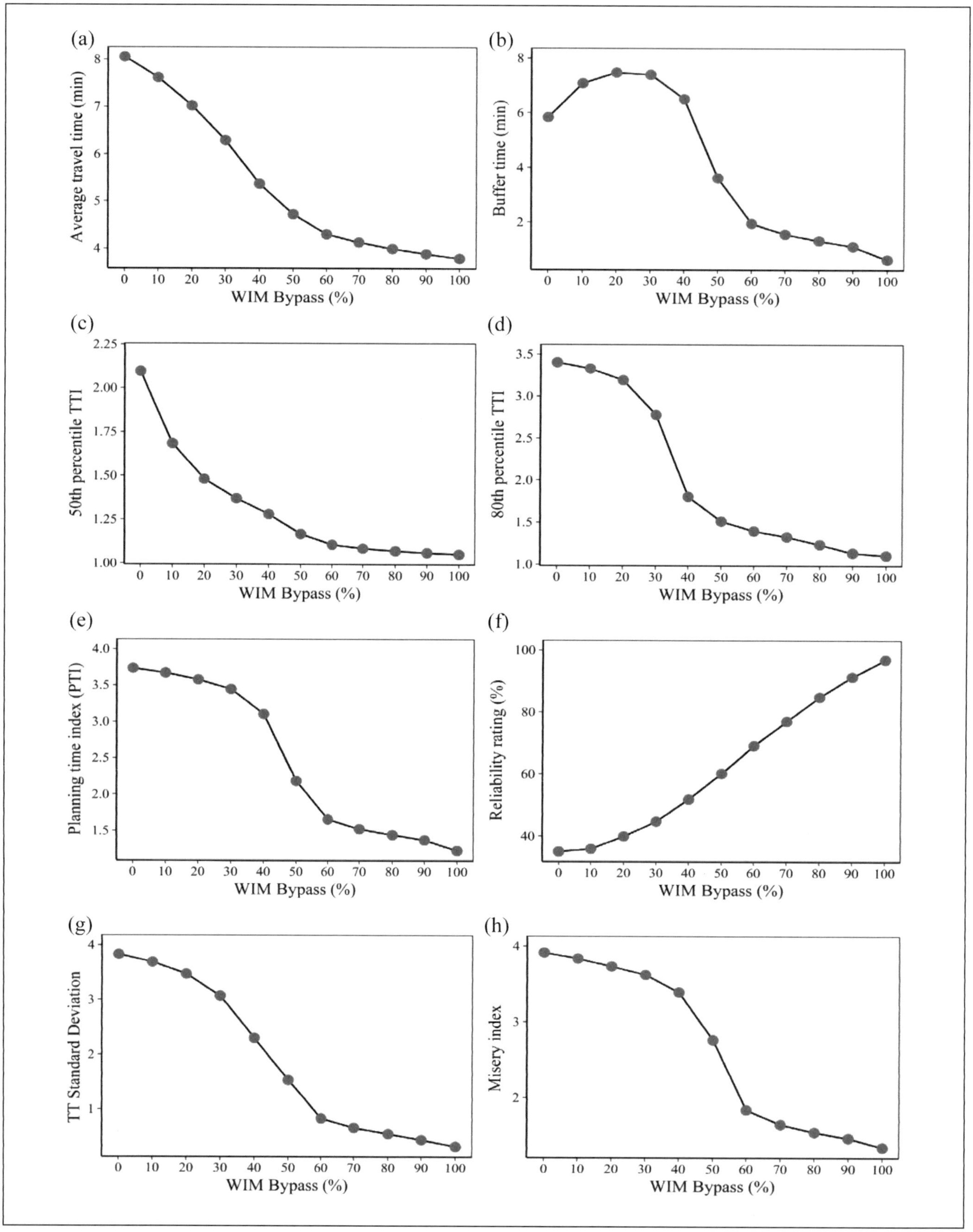

Figure 5. Impact of WIM on travel time reliability performance measures.

percentage surpasses 20%, improvement in travel time reliability performance measures become more evident. As the bypass level approaches 100%, the effects of weigh station on travel time reliability almost fully dissipates with average truck travel time getting closer to what is observed when the weigh station is closed. Overall, the results of the microsimulation model suggest that WIM technology reduces travel time variability and provides benefits to the trucking industry and enforcement agencies.

As discussed above, WIM bypass levels are directly determined by the interplay of three factors, namely WIM accuracy, percentage of transponder-equipped trucks, and percentage of overweight trucks. This study evaluated the combined impact of these factors on travel time reliability by modeling WIM bypass levels directly. Future research is needed to evaluate these factors explicitly and evaluate the relative impact of these factors on WIM bypass percentage and travel time reliability.

Author Contributions

The authors confirm contribution to the paper as follows: study conception and design: Samandar, Williams, Ahmed; data collection: Samandar, Ahmed; analysis and interpretation of results: Samandar, Williams, Ahmed; draft manuscript preparation: Samandar, Williams, Ahmed. All authors reviewed the results and approved the final version of the manuscript.

References

1. Small, K. A., R. Nolan, X. Chu, and D. Lewis. *NCHRP Report 431: Valuation of Travel-Time Savings and Predictability in Congested Conditions for Highway User-Cost Estimation.* TRB, National Research Council, Washington, D.C., 1999.
2. Benekohal, R., Y. El-Zohairy, E. Forrler, and M. Aycin. Truck Delay and Traffic Conflicts around Weigh Stations: A Case Study in Illinois. *Transportation Research Record: Journal of the Transportation Research Board*, 1999. 1653: 52–60.
3. Al-Qadi, I., H. Wang, Y. Ouyang, K. Grimmelsman, and J. Purdy. *LTBP Program's Literature Review on Weigh-in-Motion Systems.* Report No. FHWA-HRT-16-024. Federal Highway Administration, U.S. Department of Transportation, McLean, VA, 2016.
4. Benekohal, R., Y. El-Zohairy, and S. Wang. Truck Travel Time around Weigh Stations: Effects of Weigh in Motion and Automatic Vehicle Identification Systems. *Transportation Research Record: Journal of the Transportation Research Board*, 2000. 1716: 135–143.
5. Kamyab, A., T. Maze, M. Nelson, and B. McCall. Simulation Modeling of Electronic Screening at Weigh Stations. *Transportation Research Record: Journal of the Transportation Research Board*, 1998. 1613: 57–62.
6. Glassco, R. A. *Simulation and Analysis of Three Congested Weigh Stations Using Westa.* Contract No: DTFH61-95-C00040. Federal Highway Administration, U.S. Department of Transportation, McLean, VA, 1999.
7. Traffax Inc. *BlueSTATS.* http://www.traffaxinc.com/content/blustats.html. Accessed July 15, 2017.
8. North Carolina Department of Transportation. *NCPass.* https://www.ncdot.gov/programs/transponder/. Accessed July 10, 2017.
9. Johnson, S., and D. Murray. Empirical Analysis of Truck and Automobile Speeds on Rural Interstates: Impact of Posted Speed Limits. Presented at 89th Annual Meeting of the Transportation Research Board, Washington, D.C., 2010.
10. *Highway Capacity Manual: A Guide for Multimodal Mobility Analysis,* 6th ed. Transportation Research Board, Washington, D.C., 2016.
11. Schroeder, B., K. Salamati, and J. Hummer. Calibration and Field Validation of Four Double-Crossover Diamond Interchanges in VISSIM Microsimulation. *Transportation Research Record: Journal of the Transportation Research Board*, 2014. 2404: 49–58.

The Standing Committee on Freight Transportation Planning and Logistics (AT015) peer-reviewed this paper (18-06365).

Article

TRR
JOURNAL OF THE TRANSPORTATION RESEARCH BOARD

Transportation Research Record
2018, Vol. 2672(9) 130–140

DOI: 10.1177/0361198118783901
journals.sagepub.com/home/trr
SAGE

Reducing Unmet Demand and Spoilage in Cut Rose Logistics: Modeling and Control of Fast Moving Perishable Goods

Xiao Lin[1], Rudy R. Negenborn[1], Mark B. Duinkerken[1], and Gabriel Lodewijks[2]

Abstract

Fresh cut flower supply chains are aware of the need for reducing spoilage and increasing customer satisfaction. This paper focuses on a part of the cut rose supply chain, from auction house to several end customers. A new business mode is considered that would allow end customers to subscribe to florists and have a continuous supply of bouquets of roses. To make this business mode feasible, we propose to benefit from real-time information on roses' remaining vase life. First, a quality-aware modeling technique is applied to describe supply chain events and quality change of cut roses among several supply chain players. Then, a distributed model predictive control strategy is used to make up-to-date decisions for supply chain players according to the latest logistics and quality information. This approach provides a tool for multiple stakeholders to collaboratively plan the logistics activities in a typical cut rose supply chain based on roses' estimated vase life in real time. The proposed approach is compared with a currently used business mode in simulation experiments. Results illustrate that the new business mode and the planning approach could reduce unmet demand and spoilage in a cut rose supply chain.

Cut flowers and foliage are some of the most traded commodities in the world. In 2014 alone, the amount of these perishable goods imported to Europe from developing countries was worth almost €1.2 billion, according to the Centre for the Promotion of Imports from developing countries (*1*). The Netherlands is the biggest cut flower trader in Europe, with an annual total value of €2.9 billion exported to other countries.

A typical international cut flower supply chain consists of growers, forwarders, importers, retailers, and customers. In the Netherlands, Royal FloraHolland, the Dutch auction house, is the main hub for cut flower trades. In 2016, the turnover of cut roses, for instance, had a value of €746 million, making it the largest part of the cut flower trade in Royal FloraHolland (*2*).

Cut roses are highly perishable and vulnerable to high temperatures. Efficiencies of the supply chain are often affected by uncertainties from the weather-dependent perishable nature, as well as from the dynamics of supply and demand (*3*). The nature of this fragile product, the fast moving features of the supply chain, and the complexity of supply–demand interactions create the need for close coordination among different parties within the supply chain.

This paper focuses on the part of the supply chain after the auction house, including wholesalers, florists, and customers. In today's business mode, a common practice of florists and wholesalers is to maintain a certain stock of roses. However, in this business mode, uncertainties of demand and degradation rate often leads to unmet demand or spoilage. To reduce the unmet demand and spoilage, this paper considers a business mode in which customers can subscribe to florists to have a continuous fresh rose supply. When a bouquet at a customer is no longer fresh, another fresh bouquet is supplied to them. Meanwhile, with the quality information of roses made available to suppliers by sensing and communication technologies, the demand from customers can be predicted.

To benefit from the information of demand and quality, the authors propose an approach to schedule logistics activities by combining quality-aware models and distributed model predictive control (DMPC). The quality-aware model is developed to represent the logistics

[1]Department of Maritime and Transport Technology, Delft University of Technology, Delft, The Netherlands
[2]School of Aviation, University of New South Wales, Sydney, NSW, Australia

Corresponding Author:
Address correspondence to Xiao Lin: x.lin@tudelft.nl

activities and quality change that roses are going through during their life cycle. The DMPC strategy is applied to assist wholesalers and florists when making decisions on how many roses to purchase and which of the stocked roses to sell, with the consideration of quality changes due to possible disturbances from environmental factors.

The remainder of this paper is organized as follows. First, a literature review on perishable goods logistics is conducted. Then the specific objectives and contributions of this paper are highlighted. Then the problem statement adopting a system and control perspective is introduced, and the assumptions considered in this approach are explained. Then the quality-aware models to describe the considered part of the rose supply chain, using integer/mixed-integer linear programming formulations are presented. Then the DMPC strategy for achieving the coordination within the supply chain is discussed. Then the proposed modeling approach and control strategy are assessed in simulation experiments. The potential of the proposed business mode in comparison with the current mode of operation is discussed. Lastly, the study is concluded and directions are provided for future research.

Literature Review

The transport of cut roses is a race against time due to their perishable nature. Van Meeteren (*4*) points out that temperature is one of the most important factors in rose quality control, and modeling temperature and quality relations is beneficial for rose supply chains. Tromp et al. (*5*) have examined two different models for rose quality prediction, namely, a kinetics model and a time–temperature model. Through analysis of experiments, they prove that the time–temperature model has practical value in predicting the remaining vase life of roses during transport and storage. To implement this model in rose supply chain optimization and coordination, temperature is the key indicator that needs to be monitored with roses during transport. Since this information can be made available in real time by advancing sensor and communication technologies (*6*), this research focuses on how real-time quality information can be used to increase the effectiveness of supply chain activities and reduce cost and spoilage.

Limited research addresses logistics scheduling in a rose supply network with the consideration of the decreasing quality of fresh products. Verdouw et al. (*7*) made an initial step to address this topic by designing a conceptual framework for the Dutch horticultural supply chain virtualization. The paper proposed that by means of Internet-of-Things technology, a dynamic, connected, intelligent, and quality controlled supply chain can be realized. De Keizer et al. (*8*) performed a quantitative study by designing a network for flower distribution. The research applied a network flow model to represent the amount of flowers being transported from auctions to wholesalers and finally to retailers. Quality aspects were considered with an index of time–temperature summation attached to nodes (locations) and arcs (transportations) in the network model.

Beside De Keizer et al. (*8*), other research also combines flow models with quality degradation for perishable goods in general. Rong et al. (*9*) represent temperature differences using multiple nodes for the same location. Goods at the same location can then be distinguished by putting them to different nodes, resulting in different qualities. Yu and Nagurney (*10*) introduce arc multipliers to represent the change of environmental conditions and durations. De Keizer et al. (*11*) use fractions of a flow to represent goods with different quality categories in the same flow. Although flow models have been proved to be very useful in conventional transport systems (*12*, *13*), they are often not effective when including the change in product quality together with the decisions of allocating products. This is because these models usually need to attach quality attributes to network attributes. Nevertheless, the degradation of perishable goods does not only depend on the location of the goods but also on the environmental and products' internal conditions. In other words, the products' change of location and change of quality are two different series of events, and thus one should not be attached to the other.

Dabbene et al. (*14*) propose a hybrid model, combining event-driven logistics dynamics and time-driven quality dynamics. This approach separates the two characteristics of the product, namely logistics activities and product quality. The representation, however, limits the model application to a fixed supply chain without route choices, rather than a supply network.

In our previous research (*15*), a preliminary quality-aware model was developed to resolve the aforementioned problems. Instead of focusing on network flows, the proposed approach focused on a certain number of perishable goods using a state-space representation, consisting of a quality aspect and a logistics aspect. The two aspects can affect each other but are not dependent on each other. Moreover, the flexibility of the model allows different route options to be considered in the scheduling of logistics activities. In this paper the model is extended in order to describe several different players in a fast-moving supply chain.

Apart from a quality-aware modeling approach, the authors have proposed the application of a model predictive control (MPC) strategy (*16*), an online optimization-based control technique (*17*). MPC updates decision-making in response to real-time information over a given horizon. It has wide applicability in freight transportation systems (*18–20*). For perishable goods logistics, MPC also has great potential in handling the decision-making process, in which quality of goods and environmental factors (such as temperature) can change rapidly. Therefore, decisions obtained from the MPC approach in response to these changes could increase the efficiency of logistics activities and reduce the loss of perishable goods. This paper adopts this strategy to assist the decision-making process for several cut rose supply chain players as price and quality changes from time to time.

Objective and Contributions

The objective of this research is to develop a decision-making strategy for cut rose supply chain players with the consideration of rose quality. This strategy controls the supply chain in a coordinated way to reduce unmet demand and spoilage. By utilizing real time and predicted information on the vase life of roses, the strategy can assist supply chain players to make informed decisions on amounts and time of purchasing.

The contribution of this research is twofold. From the theoretical perspective, the quality-aware modeling approach is extended, so that it can effectively describe logistics activities, quality attributes, and decision-making in a cut rose supply chain. From the practical perspective, it is demonstrated that the quality-aware modeling approach and the DMPC technique are capable of handling decision-making processes for supply chain players in a changing environment.

Problem Statement and Assumptions

This paper considers a cut rose supply network from auction house to end customers. Several involved players are the auction house, a wholesaler, several florists, and customers (see Figure 1). The wholesaler purchases roses from the auction house, which are kept dry at a low temperature. The florists buy roses from the wholesaler and put them in water. Because roses are highly perishable and the demand is often not known in advance, it is difficult for both florists and the wholesaler to determine the number of roses in stock that would result in no loss of sales or spoilage of roses.

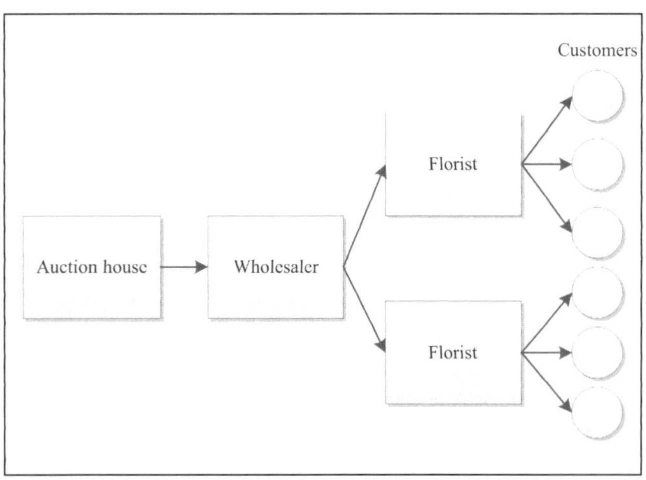

Figure 1. Supply network for cut roses considered in this study.

In this study the end customer is seen as a supply chain player if they prefer to have a continuous supply of rose bouquets. When their roses perish, they will soon need another bouquet. A future scenario is considered where supply chain players adopt sensing technology so that roses' remaining vase life can be predicted. Customers can subscribe to a local florist who keeps their houses decorated with bouquets of roses. Thus the demand can be known to the florist by estimation of the vase life of roses.

Assumptions

The following assumptions are considered in this study:

- Vase life of roses can be estimated and predicted by the wholesaler and florists;
- An auction house has an unlimited amount of flowers that are available to the wholesaler at any time;
- Wholesalers and florists discard roses with low quality according to their standards;
- Wholesalers purchase roses in boxes, and florists purchase roses in bouquets. Each box contains several bouquets;
- Wholesalers can always satisfy demands from florists, while florists only purchase on fixed days per week. A penalty is added to florists if they do not fulfill the demand from customers.

Quality-Aware Model for Cut Rose Supply Chains

This study focuses on the decisions and operations of the wholesaler and florists. Each supply chain player is described in a quality-aware modeling method, consisting of a logistics perspective and a quality perspective.

Logistics and Quality Evolution in a Quality-Aware Model

From the logistics perspective, roses are considered as units (boxes by the wholesaler, bouquets by the florists) that go through different stages of the supply chain, denoted by a directed graph $\mathcal{G} = \{\mathcal{N}, \mathcal{E}\}$. The nodes in collection $\mathcal{N}$ stand for possible locations of each unit, and the directed arcs in collection $\mathcal{E}$ stand for possible transitions between locations. Notation $l_{mi}(k) = 1$ represents that at time step k unit m is at node i. A transition from one node to another takes place if $u_{mij}(k) = 1$, where a decision is made at time step k, that unit m moves from location i to location j at the next time step. Note that every node has a self-directed arc ($i = j$), which allows units to stay at the same location for several time steps.

From the quality perspective, we consider the quality of a unit of rose q_m as its remaining vase life t_{VL}. The vase life can be predicted using a time–temperature model (5) with temperature T and time period in days for storage t_{Day}. The logistics and quality evolution of each unit $m \in \mathcal{M}$ is described as follows:

$$l_{mi}(k) = \begin{cases} 1, & \text{if unit } m \text{ is at node } i \text{ at time step } k, \\ 0, & \text{otherwise,} \end{cases} \tag{1}$$

$$u_{mij}(k) = \begin{cases} 1, & \text{if unit } m \text{ moves from node } i \text{ to node } j \text{ after time step } k, \\ 0, & \text{otherwise,} \end{cases} \tag{2}$$

$$l_{mj}(k+1) = \begin{cases} 1, & \text{if } l_{mi}(\mathrm{k}) = 1 \text{ and } u_{mij}(k) = 1, \\ 0, & \text{otherwise,} \end{cases} \tag{3}$$

$$t_{\mathrm{VL}} = A - \frac{1}{20}(T - 273.15)t_{\mathrm{Day}}, \tag{4}$$

$$\Delta q_i = \frac{1}{20}(T_i - 273.15), \tag{5}$$

$$q_m(k+1) = q_m(k) - \sum_{(i,j)\in\mathcal{E}} \Delta q_i(k) u_{mij}(k). \tag{6}$$

in which Equations 1–3 represent the evolution of logistics attributes (locations and movements) of unit m. Equations 4–6 are the evolution of quality attributes, in which Equations 5 and 6 illustrate that deterioration rate per day Δq_i is related to the temperature T_i at node i. A is the initial vase life, which is estimated by Tromp et al. (5) to be 10 days.

Equations 1–6 describe the two aspects of the dynamics (logistics and quality) of a unit of roses in a quality-aware model. Next, the model is developed specifically for the wholesaler and the florists in cut rose supply chains.

Quality-Aware Logistics Model for Florists

Each florist purchases roses from a wholesaler, stores them in water, and then supplies them to subscribed customers. The graphical representation ($\mathcal{G}_f = \{\mathcal{N}_f, \mathcal{E}_f\}$) of the connections between a florist $f \in \mathcal{F}$ and other parties is shown in Figure 2. In this graph, node $o_f = 1$ represents the location of the wholesaler; node $b = 2$ represents the location of the florist; node $c \in \mathcal{C} = \{3, 4, 5\}$ represents different customers that subscribed to this florist. Node $z_f = 6$ is a virtual node for disposing of roses that are no longer fresh.

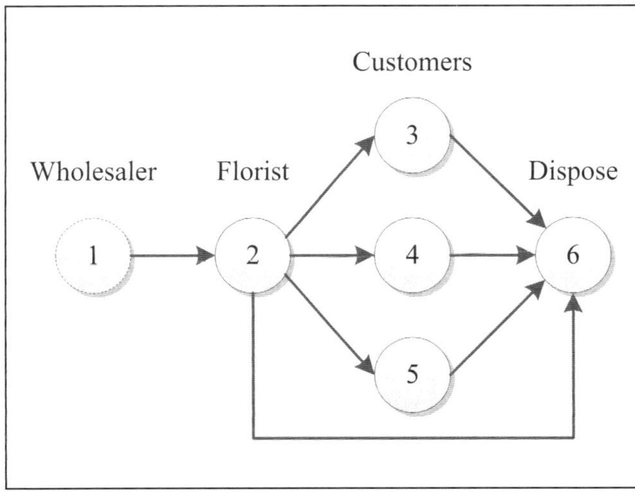

Figure 2. The graph $\mathcal{G}_f$ for a florist.

Now we consider the dynamics of each bouquet unit m in the collection of cut roses $\mathcal{M}_f$ under control of a florist f at step k, and the dynamics of all roses at the florist f at time step k as follows:

$$\sum_{(i,j)\in\mathcal{E}_f} u_{mij}(k) = 1, \forall m \in \mathcal{M}_f, k \in \{1, 2, \ldots\}, \tag{7}$$

$$\sum_{p\in P(i)\cup\{i\}} u_{mpi}(k) = \sum_{j\in S(i)\cup\{i\}} u_{mij}(k+1), \quad \forall m \in \mathcal{M}_f, i \in \mathcal{N}_f, k \in \{1, 2, \ldots\}, \tag{8}$$

$$\sum_{j\in S(i)\cup\{i\}} u_{mij}(1) = l_{mi}(1), \forall m \in \mathcal{M}_f, i \in \mathcal{N}_f, \tag{9}$$

$$\sum_{m\in\mathcal{M}_f} \sum_{i\in P(j)} u_{mij}(k) \leq C_j, \forall j \in \mathcal{N}_f, k \in \{1, 2, \ldots\}, \tag{10}$$

$$q_m(k+1) = q_m(k) - \sum_{(i,j)\in\mathcal{E}_f} \Delta q_i(k) u_{mij}(k), \quad \forall m \in \mathcal{M}_f, k \in \{1, 2, \ldots\}, \tag{11}$$

$$Q(1 - u_{mbc}(k)) \geq q_f^{\mathrm{low}} - q_m(k), \quad k \in \{1, 2, \ldots\}, c \in \mathcal{C}, \tag{12}$$

$$Q(1 - u_{mbb(k))} \geq q_f^{\mathrm{low}} - q_m(k), \forall m \in \mathcal{M}_f, k \in \{1, 2, \ldots\}, \tag{13}$$

$$Q(1 - u_{mcc}(k)) + q_m(k) \geq 0, \quad \forall m \in \mathcal{M}_f, k \in \{1, 2, \ldots\}, c \in \mathcal{C}, \tag{14}$$

$$\sum_{m\in\mathcal{M}_f} u_{mo_f b}(\tau) \leq QB_k^f(\tau), \forall \tau \in \{1, \ldots, N_{\mathrm{P}}\}, k \in \{1, 2, \ldots\}. \tag{15}$$

$$\mathbf{l}_f(k) = \{l_1(k), \ldots, l_m(k), \ldots, l_{M_f}(k)\}^{\mathrm{T}}, \tag{16}$$

$$\mathbf{q}_f(k) = \{q_1(k), \ldots, q_m(k), \ldots, q_{M_f}(k)\}^{\mathrm{T}}, \tag{17}$$

$$\Delta \mathbf{q}_f(k) = \{\Delta q_1(k), \ldots, \Delta q_i(k), \ldots, \Delta q_{N_f}(k)\}^{\mathrm{T}}, \quad (18)$$

$$\mathbf{u}_f(k) = \{u_{111}(k), \ldots, u_{mij}(k), \ldots\}^{\mathrm{T}}, \quad (19)$$

$$\mathbf{X}_f(k) = \{\mathbf{l}_f^{\mathrm{T}}(k), \mathbf{q}_f^{\mathrm{T}}(k)\}^{\mathrm{T}}, \quad (20)$$

$$\mathbf{X}_f(k+1) = g_f\left(\mathbf{X}_f(k), \mathbf{u}_f(k), \Delta \mathbf{q}_f(k)\right), \quad (21)$$

in which Equation 7 ensures that at any time step, the transition of locations of a unit/bouquet m should follow only one arc; Equation 8 makes sure that the unit always follows the directed arcs when being moved ($P(i)$ and $S(i)$ are the collection of the predecessor and successor nodes of node i, respectively); Equation 9 ensures that the decisions for movements of the upcoming time step $k+1$ should be from the current location i at time step k; Constraint 10 guarantees that the number of units at a certain location j at the same time step k does not exceed the capacity of this location C_j; Equation 11 keeps track of the quality of each unit. In the following constraints, Q is a large, positive value. Equations 12 and 13 force roses with a vase life less than q^{low} at florists to be discarded, and thus customers ($c \in \mathcal{C}$) will not receive roses with too low a quality; Constraint 14 describes that roses at customers are discarded when running out of vase life. Constraint 15 describes that the purchases made by florists from the wholesaler follow a pattern B, an array denoting on which in the N_{P} upcoming days the florists visit the wholesaler. For instance, $B_k^f = \{0, 1, 0, 0, 0, 0, 0\}$ represents whether the florist f will have purchasing plans on the following days seen from day k, and the second element in the array ($B_k^f(2) = 1$) indicates that on day $k+2$, florist f has a purchase plan. In this paper, we consider the length of the vector B related to a predictive horizon (N_{P}) determined by the proposed control strategy, which will be discussed later. In Equations 16–21, arrays $\mathbf{l}_f(k)$, $\mathbf{q}_f(k)$, $\Delta\mathbf{q}_f(k)$, $\mathbf{u}_f(k)$ represent locations, qualities, deterioration rates, and decisions of all the units at time step k, respectively. $\mathbf{X}_f(k)$ stands for the current system state at time step k. Function g_f is the state transition of the system of the future system state $\mathbf{X}_f(k+1)$ based on the current state $\mathbf{X}_f(k)$ and decisions of movements $\mathbf{u}_f(k)$. Note that the arrays in $\mathbf{X}_f(k+1)$ can be derived from Equations 3 and 6, respectively.

Quality-Aware Logistics Model for the Wholesaler

The part of the logistics system for the wholesaler w is represented by graph $\mathcal{G}_w$ in Figure 3. The wholesaler receives demand information from the florists and purchases roses from the auction house. The wholesaler also discards roses of low quality according to its own regulation. In $\mathcal{G}_w$ we represent the auction house by node $o_w = 1$, wholesaler by node $s = 2$, dispose by node $z_w = 3$.

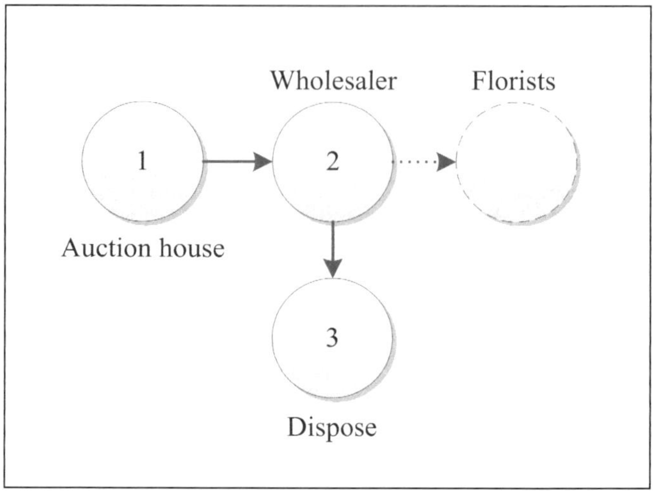

Figure 3. The graph $\mathcal{G}_w$ for the wholesaler.

Note that the wholesaler buys roses in boxes but sells them to florists in bouquets. Therefore, a box of roses can be sold to different florists. We introduce a variable $h_m(k)$, denoting the quantity (of bouquets) of roses left in a box m at the wholesaler, at time step k. Decision variable $r_m(k)$ represents the number of bouquets required by florists from the box m at time step k from the wholesaler.

We now discuss the dynamics of each box of roses $m \in \mathcal{M}_w$ under control of the wholesaler w and the dynamics of all the boxes of roses is demonstrated in a state-space fashion as follows:

$$\sum_{(i,j)\in\mathcal{E}_w} u_{mij}(k) = 1, \forall m \in \mathcal{M}_w, k \in \{1, 2, \ldots\}, \quad (22)$$

$$\sum_{p\in P(i)\cup\{i\}} u_{mpi}(k) = \sum_{j\in S(i)\cup\{i\}} u_{mij}(k+1), \quad \forall m \in \mathcal{M}_w, i \in \mathcal{N}_w, k \in \{1, 2, \ldots\}, \quad (23)$$

$$\sum_{j\in S(i)\cup\{i\}} u_{mij}(1) = l_{mi}(1), \forall m \in \mathcal{M}_w, i \in \mathcal{N}_w, \quad (24)$$

$$q_m(k+1) = q_m(k) - \sum_{(i,j)\in\mathcal{E}_w} \Delta q_i(k) u_{mij}(k), \quad \forall m \in \mathcal{M}_w, k \in \{1, 2, \ldots\}, \quad (25)$$

$$Q(1 - u_{mss}(k)) \geq q_w^{\mathrm{low}} - q_m(k), \forall m \in \mathcal{M}_w, k \in \{1, 2, \ldots\}, \quad (26)$$

$$h_m(k+1) = h_m(k) - r_m(k), \forall m \in \mathcal{M}_w, k \in \{1, 2, \ldots\}, \quad (27)$$

$$\sum_{\tau=0}^{N_{\mathrm{P}}-1} r_m(\tau) \leq h_m(k), \forall m \in \mathcal{M}_w, k \in \{1, 2, \ldots\}, \quad (28)$$

$$r_m(k) \leq Q(u_{mo_ws}(k) + u_{mss}(k)), \forall m \in \mathcal{M}_w, k \in \{1, 2, \ldots\}, \tag{29}$$

$$\sum_{m \in \mathcal{M}_w} r_m(k) = \sum_{f \in \mathcal{F}} d_f(k), \forall k \in \{1, 2, \ldots\}, \tag{30}$$

$$u_{mss}(k) \leq Q\left(h_m(1) - \sum_{\tau=1}^{k} r_m(\tau)\right), \forall m \in \mathcal{M}_w, k \in \{1, 2, \ldots\}, \tag{31}$$

$$\mathbf{l}_w(k) = \{l_{11}(k), \ldots, l_{mi}(k), \ldots\}^{\mathrm{T}}, \tag{32}$$

$$\mathbf{q}_w(k) = \{q_1(k), \ldots, q_m(k), \ldots, q_{M_w}(k)\}^{\mathrm{T}}, \tag{33}$$

$$\Delta\mathbf{q}_w(k) = \{\Delta q_1(k), \ldots, \Delta q_i(k), \ldots, \Delta q_{N_w}(k)\}^{\mathrm{T}}, \tag{34}$$

$$\mathbf{h}(k) = \{h_1(k), \ldots, h_m(k), \ldots, h_{M_w}(k)\}^{\mathrm{T}}, \tag{35}$$

$$\mathbf{u}_w(k) = \{u_{111}(k), \ldots, u_{mij}(k), \ldots\}^{\mathrm{T}}, \tag{36}$$

$$\mathbf{r}(k) = \{r_1(k), \ldots, r_m(k), \ldots, r_{M_w}(k)\}^{\mathrm{T}}, \tag{37}$$

$$\mathbf{d}(k) = \{d_1(k), \ldots, d_f(k), \ldots, d_F(k)\}^{\mathrm{T}}, \tag{38}$$

$$\mathbf{X}_w(k) = \{\mathbf{l}_w^{\mathrm{T}}(k), \mathbf{q}_w^{\mathrm{T}}(k), \mathbf{h}^{\mathrm{T}}(k)\}^{\mathrm{T}}, \tag{39}$$

$$\mathbf{X}_w(k+1) = g_w(\mathbf{X}_w(k), \mathbf{u}_w(k), \mathbf{r}(k), \Delta\mathbf{q}_w(k), \mathbf{d}(k)), \tag{40}$$

in which Equations 22 and 23 are the topology constraints that ensure units appear at one place at each time step and move along the directed arcs of the graph; Equation 24 ensures that the decisions for movements of the upcoming time step $k + 1$ should be from the current location i at time step k; Equation 25 keeps track of the quality of each unit; Inequality 26 keeps the quality of roses at the wholesaler by discarding the ones that have a quality lower than q_w^{low}; Equation 27 explains how the quantity of roses in a box m can change over time; Inequality 28 ensures that the number of bouquets taken from each box m over the horizon N_{P} should be no more than the number of bouquets that are left at each time step k; Inequality 29 ensures that florists can only purchase roses from boxes arriving or stored at the wholesaler; Equation 30 ensures that roses removed from boxes at the wholesaler equal the demand from the florists. Equation 31 ensures that once all the roses are removed from the box, the unit for the box moves to the next stage (dispose). In Equations 32–40, arrays $\mathbf{l}_w(k)$, $\mathbf{q}_w(k)$, $\Delta\mathbf{q}_w(k)$, $\mathbf{h}(k)$, $\mathbf{u}_w(k)$, $\mathbf{r}(k)$, and $\mathbf{d}(k)$ represent locations, qualities, deterioration rates, quantities left in boxes, decisions of all the movements, decisions of all the numbers of bouquets taken from each box, and demand from all the florists at time step k, respectively. $\mathbf{X}_w(k)$ is the current system state for all roses considered at the wholesaler at time step k. Function g_w is the state transition to the future state $\mathbf{X}_w(k + 1)$.

Control Strategy for Real-Time Coordination

The previous section presents the quality-aware models for rose handling at the wholesaler and florists. This section introduces the control strategy for each supply chain player to make optimal decisions, and to coordinate with others in the supply chain. Firstly, the objective functions of controllers at the wholesaler and the florists are introduced. Subsequently, a rotating units method is introduced in order to cope with a fast-moving rose supply chain. Then, control algorithms for each controller and inter-controller communication are described.

Objectives of the Controllers

The controller at each florist f aims to minimize the cost of buying roses and unmet demand. Each bouquet costs α_1 and the penalty for each unmet demand per day is α_2. Similarly, the controller at the wholesaler seeks to minimize the cost of purchasing roses. Each box of roses costs $\beta(k)$, as this price may vary from day to day. An upcoming time period $\{1, \ldots, k, \ldots, N_{\mathrm{P}}\}$ is considered. The objective functions for controllers at florists (J_f) and the wholesaler (J_w) are listed as follows:

$$\min \quad J_f = \alpha_1 \sum_{k=1}^{N_{\mathrm{P}}} \sum_{m \in \mathcal{M}_f} u_{mo_f b}(k) + \alpha_2 \sum_{k=1}^{N_{\mathrm{P}}} \sum_{j \in C} \left(1 - \sum_{m \in \mathcal{M}_f} \sum_{i \in P(j) \cup \{j\}} u_{mij}(k)\right), \tag{41}$$

$$\min \quad J_w = \sum_{k=1}^{N_{\mathrm{P}}} \sum_{m \in \mathcal{M}_w} \beta(k) u_{mo_ws}(k). \tag{42}$$

Rotating Unit Method and Communication between Controllers

In this modeling approach, the number of units in a system needs to be defined before optimization. However, roses are fast-moving goods as they come into and are consumed from the supply chain all the time. This feature requires controllers to be capable of introducing new units and disposing of consumed/spoiled units. A rotating unit method is applied: when a bouquet/box enters a supply chain player, it is registered to a unit with the new attributes (quality and location). When the bouquet/box is disposed of, it is unregistered and the unit moves to the beginning of the supply chain to register as a new bouquet/box of roses. Therefore, given a pre-determined number of units, these units can be "reused" over time, instead of trying to include every future unit in the model. Consider any unit m in the wholesaler/florist, the following operations show how this method is applied:

$$q_{m_f}(k+1) = q_{m_w}(k+1), \text{if} \quad u_{m_f o_f b}(k) = 1 \quad \text{and} \quad r_{m_w} \geq 0, \\ \forall m_f \in \mathcal{M}_f, m_w \in \mathcal{M}_w, f \in \mathcal{F}, k \in \{1, 2, \ldots, N_{\text{P}}\}, \tag{43}$$

$$q_{m_f}(k+1) = q_w^{\text{low}}, \text{if} \quad u_{m_f i z_f}(k) = 1, \\ \forall m_f \in \mathcal{M}_f, f \in \mathcal{F}, i \in P(z_f), k \in \{1, 2, \ldots, N_{\text{P}}\}, \tag{44}$$

$$l_{m_f o_f}(k+1) = 1, l_{m_f z_f}(k+1) = 0, \text{if} \quad u_{m_f i z_f}(k) = 1, \\ \forall m_f \in \mathcal{M}_f, i \in P(z_f), k \in \{1, 2, \ldots, N_{\text{P}}\}, \tag{45}$$

$$q_{m_w}(k+1) = q_{m_w}^{\text{ini}}(k+1), \text{if} \quad u_{m_w o_w s}(k) = 1, \\ \forall m_w \in \mathcal{M}_w, k \in \{1, 2, \ldots, N_{\text{P}}\}, \tag{46}$$

$$l_{m_w o_w}(k+1) = 1, l_{m_w z_w}(k+1) = 0, \text{if} \quad u_{m_w i z_w}(k) = 1, \\ \forall m_w \in \mathcal{M}_w, k \in \{1, 2, \ldots, N_{\text{P}}\}. \tag{47}$$

$$h_{m_w}(k+1) = h^{\text{ini}}, \text{if} \quad u_{m_w i z_w}(k) = 1, \\ \forall m_w \in \mathcal{M}_w, k \in \{1, 2, \ldots, N_{\text{P}}\}. \tag{48}$$

Equation 43 describes the communication of quality between the wholesaler and a florist when purchases happen: the quality of a new bouquet moving into a florist f is updated with the quality of the box from which the bouquet is taken. If the bouquets are taken from more than one box, their qualities should be updated according to the qualities of the boxes. Note that units are indexed here (as m_f and m_w) to discriminate a "box unit" of the wholesaler from a "bouquet unit" from a florist. Equations 44 and 45 show that when a bouquet is disposed of, the unit is unregistered with the bouquet, moved to node 1, and registered with a newly registered minimum acceptable quality from the wholesaler. Although the quality information of the new bouquet is not yet available to the florist before the purchase, it is guaranteed with a minimum quality (ensured by constraint 26). Similarly, wholesalers apply the same principle with the rotating unit method. In Equations 46 and 48, the quality and initial quantity is updated with information from the auction house. Equation 47 takes care of the re-registration of units to new boxes of roses.

Distributed Control Algorithms

We list the algorithms for the control strategies for the florists and the wholesaler as follows.

Algorithm for florist f:

1. Examine the current system state $\mathbf{X}_f(k)$ and the purchase pattern B_k^f.
2. Solve the binary integer linear programming problem with the objective function (Equation 41) and constraints (Equations 7–15). Compute the optimal solution over the horizon: $\left(\hat{\mathbf{u}}_f(k), \ldots, \hat{\mathbf{u}}_f(k+N_{\text{P}}-1)\right)$.
3. Send purchase plans $\left(\sum_{m \in \mathcal{M}_f} u_{m o_f b}(k)\right)$ for the time steps $k+1$ to $k+N_{\text{P}}$ to the wholesaler.
4. After receiving confirmation from the wholesaler, purchase roses from wholesaler with certain qualities, update qualities of newly purchased bouquets in $\mathbf{q}_f(k+1)$ using Equation 43.
5. Execute the decisions $\hat{\mathbf{u}}(k)$ using Equation 21, and rotate units that are disposed of, and register qualities to these units according to Equations 43–45. This results in the new system state $\mathbf{X}_f(k+1)$.

Algorithm for wholesaler w:

1. Examine the current system state $\mathbf{X}_w(k)$.
2. Receive the total demand from florists $\mathbf{d}(k)$ over the prediction horizon.
3. Solve the mixed-integer linear programming problem with the objective function (Equation 42), constraints (Equations 22–31), and the daily update of the price $\beta(k)$. Compute the optimal solution over the horizon: $(\hat{\mathbf{u}}_w(k), \ldots, \hat{\mathbf{u}}_w(k+N_{\text{P}}-1))$ and $(\hat{\mathbf{r}}(k), \ldots, \hat{\mathbf{r}}_M(k+N_{\text{P}}-1))$.
4. Sell roses to florists with certain qualities according to $\hat{\mathbf{r}}(k)$. Update the florists about the qualities of the sold items using Equation 43.
5. Execute the decisions $\hat{\mathbf{u}}_w(k)$ and $\hat{\mathbf{r}}_m(k)$; rotate units that are disposed of; and register qualities to these units according to Equations 46–48. This results in the new system state $\mathbf{X}_w(k+1)$.

In Figure 4, a flowchart is presented to illustrate how and when inter-controller communications are carried out within a control loop at time step k. When florists have generated purchase plans, they send this information to the wholesaler. The wholesaler then decides which roses to sell to the florists. The quality information is updated with the florists, as these bouquets being traded are registered with units at each florist.

Simulation Experiments

In this section, simulation experiments are carried out to compare the effectiveness of the current handling method and the proposed approach. Firstly, the current approach is explained and the parameters of the scenario are introduced. Then, results of simulations are presented and discussed.

To demonstrate the potential of the proposed approach in the business mode, its performance is compared with a simulation of a reference group without the proposed approach. The supply chain players in the reference group apply the strategy of maintaining their stocks at certain levels. This is commonly seen in today's cut flower supply chains. In this paper it is assumed that

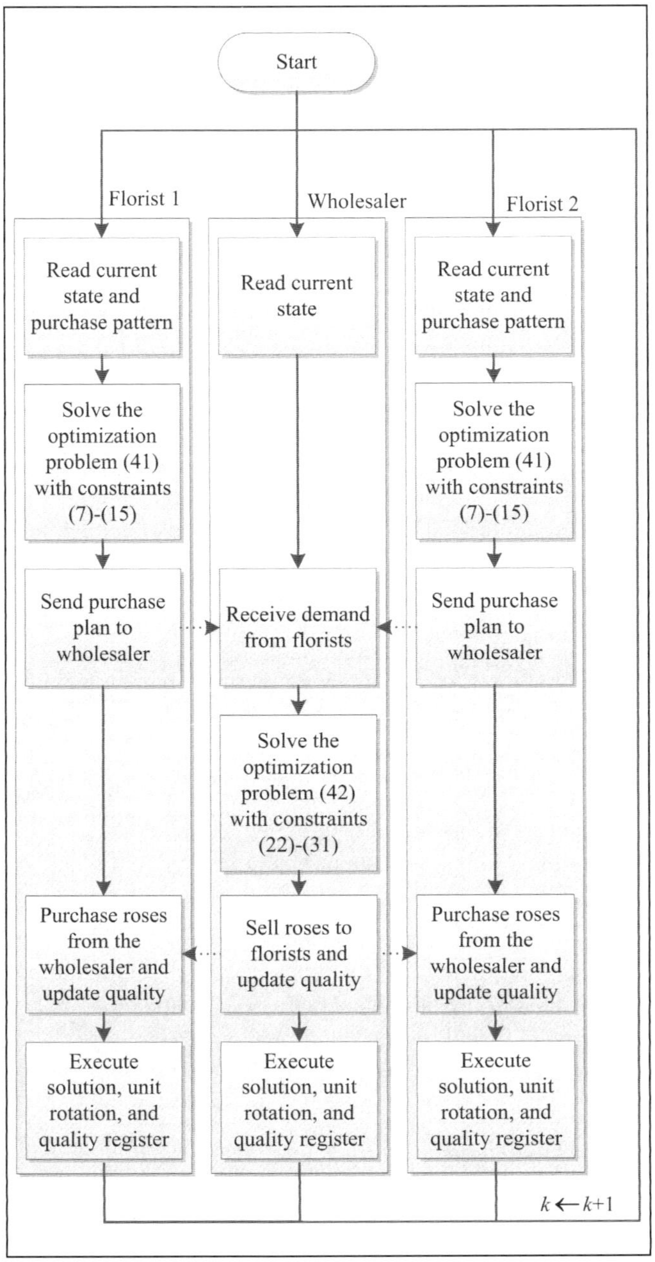

Figure 4. Flowchart of the coordinated control strategy at time step *k*.

florists 1 and 2 aim to maintains their stock at two and three bouquets, respectively; the wholesaler maintains the stock at no less than five bouquets.

Scenario Description

We consider the scenario of one wholesaler, two florists, and six customers (as shown in Figure 1). The parameters and initial system states are given in Table 1. In the scenario, we consider $N_{\text{total}} = 16$ days as the total time period. The purchase pattern of each florist each week stays the same. The price of purchasing a box of roses from auction houses can vary from day to day. We consider $\beta^t(k)$ as the price of day $t + k - 1$ seen from day t (e.g., $\beta^3(1)$ is the price of day 3 seen from day 3). Since the auction price cannot be predicted, the wholesaler uses the average price (€15) as a reference for the predictive controller. Similarly, $\Delta\mathbf{q}_f^t(k)$ and $\Delta\mathbf{q}_w^t(k)$ are the degradation rates for day $t + k - 1$ seen from day t. Initial locations of each unit $\mathbf{l}_f(1)$ and $\mathbf{l}_w(1)$ are denoted as the node numbers of the locations, instead of the binary indicators (e.g., if $l_{mi}(1) = 1$ then the m-th number in the value is i).

The optimization problems are solved by Cplex v12.5.1 in a Matlab 2015b, Windows 7 64-bit environment, on a desktop with Intel Core 2 Q8400 2.66GHz and 4GB RAM. Running the whole program takes only 19 seconds.

Results and Discussion

The results from the current approach and the proposed approach are compared in Table 2. Performance indicators are unmet demand from florists, total cost of the wholesaler buying roses from an auction house, and total number of spoiled bouquets over the whole simulation.

Compared with the current approach, florists in the proposed approach have the number of spoiled bouquets reduced to zero. Unmet demands from subscribed customers are also much lower. The wholesaler has one spoiled bouquet but has fewer costs making purchases from the auction house. Although the auction price per box of roses of the upcoming days is assumed unknown in advance, considering an average price of €15 (β in Table 1) is helpful for the wholesaler in making purchasing plans.

Sensitivity Analysis

A sensitivity analysis was carried out regarding α_1 (the cost of buying from the wholesaler), α_2 (penalty of spoiling cut roses), β (cost of buying from the auction), and uncertainty in prediction of Δq. Experiments were conducted with the adjusted parameter values in different scenarios, with results shown in Table 2. A scenario with a static bid price assumes that the cost of each box from the auction house is static ($\beta^t(k) = 15$, $\forall k \in \{1, \ldots, N_{\text{P}}\}, t \in \{1, \ldots, N_{\text{total}}\}$). A dynamic environment refers to the scenario introduced by Table 1, in which prediction of vase-life change may not be accurate. A static environment assumes that the prediction of quality change is accurate.

Table 1. Scenario Considered in the Experiments

	Parameter	Value
System	N_P	7
	N_{total}	16
	M	$M_f = 12, M_w = 8$
	N	$N_f = 6, N_w = 3$
	q^{low}	$q_f^{low} = 6, q_w^{low} = 8$
	W	1
	F	2
	C	3
	α_1	€5
	α_2	€25
	$\beta^t(k)$	$\beta^t(1) = N\sim(15, 1.5), \beta^t(k)\|_{k\geq 2} = 15, t \in \{1, \ldots, N_{total}\}$
Florist 1	$\mathbf{I}_f(1)\|_{f=1}$	$\{1, 1, 1, 1, 1, 1, 1, 1, 2, 3, 4, 5\}$
	$\mathbf{q}_f(1)\|_{f=1}$	$\{7, 7, 7, 7, 7, 7, 7, 7, 7, 3, 1, 2\}$
	$B_1^f(1)\|_{f=1}$	$\{1, 0, 0, 1, 0, 0, 0\}$
	$\Delta\mathbf{q}_f^t(k)\|_{f=1}$	$\Delta q_i^t\|_{i=1,6}(k) = 0, \Delta q_2^t(k) = 0.8, \Delta q_3^t(k) = 1, \Delta q_i^1\|_{i=4,5}(1) = 1.2$
		$\Delta q_i^{t+1}\|_{i=4,5}(1) = N\sim\left(\Delta q_i^t\|_{i=4,5}(1), 0.1\right), \forall t \in \{1, \ldots, N_{total} - 1\}$
		$\Delta q_4^t(k)\sim N\left(\Delta q_4^t(1), 0.05 + 0.003k\right), \forall k \in \{2, \ldots, N_P\}$
		$\Delta q_5^t(k)\sim N\left(\Delta q_5^t(1), 0.06 + 0.005k\right), \forall k \in \{2, \ldots, N_P\}$
Florist 2	$\mathbf{I}_f(1)\|_{f=2}$	$\{1, 1, 1, 1, 1, 1, 1, 1, 1, 3, 4, 5\}$
	$\mathbf{q}_f(1)\|_{f=2}$	$\{7, 7, 7, 7, 7, 7, 7, 7, 7, 3, 1, 2\}$
	$B_1^f(1)\|_{f=2}$	$\{0, 1, 0, 0, 1, 0, 0\}$
	$\Delta\mathbf{q}_f^t(k)\|_{f=2}$	$\Delta q_i^t\|_{i=1,6}(k) = 0, \Delta q_3^t(k) = 1.1, \Delta q_i^1\|_{i=4,5}(1) = 1.1$
		$\Delta q_i^{t+1}\|_{i=4,5}(1) = N\sim\left(\Delta q_i^t\|_{i=4,5}(1), 0.12\right), \forall t \in \{1, \ldots, N_{total} - 1\}$
		$\Delta q_4^t(k)\sim N\left(\Delta q_4^t(1), 0.05 + 0.003k\right), \forall k \in \{2, \ldots, N_P\}$
		$\Delta q_5^t(k)\sim N\left(\Delta q_5^t(1), 0.06 + 0.005k\right), \forall k \in \{2, \ldots, N_P\}$
Wholesaler	$\mathbf{I}_w(1)$	$\{1, 1, 1, 1, 1, 1, 1, 2\}$
	$\mathbf{q}_w(1)$	$\{9, 10, 10, 10, 10, 10, 10, 10\}$
	$\mathbf{h}(1)$	$\{5, 5, 5, 5, 5, 5, 5, 2\}$
	$\Delta\mathbf{q}_w^t(k)$	$\Delta q_i^t\|_{i=1,3}(k) = 0, \Delta q_2^1(1) = 0.3$
		$\Delta q_2^{t+1}(1)\sim N\left(\Delta q_2^t(1), 0.02\right), \forall t \in \{1, \ldots, N_{total} - 1\}$
		$\Delta q_2^t(k)\sim N\left(\Delta q_2^t(1), 0.01 + 0.002k\right), \forall k \in \{2, \ldots, N_P\}$

When the price for a box of roses from the auction house is static and predictable, the wholesaler only buys roses when necessary, spending less in purchasing roses and resulting in less spoilage. While in reality, the wholesaler needs to compare the current auction price with the average value and buys roses in advance if the auction offers a good deal, which can result in increased buying frequency. In a dynamic environment, when the cost of each bouquet $\alpha_1 = 0$, the wholesaler buys more roses from the auction house and thus more roses flow into this part of the supply network, which results in an increase in spoilage. When $\alpha_1 = 5$, $\alpha_2 = 0.5$, as the penalty for unmet demand is so low, florists would rather pay the penalty instead of purchasing roses to fulfill the demand from customers. If the penalty, on the other hand, is very high (e.g., $\alpha_2 = 500$), the florists will sometimes supply bouquets to the customers before the previous bouquet deteriorates to reduce the possibility of demand being unsatisfied. In addition, experiments in static circumstances result in lower unmet demand, spoilage, and cost, with accurate quality information.

In summary, the results illustrate that this approach of quality-aware modeling and control can largely benefit supply chain players. The application of the new business mode brings wholesalers, florists, and customers into closer coordination. In this way, purchased roses are better used, with reduced prices, unmet demand, and spoilage. A closer analysis shows that a higher penalty for unmet demand or lower price of bouquets may result in a higher flow volume, and more accurate prediction of vase life may help stakeholders make even better decisions in logistics operations.

Conclusions and Future Research

Cut roses are traded and transported in significant amounts across continents. Nevertheless, the highly perishable nature and complexity in supply–demand

Table 2. Comparing Results from the Current Approach with the Proposed Approach and Sensitivity Analysis

Scenarios	Parameters	Florist 1		Florist 2		Wholesaler	
		Unmet	Spoil	Unmet	Spoil	Cost	Spoil
Current approach	$\alpha_1 = 5, \alpha_2 = 25$	2	1	3	3	59	1
Proposed approach	$\alpha_1 = 5, \alpha_2 = 25$	0	0	1	0	41	1
Static bid price	$\alpha_1 = 5, \alpha_2 = 25$	0	0	1	0	30	0
Dynamic environment	$\alpha_1 = 0, \alpha_2 = 25$	0	2	1	5	113	2
	$\alpha_1 = 5, \alpha_2 = 0.5$	36	0	42	0	0	2
	$\alpha_1 = 5, \alpha_2 = 500$	0	0	1	0	57	2
Static environment	$\alpha_1 = 5, \alpha_2 = 25$	0	0	1	0	41	0
	$\alpha_1 = 0, \alpha_2 = 25$	0	2	1	4	83	0
	$\alpha_1 = 5, \alpha_2 = 0.5$	36	0	42	0	0	2
	$\alpha_1 = 5, \alpha_2 = 500$	0	0	1	0	42	0

relations often brings challenges to supply chain players in reducing waste and loss of sales at the same time. The authors consider a business mode that brings end customers into the rose supply chain together with wholesalers and florists. In this mode, customers can subscribe to a florist to receive a continuous supply of fresh bouquets of roses. This business mode can be realized by sensor and communication technologies, which makes remaining vase life known to other players in the supply chain. In order to fully benefit from this business mode, the authors propose a decision-making strategy combining a quality-aware modeling method and a distributed control approach. This strategy is designed to coordinate several players in a fast-moving perishable goods supply chain with uncertainties. Simulation experiments illustrate that the realization of this business mode via the proposed approach could significantly reduce unmet demand and spoilage in a cut rose supply chain.

This paper focuses on a part of a cut rose supply chain from an auction house to the end customers. Future research may include modeling of larger scale supply networks, impacts on long-haul transport, investigation of more detailed and practical extensions in supply chain coordinations, and validation in real-world experiments with different types of horticultural products.

Acknowledgments

This research is supported by the China Scholarship Council under Grant 201406950004, entitled "Controlled cool logistics: Real-time coordination for fresher perishable products."

Author Contributions

The authors confirm contributing to the paper as follows: study conception and design: Lin, Negenborn, Lodewijks; data collection: Lin; analysis and interpretation of results: Lin, Duinkerken, Negenborn; draft manuscript preparation: Lin, Negenborn, Duinkerken. All authors reviewed the results and approved the final version of the manuscript.

References

1. CBI. *CBI Trade Statics: Cut Flowers and Foliage*. The Centre for the Promotion of Imports from developing countries, The Hague, The Netherlands, 2016.
2. Royal FloraHolland. *Royal FloraHolland Annual Report 2016*, 2017. https://www.royalfloraholland.com/en/about-floraholland/who-we-are-what-we-do/factsand-figures/annual-reports. Accessed July 31, 2017.
3. Verdouw, C. N., A. J. M. Beulens, J. H. Trienekens, and T. Verwaart. Mastering Demand and Supply Uncertainty with Combined Product and Process Configuration. *International Journal of Computer Integrated Manufacturing*, Vol. 23, No. 6, 2010, pp. 515–528.
4. Van Meeteren, U. Causes of Quality Loss of Cut flowers - a Critical Analysis of Postharvest Treatments. *IX International Symposium on Postharvest Quality of Ornamental Plants*, Vol. 847, 2008, pp. 27–36.
5. Tromp, S., R. G. M. van der Sman, H. M. Vollebregt, and E. J. Woltering. On the Prediction of the Remaining Vase Life of Cut Roses. *Postharvest Biology and Technology*, Vol. 70, 2012, pp. 42–50.
6. Abad, E., F. Palacio, M. Nuin, A. G. De Zarate, A. Juarros, J. M. Gómez, and S. Marco. RFID Smart Tag for Traceability and Cold Chain Monitoring of Foods: Demonstration in an Intercontinental Fresh fish Logistic Chain. *Journal of Food Engineering*, Vol. 93, No. 4, 2009, pp. 394–399.
7. Verdouw, C. N., A. J. M. Beulens, and J. G. A. J. van der Vorst. Virtualisation of Floricultural Supply Chains: A Review from an Internet of Things Perspective. *Computers and Electronics in Agriculture*, Vol. 99, 2013, pp. 160–175.
8. De Keizer, M., R. Haijema, J. M. Bloemhof, and J. G. A. J. van der Vorst. Hybrid Optimization and Simulation to Design a Logistics Network for Distributing Perishable Products. *Computers and Industrial Engineering*, Vol. 88, 2015, pp. 26–38.
9. Rong, A., R. Akkerman, and M. Grunow. An Optimization Approach for Managing Fresh Food Quality Throughout the Supply Chain. *International Journal of Production Economics*, Vol. 131, No. 1, 2011, pp. 421–429.
10. Yu, M., and A. Nagurney. Competitive Food Supply Chain Networks with Application to Fresh Produce.

European Journal of Operational Research, Vol. 224, No. 2, 2013, pp. 273–282.
11. De Keizer, M., R. Akkerman, M. Grunow, J. M. Bloemhof, R. Haijema, and J. G. A. J. van der Vorst. Logistics Network Design for Perishable Products with Heterogeneous Quality Decay. *European Journal of Operational Research*, Vol. 262, No. 2, 2017, pp. 535–549.
12. Li, L., R. R. Negenborn, and B. de Schutter. Receding Horizon Approach for Container Flow Assignment in Intermodal Freight Transport. *Transportation Research Record: Journal of the Transportation Research Board*, 2014. 2410: 132–140.
13. Chen, C.-C., Y.-H. Tsai, and P. Schonfeld. Schedule Coordination, Delay Propagation, and Disruption Resilience in Intermodal Logistics Networks. *Transportation Research Record: Journal of the Transportation Research Board*, 2016. 2548: 16–23.
14. Dabbene, F., P. Gay, and N. Sacco. Optimisation of Fresh-food Supply Chains in Uncertain Environments, Part I: Background and Methodology. *Biosystems Engineering*, Vol. 99, No. 3, 2008, pp. 348–359.
15. Lin, X., R. R. Negenborn, and G. Lodewijks. Towards Quality-Aware Control of Perishable Goods in Synchromodal Transport Networks. *Proc., 5th IFAC Conference on Sensing, Control and Automation Technologies for Agriculture*, Seattle, Washington, 2016, pp. 132–137.
16. Lin, X., R. R. Negenborn, and G. Lodewijks. Survey on Operational Perishables Quality Control and Logistics. *Proc. of the 6th International Conference on Computational Logistics*, Delft, The Netherlands, 2015, pp. 398–421.
17. Negenborn, R. R., and J. M. Maestre. Distributed Model Predictive Control: An Overview and Roadmap of Future Research Opportunities. *IEEE Control Systems*, Vol. 34, No. 4, 2014, pp. 87–97.
18. Zheng, H., R. R. Negenborn, and G. Lodewijks. Closed-Loop Scheduling and Control of Waterborne AGVs for Energy-Efficient Inter Terminal Transport. *Transportation Research Part E: Logistics and Transportation Review*, Vol. 105, 2016, pp. 261–278.
19. Nabais, J. L., R. R. Negenborn, R. B. Carmona-Benítez, L. F. Mendonça, and M. A. Botto. Hierarchical MPC for Multiple Commodity Transportation Networks. In J. Maestre, R. R. Negeborn, eds. *Distributed Model Predictive Control Made*. Springer, Dordrecht, 2013, pp. 535–552.
20. Xin, J., R. R. Negenborn, and G. Lodewijks. Energy-efficient Container Handling using Hybrid Model Predictive Control. *International Journal of Control*, Vol. 88, No. 11, 2015, pp. 2327–2346.

The Standing Committee on Freight Transportation Planning and Logistics (AT015) peer-reviewed this paper (18-06644).

TRR
JOURNAL OF THE TRANSPORTATION RESEARCH BOARD

Article

Transportation Research Record
2018, Vol. 2672(9) 141–151

DOI: 10.1177/0361198118788185
journals.sagepub.com/home/trr
SAGE

Evaluation of Commercial Truck Parking Detection for Rest Areas

Wei Sun[1], Ethan Stoop[1], and Scott S. Washburn[1]

Abstract

Florida's interstate rest areas are heavily utilized by commercial trucks for overnight parking. Many of these rest areas regularly experience 100% utilization of available commercial truck parking spaces during the evening and early-morning hours. Being able to communicate availability of commercial truck parking space to drivers in advance of arriving at a rest area would reduce unnecessary stops at full rest areas as well as driver anxiety. In order to do this, it is critical to implement a vehicle detection technology to reflect the parking status of the rest area correctly. The objective of this project was to evaluate three different wireless in-pavement vehicle detection technologies as applied to commercial truck parking at interstate rest areas. This paper mainly focuses on the following aspects: (a) accuracy of the vehicle detection in parking spaces, (b) installation, setup, and maintenance of the vehicle detection technology, and (c) truck parking trends at the rest area study site. The final project report includes a more detailed summary of the evaluation. The research team recorded video of the rest areas as the ground-truth data and developed a software tool to compare the video data with the parking sensor data. Two accuracy tests (event accuracy and occupancy accuracy) were conducted to evaluate each sensor's ability to reflect the status of each parking space correctly. Overall, it was found that all three technologies performed well, with accuracy rates of 95% or better for both tests. This result suggests that, for implementation, pricing, and/or maintenance issues may be more significant factors for the choice of technology.

Rest areas along Florida's interstate routes are heavily utilized by commercial trucks for overnight parking. Many of the rest areas regularly experience over-capacity utilization of the commercial truck parking spaces during the evening and early-morning hours (*1*). Currently, information on the number of available commercial truck parking spaces at Florida rest areas is not available remotely. Thus, aside from obtaining this information through direct communication with someone already on site, a truck driver's only sure way of determining if parking is available at a rest area is to stop at the rest area. Being able to communicate the availability of commercial truck parking space to drivers in advance of arriving at a rest area would reduce unnecessary stops at full rest areas as well as driver anxiety. In order to do this, it is critical to implement vehicle detection technology to reflect the parking status of the rest area correctly.

The objective of this project was to evaluate available vehicle detection technologies as applied to commercial truck parking at interstate rest areas in the state of Florida. This evaluation addressed the following aspects: (a) accuracy of the vehicle detection in parking spaces, and (b) installation, setup, and maintenance of the vehicle detection technology. Also described in this paper are trends in truck parking at the rest area study site. The final project report includes a more detailed summary of the evaluation (*2*).

In 2011, the Federal Motor Carrier Safety Administration (FMCSA) conducted field operations tests on parking detection technologies for commercial truck parking, which evaluated the performance of magnetic sensors (*3*). With the development of vehicle parking detection technologies, radar technology has begun to be used in vehicle parking detection sensors. In 2014, the San Francisco Municipal Transportation Agency conducted a parking sensor technology performance evaluation focusing on on-street parking vehicles, which evaluated the performance of sensors of various technologies, including magnetometer, radar, infrared, and image recognition (*4*). However, parking detection sensor applications for passenger vehicles do not necessarily fully translate to commercial vehicle parking situations.

The testing performed for the subject study was conducted during the summertime. Thus, the high heat and rainy conditions of Florida's summer months provided test conditions on one end of the weather spectrum.

[1]Department of Civil and Coastal Engineering, University of Florida, Gainesville, FL

Corresponding Author:
Address correspondence to Wei Sun: w.sun2014@ufl.edu

Vehicle Detection Technologies

Several vendors sell a product, or products, that are designed solely for the purpose of vehicle parking detection or a multipurpose detector that they claim can be applied to parking situations. After reviewing the literature and vehicle detection technology vendor web sites, as well as discussing with vendor representatives the feasibility of participation, the following three vendors/products were included in the project:

1. NEDAP/IPsens/SENSIT (*5*, *6*)
2. Sensys (*7*)
3. CivicSmart (*8*)

The vehicle detection technologies above feature wireless in-pavement sensors, which are placed in the pavement in an array at each parking space and detect the presence of a vehicle as it parks within a boundary around them. The sensor technologies and operating frequencies for the three vendors are:

- SENSIT has magnetic and infrared sensors with operating frequency of 902–928 MHz
- Sensys has microwave radar sensors with operating frequency of 2.400–2.4835 GHz
- CivicSmart has microwave radar sensors with operating frequency of 2.405–2.480 GHz

Field Testing of the Parking Detection Technologies

Test Site

The vehicle parking technology was tested at two rest areas within Florida Department of Transportation (FDOT) District 2. The sites are located in FDOT rest area facilities 20161 (I-75 northbound) and 20162 (I-75 southbound) in Columbia County, Florida, at milepost 413 of I-75 (GPS coordinates: 29.978335, −82.57862). Aerial views of this site are provided in Figure 1.

The truck parking spaces at the study site are 82 ft long and 15 ft wide, and oriented at a 30-degree angle. The detection of truck parking required at least two sensors per parking space, due to the various lengths of trucks.

Installation of Vehicle Detection Technologies

Each of the three vendors was assigned a 10-parking space section on either the northbound side or southbound side of the rest area. To avoid possible interference among the signals of the different vendors, vendors with similar operating frequencies were not located on the same side. The number of sensors to install per parking space was left to the vendors' judgement (CivicSmart: 2 sensors per space, SENSIT and Sensys: 3 sensors per space).

The installation process for all three vehicle detection technologies was similar: (a) determine the desired sensor locations, (b) drill holes the size of the sensors into the pavement and place sensors in them, (c) apply epoxy glue to secure the sensors, (d) mount the relay node and data collector devices to lighting poles in the parking lot. Figures 2 and 3 show the installed wireless in-pavement sensors and supporting devices. The data collector devices transmit the sensor data to the vendors' backend systems, which had the capability to monitor sensor status, store and analyze the data, and provide alerts on detector failures and/or anomalous readings.

Video Data Collection

This project used video to record the truck parking area as ground-truth data—one camera for each vendor's section of 10 parking spaces. The cameras were mounted to the top of light poles in the rest area. Figures 4 and 5 show two of the respective camera fields of view.

Once the video equipment was installed, and the vendors had finished calibrating their sensors, the data collection commenced. Since the start dates of the vendors were different, the amount of data collected for each vendor was different, but at least one month of video data for each vendor was collected. The amount of data used for the accuracy tests was a function of several factors: the truck parking demand, the accuracy rate of the parking detection equipment, and the desired statistical confidence range of the measured accuracy rate. Video was recorded on the following dates:

- Northbound rest area (SENSIT and CivicSmart): August 10–October 5, 2016
- Southbound rest area (Sensys): August 24–October 5, 2016

Video Data Reduction

The ingress and egress movements of each vehicle were recorded manually from the video data. These data items included the following information for each entering/exiting truck:

1. parking space number
2. time entered space
3. time exited space
4. vehicle type (truck with no trailer, truck with vehicle trailer, truck with flatbed trailer, truck with closed trailer, truck with double closed trailer, truck with tanker trailer, single unit truck, recreational vehicle, and other: usually personal car).

Data Analysis Software Tool

The research team developed a data analysis software tool to process and analyze the data. The manually recorded video

Figure 1. Aerial view of the two rest areas: (*a*) southbound rest area and (*b*) northbound rest area.
(Source: Google Maps).

Figure 2. Installed in-pavement sensors (from left to right: SENSIT, Sensys, and CivicSmart).

Figure 3. Installed relay node and data collection devices.

Figure 4. Camera view of northbound rest area.

data were saved into comma-separated value (CSV) formatted files. The vendors' parking sensor data was provided to the research team also in CSV-file format. The software tool could load the CSV-formatted files of both video and parking sensor data and conduct the accuracy tests.

As shown in Figures 6–9, the software tool could generate four kinds of charts for truck parking analysis: (a) parking status chart, (b) interval number of parking chart, (c) interval occupancy chart, and (d) parking duration chart. The parking status chart shows the parking space occupancy status for any specific time. The interval number of parking chart shows the number of vehicles parked at each parking space for any time interval. The interval occupancy chart shows the occupancy percentage of each parking space for any time interval. The parking duration chart shows the occupancy status of each parking space for any day (24 h).

Analysis of Ground-Truth Data

The research team conducted a limited analysis of the ground-truth data regarding the parking vehicle composition, time of day of parking space utilization, and parking duration. Note that the following analysis was based on the 30 parking spaces studied, not the entire rest area.

As shown in Figure 10, approximately 88% of the vehicles parking in the truck parking area were trucks, approximately 5% were RVs, and approximately 7% were other vehicles, such as personal cars, motorcycles, etc.

Figure 11 shows the parking space utilization of one specific week, and Figure 12 shows the parking duration for overnight parking and daytime parking. The rest area was highly occupied during the nighttime, with an average of approximately 90% parking space utilization for most days from 10:00 p.m. to 7:00 a.m. the next morning. The average

Figure 5. Camera view of southbound rest area.

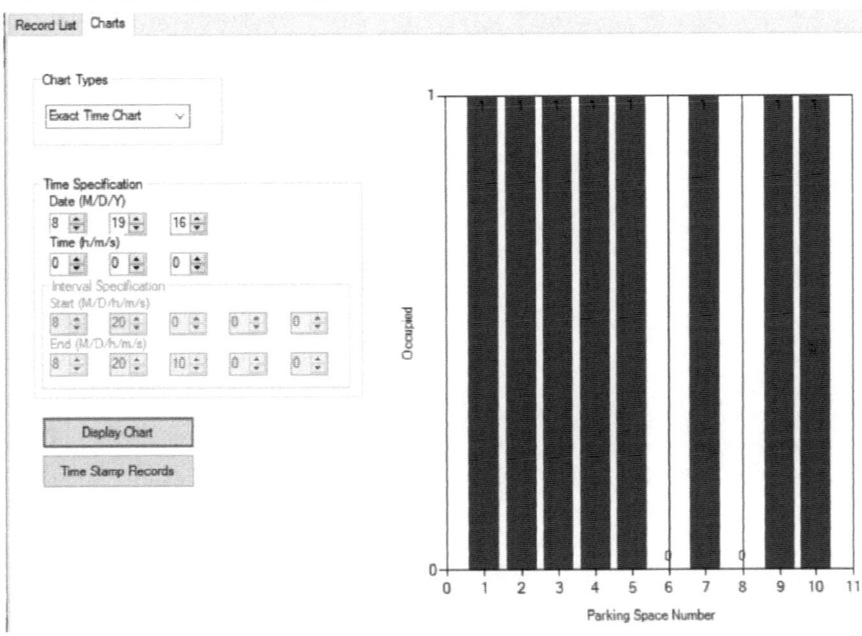

Figure 6. Parking status chart.

duration of overnight parking per truck was approximately 10 hours. During the daytime, the rest area was less occupied, with an average of approximately 30% parking space utilization and the parking events were mostly short-term parking, with an average parking duration of approximately 30 minutes.

Accuracy Tests

The research team obtained raw data collected by the vehicle sensors and compared them with the ground-truth data for the accuracy tests. SENSIT sensor data were collected from the company's backend system, the Truck Parking Manager. Sensys sensor data were collected from the company's backend system, the Sensys Networks Archive Proxy and Statistics (SNAPS) system. CivicSmart sensor data were collected from the company's backend system, the Parking Enterprise Management System (PEMS). These CSV-formatted raw data files from the sensors can be loaded into the data analysis software tool developed by the research team.

Two accuracy tests were applied for evaluation of the parking detection technologies: turnover accuracy test and occupancy accuracy test. The turnover accuracy test evaluated the sensor's ability to identify parking events (parking

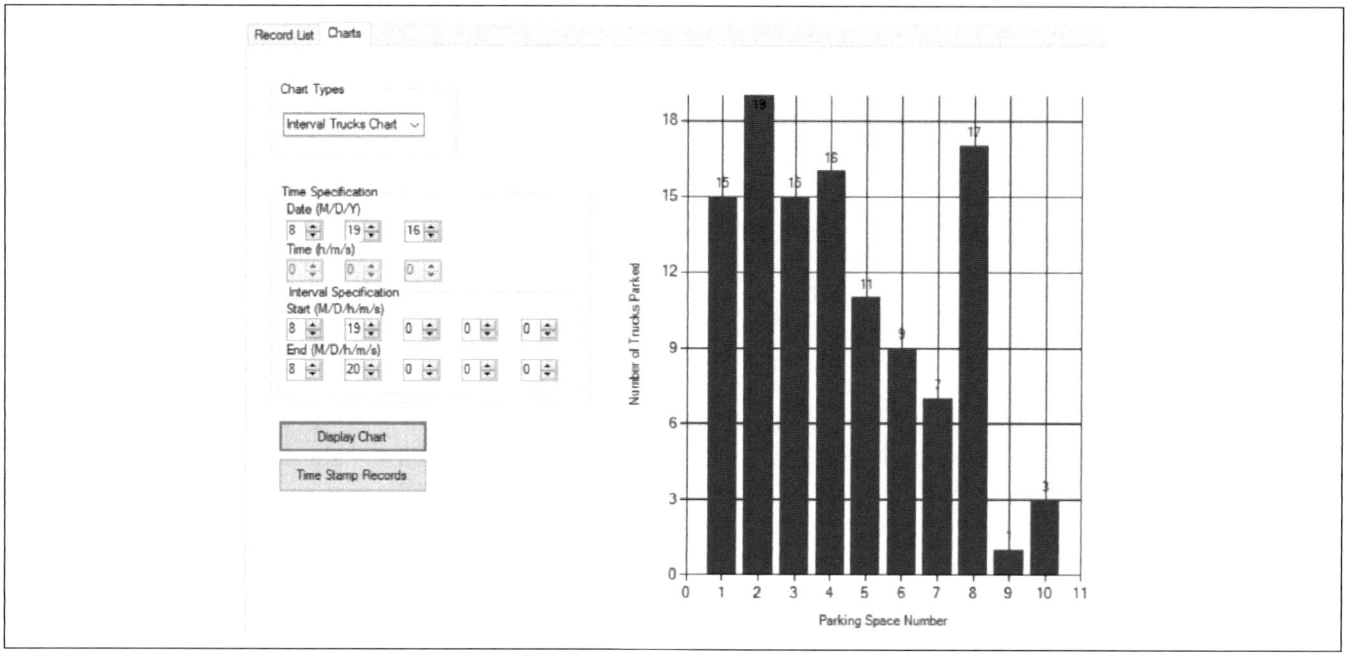

Figure 7. Interval number of parking chart.

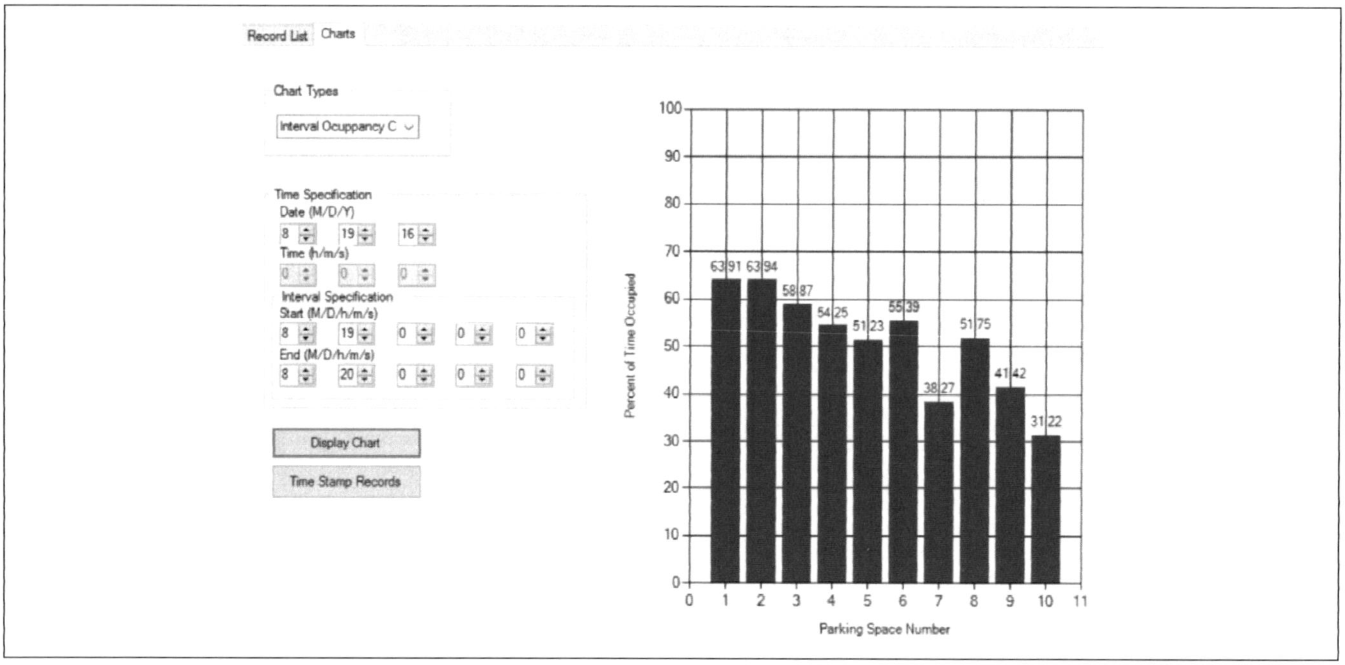

Figure 8. Interval occupancy chart.

ingress or egress), the accuracy rate was the percentage of ground-truth parking events that can be captured by the sensors. The occupancy accuracy test evaluated the percentage of time in which the sensors report the status of the parking spaces (vacant or occupied) correctly.

In this project, each parking space had two or three sensors. For all vendors, the method used to identify an occupied parking space was that at least one of the detectors was at "on" status (vehicle over or near the detector). Conversely, if all the sensors in the space were in the "off" position, the parking space was considered vacant. The parking space status obtained from the sensor was compared with the ground-truth data for accuracy tests. In this case, the accuracy included turnover/occupancy by any type of vehicle parked in the parking space.

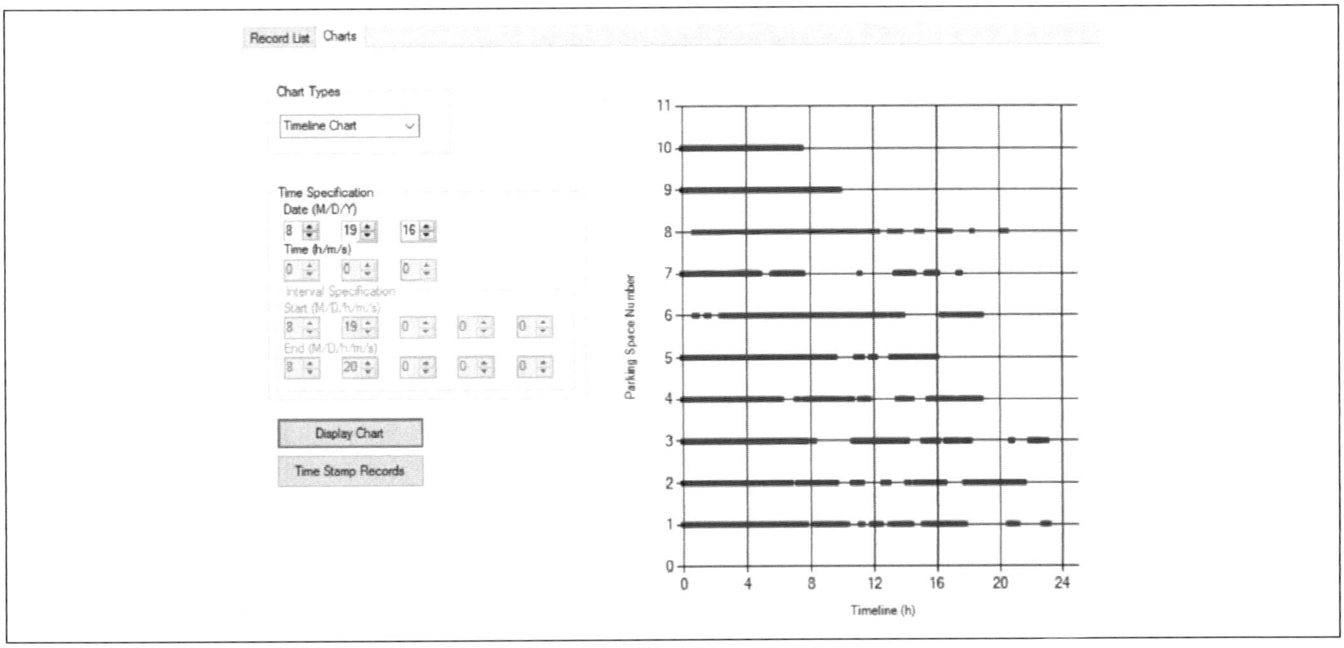

Figure 9. Parking duration chart.

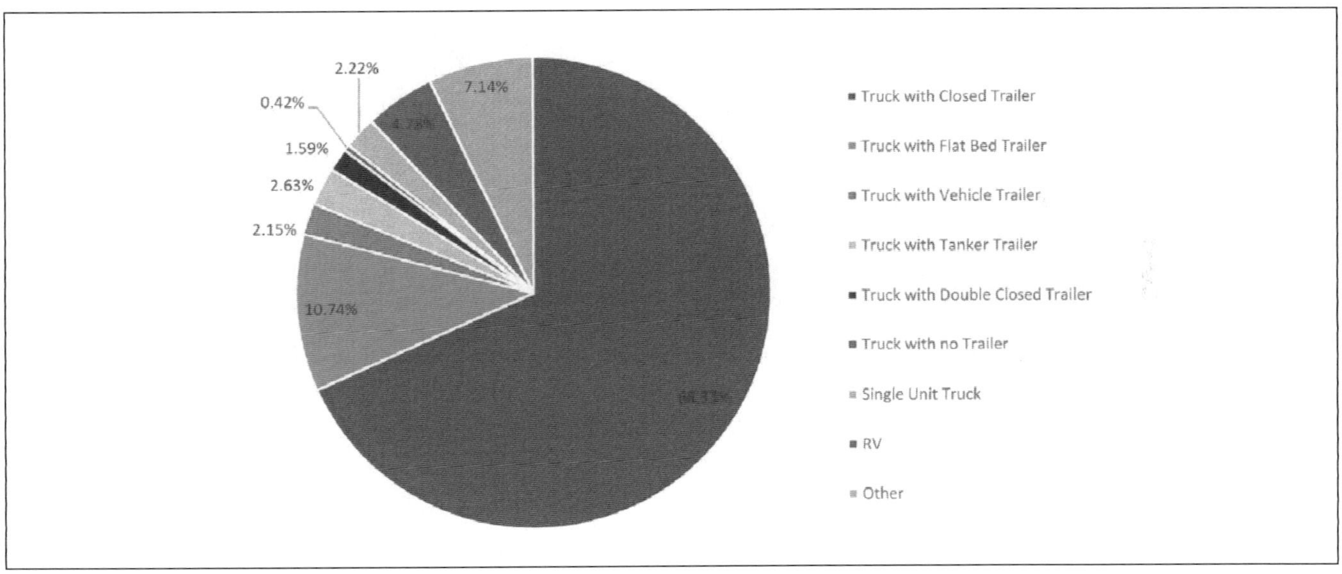

Figure 10. Parking vehicle composition.

Table 1 shows the accuracy test results for the three vendors' sensors. Overall, all three sensors performed well, with an average of over 95% turnover accuracy and 97% occupancy accuracy. The turnover ratio (average number of parking events per hour) was also calculated for each vendor, as an informational item to give an indication of the amount of parking activity in each vendor's set of assigned spaces. This measure should not be interpreted as a reflection of the accuracy of the respective systems. Given that truck drivers typically park in the closest available space to the entrance to the rest area, these turnover ratio values matched our expectations since the SENSIT detectors were closer to the entrance of the rest area than the Sensys and CivicSmart detectors.

Table 2 shows the 95% confidence interval of the turnover accuracy for each detection technology. The confidence interval was calculated based on the number of parking events (n) and the turnover accuracy ($\hat{p}$), following a binomial distribution, as shown in Equation 1.

$$\text{Confidence Interval} = \hat{p} \pm z_{\alpha/2}\sqrt{\frac{\hat{p}\left(1-\hat{p}\right)}{n}}, \alpha = 5\% \quad (1)$$

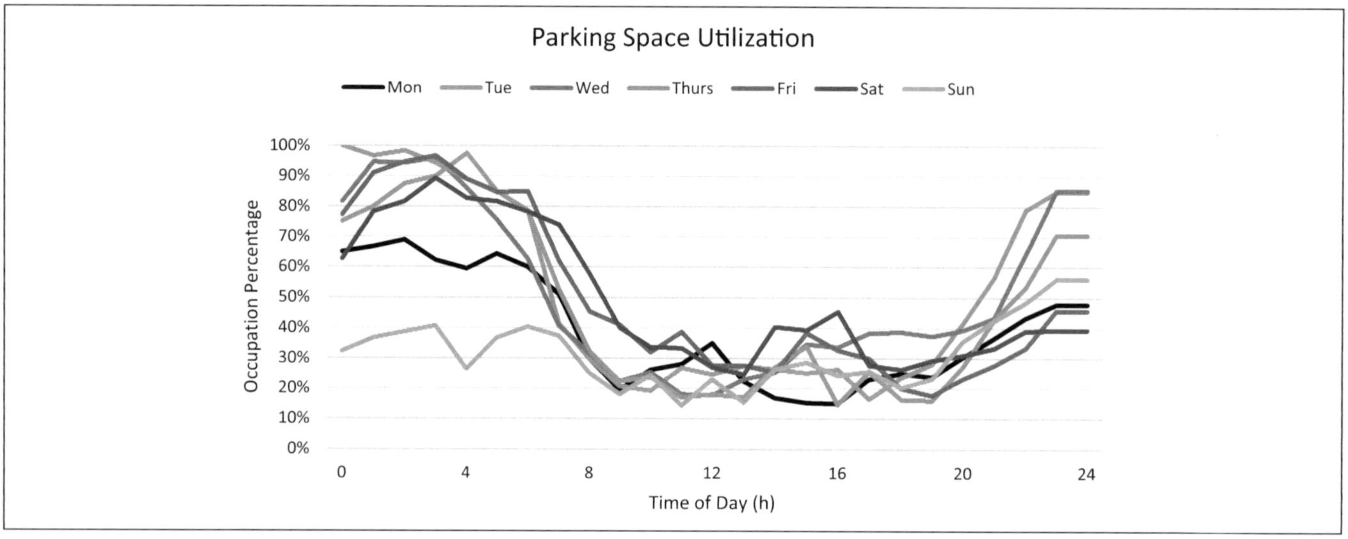

Figure 11. Time of day parking space utilization.

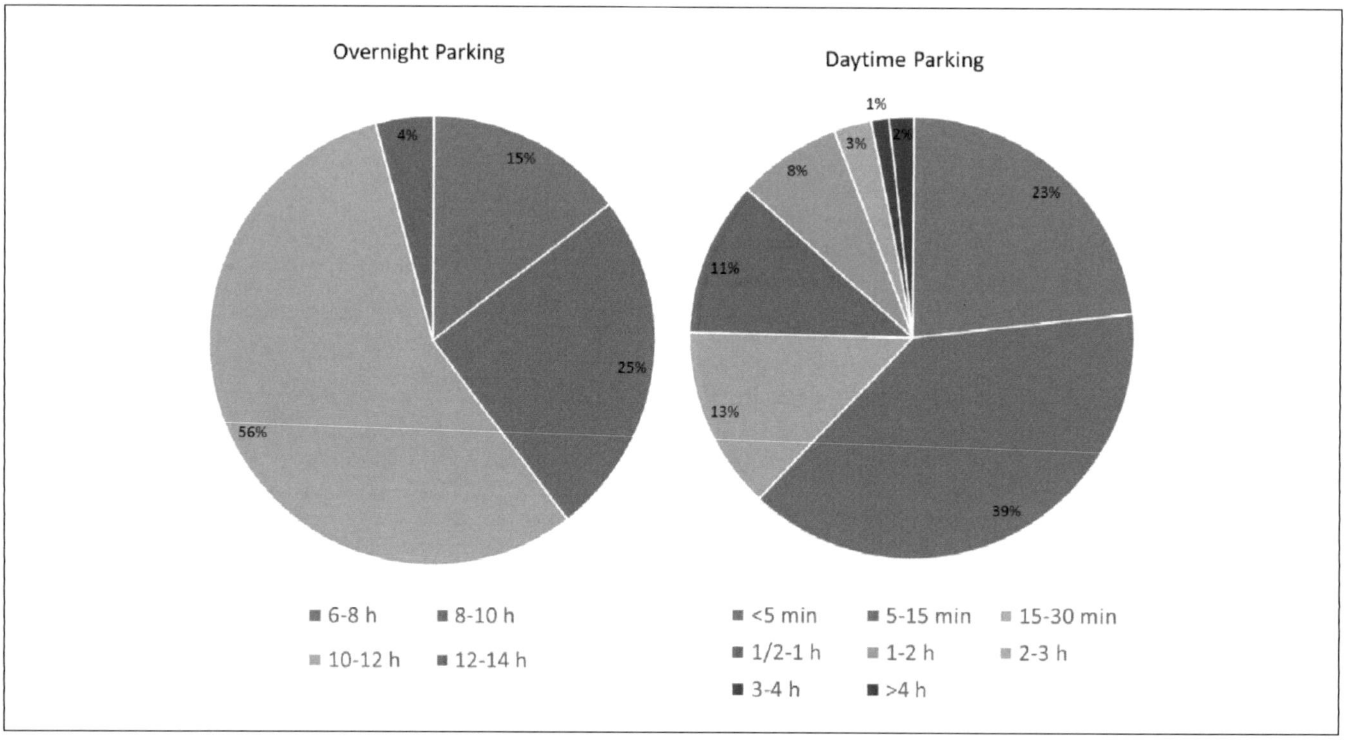

Figure 12. Parking duration.

Where:

α is the desired significance level, in this case, 5%

n is the number of parking events for a specific vendor

$\hat{p}$ is the turnover accuracy for a specific vendor

With respect to the difference in the number of parking events between the vendors, the following should be noted:

- The confidence interval values would not change for Sensys if it had the same number of parking events as SENSIT, and the $\hat{p}$ value remained unchanged.
- The confidence interval for CivicSmart would change to 95–97% if it had the same number of parking events as SENSIT, and the $\hat{p}$ value remained unchanged.

Table 1. Accuracy Test Results

Vendor	Number of parking events	Turnover accuracy (%)	Turnover ratio (parking events/h)	Hours of data analyzed	Occupancy accuracy (%)
SENSIT	1073	95.25	8.01	134	97.36
Sensys	776	97.94	4.62	168	99.15
CivicSmart	488	96.11	4.45	112	97.20

Table 2. Turnover Accuracy Confidence Interval

	SENSIT	Sensys	CivicSmart
Number of parking events (n)	1073	776	488
Turnover accuracy ($\hat{p}$)	95.25%	97.94%	96.11%
95% confidence interval	94%–97%	97%–99%	94%–98%

Table 3. Accuracy Test Results under Dry Condition

Vendor	Analysis period	Number of parking events	Turnover accuracy (%)	Number of hours	Occupancy accuracy (%)	General weather condition
SENSIT	8/18/16 8:00–23:59	208	93.27	16	97.89	Dry
	8/20/16 0:00–17:00	158	100	24	98.99	Dry
	Sum	366	96.18	40	98.55	
Sensys	9/06/16 0:00–23:59	105	100.00	24	99.72	Dry
	9/09/16 0:00–23:59	109	96.33	24	99.52	Dry
	Sum	214	98.13	48	99.62	
CivicSmart	8/20/16 6:00–20:00	63	98.41	14	99.19	Dry
	8/24/16 16:00–20:00	23	91.3	4	89.95	Dry
	8/25/16 6:00–13:00	38	100	7	99.14	Dry
	8/26/16 16:00–20:00	18	100	4	99.41	Dry
	8/27/16 6:00–20:00	41	95.12	14	89.99	Dry
	8/28/16 6:00–15:00	24	100	9	96.97	Dry
	9/09/16 6:00–20:00	80	97.5	14	98.91	Dry
	Sum	287	97.56	66	96.32	

Despite the different number of parking events across vendors, the relatively narrow range of the 95th percentile confidence interval values for each vendor shows that the accuracy results are reasonably stable.

Other Issues

Intermittent Detector Status Change Events. The SENSIT and CivicSmart sensors occasionally displayed short periods of the detector status oscillating between on (vehicle parked) and off (vehicle not parked) while a vehicle was parked in a space. For example, the ground-truth data may have shown that a truck had been parked in a space continuously for two hours, but the sensors might report several status change events during that period, with each status change event lasting only 10–40 seconds. However, the vendors' parking status algorithms account for this issue by employing a short lag interval (1–2 minutes) before registering a change of parking status in their database. For this type of application, such a lag in registering parking space status changes is not a significant concern.

Performance under Raining Conditions. Since the project was conducted during the rainy season of Florida, rain and thunderstorms occurred during the analysis period. Tables 3 and 4 show the accuracy test results under dry and rainy conditions, respectively. Generally, there was a slightly drop of accuracy during rainy conditions. The Sensys sensors appeared to be the least influenced by rainy conditions—the decrease in accuracy for both tests was less than 1%. The SENSIT sensors' accuracy decreased approximately 1.5% for both tests. While the occupancy accuracy for CivicSmart appeared to increase by 2% relative to dry conditions, its turnover accuracy decreased by more than 3%. While the analysis of the influence of rain on the parking detection technologies was limited and should not be considered conclusive, the research team suspects that heavy rain that causes large amounts of sheeting water over the parking spaces

Table 4. Accuracy Test Results under Raining Conditions

Vendor	Analysis period	Number of parking events	Turnover accuracy (%)	Number of hours	Occupancy accuracy (%)	General weather conditions
SENSIT	8/19/16: 0:00–23:59	218	97.71	24	97.88	Rain, Thunderstorm
	8/31/16: 0:00–22:00	206	95.15	22	98.27	Rain
	9/05/16: 0:00–23:59	131	93.13	24	96.24	Rain
	9/11/16: 0:00–23:59	152	91.45	24	95.12	Thunderstorm
	Sum	707	94.77	94	96.85	
Sensys	9/01/16: 0:00–23:59	114	97.37	24	98.63	Rain, Thunderstorm
	9/03/16: 0:00–23:59	110	93.64	24	98.01	Rain, Thunderstorm
	9/05/16: 0:00–23:59	104	99.04	24	99.20	Rain
	9/11/16: 0:00–23:59	106	100.00	24	99.49	Thunderstorm
	9/14/16: 0:00–23:59	128	99.22	24	99.49	Rain, Thunderstorm
	Sum	562	97.87	120	98.96	
CivicSmart	8/31/16: 6:00–15:00	84	97.62	9	98.5	Rain
	9/01/16: 6:00–20:00	57	89.47	14	96.31	Rain, Thunderstorm
	9/04/16: 11:00–20:00	27	85.19	9	99.61	Rain, Thunderstorm
	9/05/16: 6:00–20:00	33	100	14	99.87	Rain
	Sum	201	94.03	46	98.47	

could slightly affect the performance of the sensors. Note that the comparison made here was based on general weather conditions and did not explicitly consider water depth; this issue should be considered in future research.

Vehicle Composition. Even though the parking spaces studied were within the truck parking zone, approximately 7% of the vehicles parking there were personal cars and motorcycles. The research team observed many cases where a small personal vehicle parked squarely between two sensors without triggering any of them, or a motorcycle parked at the far end of a parking space without triggering any of the sensors, which ended up affecting the accuracy tests. In this case, the vehicle composition of the truck parking area would affect the number and layout of sensors needed in the parking space in order to report the occupancy status correctly.

Durability/Maintenance. Over the course of numerous site visits during the data collection period, some brief observations were made of the parking sensors. No damage to the various detectors was observed visually. Over the data collection period, only one sensor (CivicSmart) suffered a failure. This was determined to have been the result of a faulty sensor rather than due to any physical damage. The sensor was quickly replaced in the field by CivicSmart maintenance personnel. It should be kept in mind, however, that the duration of this project was relatively short.

Summary and Recommendations

Overall, it was found that all three technologies performed well, with accuracy rates of 95% or better for both tests (turnover and occupancy). This result suggests that, for implementation, pricing, and/or maintenance issues may be more significant factors for the choice of technology.

When considering truck parking detection technology at rest areas, the following observations and recommendations are offered, with respect to the findings of this study:

1. Currently, no conclusions can be made with regard to the influence of rain or snow on the performance of parking detection technologies. Future research should consider studying the possible influence of various depths of surface water or snow on the performance of the sensors.
2. The actual vehicle composition of the truck parking area should be considered to determine the number and layout of sensors needed in the parking spaces for research or implementation. If the percentage of small personal cars and motorcycles using the truck parking spaces is high, more sensors should be installed in the parking spaces in order to report the occupancy status correctly.
3. When a false detection error occurs, it will only be corrected with the next correctly detected parking event. In this case, during the implementation of the parking detection technologies, manual updates of the status of the sensors by checking the live video of the rest area or by maintenance personnel on site could be helpful. This is especially so in the late-night to early-morning time period, when there are much fewer parking events than during the daytime, and thus false detection errors, once occurring, are less likely to be corrected by sensors in a timely manner.

Acknowledgments

This research was sponsored and supported by the Florida Department of Transportation (FDOT). The authors thank Marie Tucker (FDOT Central Office) and Pradeep Rao (HNTB) for their assistance with this project.

Author Contributions

The authors confirm contribution to the paper as follows: study conception and design: Wei Sun, Scott S. Washburn; data collection: Ethan Stoop, Wei Sun, Scott S. Washburn; analysis and interpretation of results: Ethan Stoop, Wei Sun, Scott S. Washburn; draft manuscript preparation: Ethan Stoop, Wei Sun, Scott S. Washburn. All authors reviewed the results and approved the final version of the manuscript.

References

1. Bayraktar, M. E., Y. Zhu, and F. Arif. *Commercial Motor Vehicle Parking Trends at Rest Areas and Weigh Stations*. Report BDK80-977-14. Florida Department of Transportation Research, Tallahassee, 2012.
2. Washburn, S. S., W. Sun, and E. Stoop. *Commercial Truck Parking Detection Technology Evaluation for Columbia County Rest Areas*. Report BDV31-977-56. Florida Department of Transportation Research, Tallahassee, 2016.
3. Fallon, J., and K. Howard. *SmartPark Truck Parking Availability System: Magnetometer Technology Field Operational Test Results*. No. FMCSA-RRT-10-041. Federal Motor Carrier Safety Administration, Washington, D.C., 2011.
4. SFpark. *Parking Sensor Technology Performance Evaluation*. 2014. http://sfpark.org/resources/parking-sensor-technology-performance-evaluation/. Accessed July 28, 2017.
5. Nedap Identification Systems. *SENSIT Products*. 2016. http://www.nedapidentification.com/products/sensit/. Accessed July 28, 2017.
6. IPsens, LLC. *Advanced Maintenance Monitoring Application & Services Pre-release Documentation*. IPsens, LLC, Branson, Mo., 2016.
7. Sensys Networks. *Sensys Products*. 2016. http://www.sensys-networks.com/products. Accessed July 28, 2017.
8. CivicSmart. *Wireless Vehicle Detection Sensors*. 2016. http://www.civicsmart.com/wireless-vehicle-detection-sensors/. Accessed July 28, 2017.

The Standing Committee on Trucking Industry Research (AT060) peer-reviewed this paper (18-01082).

The opinions, findings, and conclusions expressed in this publication are those of the author(s) and not necessarily those of the Florida Department of Transportation or the U.S. Department of Transportation.

Article

TRR
JOURNAL OF THE TRANSPORTATION RESEARCH BOARD

Transportation Research Record
2018, Vol. 2672(9) 152–163

DOI: 10.1177/0361198118784175
journals.sagepub.com/home/trr
SAGE

Comprehensive Exploratory Analysis of Truck Route Choice Diversity in Florida

Trang D. Luong[1], Divyakant Tahlyan[1], and Abdul R. Pinjari[2]

Abstract

This study presents a comprehensive exploratory analysis of truck route choice diversity in the state of Florida, for both long-haul and short-haul truck travel segments. It employs six metrics to measure three different dimensions of diversity in truck route choice between any given origin–destination (OD) pair. These dimensions are: (a) number of distinct routes used to travel between the OD pair, (b) the extent of overlap (or lack thereof) among the routes, and (c) the evenness (or dominance) in the usage of different unique routes. The diversity metrics were applied to a large database of 73,000 truck routes derived from 200 million GPS records. Descriptive analysis and statistical modeling of the diversity metrics offered insights into the determinants of various dimensions of truck route choice diversity between any OD pair. The results are useful for improving choice set generation algorithms for truck route choice modeling and in truck route policies and investments.

An essential step toward enhancing highway freight mobility is to improve our understanding of freight-truck route choice behavior. Specifically, analysis of the routes that trucks use to travel between different origins and destinations can support the design of truck routing policies aimed at mitigating congestion, improving travel time reliability, and facilitating truck movement during network disruptions. However, research on truck route choice has been limited due to insufficient data on truck movements. The recent availability of global positioning systems (GPS) data has encouraged research on truck route choice modeling (*1–3*) and highway freight performance measures (*4, 5*). Yet little attention has been paid to exploring the diversity or variability of truck route choice between travel origins and destinations.

An improved understanding of truck route diversity has significance in both freight modeling and planning. For modeling applications, understanding truck route choice diversity can help determine the number and structure (i.e., extent of overlap) of route alternatives to be used in route choice models and traffic assignment procedures (*6*). For planning applications, analyzing the diversity of truck route choices observed in the field can help inform truck routing decisions during regular and emergency situations. For example, identifying origin–destination (OD) pairs with high travel demand but low diversity in the routes used (e.g., a single route used) can help identify critical segments of the network and inform routine infrastructure maintenance scheduling as well as re-routing efforts during emergency recovery. Also, one can apply route diversity measures to evaluate the redundancy of (or lack of) truck routes in existing transportation infrastructure to justify long-term investment in truck corridors to increase network redundancy.

This research presents a comprehensive exploratory analysis of truck route choice diversity in the State of Florida for both long-haul and short-haul travel segments. Specifically, the paper addresses two broad questions: How to measure the degrees of diversity in the routes trucks use to travel between an OD pair? and, What factors influence the diversity of truck route choice between an OD pair? To this end, six metrics are utilized to measure the following three different dimensions of diversity in route choice between a given OD pair: (a) number of different routes used between the OD pair, (b) extent of overlap (or lack thereof) among the routes, and (c) evenness (or dominance) of the use of different unique routes between that OD pair. These metrics are applied to quantify truck route choice diversity using a database of about 73,000 routes derived from more than 200 million truck-GPS records. Next, statistical models are estimated to explore the influence of various

[1]Department of Civil and Environmental Engineering, University of South Florida, Tampa, FL
[2]Department of Civil Engineering, Center for Infrastructure, Sustainable Transportation and Urban Planning (CiSTUP), Indian Institute of Science (IISc), Bangalore, India

Corresponding Author:
Address correspondence to Abdul R. Pinjari: abdul@iisc.ac.in

determinants on the three dimensions of route choice diversity between hundreds of OD pairs. The models provide insights into the influence of the characteristics of truck travel demand, OD location, and network structure on the diversity of truck route choice between an OD pair. Potentially, these insights can help travel modelers in improving choice set generation algorithms for modeling truck route choice and help planners in devising truck routing policies.

The next section reviews past studies on variability in route choice behavior. The following section summarizes the truck-GPS data used for this study. The metrics used to quantify diversity in truck route choice are then elaborated. The next section presents the statistical modeling methodologies used in this study. Subsequently, empirical results are presented and discussed. The last section concludes the paper.

Literature

Few studies in the literature have analyzed route diversity. Jan et al. (*6*) were among the first to analyze variability in observed routes, using GPS data from a household survey in Lexington, Kentucky. They concluded that most chosen routes differed from the shortest-time path and that the extent of variability across trips made by different households was greater than that across trips made within the same household. In another study, Papinski et al. (*7*) estimated that up to 20% of travelers deviated from their planned route. They found that the observed routes are longer than both shortest-time and shortest-distance routes because travelers consider many route characteristics in addition to time and distance. Additionally, Spissu et al. (*8*) examined the intra-personal and inter-personal variability in route choice using GPS data of 679 routes chosen by a sample of 12 university students. Their results suggested higher levels of intra-individual variability (than inter-individual variability) for discretionary trips and higher levels of inter-individual variability for work or study trips. Finally, Zhu and Levinson (*9*) utilized 13-day GPS data of 143 individuals in the Minneapolis–St. Paul Twin Cities to analyze the diversity of commuting trips to investigate their hypothesis that people choose a portfolio of routes over time.

In all these studies, the datasets were limited to only tens or hundreds of observed routes. The metrics used by these studies for analyzing route variability cannot be used to measure the amount of overlap among routes or dominance of usage of routes. Furthermore, none of the above studies are focused on truck route choice. The current paper fills such gaps in research through a comprehensive analysis of truck route diversity using six different metrics of diversity applied to a large database of truck routes. In addition, this is perhaps the first empirical study that examines the influence of truck travel demand, OD location, and network structure on diversity of truck route choice. A comparison of route diversity patterns between long-haul and short-haul travel segments is also presented.

Data

The truck-GPS data used in this study was obtained from the American Transportation Research Institute (ATRI). The database of long-haul truck trips (trips longer than 50 miles) comprised 145 million GPS records collected from nearly 50,000 freight trucks that traveled in the state of Florida from March to June 2010. The data used to derive short-haul trips (trips shorter than 50 miles) comprised about 96 million GPS records corresponding to 110,000 freight trucks and spanned six counties of the Tampa Bay region in Florida. Temporally, the short-haul data corresponded to the first 15 days in October 2015, December 2015, April 2016, and June 2016.

The raw GPS data were first converted into a database of truck trips using algorithms developed by Thakur et al. (*10*) and later refined by Pinjari et al. (*11*). To derive the chosen route for each trip, raw GPS records corresponding to each trip were map-matched to high-resolution NAVTEQ roadway networks provided by the Florida Department of Transportation (FDOT), using the procedure developed by Kamali et al. (*12*). The 2010 NAVTEQ network used to derive long-haul routes comprised 1.5 million links and 5.8 million nodes whereas the 2015 NAVTEQ network used for short-haul routes comprised 1.8 million links and 6.9 million nodes. After the map-matching and validation process, a database of about 78,000 long-haul trips and 225,000 short-haul trips was retained.

To analyze route choice and the diversity therein, it is useful to aggregate trip end locations to larger spatial units. This spatial aggregation allows analysts to observe a sufficient number of trips to gain an uncensored view of the various routes trucks choose between an OD pair. Hence, all trip end locations in this study were aggregated to the traffic analysis zones (TAZs) defined in the Florida Statewide Travel Demand Model. TAZ OD pairs with at least 50 long-haul trips and at least 30 short-haul trips were selected, since OD pairs with fewer trips might not offer a complete picture of truck route diversity. The final long-haul dataset used in this analysis comprises 277 TAZ OD pairs with a total of 30,263 trips longer than 50 miles. The short-haul dataset comprises 527 TAZ OD pairs with a total of 42,884 trips of 5 to 50 miles length. Trips shorter than 5 miles were not considered in this analysis because such trips would typically not have many route choice options.

Diversity Metrics

To measure diversity in truck route choice between any given OD pair, we employ the following six metrics: (a) number of unique routes, (b) average commonality factor, (c) average path size, (d) non-overlapping index, (e) standardized variance of route usage, and (f) standardized Shannon entropy of route usage. The first metric measures the number of distinct or "unique" routes traveled by trucks between an OD pair. The next three metrics measure the extent of overlap (or lack thereof) among the unique routes identified between that OD pair. The last two metrics measure the evenness (or, otherwise, dominance) in the use of unique routes between the OD pair. These three dimensions together provide a complete picture of the diversity in truck route choice between any OD pair. All six metrics are defined next.

Number of Unique Routes

Many routes chosen between an OD pair overlap substantially with each other and differ only by a few links. To determine a set of distinct or unique routes based on the amount of overlap, we used the commonality factor (C_{ij}) between routes i and j, proposed by Cascetta et al. (*13*) as follows:

$$C_{ij} = l_{ij} / \sqrt{L_i L_j} \tag{1}$$

where

L_i and L_j = the lengths of routes i and j, respectively; and

l_{ij} = the length of the shared portion between the two routes.

The two routes are defined as unique if the commonality factor between them is less than 0.95. To determine the number of unique routes observed between an OD pair, all routes between that OD pair are arranged in ascending order of route length. The shortest route is the first unique route. The commonality factor of each subsequent route is computed with respect to all preceding unique routes to determine whether it is a unique route (if all C_{ij} values for that route are less than 0.95). The result of this process is a set of unique routes between an OD pair, where the C_{ij} value between any two unique routes is less than 0.95. The size of this unique route set represents the number of unique routes used between that OD pair.

Average Commonality Factor

Since the number of unique routes metric does not quantify the extent of overlap (or lack thereof) among the unique routes observed between an OD pair, we derive the average commonality factor metric to measure the degree of overlap between all pairs of unique routes between the OD pair, as shown below:

$$\overline{\overline{CF}} = \frac{\sum_1^{\binom{K}{2}} C_{ij}}{K} \tag{2}$$

where

C_{ij} = the commonality factor computed for two unique routes i and j between an OD pair;

K = the number of unique routes between the OD pair; and

$\binom{K}{2}$ = the number of pairs of unique routes.

Ranging between 0 and 1, an average commonality factor value closer to 0 represents low overlap between the unique routes. Therefore, for a given number of unique routes between an OD pair, a lower value of $\overline{\overline{CF}}$ suggests a greater extent of diversity (i.e., less overlap) among the routes.

Average Path Size

Proposed by Ben-Akiva and Bierlaire (*14*), path size (in Equation 3, PS) is a metric commonly used in the route choice modeling literature to measure the degree of overlap of two routes between an OD pair. The PS for each unique route i is defined as follows:

$$PS_i = \sum_{a \epsilon \Gamma_i} \left(\frac{l_a}{L_i}\right) \frac{1}{\sum_{j \epsilon \, unique \, routes} \delta_{aj}} \tag{3}$$

where

Γ_i is the set of all links composing unique route i;

l_a is the length of link a;

L_i is the length of route i; and

δ_{aj} is equal to 1 if route j belonging to the set of K unique routes uses link a, 0 otherwise.

The maximum possible value of PS is 1 and the minimum value tends to 0. A route with no overlap with any other routes has a PS value of 1. Average PS in an OD pair is the mean value of PS across all unique routes between that OD pair. For a given number of unique routes between an OD pair, the closer to 1 is the average PS value, the greater is the diversity (i.e., lower overlap) among routes.

Non-overlapping Index

Complementary to the above two metrics, the degree of non-overlap among the unique routes between an OD pair is quantified using the non-overlapping index. This index is measured as the ratio between the total length of links (on unique routes) that were used only once and the total length of all links (on unique routes) that were used at least once. This index ranges between 0 and 1, where a value closer to 1 represents low overlap among unique routes.

Standardized Variance of Route Usage

Another dimension of diversity is based on the evenness in the usage of different unique routes between an OD pair. The most even usage is when all observed trips are equally distributed among the observed unique routes between that OD pair. A complementary concept is the degree of dominance, when most trips are observed to have taken only one or a few unique routes.

To measure the degree of evenness of route usage, the distribution of N trips among K different unique routes between a given OD pair may be characterized as a multinomial distribution, with each trip being allocated to any one of the K different unique routes. If the random variable $X_k (k = 1, 2, 3, \ldots, K)$ indicates the number of trips choosing route k and p_k is the proportion of trips allocated to route k, vector $X = (X_1, X_2, \ldots, X_K)$ follows a multinomial distribution with parameters N and p, where $p = (p_1, p_2, \ldots, p_K)$. The variance of such multinomial-distributed random variables may be written as $Var(X_k) = N^* p_k^* (1 - p_k)$.

The variance of route usage between an OD pair is defined as the sum of variances of usage frequency for each route, as: $N^* \sum_1^K p_k^* (1 - p_k)$. This metric is influenced by three factors that vary across OD pairs: (a) number of observed trips (the more trips, the higher the variance); (b) number of unique routes (the more routes, the higher the variance); and (c) evenness of the distribution of the observed trips among various unique routes. For a given OD pair with N observed trips and K unique routes, the maximum possible value of variance of route usage, when all trips are evenly distributed among all unique routes, is determined as follows:

$$\begin{aligned} &Max\ variance\ of\ route\ usage \\ &= N^* K^* (1/K)^* (1 - 1/K) = N^* (1 - 1/K) \end{aligned} \tag{4}$$

where

N = the number of observed trips between an OD pair; and

K = the number of unique routes between the OD pair.

To measure solely the nature of trip distribution without being influenced by the number of observed trips (N) or unique routes (K), this metric may be standardized as the ratio of the variance of usage to the maximum possible variance as shown in Equation 5:

$$Standardized\ variance\ of\ route\ usage = \frac{\sum_1^K p_k^* (1 - p_k)}{\left(1 - \frac{1}{K}\right)} \tag{5}$$

where

K = the number of unique routes between an OD pair; and

p_K = proportion of trips taking the kth unique route.

The closer this metric is to its maximum possible value 1, the more evenly distributed the observed trips are among various unique routes. For example, if 100 trips use two unique routes between an OD pair, the standardized variance of usage for that OD pair would be 1 if 50 trips take the first route and the other 50 trips take the second route.

Standardized Shannon Entropy of Route Usage

Shannon entropy (*15*) is a metric typically used to measure the evenness of distribution of different entities among a given number of categories. Proposed in the field of information science, the concept of entropy has been applied widely in different fields, such as geodiversity and biodiversity, and is used in transportation to measure land-use diversity. We apply the same concept for route choice diversity to measure the extent of distribution of all trips among different unique routes. The Shannon entropy of usage of K unique routes between an OD pair is defined as $\sum_1^K p_k ln(p_k)$. The maximum value of the Shannon entropy of usage is $K^*(1/K)^* \ln(1/K) = \ln(1/K)$ when all trips are equally distributed among the identified unique routes between an OD pair. To eliminate the effect of number of unique routes, the standardized Shannon entropy (SE in Equation 6) of route usage is computed as below:

$$Standardized\ SE\ of\ route\ usage = \frac{\sum_1^K p_k ln(p_k)}{\ln(1/K)} \tag{6}$$

The maximum possible value standardized Shannon entropy is 1, when all trips are evenly distributed among all unique routes.

An Illustration

Figure 1 illustrates the application of the above diversity metrics for two different OD pairs. The first OD pair is selected from the long-haul data, with eight unique routes which range from 62 to 84 miles long. Note that many of the eight unique routes overlap considerably with each other. Such overlap is measured by the average commonality factor and average path size. Out of the 66 trips observed between this OD pair, 53 use the first unique route, indicating the dominance of the first unique route. The second OD pair is from short-haul data with 32 observed trips which are more evenly distributed among the different routes than those between the first OD pair. Such differences in dominance (or evenness) of route usage are measured by the two metrics: standardized

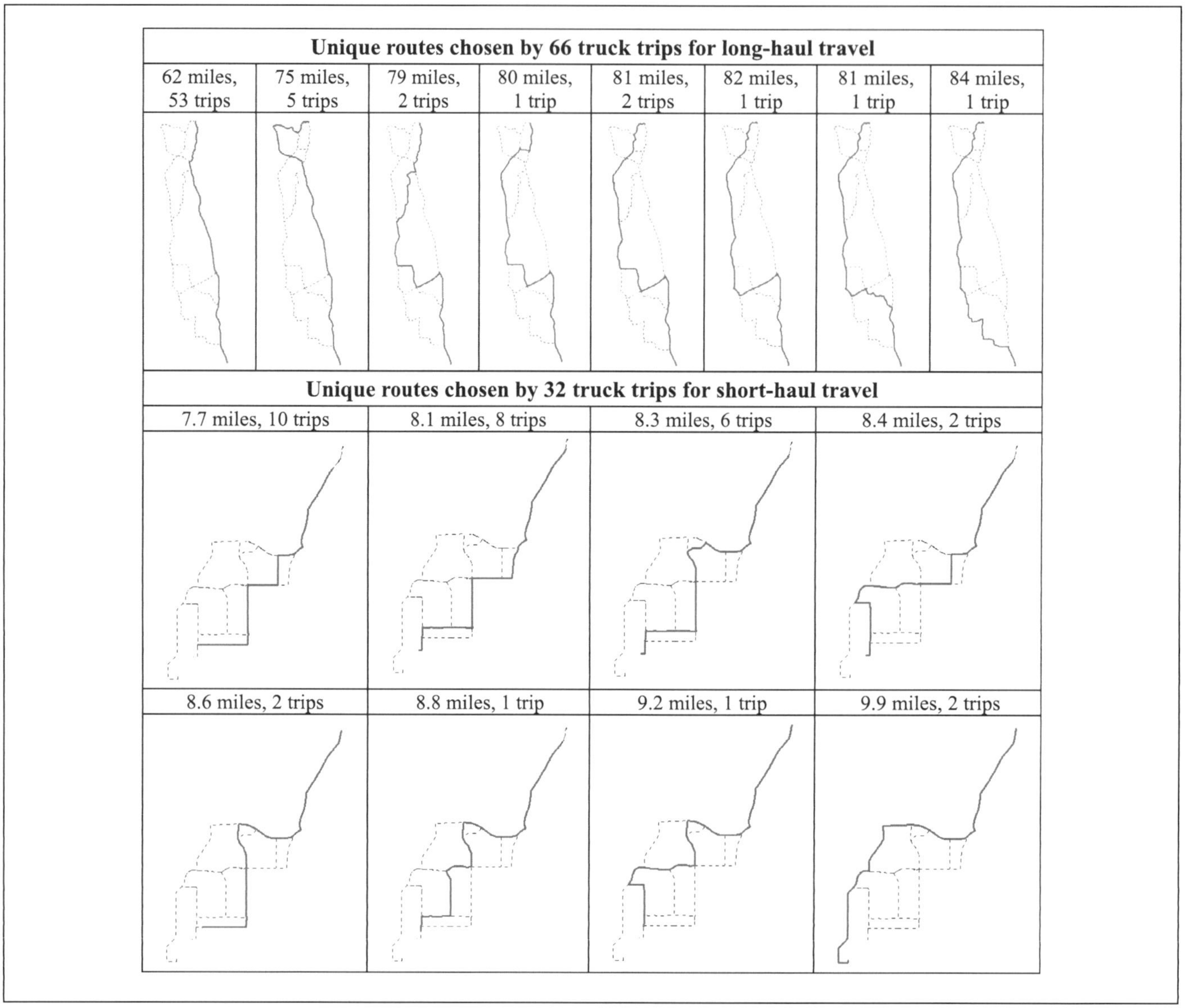

Figure 1. Examples of unique routes for long-haul and short-haul OD pairs.

variance of route usage and the standardized Shannon entropy of route usage.

Modeling Methodology

In this section we explain the statistical model structures used to analyze the determinants of the following three of the six metrics developed for each OD pair: (a) number of unique routes, (b) average path size of unique routes, and (c) standardized Shannon entropy of route usage.

Count Data Models for Number of Observed Unique Routes

Negative binomial (NB) regression is an appropriate choice to model count data given by the number of observed unique routes in this research. Typically, Poisson regression is preferred if the mean of the count process is equal to the variance. If there is a significant difference between the mean and the variance of the count process, the data are said to be over-dispersed and NB regression is preferred. Our empirical data supported the use of NB regression over Poisson regression, because of over-dispersion in the data.

In NB regression, the probability $P(y_i)$ of an OD pair i having y_i number of unique routes is defined as follows:

$$P(y_i) = \frac{\Gamma\left(\frac{1}{\alpha} + y_i\right)}{\Gamma\left(\frac{1}{\alpha}\right) y_i!} \left(\frac{1/\alpha}{\left(\frac{1}{\alpha}\right) + \lambda_i}\right)^{1/\alpha} \left(\frac{y_i}{\left(\frac{1}{\alpha}\right) + \lambda_i}\right)^{y_i} \tag{7}$$

where

$\Gamma(\cdot)$ = the gamma function:

$\lambda_i = exp(\beta X_i + \varepsilon_i)$;

X_i = a vector of explanatory variables;

β = a vector of parameters to be estimated; and

$exp(\varepsilon_i)$ is a Gamma-distributed disturbance term with unit mean and variance given by the dispersion parameter α.

The model parameters can be estimated using maximum likelihood estimation technique. Depending on the count process being modeled, the regression can be right, left, or two-side truncated. To model the number of unique routes between an OD pair, our count data models were left-truncated at 1, because any OD pair in the data has at least one unique route.

Fractional Response Models for Average Path Size and Standardized Shannon Entropy of Usage

It is worth noting that all diversity metrics proposed in this study, except the number of unique routes, range between 0 and 1. The fractional response model structure proposed by Papke and Wooldridge (*16*) may be used to model such quantities whose values lie between 0 and 1. Although proportion data may be modeled by logit transformation of the dependent variable [*i.e.*, $\ln(y_i/(1-y_i) = \beta X_i$] followed by ordinary least squares regression, this transformation cannot be used when the dependent variable might take values of zero or one. This issue can be resolved with the fractional response model (*16*) whose expected value of the dependent variable is: $E(y_i|x_i) = G(x_i\beta)$. Two possible functional forms for $G(z)$ are: (a) logistic function, $G(z) = exp(z)/(1 + exp(z))$ and (b) cumulative density function of a standard normal distribution. According to this model, the quasi likelihood of an OD pair with an observed value y_i is given as below:

$$\mathcal{L}_i(\beta) = y_i{*}\log[G(x_i\beta)] + (1-y_i){*}\log[1 - G(x_i\beta)] \quad (8)$$

where

$G(x_i\beta)$ = a known function with $0 < G(z) < 1 \forall z \in \mathbb{R}$.

The parameters are estimated using maximization of the quasi log-likelihood function.

Descriptive Analysis

The first set of rows in Table 1 summarizes the mean and standard deviation (SD) values of all diversity metrics calculated for our long-haul and short-haul datasets. There are 19 long-haul OD pairs and 22 short-haul OD pairs that have one single observed unique route. Except for the number of unique routes metric, all other diversity metrics reported were computed after excluding such OD pairs with only one unique route.

It is worth noting that the short-haul routes exhibit greater diversity than long-haul routes with higher average values for observed unique routes, non-overlapping index, standardized variance of usage, and standardized Shannon entropy of usage. The SDs are also higher for short-haul data, suggesting a greater incidence of higher values. In other words, short-haul routes are more diverse than long-haul routes from the standpoint of lower overlap as well as lower dominance in their usage.

Potential Determinants of Diversity

To explore the correlates of diversity in truck route choice, we extracted a variety of factors describing observed travel demand, OD locations, and network structure between all OD pairs. These variables are presented in the second set of rows in Table 1 and discussed next.

Trip Characteristics. The first category of variables includes the number of trips observed for each OD pair and the number of trucks taking those trips (a measure of truck travel demand), spatial separation (straight-line distance or direct distance) between the OD locations, and travel conditions measured between the OD pair (particularly on the most used route). In terms of travel conditions of the most used route, we measure travel time variability and level of route circuity (defined as the ratio of route length to the direct OD distance).

OD Location Characteristics. Characteristics of origin and destination TAZs include land-use descriptors (employment densities, TAZ size, urban/rural classification) and spatial dispersion of freight activity centers (defined as the average distance from all trip ends' centroids to each trip end location).

Network Structure. To explore the impact of network structure on the diversity of observed routes, we hypothesized two different areas of influence between OD pairs. In the first hypothesis, the diversity of route choice between an OD pair is provided by the entire road network inside an elliptical area of influence connecting that OD pair, referred to as the long ellipse. The long ellipse's major axis is assumed to be the same distance and orientation of the straight line connecting the centroids of origin and destination TAZs. Its minor axis is set to be one-third of the major axis length. In the second hypothesis, the diversity of route choice between an OD pair is differentially impacted by two different areas of influence. The first category of influence areas are circular areas around the origin and destination TAZ centroids, referred to as circular buffers. The buffer radii explored are 1, 2, and 5 miles for direct distances of 5 to 10, 10 to 20, and more than 20 miles, respectively. The

Table 1. Descriptive Statistics of All Data Used for Route Diversity Analysis

No.	Diversity metrics		Long-haul Average	Long-haul SD	Short-haul Average	Short-haul SD
1	Number of unique routes		8.61	6.54	9.03	6.51
2	Average commonality factor		0.69	0.17	0.68	0.18
3	Average path size		0.28	0.12	0.29	0.14
4	Non-overlapping index		0.26	0.13	0.32	0.15
5	Standardized variance of usage		0.62	0.26	0.65	0.25
6	Standardized Shannon entropy of usage		0.57	0.21	0.61	0.21
No.	**Potential determinants of diversity**		**Long-haul Average**	**Long-haul SD**	**Short-haul Average**	**Short-haul SD**
1	No. of trips observed for an OD pair		109.3	99.7	81.1	123.7
2	No. of trucks observed for an OD pair		19.7	16.6	68.3	113.5
3	OD airway distance (miles)		109.8	69.6	21.9	11.0
4	Travel time of trips taking the most used route (minutes)	SD	7.5	3.6	4.8	18.4
5		Average	146.1	73.9	31.4	13.0
6		95th percentile	158.5	77.4	35.9	14.3
7		5th percentile	136.9	71.7	27.8	12.1
8	Ratio of the most used route length to airway OD distance		1.2	0.1	1.2	0.4
9	Employment density of OD TAZs (1000 jobs/sq. mile)	All types	7.0	5.5	5.7	4.0
10		Industrial	1.5	1.1	0.9	0.7
11		Service	3.4	3.1	3.0	2.8
12		Commercial	2.1	1.9	1.8	1.3
13	Average area of OD TAZs (miles^2)		2.2	2.8	2.3	3.4
14	Indicator if both OD TAZs are urban		0.8	0.3	0.9	0.3
15	Average distance from centroid of all trip ends to each trip end (miles)		0.4	0.5	0.8	1.0
16	Average distance from TAZ centroid to major arterials (miles)		6.0	3.5	4.7	3.2
17	Length of major arterials (miles)*	Long ellipse	331.8	411.8	27.1	27.9
18		Short ellipse	274.2	386.1	15.7	17.5
19		Ending buffers	29.0	23.6	8.8	10.0
20	Length of minor arterials (miles)*	Long ellipse	621.4	855.4	42.6	49.4
21		Short ellipse	502.6	800.1	18.2	25.9
22		Ending buffers	62.8	60.2	17.8	24.0
23	Length of collectors (miles)	Long ellipse	1276.6	1561.8	110.5	114.3
24		Short ellipse	1031.3	1446.2	48.7	62.2
25		Ending buffers	137.9	96.0	48.1	56.7
26	Length of local roads (miles)	Long ellipse	10155.1	14471.0	673.1	639.2
27		Short ellipse	9867.2	11496.5	311.3	349.2
28		Ending buffers	1212.1	768.9	259.8	279.2
29	Length of toll roads (miles)	Long ellipse	81.4	117.6	5.8	8.2
30	No. of rest stops	Long ellipse	9.2	12.1	1.0	1.4
31		Short ellipse	8.1	11.6	0.6	1.0
32		Ending buffers	0.3	0.9	0.4	0.8
33	No. of interchanges	Long ellipse	84.4	124.6	9.3	12.1
34		Short ellipse	60.5	107.6	4.0	6.7
35		Ending buffers	18.1	24.3	4.2	7.1
36	No. of traffic signals	Long ellipse	728.6	1043.8	59.2	86.3
37		Short ellipse	532.9	901.4	25.0	47.4
38		Ending buffers	136.2	168.4	29.1	53.0

*The lengths of major and minor arterials include both toll roads and toll-free roads.

second area of influence is elliptically shaped, referred to as the short ellipse, with the major axis as the difference of straight-line distance and radius of the circular buffers on each end.

Within these hypothesized areas of influence for each OD pair, we computed densities of various road types (major arterials, minor arterials, collectors, and local roads) to characterize the network structure between the OD pair. In addition, other facilities along the roadway, such as traffic signals, intersections, interchanges, and truck rest stops, were counted within long and short ellipses and circular buffers.

Table 2. Estimation Results of Truncated Negative Binomial Regression of Number of Unique Routes for Long-Haul and Short-Haul Datasets

Variable description	Long-haul data		Short-haul data	
	Coefficient	t-stat	Coefficient	t-stat
Trip characteristics				
Logarithm of number of truck IDs	0.293	5.53	0.358	10.93
Logarithm of number of trips	0.379	4.97	–	–
Trip indicator (1 if more than 150 trips, 0 otherwise)	–	–	0.159	1.80
Travel time variability on the most used route indicator (1 if the difference of 95th and 5th percentile of travel time greater than 15 minutes, 0 otherwise)	0.172	1.86	–	–
Ratio of length of the most used route to direct OD distance (miles/mile)	1.486	3.51	–	–
OD characteristics				
Average area of OD TAZs (mile^2)	–0.040	–1.85	–0.036	–3.85
Indicator if both OD TAZs are urban zones	–	–	0.277	3.83
Indicator if direct OD distance 50–200 miles	0.381	3.59	N/A	N/A
Indicator if direct OD distance 10–20 miles	N/A	N/A	0.128	2.84
Indicator if direct OD distance more than 40 miles	N/A	N/A	–0.671	–6.52
Industrial employment density (1,000 jobs/mile^2)	–	–	0.215	6.41
Commercial employment density (1,000 jobs/mile^2)	–	–	0.059	2.95
Average distance from centroid of all trip ends to each trip end (miles)	–	–	0.283	12.54
Average distance from TAZ centroids to nearest major or minor arterials (miles)	–	–	0.042	4.87
Network structure				
Ratio of toll roads to major arterials in long ellipse (miles/mile)	1.283	3.09	–	–
Density of major and minor arterials in 5-mile buffers around both endings (miles/mile^2)	–0.583	–2.87	–	–
Density of collectors in 5-mile buffers around both endings (miles/mile^2)	0.340	2.29	–	–
Density of major, minor arterials and collectors in short ellipse (miles/mile^2)	0.231	2.12	–	–
Density of minor arterials and collectors in long ellipse (miles/mile^2)	–	–	0.108	2.70
Proportion of major arterials to total length of major, minor arterials and collectors in long ellipse (miles/mile)	–	–	–0.534	–2.41
Proportion of minor arterials and collectors to total length of major, minor arterials and collectors in short ellipse (miles/mile)	1.375	2.24	–	–
Constant	–3.997	–5.06	–0.254	–1.41
Dispersion parameter	0.235	5.78	0.075	5.80
Number of observations (OD pairs)	277		527	
Log likelihood at convergence	–782.90		–1372.45	
Log likelihood for constant-only model	–842.92		–1617.90	
Adjusted ρ^2 with respect to constant-only model	0.056		0.141	

Note: For variables that have a significant influence in one model but not in the other model, "–" appears in place of parameter estimate and t-stat for that variable in the latter model. "N/A" is used when variable is not applicable to a specific model.

Statistical Modeling Results

Statistical models were estimated separately for long-haul and short-haul datasets to analyze the determinants of diversity metrics: number of unique routes, average path size, and standardized SE. Such empirical model results are discussed here.

NB Regression Model for Number of Unique Routes

Table 2 presents the NB regression estimation results for the number of unique routes, separately for long-haul and short-haul travel segments. Both model results indicate that OD pairs with a higher number of observed trucks are likely to have more unique routes. This is an expected result because more trucks traveling between an OD pair may lead to greater diversity in route choice due to heterogeneity in preferences of truck drivers, operators, and the businesses they serve. Similarly, OD pairs with more observed trips have more unique routes, in both travel segments (specifically, when there are more than 150 trips in the short-haul segment). More trips suggests a greater demand for travel and may lead to greater diversity in route choices as well. The next variable in the long-haul model, indicating high travel time variability (when the difference between 95th and 5th percentile travel time on the most used route is greater than 15 minutes), suggests more unique routes since the high variability (or low reliability) in travel conditions causes truckers to prefer alternative routes. Specific to the long-haul model, deviation of the most used route from the straight-line OD distance (measured as the ratio of the most used route length to straight-line distance) has a positive influence on the number of observed unique routes. When the most used route is more circuitous, more available routes in the network may exist and increase truckers' considerations for more alternate routes. Interestingly, neither travel time variability nor route circuity had a significant influence in the short-haul segment.

In the context of OD location characteristics, OD pairs with larger OD TAZs are likely to have a lower number of unique routes in both travel segments, perhaps because those TAZs are typically in areas with sparse network, population, and employment density, and therefore fewer network options to travel. For the same reason, both OD locations being in an urban zone is associated with a higher number of unique routes in the short-haul model. In the context of direct distance, OD pairs that are more than 200 miles apart tend to have fewer unique routes. This finding is consistent with the limited options the roadway network of Florida offers for long-haul travel.

We now discuss the influence of OD location variables that are significant only in the short-haul model. The number of unique routes increases as spatial separation increases from small (<10 miles) to moderate (10–20 miles), and then decreases in the longest segment. This may be because the road network in the Tampa Bay region does not offer many route options for local (<10 miles) or cross-county (>40 miles) travel. Employment densities in the OD TAZs are positively correlated with the number of unique routes; perhaps because a greater employment density is a surrogate for the heterogeneity of businesses served by freight trucks, which leads to a greater diversity in route choice. Similarly, average distance between the TAZ centroid of all trip ends to each trip end (a measure of spatial dispersion of the freight activity generators in the OD TAZs) is positively associated with the number of unique routes. The average distance from TAZ centroids to the nearest major arterial is a surrogate for how far trucks need to travel to reach a major arterial, which is negatively correlated with the number of unique routes.

In the context of network structure, long-haul OD pairs with higher ratio of toll roads to major arterials captured in the long ellipse are likely to have more unique routes; which can be explained by truck operators' deciding to look for alternative routes to avoid tolls (*2*). This variable is, however, insignificant in the short-haul model, mostly because our study region for the short-haul segment does not have many toll roads. In the long-haul model, OD locations with higher density of major and minor arterials in circular buffers around trip ends are likely associated with lower number of unique routes while OD locations with higher density of collectors are likely associated with more unique routes. This may be because the availability of more major and minor arterials around origins or destinations reduces the need to search for alternative routes. On the other hand, OD pairs with higher density of major and minor arterials and collectors in the short ellipse are likely to have more unique routes; probably because of an increased number of route options. For similar reasons, OD pairs with a greater proportion of minor arterials and collectors (with respect to major and minor arterials and collectors) in the short ellipse are likely to have a greater number of observed unique routes in the long-haul model. In the short-haul model, whereas the density of the minor arterials and collectors in the long ellipse has a positive influence on the number of observed unique routes, the influence of the proportion of major arterials (with respect to major, minor arterials and collectors) is negative. All these results highlight subtle but notable differences in the influence of network structure on the diversity of truck route choice between long-haul and short-haul travel segments.

Fractional Response Models for Average Path Size and Standardized Shannon Entropy of Usage

Table 3 presents the fractional response model estimation results for average path size (first set of rows) and standardized Shannon entropy of usage (second set of rows), estimated for OD pairs with at least two observed unique routes.

The average path size models have the results expected for both long-haul and short-haul segments. OD pairs with more observed unique routes are likely to have lower average path size (i.e., greater overlap). OD pairs with a higher proportion of trips on the most used route are likely to have higher average path size (i.e., lower overlap). The presence of a dominant route may imply the presence of other longer routes that do not overlap much and are less preferable. A greater spatial separation

Table 3. Estimation Results of Fractional Response Models for Average Path Size and Standardized Shannon Entropy of Usage

	Long-haul data		Short-haul data	
Variables in average path size model	Coefficient	t-stat	Coefficient	t-stat
Number of unique routes	−0.068	−9.66	−0.056	−9.74
Proportion of trips on the most used route	0.534	3.72	0.705	6.25
Direct OD distance (miles)	–	–	−0.012	−5.71
Direct OD distance indicator (1 if more than 200 miles, 0 otherwise)	−0.175	−1.95	–	–
Constant	−0.665	−4.73	−0.539	−4.70
Number of observations (OD pairs)	258		505	
Log pseudo likelihood at convergence	−101.08		−202.11	
Log pseudo likelihood for constant-only model	−106.37		−211.61	
Rho-square with respect to constant-only model	0.050		0.045	
	Long-haul data		**Short-haul data**	
Variables in standardized Shannon entropy model	Coefficient	t-stat	Coefficient	t-stat
Number of unique routes	0.059	5.76	0.052	6.37
Number of trips	−0.003	−4.83	−0.002	−3.74
Average path size	−1.345	−1.78	−1.667	−4.31
Average distance from centroid of all trip ends to each trip end (miles)	–	–	0.255	4.60
Constant	0.498	1.81	0.462	2.80
Number of observations (OD pairs)	258		505	
Log pseudo likelihood at convergence	−119.19		−224.05	
Log pseudo likelihood for constant-only model	−126.39		−241.98	
Rho-square with respect to constant-only model	0.057		0.074	

of OD pairs is associated with a lower value of path size (i.e., greater overlap) of the different unique routes; perhaps because an increase in spatial separation may reduce the number of travel routes offered by the network.

As expected for both long-haul and short-haul travel, when modeling standardized Shannon entropy, OD pairs with more observed unique routes are likely to have a higher Shannon entropy (i.e., more even distribution of trips among unique routes). OD pairs with more observed trips are likely to have a more even usage of the routes. OD pairs with a higher average path size value (or lower overlap) among unique routes demonstrate a more uneven usage of different routes. Such OD pairs with fewer overlapping routes are likely to have one or few dominant routes that are largely preferred over other routes. In the short-haul model, OD pairs with a greater average distance from the centroid of the trip end TAZs to all trip ends (i.e., greater spatial dispersion of freight activity generators) are likely to be associated with a more even distribution of trips among different unique routes. This suggests the influence of heterogeneity or spatial dispersion in trip ends on the heterogeneity of preferences for truck routes.

Summary and Conclusion

This study presents a comprehensive exploratory analysis of truck route choice diversity in Florida, for both long-haul and short-haul travel segments. To measure the diversity in truck routes between any OD pairs, we develop the following six metrics: (a) number of unique routes, (b) average commonality factor, (c) average path size, (d) non-overlapping index, (e) standardized variance of route usage, and (f) standardized Shannon entropy of route usage. The first metric measures the number of distinct truck routes between an OD pair. The next three metrics measure the extent of overlap (or lack thereof) among the routes. The last two metrics measure the evenness (or, otherwise, the dominance) of the usage of the routes between the OD pair. These three dimensions together provide a complete picture of the diversity in truck route choice between an OD pair. The diversity metrics were utilized to describe truck route choice diversity in Florida from a database of about 73,000 truck trips derived from more than 200 million GPS records. We produced a rich database of diversity metrics for 277 TAZ OD pairs for long-haul travel (trips longer than 50 miles) in the State of Florida and 527 TAZ OD pairs for short-haul travel (trips between 5 and 50 miles) in the Tampa Bay region. In addition, we compiled an extensive set of variables characterizing the truck travel, OD location, and network structure between these OD pairs that could potentially influence the extent of route choice diversity. Negative binomial regression models were estimated to explore the influence of these variables on the number of unique routes traveled between an OD pair. Further, fractional response models were estimated to explore the determinants of average path size (overlap among routes) and standardized Shannon entropy (evenness) of route usage.

The analysis suggests that short-haul truck travel exhibits greater diversity in route choice than long-haul travel, in terms of number of unique routes observed, the extent of non-overlap between unique routes, as well the evenness of usage of different unique routes. Within the long-haul segment, OD pairs that are more than 200 miles apart exhibit lower diversity than those that are closer. Among the short-haul OD pairs, local (< 10 miles) and cross-county (>40 miles) travel exhibit lower diversity than medium-distance travel. OD pairs in urban zones are associated with a greater diversity in route choice, because urban areas offer more network options for route choice. OD pairs with a greater number of trips and/or trucks observed (i.e., greater demand for travel) are associated with a higher number of unique routes. OD pairs with more variability in travel conditions (e.g., travel time) or routes that deviate more from a straight line have more diverse traveled routes. In addition, network structure between OD pairs has a considerable influence on the diversity of truck route choices. Another important finding is that the determinants and their extent of influence differs between short-haul and long-haul travel segments. For example, OD TAZ land use (employment density and diversity of freight activity locations) has a significant influence on route choice diversity only in the short-haul segment. Furthermore, network structure variables have differential impacts on route diversity between the two segments. Finally, OD pairs with a higher number of observed unique routes have greater overlap (i.e., lower average path size) and lower dominance of route usage, whereas OD pairs with fewer overlapping routes exhibit greater dominance of usage.

The findings from this study can be applied to improve algorithms for generating choice sets for truck route choice modeling. Specifically, route choice set generation algorithms can be customized based on truck travel demand, OD location, and network structure characteristics found to be influential in this analysis. Furthermore, the count models of number of unique routes can be used to determine the appropriate number of choice alternatives to be generated for route choice modeling and traffic assignment purposes.

An enhanced understanding of route diversity can also help improve truck routing policies and determine routing decisions during emergency situations. For example, OD pairs with only one unique route but high travel demand tend to be vulnerable in the event of roadway disruptions. Thus transportation planners can focus on such OD pairs to build a proactive plan for emergency detours as well as better maintenance of critical road segments. Furthermore, such OD pairs can be priority candidates for future roadway expansion to create additional route options.

Acknowledgments

This study was funded by FDOT and the Center for Advanced Infrastructure and Transportation (CAIT), a USDOT University Transportation Center. Three anonymous reviewers provided constructive comments on an earlier manuscript. The authors are grateful to the American Transportation Research Institute (ATRI) for providing the truck-GPS data used in this study.

Author Contributions

The authors confirm contribution to the paper as follows: study conception and design: Abdul Pinjari; analysis and interpretation of results: Trang Luong, Divyakant Tahlyan, Abdul Pinjari; draft manuscript preparation: Trang Luong, Divyakant Tahlyan, Abdul Pinjari. All authors reviewed the results and approved the final version of the manuscript.

References

1. Knorring, J. H., R. He, and A. L. Kornhauser. Analysis of Route Choice Decisions by Long-haul Truck Drivers. *Transportation Research Record: Journal of the Transportation Research Board*, 2005. 1923: 46–60.
2. Wang, Z., and A. Goodchild. GPS Data Analysis of the Impact of Tolling on Truck Speed and Routing: Case Study in Seattle, Washington. *Transportation Research Record: Journal of the Transportation Research Board*, 2014. 2411: 112–119.
3. Hess, S., M. Quddus, N. Rieser-Schüssler, and A. Daly. Developing Advanced Route Choice Models for Heavy Goods Vehicles Using GPS Data. *Transportation Research Part E: Logistics and Transportation Review*, Vol. 77, 2015, pp. 29–44.
4. Wang, Z., A. Goodchild, and E. McCormack. Measuring Truck Travel Time Reliability Using Truck Probe GPS Data. *Journal of Intelligent Transportation Systems*, Vol. 20, No. 2, 2016, pp. 103–112.
5. Liao, C. F. Generating Reliable Freight Performance Measures with Truck GPS Data: Case Study in Twin Cities Metropolitan Area, Minnesota. *Transportation Research Record: Journal of the Transportation Research Board*, 2014. 2410: 21–30.
6. Jan, O., A. Horowitz, and Z. R. Peng. Using Global Positioning System Data to Understand Variations in Path Choice. *Transportation Research Record: Journal of the Transportation Research Board*, 2000. 1725: 37–44.
7. Papinski, D., D. Scott, and S. Doherty. Exploring the Route Choice Decision-making Process: A Comparison of Planned and Observed Routes Obtained using Person-based GPS. *Transportation Research Part F: Traffic Psychology and Behaviour*, Vol. 12, No. 4, 2009, pp. 347–358.
8. Spissu, E., I. Meloni, and B. Snjust. Behavioral Analysis of Choice of Daily Route with Data from Global Positioning System. *Transportation Research Record: Journal of the Transportation Research Board*, 2011. 2230: 96–103.

9. Zhu, S., and D. Levinson. A Portfolio Theory of Route Choice. *Transportation Research Part C: Emerging Technologies*, Vol. 35, 2013, pp. 232–243.
10. Thakur, A., A. B. Zanjani, A. R. Pinjari, J. Short, V. Mysore, and S. F. Tabatabaee. Development of Algorithms to Convert Large Streams of Truck GPS Data into Truck Trips. *Transportation Research Record: Journal of the Transportation Research Board*, 2015. 2529: 66–73.
11. Pinjari, A. R., M. Kamali, T. D. Luong, and S. Ozkul. *GPS Data for Truck-Route Choice Analysis of Port Everglades Petroleum Commodity Flow*. Report BDK25-977-17. Florida Department of Transportation, 2015.
12. Kamali, M., A. Ermagun, K. Viswanathan, and A. R. Pinjari. Deriving Truck Route Choice from Large GPS Data Streams. *Transportation Research Record: Journal of the Transportation Research Board*, 2016. 2563: 62–70.
13. Cascetta, E., A. Nuzzolo, F. Russo, and A. Vitetta. A Modified Logit Route Choice Model Overcoming Path Overlapping Problems. Specification and Some Calibration Results for Interurban Networks. *Proc., 13th International Symposium on Transportation and Traffic Theory*, Lyon, France, 1996, Permagon Press, pp. 697–711.
14. Ben-Akiva, M., and M. Bierlaire. Discrete Choice Methods and Their Applications to Short Term Travel Decisions. In *Handbook of Transportation Science*, Vol. 23, Springer, Boston, Mass., 1999, pp. 5–33.
15. Shannon, C. E. A Mathematical Theory of Communication. *Bell System Technical Journal*, Vol. 27, 1948, pp. 379–423.
16. Papke, L. E., and J. Wooldridge. Econometric Methods for Fractional Response Variables with an Application to 401 (K) Plan Participation Rates. *Journal of Applied Econometrics*, Vol. 11, No. 6, 1996, pp. 619–632.

The Standing Committee on Trucking Industry Research (AT060) peer-reviewed this paper (18-05667).

The opinions, findings, and conclusions expressed in this paper are not necessarily those of FDOT or USDOT.

Article

TRR
JOURNAL OF THE TRANSPORTATION RESEARCH BOARD

Transportation Research Record
2018, Vol. 2672(9) 164–172

DOI: 10.1177/0361198118797508
journals.sagepub.com/home/trr
SAGE

Evaluating the Impact of Real-Time Mobility and Travel Time Reliability Information on Truck Drivers' Routing Decisions

Xiaoqiang Kong[1], William L. Eisele[2], Yunlong Zhang[1], and Daren B. H. Cline[3]

Abstract

This study represents the first research to investigate the impacts of two critical determinants—level of congestion and travel time reliability—on routing decisions with two groups of truck drivers having different levels of awareness of the real-time and the historical traffic conditions on available routes. The research analyzed 14,538 global positioning system devices recording trips on the I-495 crossing through Maryland, Virginia, and Washington, DC, and 2,166 trips in the Dallas area, to explore how truck drivers make routing decisions based on real-time travel time and reliability information by applying a binary logistic regression model. Researchers found that for truck drivers who are not familiar with the historical traffic and travel time conditions on available routes, real-time congestion information is a significant factor in their routing decision-making process, while travel time reliability is not a major consideration. For frequent truck drivers who are familiar with the historical traffic and travel time conditions on available routes, travel time reliability is a significant factor in their routing decision-making process, and traffic congestion information is not a significant factor. These results bring more accuracy to travel time prediction and provide valuable insights into traffic management and reliability performance measures. Moreover, this research provides statistical evidence proving the potential value of delivering travel time reliability information to drivers, traffic management agencies, and navigation map developers.

Route selection is a classic and fundamental transportation topic. Researchers have tried to predict routing decisions with stochastic and deterministic models (*1*). More accurate route prediction estimates help transportation planners or operators to better manage the transportation system and tackle congestion issues. Previous studies on travel behaviors have identified that certain determinants can generate significant impacts on routing selection. The determinants include the availability of alternate routes, travel time, travel time reliability, level of congestion, and concern for safety (*2–5*).

Research has shown that congestion is a significant contributor to routing decisions. Travel time reliability, defined as the variability of travel time, has been ignored in most routing studies and practices. Congestion and travel time reliability may play a role in passenger and freight vehicle drivers' route decision-making processes. However, assuming alternative routes exist, these two variables would only influence routing selection if drivers had a sense of traffic conditions, including the level of congestion and travel time reliability from historical traffic conditions or travel experiences, on the available routes. In other words, there would be no rational routing selection if the drivers were not familiar with the traffic conditions on alternate routes. Drivers usually follow the recommended route from map applications on their navigation devices. However, the recommended routes are normally calculated with current traffic congestion information. Most existing navigation devices or map applications fail to consider the travel time reliability. An experienced driver may prefer a route with higher travel time reliability but a longer average travel time

[1]Zachry Department of Civil Engineering, Texas A&M University, College Station, TX
[2]Texas A&M Transportation Institute, College Station, TX
[3]Department of Statistics, Texas A&M University, College Station, TX

Corresponding Author:
Address correspondence to Xiaoqiang Kong: jackxqkong@gmail.com

over the one with a shorter travel time suggested by the navigation device.

This research focused on the routing behavior of truck drivers. Routing behavior of passenger vehicle drivers has been extensively studied. A paper by Bitzios and Ferreira in 1993 suggested that fleet and passenger vehicle drivers could have different travel behaviors (*6*). In 2001, Eisele et al. proved that the travel time distribution between passenger vehicles and commercial vehicles is statistically different, and commercial vehicles show more variability in travel time (*7*).

Kittelson and Associates, and Rashidi Moghaddam et al. pointed out that the routing decisions of passenger travelers and freight movers are greatly affected by congestion level and travel time reliability (*8, 9*). The SHRP2 program addressed travel time reliability as a particular operational characteristic of the highway system. Strategically applied travel time reliability information could significantly reduce overall congestion and travel delay (*10*). Abdel-Aty et al. stated that the impact of travel time reliability on routing selection varies across individuals and that commuters prefer to minimize travel time when the travel time variation is consistent (*11*). Tilahun and Levinson demonstrated that route preference is decided through momentary recollections of past experience (*12*). Schneider et al. found that the efficiency of vehicle routes improved when drivers were familiar with the region. Schneider et al. also pointed out that this efficiency can be intensified with a higher level of familiarity (*13*).

The data collection methods used in past routing behavior research include stated preference surveys (SPs), travel logs, licenses plate studies, and Bluetooth® and global positioning systems (GPS) (*14*). The stated preference survey (SP) was one of the most frequently used methods before advanced data collection technologies such as Bluetooth and GPS emerged. By applying the SP method, researchers can easily control the experiment environment and make all the assumptions to simulate the real-world conditions. With the increasing accessibility and affordability of GPS devices, detailed disaggregated data have become more accessible for researchers to study travel behavior. The biggest advantage of using GPS data to study routing behavior is that those data reflect the real routing decisions made by travelers in the real world. Using empirical GPS data to conduct routing behavior research can overcome the weaknesses of the stated preference method, including the uncertainty of the stated routing choices made by participants in the laboratory reflecting their actual routing decisions. This is the biggest criticism of the widely used SP method.

The increasing accessibility of detailed disaggregated GPS data greatly improves the accuracy of routing behavior studies. However, this is not readily the case for freight data because freight companies want to keep their detailed disaggregated data confidential. This is the main reason routing behavior studies of truck drivers are not as extensive as those of passenger car drivers to date.

Objective

Two objectives of this research were identified: (a) to identify the impacts of congestion and travel time reliability on the routing behavior of truck drivers who are not familiar with available routes, and (b) to identify impacts of congestion level and travel time reliability on the routing behavior of truck drivers who have more familiarity with available routes.

To achieve the objectives, the authors applied the logistic regression model using empirical freight data from Maryland and the Dallas–Fort Worth (DFW) area of Texas. To the best of the knowledge of the authors, this is the first study to evaluate the effects of congestion level and travel time reliability on the routing behavior of truck drivers by using detailed disaggregate freight data with geospatial information.

Data Description

The Maryland Department of Transportation's State Highway Administration and the Texas Department of Transportation Planning and Programming Division provided two INRIX origin-destination datasets.

INRIX collected 400 gigabytes of data during February, June, July, and October 2015, including 19,690,402 trips and 1,376,720,203 waypoints in Maryland, Virginia, and Washington, DC, and collected 340 gigabytes of data from January to April 2016, including 29,138,492 trips and 1,608,100,000 waypoints in the DFW area. A waypoint is a geospatial point recorded by GPS devices that represents the location of the recorded vehicle.

These detailed disaggregate traffic data are composed of trip information and recorded waypoints with geospatial information. Trip information includes data provider type, device ID, geospatial type, provider driving profile, and vehicle weight class. Waypoint data provide geospatial information with longitude and latitude and the capture time of each waypoint. There are three data sources: consumer, fleet, and mobile. This research only used data from the fleet source.

Study Area Selection

To evaluate the impact of real-time traffic information and travel time reliability on drivers with different levels of familiarity with the alternative routes, it is important to eliminate other factors that could largely influence routing selection, such as toll or travel distance and

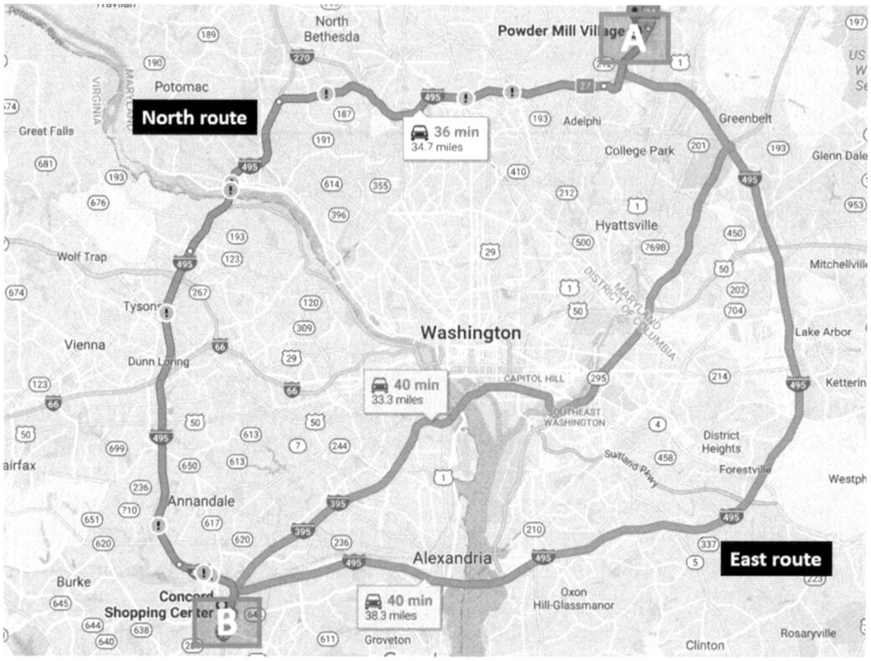

Figure 1. Selected research area in Maryland.

functional class of the available routes. Two identical or similar routes were necessary for this study.

In the Maryland, Virginia, and Washington, DC, area, the I-495 east route and I-495 north route were chosen to fulfill the requirements, as shown in Figure 1. The chosen routes provided a sufficient number of observations. As Figure 1 illustrates, from area A to B, the distance was 34.7 miles for the north route and 38.3 miles for the east route. The estimated driving time during free-flow speeds was 36 minutes for the north route and 40 minutes for the east route. The trips passing areas A and B were filtered from the original dataset for research purposes. The exported data were divided into two datasets based on the directional information: northbound trips and southbound trips. Trips that passed area A first and then passed area B were in the southbound dataset, and trips that passed area B first and then passed area A were in the northbound dataset.

In Dallas, Texas, a part of the I-30 and I-35E and a part of the I-30 were selected as the research areas. The north point A was defined as the interchange of I-30 and I-20 east of Dallas, and the south point B was defined as the interchange of I-35E and I-20 south of Dallas. The north route was a part of I-30 and I-35E, and went through downtown Dallas. The east route was a part of I-20. Their lengths were about 22.5 miles and 23.9 miles, respectively. Filtered trips were subset into two directional datasets: southbound dataset and northbound dataset.

Data Cleaning

The data-cleaning step involved extracting data of fleet trips passing points A and B from the raw datasets. Secondly, it was important to eliminate those trips passing points A and B but using other minor routes instead of the north route and the east route shown in Figures 1 and 2. Another necessary step was to eliminate the local delivery trips occurring between points A and B because the routing of those trips was dominated by the destinations of their deliveries in the regions. After the data cleaning process, the Maryland freight dataset was comprised of 14,538 fleet trips and 750,608 waypoints, and the Dallas freight dataset had 2,166 trips and 726,415 waypoints.

Frequent Trip Dataset and Non-Frequent Trip Dataset

The INRIX data provided a unique device ID for every recorded trip, which represented the identification number of each data collection device. This unique identification of the recording device allowed for categorizing trips by the frequency of the device ID. Trips with high occurrences of the same device ID were filtered out. It is highly possible that drivers assigned to those trips were frequent users of the selected research routes. Those drivers were possibly intuitively aware of the travel time reliability of the selected routes, that is, they had a higher level of familiarity with the traffic conditions on all available routes. Accordingly, in the Maryland dataset, trips

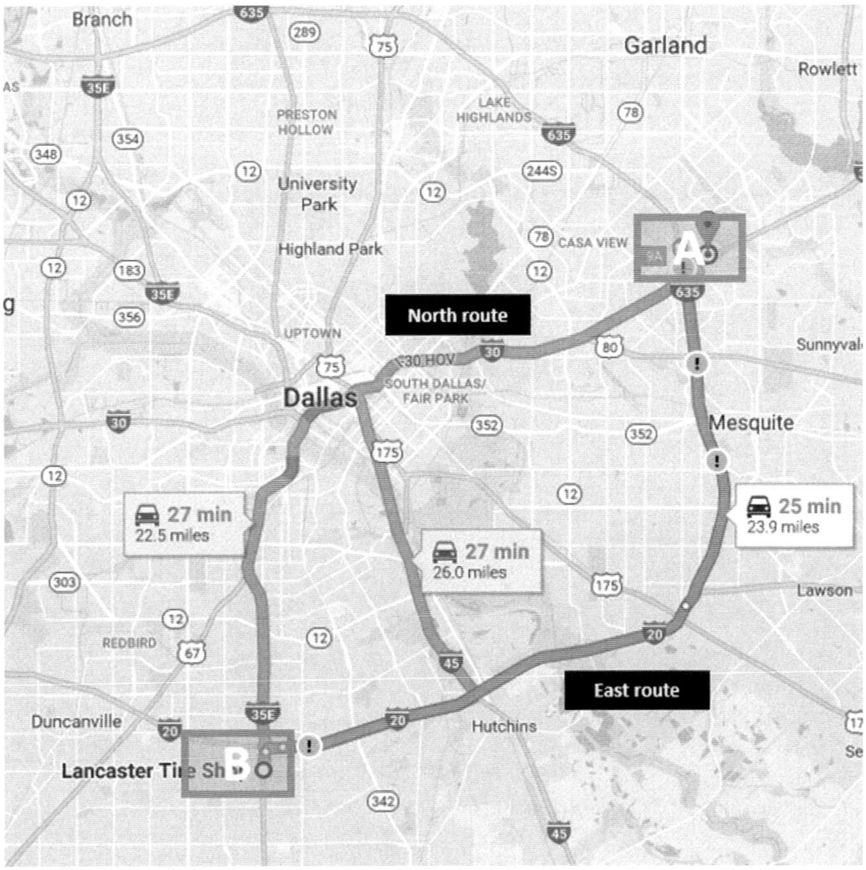

Figure 2. Selected research area in Dallas.

with frequent drivers had 295 observations. In Dallas, there were 132 trips with frequent drivers. The rest of the trips in the two datasets were defined as trips with non-frequent drivers.

Methodology

This research focused on two critical measures: the ratio of the congestion level on each route and the ratio of travel time reliability on each route. This study used real-world route choices to evaluate the effects of these two measures on truck drivers' routing decisions and to compare their impact on the routing decisions of truck drivers with various levels of familiarity with the traffic conditions on the alternative routes.

The routing decision for a truck driver to decide to travel on the north route or the east route was a discrete binary response. Thus, the binary response logistic regression was suitable to identify a relationship between routing choices and independent variables.

Travel Time Index—TTIRATIO

Transportation researchers use travel time index (TTI) to measure congestion intensity. The common TTI definition is the average travel time divided by the free-flow travel time, as shown in Equation 1 (*15*).

To compare the congestion level of the two routes, this research used TTIRATIO to represent the ratio of the congestion conditions on both routes. For consistency, TTIRATIO was always defined as the TTI of the north route over the TTI of the east route, as shown in Equation 2:

$$\text{TTI} = \frac{\textit{mean of travel time of all observations during one hour on selected route}}{\textit{free flow travel time on this route}} \quad (1)$$

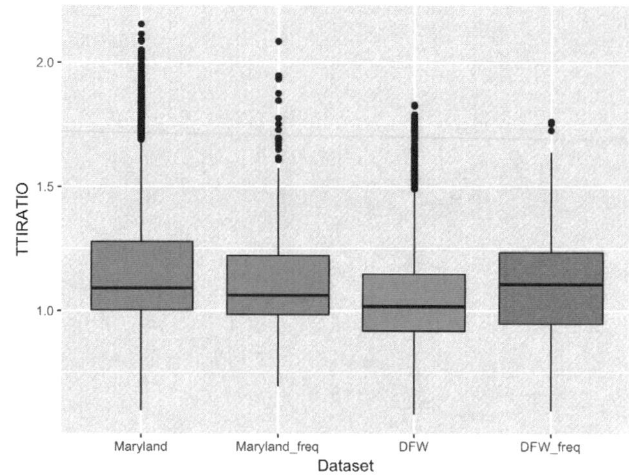

Figure 3. Boxplot of TTIRATIO in different datasets.

$$\text{TTIRATIO} = \frac{TTI\ of\ north\ loop}{TTI\ of\ east\ loop} = \frac{mean\ of\ travel\ time\ on\ north\ loop / free\ flow\ travel\ time\ on\ north\ loop}{mean\ of\ travel\ time\ on\ east\ loop / free\ flow\ travel\ time\ on\ east\ loop} = \frac{mean\ of\ travel\ time\ on\ north\ loop * free\ flow\ travel\ time\ on\ east\ loop}{mean\ of\ travel\ time\ on\ east\ loop * free\ flow\ travel\ time\ on\ north\ loop} \quad (2)$$

For each directional dataset for both Maryland and Dallas, the boxplot in Figure 3 illustrates that the majority of TTIRATIO values were concentrated between 1.0 and 1.4. Some extreme values did exist in the dataset.

Planning Time Index—PTIRATIO

The planning time index (PTI) is a measure of travel time reliability. The simplest method to measure PTI is a ratio of the 95th percentile peak-period travel time divided by the free-flow travel time, as shown in Equation 3 (*15*).

To compare the travel time reliability of the two routes, PTIRATIO was defined as the PTI of the north route divided by the PTI of the east route, as shown in Equation 4:

$$\text{PTI} = \frac{95\text{th}\ percentile\ of\ travel\ time\ of\ all\ observations\ during\ one\ hour\ on\ selected\ route}{free\ flow\ travel\ time\ on\ this\ route} \quad (3)$$

$$\text{PTIRATIO} = \frac{PTI\ of\ north\ loop}{PTI\ of\ south\ loop} = \frac{95\text{th}\ percentile\ of\ travel\ time\ of\ north\ loop / free\ flow\ travel\ time\ of\ north\ loop}{95\text{th}\ percentile\ of\ travel\ time\ of\ east\ loop / free\ flow\ travel\ time\ of\ east\ loop} = \frac{95\text{th}\ percentile\ speed\ of\ north\ loop * free\ flow\ travel\ time\ of\ east\ loop}{95\text{th}\ percentile\ of\ travel\ time\ of\ east\ loop * free\ flow\ travel\ time\ of\ north\ loop} \quad (4)$$

The PTIRATIO boxplot in Figure 4 shows that most of the values were between 1.0 and 1.4.

Logit Model

Logistic regression is a regression model that can establish a relationship between discrete results with various given independent variables (*16*, *17*). In this case, the binary response (0 and 1) represented the drivers' routing choices (1 = north route, 0 = east route). In a logistic regression model, the log odds of the response are modeled as a linear function:

$$Logit(p_i) = \log\left[\frac{p_i}{1-p_i}\right] = \alpha + \beta_i X_i \quad (5)$$

In which

p_i = probability of the response with given explanatory variables;

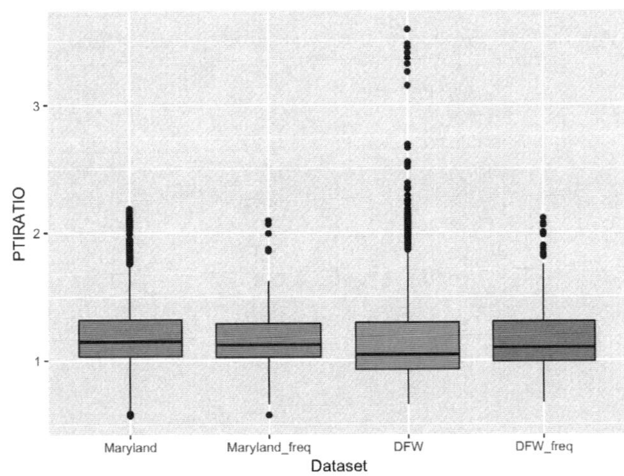

Figure 4. Boxplot of PTIRATIO in different datasets.

α = intercept parameter;

$\frac{p_i}{1-p_i}$ = odds of selecting the north route;

β_i = coefficient parameter of the vector of the explanatory variable TTIRATIO and PTIRATIO, which represents how much the explanatory variable can affect the log odds; and

X_i = explanatory variables.

For this analysis, researchers applied three models, as shown in Equations 6–8.

Model 1: logistic model with TTIRATIO variable only:

$$Logit(p_i) = \alpha + \beta_i \text{TTIRATIO} \quad (6)$$

Model 2: logistic model with PTIRATIO variable only:

$$Logit(p_i) = \alpha + \beta_j \text{PTIRATIO} \quad (7)$$

Model 3: logistic model with both variables:

$$Logit(p_i) = \alpha + \beta_i \text{TTIRATIO} + \beta_j \text{PTIRATIO} \quad (8)$$

Results

The odds ratio is used to explain the meaning of estimated coefficients of explanatory variables to better communicate the logit model results.

Figures 3 and 4 show that most of the TTIRATIO and PTIRATIO values were between 1.0 and 1.4. The traditional approach to explaining the coefficient of a continuous variable is that each estimated coefficient represents the expected change in the log odds ratio of driving on the north route with a 1-unit increase in TTIRATIO or PTIRATIO. However, in this case, increasing 1 unit in TTIRATIO or PTIRATIO was not practically meaningful. In this research, the expected change in the odds of using the north route was calculated using a 0.1-unit increment in the corresponding value. For example, a 0.1-unit increase in TTIRATIO meant that the average travel time on the north route was 10% greater than the average travel time on the east route.

Non-Frequent Trips

Table 1 shows the binary logistic regression results of the Maryland and Dallas datasets. In the Maryland dataset, the results of Model 1 (Equation 6) and Model 2 (Equation 7) show that TTIRATIO and PTIRATIO had negative influences on the route decision-making process. It is reasonable that truck drivers switch to an alternative route when they find their route is more congested or less reliable. There was a 7% decrease in the odds of choosing the north route. There was a 4% expected decrease in the odds of driving on the north route when PTIRATIO increased by 0.1 unit. Truck drivers tend to make rational choices to avoid more-congested and less-reliable routes.

The authors also wanted to investigate collinearity between TTIRATIO and PTIRATIO. Generally, higher congestion levels lead to lower travel time reliability. Model 3 (Equation 8) in Table 1 proves that the collinearity between TTIRATIO and PTIRATIO was present because the coefficient of PTIRATIO was positive. The

Table 1. Binary Logistic Regression Results of Maryland and Dallas Datasets

Dataset	Model	Variable	Coefficient	Significance	Odds ratio	Odds change
Maryland non-frequent trips	1	TTIRATIO	−0.70193	0.000	0.932	−7%
	2	PTIRATIO	−0.36530	0.000	0.964	−4%
	3	TTIRATIO	−0.73894	0.000	0.929	−7%
		PTIRATIO	0.07915	0.240	1.008	1%
Dallas non-frequent trips	1	TTIRATIO	−0.86460	0.000	0.917	−8%
	2	PTIRATIO	0.04844	0.330	1.005	0%
	3	TTIRATIO	−1.14330	0.000	0.892	−11%
		PTIRATIO	0.32000	0.006	1.033	3%

Note: Odds ratio = odds change with a 0.1-unit increase in the corresponding variable.

Table 2. Correlation between TTIRATIO and PTIRATIO

Trip type	Location	Correlation	Significance
Non-frequent trips	Maryland	0.52***	0.000
	Dallas	0.44***	0.000

$^{***}p < 0.01$.

odds ratio showed that the expected change in the odds of selecting the north route increased by 1% when the north route became less reliable. As shown in Table 2, the correlation between TTIRATIO and PTIRATIO was 0.52. Even this is not a very large correlation value; with a large dataset and a binary response, the correlation between these two variables could lead to questionable results.

In the Dallas dataset, the results of Model 1 illustrate that the impact of TTIRATIO on the routing selection was significant and had a negative value. Every 0.1-unit increase in TTIRATIO caused an 8% decrease in the odds of using the north route in the selected study area. In contrast to the results of the Maryland dataset in Model 2, the impact of PTIRATIO was positive, even with only the PTIRATIO variable in the model. The expected change in odds was almost zero with a 0.1-unit change in PTIRATIO. In Model 3, the correlation between TTIRATIO and PTIRATIO led to more questionable results. PTIRATIO became a significant factor with a positive value on the routing selection based on the results of Model 3.

As Table 1 illustrates, for both datasets, the results of TTIRATIO in Model 1 were consistent, whereas the results of PTIRATIO were not. The Dallas datasets showed the unintuitive result of PTIRATIO in Model 2, even excluding the collinearity issue with TTIRATIO.

Rational drivers would be more likely to avoid the routes with higher congestion levels and higher variation in travel times. However, truck drivers can only make a rational routing choice if they are aware of the information on congestion and travel time reliability. To the best of the knowledge of the authors, information on congestion levels via average travel times is provided by various navigation devices, but information on travel time reliability is not typically available. The awareness of travel time reliability requires route users to have a sophisticated understanding of the historical traffic conditions of those alternative routes and to have driven on those routes frequently. The travel time reliability information would affect routing selection only when the drivers were familiar with all available routes.

Frequent Trips

The results presented in Table 3 for frequent trips in the study areas verify the assumption that frequent fleet drivers are more likely to be aware of the travel time reliability of available routes and support the statement that PTIRATIO has a significant impact on routing selection. The results from the two datasets were consistent. As shown in Table 3, the estimated coefficients of all models for both datasets were negative. In both datasets, TTIRATIO was not a significant factor in Model 1. The p-values were 0.107 and 0.102, which were slightly larger than the 0.1 significance threshold. With a 0.1-unit increase in TTIRATIO, the odds of selecting the north route decreased by 9% in the Maryland frequent trip dataset and 10% in the Dallas frequent trip dataset. PTIRATIO proved to be a highly significant factor in Model 2 for both frequent trip datasets; the p-values were smaller than 0.05. In Maryland, a 0.1-unit increase in the PTIRATIO caused a 22% decrease in the odds of using the north route. In Dallas, a 0.1-unit increase in PTIRATIO caused an 11% decrease in the odds of choosing the north route. With a 0.1-unit increase, for the frequent trips, the impact of PTIRATIO on routing selection was greater than the impact of TTIRATIO.

The results of Model 3 for both datasets were not consistent. TTIRATIO was not statically significant, while PTIRATIO was significant. Table 4 shows that the correlation between TTIRATIO and PTIRATIO decreased

Table 3. Binary Logistic Regression Results of Frequent Truck Drivers

Dataset	Model	Variable	Coefficient	Significance	Odds ratio	Odds change
Maryland frequent trips	1	TTIRATIO	−0.98430	0.107	0.906	−9%
	2	PTIRATIO	−2.51050	0.007	0.778	−22%
	3	TTIRATIO	−0.01175	0.490	0.999	0%
		PTIRATIO	−2.50306	0.016	0.779	−22%
Dallas frequent trips	1	TTIRATIO	−1.03190	0.104	0.902	−10%
	2	PTIRATIO	−1.21420	0.031	0.886	−11%
	3	TTIRATIO	−0.66380	0.214	0.936	−6%
		PTIRATIO	−1.07280	0.052	0.898	−10%

Note: Odds ratio = odds change with a 0.1-unit increase in the corresponding variable.

Table 4. Correlation between TTIRATIO and PTIRATIO

Trip type	Location	Correlation	Significance
Frequent trips	Maryland	0.48***	0.000
	Dallas	0.23***	0.006

***$p < 0.01$.

in both frequent trip datasets. In the Maryland frequent trip dataset, Model 3 demonstrates that with a 0.1-unit increase in TTIRATIO, the odds remained almost the same, while PTIRATIO was unchanged. If TTIRATIO remained the same, a 0.1-unit increase in PTIRATIO led to a 22% decrease in the odds of north route usage. Similar results occurred in the Dallas frequent trip dataset for Model 3, where a 0.1-unit increase in TTIRATIO decreased the odds by 6%. In addition, a 0.1-unit increase in PTIRATIO lowered the odds by 10%.

Conclusion

This study evaluated the impacts of congestion level and travel time reliability on non-frequent truck drivers' and frequent truck drivers' route decision-making process by using detailed disaggregate freight trip data. Analyzing routing behavior using empirical data overcomes a major drawback of the traditional SP method because the data collected by embedded GPS recording systems in freight vehicles represents real routing decisions made by truck drivers in real traffic conditions.

The regression results in this paper from empirical data of frequent trip drivers verify the assumption that travel time reliability might not even be a concern for drivers who have a low level of familiarity on available routes. For truck drivers who were familiar with the available routes, travel time reliability was a major concern, and its influence on routing selection was significant. On the contrary, congestion level had only a slight impact on the route decision-making process of frequent truck drivers, and it was not significant. Moreover, with a 0.1-unit increase in TTIRATIO and PTIRATIO, the changes in the odds of using the north route were different. Comparison of the two frequent trip datasets proved that an increase in PTIRATIO caused more of a decrease in the odds of selecting the north route than that of an increase in TTIRATIO. This finding reveals that frequent fleet drivers value travel time reliability more than congestion level.

The results from non-frequent trip drivers illustrated that congestion level significantly influenced truck drivers' route selection, which is consistent with previous studies. However, the results failed to prove that travel time reliability was a significant factor in route selection. This result is not in accordance with most existing studies. Previous SPs showed that travel time reliability could significantly influence routing decisions. These studies are based on the assumption that participants can access travel time reliability information before making routing decisions. This assumption is often invalid in the real world because truck drivers who are not familiar with the available routes cannot generate a sense of travel time reliability for those routes.

Another finding was that the correlation between TTIRATIO and PTIRATIO decreased in the frequent trip datasets, and the collinearity between these two variables did not cause a problem when applied to the logistic regression model in the frequent trip datasets.

This study shows that the influences of travel time reliability and congestion level on truck drivers' routing selection could be significantly different when those drivers have different levels of familiarity with available routes. Future work should isolate the issue by connecting GPS devices to individual drivers and could extend this research to access the impacts of these two variables (TTIRATIO and PTIRATIO) on routing decisions of truck drivers of freight vehicles with different weight classes or different truck types. Future work should also apply the same methodology to passenger vehicle datasets and compare the results with freight vehicle datasets.

Acknowledgments

The authors are very grateful to Dr David Schrank of the Texas A&M Transportation Institute for his help throughout this research. Special thanks are given to the Maryland Department of Transportation's State Highway Administration and the Texas Department of Transportation's Transportation Planning and Programming Division for the invaluable freight datasets used in this study. Thanks to Dr Lingtao Wu for his helpful comments and suggestions on the research methods and for sharing his expertise in R programming.

Author Contributions

The authors confirm contribution to the paper as follows: study conception and design: XK, WLE, YZ; data collection: INRIX; analysis and interpretation of results: XK, WLE, YZ, DBHC; draft manuscript preparation: XK. All authors reviewed the results and approved the final version of the manuscript.

References

1. Tatineni, M., D. Boyce, and P. Mirchandani. Comparisons of Deterministic and Stochastic Traffic Loading Models. *Transportation Research Record: Journal of the Transportation Research Board*, 1997. 1607: 16–23.
2. Knorring, J., R. He, and A. Kornhauser. Analysis of Route Choice Decisions by Long-Haul Truck Drivers. *Transportation Research Record: Journal of the Transportation Research Board*, 2005. 1923: 46–60.
3. Hunt, J. D., and J. E. Abraham. Stated Response Examination of Factors Influencing Commercial Movement Route Choice Behavior. In *Applications of Advanced Technologies in Transportation Engineering*, 2004, pp. 72–77.
4. Liu, H. X., W. Recker, and A. Chen. Uncovering the Contribution of Travel Time Reliability to Dynamic Route Choice using Real-Time Loop Data. *Transportation Research Part A: Policy and Practice*, Vol. 38, No. 6, 2004, pp. 435–453.
5. Zhou, H., R. J. Benz, A. Voigt, and A. C. Mao. Bluetooth Travel Time/Speed Data Analysis for Houston-Galveston Regional Transportation Study. Presented at 95th Annual Meeting of the Transportation Research Board, Washington, D.C., 2016.
6. Bitzios, D., and L. Ferreira. Factors Affecting Route Choice of Commercial Vehicle Drivers. *Proc., 18th Australasian Transport Research Forum*, Gold Coast, Australia, 1993.
7. Eisele, W., L. Rilett, K. Mhoon, and C. Spiegelman. Using Intelligent Transportation Systems Travel-Time Data for Multimodal Analyses and System Monitoring. *Transportation Research Record: Journal of the Transportation Research Board*, 2001. 1768: 148–156.
8. Kittelson, and Associates. *SHRP 2 Report, Issue S2-L11-RR-1: Evaluating Alternative Operations Strategies to Improve Travel Time Reliability*. Transportation Research Board of the National Academies, Washington, D.C., 2013.
9. Rashidi Moghaddam, Z., M. Jeihani, and S. Peeta. Understanding the Roles of Traveler Perception of Travel Time Information and Its Reliability on Route Choice Behavior using a Driving Simulator. Presented at 96th Annual Meeting of the Transportation Research Board, Washington, D.C., 2017.
10. Council, N. R. *TRB Special Report 296: Implementing the Results of the Second Strategic Highway Research Program: Saving Lives, Reducing Congestion, Improving Quality of Life*. Transportation Research Board of the National Academies, Washington, D.C., 2009.
11. Abdel-Aty, M. A., R. Kitamura, and P. P. Jovanis. Investigating Effect of Travel Time Variability on Route Choice using Repeated-Measurement Stated Preference Data. *Transportation Research Record: Journal of the Transportation Research Board*, 1995. 1493: 39–45.
12. Tilahun, N. Y., and D. M. Levinson. A Moment of Time: Reliability in Route Choice using Stated Preference. *Journal of Intelligent Transportation Systems*, Vol. 14, No. 3, 2010, pp. 179–187.
13. Schneider, M., C. Doppstadt, B. Sand, A. Stenger, and M. Schwind. *A vehicle Routing Problem with Time Windows and Driver Familiarity*. Publications of Darmstadt Technical University, Institute for Business Studies (BWL), Department of Business Administration, Economics, and Law, Darmstadt, Germany, 2010.
14. Wolf, J., S. Hallmark, M. Oliveira, R. Guensler, and W. Sarasua. Accuracy Issues with Route Choice Data Collection by Using Global Positioning System. *Transportation Research Record: Journal of the Transportation Research Board*, 1999. 1660: 66–74.
15. Schrank, D., B. Eisele, T. Lomax, and J. Bak. *2015 Urban Mobility Scorecard*. Texas A&M Transportation Institute, College Station, TX, 2015.
16. Nowakowska, M. *Logistic Models in Crash Severity Classification Based on Road Characteristics. Transportation Research Record: Journal of the Transportation Research Board*, 2010. 2148: 16–26.
17. Dissanayake, S., and J. Lu. Analysis of Severity of Young Driver Crashes: Sequential Binary Logistic Regression Modeling. *Transportation Research Record: Journal of the Transportation Research Board*, 2002. 1784: 108–114.

The Standing Committee on Trucking Industry Research (AT060) peer-reviewed this paper (18-05155).

Article

TRR
JOURNAL OF THE TRANSPORTATION RESEARCH BOARD

Transportation Research Record
2018, Vol. 2672(9) 173–183

DOI: 10.1177/0361198118798730
journals.sagepub.com/home/trr
SAGE

Delivery Process for an Office Building in the Seattle Central Business District

Haena Kim[1], Linda Ng Boyle[2], and Anne Goodchild[1]

Abstract

Movement of goods within a central business district can be very constraining with high levels of congestion and insufficient curb spaces. Pick-up and delivery activities encompass a significant portion of urban goods movement, and inefficient operations can negatively impact the already highly congested areas and truck dwell times. Identifying and quantifying the delivery processes within the building is often difficult. This paper introduces a systematic approach to examine freight movement, using a process flow map with quantitative delivery times measured during the final segment of the delivery process. This paper focuses on vertical movements such as unloading/loading activities, taking freight elevators, and performing pick-up/delivery operations. This approach allows visualization of the components of the delivery process and identification of the processes that consume the most time and have greatest variability. Using this method, the delivery process for an office building in downtown Seattle was observed, grouped into three major activities (or steps): 1. Entering, 2. Delivering, 3. Exiting. This visualization tool provides researchers and planners with a better understanding of the current practices in the urban freight system, and helps identify the non-value-added activities and time that can unnecessarily increase the overall delivery time.

The demand for goods and services is rapidly increasing in cities, in part because of the rise in online shopping and more varied delivery options. Package delivery services are a large portion of the logistic sector (*1*). Apart from long-distance intercity freight movements, the final leg of urban freight delivery involves various activities from loading/unloading goods to pick-up and delivery operations, serving the end customers (*2*). This final leg can be complex and costly, accounting for up to 28% of a product's total transportation cost (*3*).

The objective of this paper is to understand the process associated with the final leg of the urban freight delivery system. The focus of most urban freight research has been on vehicle mobility such as freight traffic and parking management. This has led to a lack of understanding of fundamental aspects of the urban goods movement, such as pick-up and delivery activities within the building (vertical movement). The process time spent outside of the vehicle can be much longer than the driving time, as much as 87% of the entire urban freight delivery process (*4*, *5*). However, analysis and documentation of the off-vehicular activities are limited, and there is sparsity of data to enable researchers to examine the overall system. An in-depth analysis of the driver's delivery process and performance for the final leg of the delivery process plays a vital role in understanding and improving urban freight delivery.

Understanding the vertical goods movement within the building is important because it can directly influence the roadway capacity and performance. The lack of curbside space, because of excessively long stays by delivery workers, could increase urban congestion as other delivery vehicles circle the city blocks while looking for parking spaces (*6*). Vertical movements can also encompass non-value-added time or time that unnecessarily increases the overall delivery time with no corresponding benefit to the customer (*7*). These factors can cause negative cascading impacts on road congestion, adding costs and pressures to the trucking industry, building management, and city officials.

This paper introduces the lean philosophy and value stream mapping (VSM) approach to identify areas of improvements within the delivery process flow of an office building in downtown Seattle (*7*). This approach provides a way to measure the delivery time for each activity within a freight delivery process. This is especially useful when the delivery process includes several

[1]Civil and Environmental Engineering, University of Washington, Seattle, WA
[2]Industrial & Systems Engineering, University of Washington, Seattle, WA

Corresponding Author:
Address correspondence to Linda Ng Boyle: linda@uw.edu

activities and tasks that differ by carriers, types of goods, and types of delivery vehicles. With the VSM approach, dwell times and failed deliveries can be better understood as it decomposes the delivery process at the micro level.

The study begins with the creation of a process flow map of an office building in a central business district, which provides information on each delivery task and identifies areas where bottlenecks and non-value-added times could occur. This map allows one to visualize the components of the delivery process as well as those tasks that are conducted by all carriers and those that are not. Identifying the processes that consume the most non-value added time and the greatest variability will help to identify strategies to improve the overall urban freight system and be better accountable for extended truck dwell times and failed deliveries.

Literature Review

Several studies regarding the "last mile" of urban freight deliveries have been conducted. Although urban freight delivery may vary by the characteristics of each city, there are activities that are observed regardless of the delivery type (e.g., loading goods, checking in, maneuvering within buildings). It is therefore important to understand how operations across various urban freight deliveries can contribute to congestion and affect the overall quality of life for residents, retailers, freight carriers, and government agencies.

Allen et al. conducted a comprehensive review of 162 freight studies (from the 1960s to 2008) in 18 countries (*8*). The data were based on freight operations from the UK, U.S., the Netherlands, Germany, and Italy, indicating active efforts worldwide on improving urban freight systems. The review noted three primary purposes of these freight studies (*8*):

1. to gain an understanding of urban freight operations,
2. for policy and decision-making, and
3. for use in urban freight modeling.

The review also highlights a need for a systems approach to measure inefficiencies and to provide better communication between public and private entities for the development and implementation of freight plans and policies (*5, 9*). A more quantitative approach can be achieved with the study method presented here, which can also provide a useful tool for highlighting the impact of freight transport movements to stakeholders, either directly (receivers, shippers, and carriers) or indirectly (city authorities, and residents). Rhodes et al. also state that quantifying and addressing both horizontal and vertical "last mile" inefficiencies are important from a planning perspective (*6*).

VSM is an effective tool for identifying system efficiencies and has been used in industries related to manufacturing and health care services (*10–12*). The urban freight delivery process consists of many activities and parties, with few standardized processes. Using a systems approach provides insight on dwell times and failed deliveries by decomposing the delivery process. Cherrett et al. emphasized the importance of understanding freight vehicle dwell times (i.e., the times the vehicle remains stationary) because shorter dwell times could reduce traffic delays and minimize environmental impacts of freight (*13*). A more in-depth understanding of vehicle dwell time was proposed by Allen et al., with 12 steps of activities performed by a goods vehicle driver when making a delivery (*4*). This paper examines the final segment of the delivery process and considers the many steps associated with the delivery tasks.

One factor that affects dwell time is the parking location. The parking options can be classified as on-street, off-street, and alternative options such as double parking or illegal parking (*14*). The decision of where to park may be influenced by the package size and weight, and distance to the recipient's location (*14*). The existence of off-street loading facilities does not necessarily mean they are always used (*13*). According to the Cherrett et al.'s review of the recent UK studies, the proportion of on-street and off-street parking varied by the type of location served (*13*). Deliveries made in shopping centers tend to include a higher percent of off-street parking facilities, whereas deliveries made to local shops on the street use more on-street parking (*13*). Based on the parked location, the levels of conflicts with pedestrians, bicyclists, and other vehicles can differ, which may cause extra delivery time.

Understanding the total delivery time as well as the time for each delivery task is important when imposing time restrictions for parking and freight loading facilities (*9, 15*). Too little time given at the loading facilities may lead to excessive enforcements using fines for parking/loading, clamps, and towing-away. These can affect delivery workers' operation significantly (*4*). Too much dwell time can be an indicator of an inefficient process with fewer on-time deliveries.

Another factor that may influence dwell time is the time associated with using elevators. Pivo et al. state that drivers would worry less about the congestion if slower traffic could be offset with faster elevator service (*16*). Delivery workers are required to use the freight elevators in many office buildings no matter the size of the delivery. The bottleneck may occur because the number of freight elevators in the office buildings is limited to one or two (*16*). Morris points out the lack of requirements regarding the number of freight elevators in commercial buildings of many American cities, including Atlanta, Boston, Chicago, Dallas, New York, and Seattle (*2*). Even though each building has different freight elevators, this study can provide insights on how much time associated with elevators can take up in the total delivery time for similar office buildings in other urban areas.

Failed deliveries is another central issue in the urban freight system. Failed deliveries are very costly as the driver needs to return (sometimes multiple times) before a successful delivery. A 2016 Interactive Media in Retail Group report in the UK showed that failed deliveries can cost up to £780 million (equivalent to $1 billion US dollars) (*17*). The cost burden for failed deliveries has prompted interest in solutions that can help streamline the final segment of the delivery process.

Data Collection

The selected office building in downtown Seattle has 62 floors with approximately 5,000 tenants, including gift shops, restaurants, and coffee shops. Each floor has a unique floor configuration, which allowed the research team to capture various delivery processes. Types of observed pickups and deliveries include office supplies, parcels, food items, assorted mail, recycling, and furniture.

The building is surrounded by four one-way streets (see Figure 1). There are seven 30-minute commercial loading zones and four mixed zones combined with 30-minute commercial loading zones and passenger drop-off zones. The loading bay has seven parking spaces with a 30-minute limit. The security booth at the loading bay has a full-time security guard and is open between 6 a.m. and 6 p.m. Inside the loading bay, there are two freight elevators which require a security fob to use. Delivery

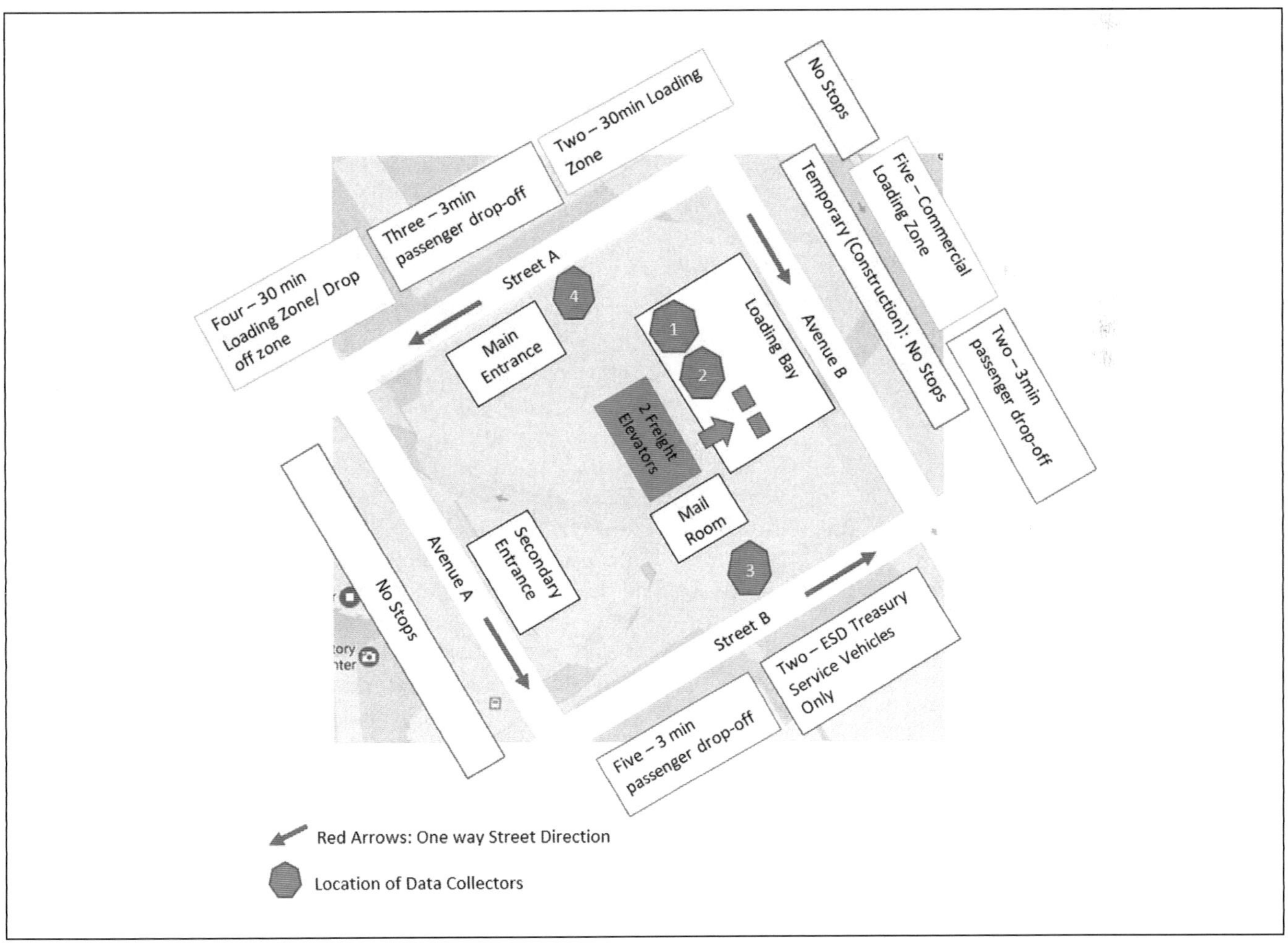

Figure 1. Configuration of the observed office building.

workers obtain a freight elevator fob from the security guard by handing in their government-issued identification card for security.

A mobile application for use in a tablet computer (Apple iPad) was developed for collecting real-time data on the delivery process. The predefined options included load/unload, waiting for/taking elevators, signing for deliveries, and much more. To identify the start of a task, the data collector taps a task button. This immediately begins recording the tasks in a web-based database and can be stopped once by tapping a sub-button once the task is finished. With this approach, each delivery task is time-stamped and the duration of each task is accurately computed even when the tasks are executed concurrently. Tasks that were not predefined could also be entered manually in the application. Other information that was recorded included whether or not a package was successfully delivered and other data-collection notes.

The data-collection process took place over five business days between January 31 and February 4, 2017, between the hours of 9:00 a.m. and 4:00 p.m. The data-collection team consisted of four people, who were trained to observe and collect data using the tablet application. The data collectors would wait until they observed a truck parking in either the loading bay or the street curbs near the building. They would then approach the delivery worker and ask permission to shadow and observe his or her delivery process. Given the observational nature of the data collection and where researchers approached the worker, these deliveries were most likely not express deliveries. Data from the tablet were then used to construct the delivery process flow map that showed the detail task durations and delivery sequences.

Process Flow Map

The process flow map is focused on the final segment of the delivery process, which is sometimes referred to as the final 50 feet (*18*, *19*). This segment includes out-of-vehicle activities and begins with the driver parking the vehicle and ends at the point when the driver drives away from the building. There are three major steps in this segment and they are further subdivided into subtasks:

1. Entering (e.g., parking vehicles, security check-in, unloading goods, waiting for elevators to go to the destination)
2. Delivering (e.g., taking an elevator to the destination, delivery or pick-up actions, waiting for elevators to go back to truck)
3. Exiting (e.g., taking an elevator to go back to truck, loading a hand truck back to truck, security check-out)

This process flow map (see Figure 2) shows the delivery activities and subtasks that can be performed in parallel and those that require a sequence of events for task completion. The square boxes represent the set of actions, and the diamonds represent the decisions made along the processes. Based on the collected data, the most shared common delivery subtasks at the study location was identified. Although each delivery person can generate many paths, the common delivery subtasks provide insights for areas where more effective delivery strategies can be deployed.

Table 1 summarizes the time duration of each subtask, in the same order shown in Figure 2. The ratio of standard deviation (sd) to the mean is used to identify processes that have the greatest variation. Those ratios greater than one are activities that were further examined.

Entering

Data collection began as soon as a truck parks at any of the designated on-street or off-street (loading bay) commercial loading zones. Drivers can enter the building through the loading bay or the main entrance on Street A or secondary entrance on Avenue A (see Figure 1).

In this study, 90% of the drivers (28 out of 31) parked in the loading bay to unload goods. Large volumes of office supplies could be a big contributor to this result. The mean duration for the "parking at loading bay" process (40 seconds) was slightly longer than for "parking at the street curb" (33 seconds). In tight spaces such as a loading bay, the drivers' maneuvering ability was limited, and several forward and backward maneuvers were necessary, as expected (*12*). During parking activities, conflicts may also occur with pedestrians, bicyclists, and other passing-by vehicles.

Depending on the location of the parked vehicle, the driver would leave the cargo compartment open or closed. In most cases, drivers at the loading bay would leave the door open because the security guard was always present. Drivers who parked on-street tended to keep the cargo compartment closed when they left the truck for delivery. Two types of cargo compartment doors were observed: rolling and swing doors. Some heavy duty trucks had a lift that goes up and down at the back of their cargo compartment to assist the driver with entering and exiting the cargo compartment. When parking, the drivers had to allow extra space if they had swing doors or the lift. Some drivers had to lock the door after closing the cargo compartment. The wait time for the lift or locking the cargo compartment can add to the total truck dwell time.

Once a delivery worker exited the truck, he or she would walk to either the security booth to check-in or the cargo compartment of the truck to unload. Several office buildings in downtown Seattle have their own unique security check-in processes. At this office building, the

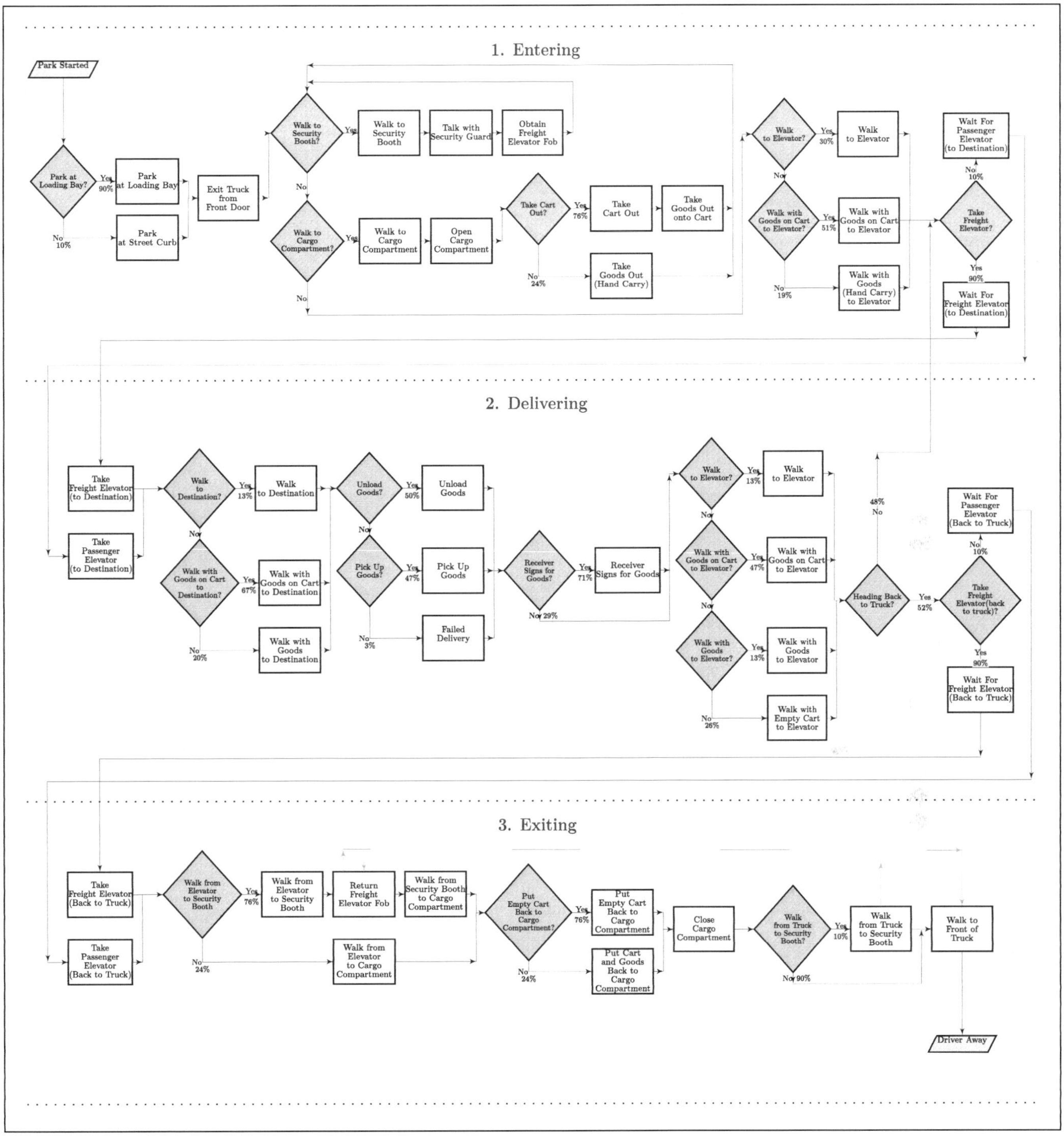

Figure 2. Delivery process flow map ($n = 31$).

delivery workers were required to check-in with a security guard to obtain a freight elevator fob by exchanging their government-issued identification cards. The duration of the check-in process could vary depending on the familiarity of the drivers with the security guard. If the driver made regular deliveries to the building and was familiar with the security guard, the check-in process would be fairly quick. However, the delivery person may also take additional time to converse with the security guard. Depending on the time of day, a bottleneck could occur if multiple delivery workers arrived at the same time for check-in.

Drivers would often carry goods by hand for small and light deliveries, and by a hand truck or dolly for large and heavy deliveries. The most common method to

Table 1. Duration (in Seconds) of Tasks within the Delivery Process Flow for Common Paths ($n = 31$)

Tasks	Subtasks	mean	sd	$\frac{sd}{mean}$	min	max	mode
1. Entering	a. Parking ended at loading bay	40	40	1.01	9	165	12
	b. Parking ended at street curb	33	30	0.89	12	67	12
	c. Exit truck from front door	14	18	1.28	3	90	3
	d. Walk from truck to security booth	27	21	0.75	4	102	8
	e. Talking with security guard	83	83	1	5	242	5
	f. Obtain freight elevator fob	34	21	0.63	2	77	12
	g. Walk to cargo compartment—Entry	17	16	0.93	2	67	2
	h. Open cargo compartment	20	20	1.03	3	75	7
	i. Take cart out	27	30	1.14	1	124	1
	j. Take goods out and place on cart	54	56	1.04	3	202	12
	k. Take goods out	51	23	0.45	26	84	26
	l. Walk to elevator	54	24	0.45	26	105	44
	m. Walk with goods on cart from truck to elevator	44	28	0.64	9	129	29
	n. Walk with goods from truck to elevator	51	29	0.57	17	105	29
	o. Wait for freight elevator (to destination)	31	49	1.56	3	193	10
	p. Wait for passenger elevator (to destination)	52	22	0.42	32	76	32
2. Delivering	a. Took freight elevator (to destination)	75	93	1.23	4	486	35
	b. Took passenger elevator (to destination)	67	50	0.76	36	126	36
	c. Walk from elevator to destination	55	19	0.35	36	81	36
	d. Walk with goods from elevator to destination	87	65	0.74	26	196	26
	e. Walk with goods on cart from elevator to destination	49	50	1.03	10	200	10
	f. Unload goods	57	59	1.04	11	221	21
	g. Pick up	37	11	0.31	23	58	35
	h. Receiver signs for goods	55	77	1.41	3	404	11
	i. Walk from destination to elevator	38	16	0.43	25	64	25
	j. Walk with goods on cart from destination to elevator	40	50	1.23	3	193	10
	k. Walk with goods from destination to elevator	44	14	0.32	25	56	56
	l. Walk with empty cart from destination to elevator	39	52	1.34	2	180	2
	m. Wait for freight elevator (back to truck)	63	35	0.55	20	124	59
	n. Wait for passenger elevator (back to truck)	NA	NA	NA	NA	NA	NA
3. Exiting	a. Took freight elevator (back to truck)	148	155	1.05	2	635	36
	b. Took passenger elevator (back to truck)	78	36	0.47	54	120	54
	c. Walk from elevator to security booth	27	29	1.07	3	97	18
	d. Return freight elevator fob	44	38	0.84	5	156	6
	e. Walk from security booth to cargo compartment	34	23	0.67	9	70	9
	f. Walk from elevator to cargo compartment	30	19	0.64	9	60	9
	g. Put empty cart back into cargo compartment	42	26	0.62	5	98	33
	h. Put goods and empty cart back into cargo compartment	36	9	0.24	28	47	28
	i. Close cargo compartment—Exit	17	16	0.97	5	54	5
	j. Walk from truck to security booth—Exit	6	5	0.76	2	11	2
	k. Walk to front of truck	22	40	1.8	4	210	8
	l. Enter truck from front door	28	32	1.13	1	124	7

unload goods was by hand, but in the case of heavy deliveries, special equipment such as a forklift or pallet jack was used.

Doors of the cargo compartment can be located either at the back or side of the truck. Of the drivers observed, 76% carried goods on dollies or hand trucks and 24% hand-carried goods. In Figure 2 and Table 1, hand trucks or dollies are represented as "cart." For pickups, the drivers skipped unloading activities and walked to the elevator directly after the security check-in.

The loading bay was located inside the building's parking facilities where two freight elevators were accessible next to the loading dock. However, the passenger elevators were located further away from the loading dock but very close to the lobby, next to the main entrance. Therefore, the delivery workers who entered the building through the main or secondary entrances were more likely to use passenger elevators. Although the passenger elevators were approximately 2 times faster than the freight elevators, the passenger elevators have a higher volume of frequent riders. This is reflected in the mean wait time (52 seconds) for the passenger elevators.

The mean wait time for the freight elevators to go from the loading bay to upper-level floors was 31 seconds, but the range in wait time was quite large (from 3 to 193 seconds) This is much greater than the wait time for the freight elevator from destination back to the loading bay. This is not surprising as the delivery person at the loading bay may have to wait a long time for the elevator if it is at the top-most floors.

Delivering

Having a unique floor configuration and delivery policy for each office made it challenging for the delivery workers who visited the building for the first time. Some offices required the delivery workers to use an interphone to enter the office suites, some were open to the public, and some had a receptionist who received and signed for goods on behalf of other office workers. If the office did not have a receptionist, the delivery worker had to find an individual receiver to deliver the goods.

Once the delivery workers arrive on the floor of their destination, they performed either delivery or pick-up activities. The mean time spent for pick-up (37 seconds) was much shorter than the mean time for delivering goods (57 seconds), which often involved unloading activities. The high volume of goods could lead to a longer time when delivery workers are required to unload each item one at a time by hand. If the high volume of goods can be unloaded in bulk, a shorter unload time can be achieved.

In total, 3% of the observed deliveries failed (or were not delivered). Each company has different policies on failed deliveries: most delivery workers look for an alternate person to sign for goods. Some delivery workers can drop off goods on the receiver's desks without obtaining a signature from anyone. Some companies allowed the delivery workers to leave the site after sending a picture(s) of the dropped-off goods and locations to the clients remotely. Company policies can also vary by the types of goods. Better communication between the delivery workers and the receivers could help reduce the failed first delivery. A simple notification system could also allow both the delivery workers and the receivers to share information on estimated arrival time or the wayfinding instructions. When the receivers are notified before the delivery arrivals, the chance of failed deliveries may be reduced. When the delivery workers are well informed about the building layouts, the chances of being lost in the building could be decreased as well.

As expected, the average time for walking with goods (44 seconds) or goods on the cart (40 seconds) was longer than the average time for those walking without any goods (38 seconds) or with an empty cart (39 seconds). For multiple deliveries, the drivers would repeat delivery and pick-up activities within the building.

The mean wait time for the freight elevator to go back to the loading bay was 63 seconds. To avoid wait time for the elevator, some delivery workers would hold the freight elevator open by blocking the elevator door until he or she comes back after completing deliveries. These delays can compound and create a continuous delay of deliveries for other drivers who are waiting for the freight elevator to other floors. Lastly, freight elevators were used by individuals that did not have any goods or freight. These individuals chose not to use the passenger elevators for their convenience, which added additional and unnecessary stops. In general, elevator bottlenecks have a significant impact on office buildings with many floors.

Exiting

The mean time in the freight elevator to go back to the loading bay was 148 seconds. Once the driver returns to the loading bay or main lobby after completing deliveries or pickups, he or she can either walk to the security booth or go back to the truck. In this study, 76% of the drivers walked from the elevator to the security booth first. At the security booth, the drivers return the elevator fob to the security guard, where they would get their identification card back.

During the peak delivery hours (10:00 a.m. to noon), the security guard experienced difficulty in accommodating all drivers and a queue began to form. In these situations, a securely automated check-in and check-out kiosk could be set up to help expedite the process. The building had a pre-screening program where some delivery workers can obtain the freight elevator fob in advance and use it without the check-in process.

The mean time for loading the empty cart or placing picked-up goods in the cargo compartment was 36 seconds with low variations (SD = 9). Closing cargo compartment was also fairly quick, with a mean time of 17 seconds. Once the drivers enter the vehicle, some of the drivers wait inside the truck to complete their paperwork, with or without the engine on. Some drivers can avoid paperwork by using a digital device that helps provide real-time paperless communication between the field and office workers.

Results

All delivery workers made at least one delivery to the building, with 26% (8 out of 31 delivery workers) making multiple deliveries. For those who visited more than one floor, the maximum number of deliveries observed was seven (see Figure 3).

Total Delivery Time

The tablet application allowed collection of data on dwell time for the three main delivery steps, the subtasks, as

Figure 3. Number of deliveries per truck.

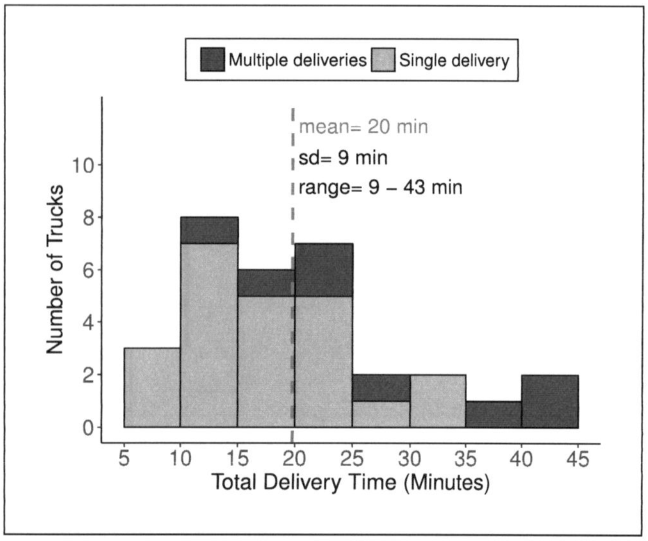

Figure 4. Distribution of total delivery time (n = 31).

well as total delivery time. Figure 4 summarizes the delivery time measured for each delivery truck by single and multiple deliveries. The mean total delivery time was 20 minutes. This is reasonable, as the parking time limit at the studied location was 30 minutes. The delivery times ranged from 9 minutes to 43 minutes. The range of total delivery times is comparable to that found in a previous study by Cherrett et al., which indicated 9 and 8 minutes as the shortest mean van dwell times according to 2001 and 2008 Winchester surveys (*13*).

As shown in Figure 5, the times for each of the three delivery steps are 7, 8, and 5 minutes for entering, delivering, and exiting, respectively. The percent of total delivery times are 35% for entering, 40% for delivering, and 25% for exiting. Unloading and organizing goods before deliveries encompass a great deal of the time in the delivery process. The highest variation is shown in the delivering step at the final destination. Differences in delivery workers' experience and familiarity with the building, and the level of interaction with the receptionist are some of the contributors to these variations.

Variation in Delivery Time

The subtasks in Table 1 are visualized in Figure 6, which shows the distribution of delivery time based on the ratio of the standard deviation and the mean (sd/mean). The variation (sd) for most processes was close to the mean. The highest variations (sd/mean > 1.5) was observed in the activities, "walk to front of truck" and "wait for freight elevator (to destination)." The six tasks with the largest sd/mean are discussed further in this section.

Task 3k (sd/mean = 1.8): Walk to Front of Truck. The high variation in the activity "Walk to front of truck" was caused by one specific case, which may not be as common in other delivery processes. A delivery worker failed to deliver the goods but spent 210 seconds lingering in the loading bay, walking back and forth between the front and the end of the truck. Although this is not common, it is important to examine because it is associated with a failed delivery.

Task 1o (sd/mean = 1.56): Waiting for Freight Elevator (to Destination). Time spent while waiting for the elevators increased the overall truck dwell time. The wait time at the bottom floor (loading bay or main lobby) may include elevator travel distance over the entire building and is affected by use during peak periods. Wait times from the office floor can be much quicker for deliveries to the middle floors of the building.

High variation in the wait time for freight elevator could also be related to the elevator age. Both freight elevators in the observed office building were installed in 1990 (27 years ago). The frequent breakdowns and slow speeds can contribute to the bottlenecks observed. Weather was also an observed factor. Strong winds from outside the building would come in through the wide opening of the loading bay entrance, and prevent the freight elevator doors from being fully closed, causing delays on the loading bay level. In these situations, a security guard would request that the delivery workers press the "close" button until the door was fully closed. However, those delivery workers that were not familiar with this defect may instead wait an excessively long time for the elevator to automatically close.

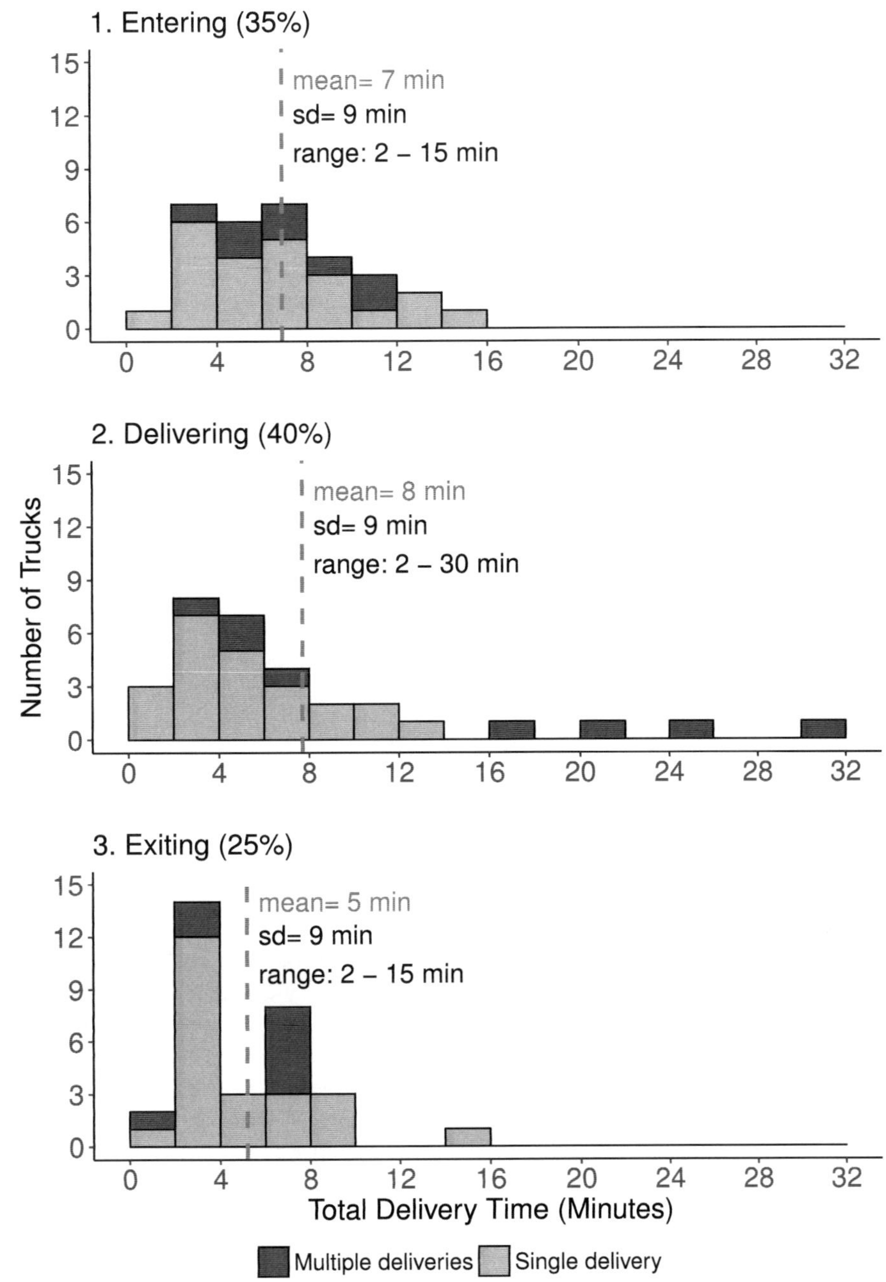

Figure 5. Delivery time by main delivery process steps (n = 31).
Note: The letters and numbers within the horizontal bars denote the tasks and subtasks identified in Table 1. For example, *3k* represented Task 3 (Exiting) and Subtask k (Walk to front of truck).

Task 2h (sd/mean = 1.41): Receiver Signs for Goods. Out of all the deliveries, 71% of the drivers were required to obtain a signature from the receiver. The time it took for the receiver to sign for the received goods varied greatly and depended on the quantity and type of received goods. For regular deliveries, the receivers anticipate certain types and amounts of goods being delivered, resulting shorter time in signing for goods. When the multiple types of delivered goods are not organized before the delivery, the receiver took a long time to sort and count each item, increasing the total dwell times for the delivery workers at the final destination.

Task 2l (sd/mean = 1.34): Walk with Empty Cart from Destination to Elevator. The walk time within the final

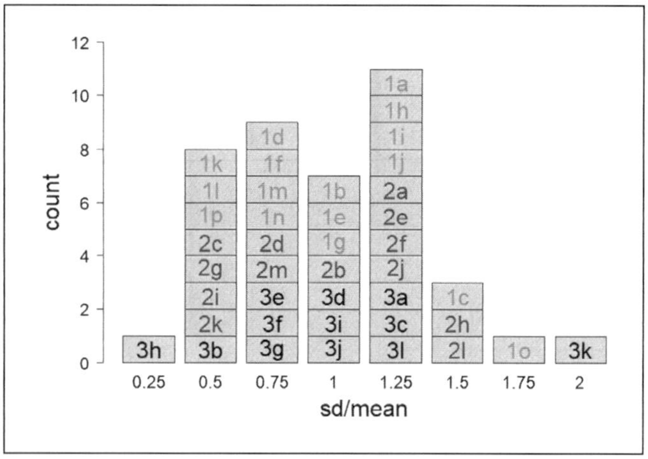

Figure 6. Ratio of standard deviation and mean (sd/mean) of delivery time.

destination also showed high variation, especially for those who walked with hand trucks or dollies, referred to as "cart" in Table 1. Depending on the size and types of cart, maneuvering the office areas with dollies and hand trucks could be time-consuming because the delivery workers required extra time to hold the office doors for hand trucks or dollies. Sharing the building infrastructure information, such as the size limits of the hallways or office doors, could help the delivery workers to plan out the deliveries ahead of time. Using standardized carts for deliveries could be another way to expedite the delivery process, avoiding any undesirable situations such as being stuck in doors or hallways.

Task 1c (sd/mean = 1.28): Exit Truck from Front Door. The high variation in "exit truck from front door" could be because of the time inside the vehicle to complete paperwork or review receiver lists while the vehicle's door was open. In some cases, the drivers were eating or using their cell phone while exiting the truck. However, most drivers did not take long to exit the vehicle (mean duration was 14 seconds).

Discussion

Freight movement is changing rapidly and it is essential to understand the process flow of goods globally, regionally, and locally. This paper focuses on the movement of goods locally, and more specifically within an office building in the Seattle central business district. There has been increasing demand for deliveries in central business districts, but there is limited space in which to move, both structurally and operationally.

The final 50 feet of the supply chain extensively involves a vertical movement of the delivery process, as deliveries and pick-up activities occur mostly while the drivers are out of the vehicle from the loading zone to the end customer. This paper introduces a systems approach to measure and observe detail tasks of the current final 50 feet of the supply chain by using a unique tablet application and a process flow map. An office building in downtown Seattle was observed by using this approach. The process flow map decomposes actions of the delivery workers, which helps the researchers identify bottlenecks in the current delivery process and where improvements can be made. The improvements can easy-to-implement solutions, such as an information board to notify delivery workers of imperfections in the freight elevator, to more high cost solutions such as a building redesign.

Although the study included only 31 observations, it still provides substantial insights on the variations that can occur for a one-week period within an office building, while also demonstrating that some steps are consistent regardless of carrier type. A future goal is to be able to compare the variations observed in this building with other building types and operations (e.g., shopping center, hotel, residential building). It would also be of interest to examine different operation types. Future process flow maps could also showcase temporal differences with respect to seasons, holidays, and weekend versus weekdays.

The scope of this study was also limited to the most common paths of the delivery process performed at one office building in downtown Seattle. However, this study can provide insights on the average delivery durations for other similar-sized office buildings in urban areas. Also, the focus of this paper is to understand the overview of the final leg of the delivery and pick-up activities by using the new systems approach of process flow maps with quantitative measures on dwell times. The quantitative measures of delivery time for each delivery task can enable researchers to identify those tasks with a high coefficient of variation value, being bigger than 1. This provides insights on the tasks that can be performed faster by others, which can be improved for other workers with a better understanding of the current delivery process flows. Further research on the final 50 feet of the pick-up and delivery process in different types of buildings could capture unique characteristics of different delivery procedures.

Acknowledgments

The work was conducted in conjunction with a project for the Seattle Department of Transportation (SDOT).

Author Contributions

The authors confirm contribution to the paper as follows: study conception and design: Haena Kim (HK), Linda Ng Boyle (LNB), Anne Goodchild (AG); data collection: HK; analysis and interpretation of results: HK, LNB, AG; draft manuscript preparation: HK, LNB, AG. All authors reviewed the results and approved the final version of the manuscript.

References

1. Kearney, A. T. *Annual State of Logistics Report: Accelerating into Uncertainty*. Council of Supply Chain Management Professionals, Washington, D.C., 2017.
2. Morris, A. G. *The Last Mile: Developing Efficient Freight Operations for Manhattan's Buildings*. The Steven L. Newman Real Estate Institute, Baruch College, New York, NY, 2009.
3. Goodman, R. W. Whatever You Call It, Just Don't Think of Last-Mile Logistics Last. *Global Logistics & Supply Chain Strategies*, Vol. 9, No. 12, 2005, pp. 46–51.
4. Allen, J., S. Anderson, M. Browne, and P. Jones. *A Framework for Considering Policies to Encourage Sustainable Urban Freight Traffic and Goods/Service Flows*. University of Westminster, London, United Kingdom, 2000.
5. Álvarez, E., and A. De la Calle. Sustainable Practices in Urban Freight Distribution in Bilbao. *Journal of Industrial Engineering and Management*, Vol. 4, No. 3, 2011, pp. 538–553.
6. Rhodes, S. S., M. Berndt, P. Bingham, R. Weisbrod, J. Bryan, T. J. Cherrett, and P. Plumeau. *NCHRP Report 14: Guidebook for Understanding Urban Goods Movement*. Transportation Research Board of the National Academies, Washington, D.C., 2012. http://dx.doi.org/10.17226/14648.
7. Dickson, E. W., S. Singh, D. S. Cheung, C. C. Wyatt, and A. S. Nugent, Application of Lean Manufacturing Techniques in the Emergency Department. *The Journal of Emergency Medicine*, Vol. 37, No. 2, 2009, pp. 177–182, http://dx.doi.org/10.1016/j.jemermed.2007.11.108.
8. Allen, J., M. Browne, and T. Cherrett, Survey Techniques in Urban Freight Transport Studies. *Transport Reviews*, Vol. 32, No. 3, 2012, pp. 287–311.
9. MDS Transmodal Limited and Centro di ricerca per il Trasporto ela Logistica (CTL). *Study on Urban Freight Transport, Final Report*. Directorate-General for Mobility and Transport (DG MOVE) European Commission, Brussels, Belgium, 2012.
10. Serrano Lasa, I., R. D. Castro, and C. O. Laburu. Extent of the Use of Lean Concepts Proposed for a Value Stream Mapping Application. *Production Planning & Control*, Vol. 20, No. 1, 2009, pp. 82–98.
11. Ries, B. S., L. N. Boyle, M. S. Vavilala, N. Kannan, H. Saxe, M. A. Kernic, F. P. Rivara, D. F. Zatzick, M. J. Bell, M. Wainwright, J. I. Groner, C. C. Giza, R. G. Ellenbogen, P. M. J. Wang, and R. Mink. Assessing Clinical Care using Interactive Value Stream Mapping. *Proc., Human Factors and Ergonomics Society 57th Annual Meeting*, San Diego, CA, 2013, pp. 1536–1540.
12. Wu, Y., L. N. Boyle, D. V. McGehee, L. S. Angell, and J. Foley. Task Analysis of Vehicle Entry and Backing. *Proc., 4th International Conference on Automotive User Interfaces and Interactive Vehicular Applications*, Portsmouth, NY, 2012, pp. 195–200.
13. Cherrett, T., J. Allen, F. McLeod, S. Maynard, A. Hickford, and M. Browne. Understanding Urban Freight Activity - Key Issues for Freight Planning. *Journal of Transport Geography*, Vol. 24, 2012, pp. 22–32.
14. Butrina, P., G. D. C. Giron-Valderrama, J. L. Machado-León, A. Goodchild, and P. C. Ayyalasomayajula. From the Last Mile to the Last 800 Feet: Key Factors in Urban Pick-up and Delivery of Goods. *Transportation Research Record: Journal of the Transportation Research Board*, 2017. 2609: 85–92.
15. McLeod, F., and T. Cherrett. Loading Bay Booking and Control for Urban Freight. *International Journal of Logistics: Research and Applications*, Vol. 14, No. 6, 2011, pp. 385–397. http://dx.doi.org/10.1080/13675567.2011.641525.
16. Pivo, G., D. Carlson, M. Kitchen, and D. Billen. Learning from Truckers: Truck Drivers' Views on the Planning and Design of Urban and Suburban Centers. *Journal of Architectural and Planning Research*, Vol. 19, No. 1, 2002, pp. 12–29.
17. Morrell, L. *Failed Deliveries are Costing up to £780 Million, Says IMRG*. eDelivery, 2016, http://edelivery.net/2016/10/failed-deliveries-costing-780-million-says-imrg. Accessed October 5, 2016.
18. United Parcel Service. *Schedule Worldwide Express Freight Pickup/Drop off during Shipment Processing*. 2017. https://www.ups.com/worldshiphelp/WS17/ENU/AppHelp/INTERNATIONAL/Schedule_Worldwide_Express_Freight_Pickup_Drop_Off_During_Shipment_Processing.htm. Accessed on July 1, 2017.
19. FedEx. *Inside Pickup and Delivery*. 2017. https://www.fedex.com/us/developer/WebHelp/ws/2014/dvg/WS_DVG_WebHelp/15_1_5_Inside_Pickup_and_Delivery.htm. Accessed on June 15, 2017.

The Standing Committee on Urban Freight Transportation (AT025) peer-reviewed this paper (18-06201).

Any opinions, findings, conclusions or recommendations expressed in this paper are those of the authors and do not necessarily reflect the views of SDOT.

Article

TRR
JOURNAL OF THE TRANSPORTATION RESEARCH BOARD

Transportation Research Record
2018, Vol. 2672(9) 184–193

DOI: 10.1177/0361198118794535
journals.sagepub.com/home/trr
SAGE

The Scope for Pavement Porters: Addressing the Challenges of Last-Mile Parcel Delivery in London

Julian Allen[1], Tolga Bektas[2], Tom Cherrett[3], Oliver Bates[4], Adrian Friday[4], Fraser McLeod[3], Maja Piecyk[1], Marzena Piotrowska[1], ThuBa Nguyen[2], and Sarah Wise[5]

Abstract

The UK parcel sector generated almost £9 billion in revenue in 2015, with growth expected to increase by 15.6% to 2019 and is characterized by many independent players competing in an "everyone-delivers-everywhere" culture leading to much replication of vehicle activity. With road space in urban centers being increasingly reallocated to pavement widening, and bus and cycle lanes, there is growing interest in alternative solutions to the last-mile delivery problem. We make three contributions in this paper: firstly, through empirical analysis using carrier operational datasets, we quantify the characteristics of last-mile parcel operations and demonstrate the reliance placed on walking by vehicle drivers with their vans being parked at the curbside for on average 60% of the total vehicle round time; secondly, we introduce the concept of "portering" where vans rendezvous with porters who operate within specific geographical "patches" to service consignees on foot, potentially saving 86% in driving distance on some rounds and 69% in time; finally, we highlight the wider practical issues and optimization challenges associated with operating driving and portering rounds in inner urban areas.

The UK parcel sector generated almost £9 billion in revenue in 2015, a 6% increase on the previous year, with growth expected to increase by 15.6% to 2019 (*1*). With over 1.7 billion parcels being delivered domestically per annum (*2*), light goods vehicles (LGVs, up to and including 3.5 metric tonnes (3.85 U.S. ton) gross weight) have seen the greatest growth with 3.6 million licensed in the UK (2015), a 23% increase relative to heavy goods vehicles since 1995 (*3–5*). The parcel distribution sector is characterized by many independent players competing in an "everyone-delivers-everywhere" culture leading to much replication of vehicle activity (*6*). This in turn negatively impacts on congestion and the need to reduce emissions in cities which is a central requirement of EU legislation (*7*).

The UK parcels market consists of three subsectors where transactions take place between different entities: business-to-business (B2B); business-to-consumer (B2C), and consumer-to-all-parties (C2X). In the UK, B2B accounted for 38%, B2C, 56%, and C2X, 6% of the parcel market in 2012 (*8*) with forecasts suggesting that volumes in the B2C and C2X subsectors will grow at approximately 4.5% to 5.5% per annum in the medium term (*9*).

Parcel carriers offer consignees a wide range of delivery options from immediate to same day, next day, or a delivery anytime within a set period of days. "Express" usually refers to services with a specified day of delivery (e.g., next day or two-day) and time of delivery (e.g., before 09:00, before 10:00). "Courier" services are usually the most time sensitive, often guaranteeing same-day delivery, or delivery before a certain time. The market for courier services is much more fragmented than for express and parcel services comprising many small owner operators. Data has suggested that next-day services

[1]Architecture and the Built Environment, University of Westminster, London, UK
[2]Southampton Business School, University of Southampton, Southampton, UK
[3]Engineering and Physical Sciences, University of Southampton, Southampton, UK
[4]School of Computing and Communications, Lancaster University, Lancaster, UK
[5]The Built Environment, University College London, London, UK

Corresponding Author:
Address correspondence to Tolga Bektas: t.bektas@soton.ac.uk

accounted for 56% of all UK domestic volumes in 2014–15 and 70% of total parcel revenues (*10*).

In order to meet customer needs, carriers have developed different logistics strategies and networks. Couriers offering immediate and same-day services typically operate a door-to-door service between the consignor and consignee. A parcel carrier based wholly within one city is likely to make use of a single depot from which multi-drop vehicle rounds are performed whereas a national or international carrier will typically make use of a hub-and-spoke network. In the case of the latter, central hubs and regional/local distribution centers may be operated, with large fully loaded vehicles operating between the hubs and other distribution centers, and smaller vehicles performing multi-drop rounds for last-mile delivery operated from several local depots in the case of large cities. In addition to this, parcel carriers are using "lifestyle" couriers (self-employed owner-drivers working on a free-lance basis) to manage local last-mile deliveries, the handling of failed first-time deliveries and customer returns. With the plethora of different operators and services, it is estimated that the UK parcel market is approximately 20% overcapacity (*11*). Given that road space in urban centers is being reallocated to pavement widening, and bus and cycle lanes (*12*), and with Transport for London predicting that traffic congestion in central London will increase by 60% by 2031 (*13*), there is growing interest in alternative solutions to the last-mile problem.

We make three contributions in this paper: firstly, through empirical analysis using carrier operational datasets, we quantify the characteristics of last-mile parcel operations and demonstrate the reliance placed on vehicle curbside parking and walking as an integral component in the last 100 m transaction; secondly, we introduce the concept of "portering" as a potentially viable option for improving the efficiency of last-mile van operations using a case study example; finally, we highlight the wider issues and challenges associated with operating and optimizing driving and portering rounds in inner urban areas.

Characteristics of Multi-Drop Operations and Their On-Street Impacts

A detailed study of 25 vehicle rounds operated by two parcel carriers making deliveries and collections across three postcodes in the West End of central London (WC1, WC2 and W1) was undertaken. This was done over three days in October 2016 and involved: i) GPS tracking of both the vehicle and the driver, ii) surveyors accompanying drivers to verify round timings, parking places used, and delivery/collection locations served, and iii) analysis of the daily manifest data for each vehicle round.

All the vehicles used were vans with a carrying capacity of up to 1 metric tonne (1.1 U.S. ton) and up to 6 m^3 (1,320 gallons) in volume. Parcels for delivery and collection were allocated to drivers each day based on predetermined and largely fixed vehicle-round structures. Parcel deliveries accounted for 94% of all activity with the transaction order being left to the driver's discretion. Drivers were responsible for selecting the route, parking locations and the clusters of consignees to service from each stopping point. The vehicle rounds studied took place in the "West End" of central London in the area of Oxford Street, Regent Street, Covent Garden, Soho, Mayfair and Piccadilly. The area has approximately 2,000 shops, 2,500 restaurants and cafes, 3,000 licensed premises, 40 theatres, 20 cinemas, 30 museums and galleries as well as 40,000 residents, and accounts for 65,000 employees generating 15% of London's total gross value added (GVA) (*14*).

The rounds emanated from three depots which had stem mileages of: Depot A (2 km (1.24 miles)); Depot B (4 km (2.5 miles)); and Depot C (11 km (6.8 miles)). The average round duration, defined as the difference in time from leaving the depot and returning, excluding time spent in the depot, was 7.3 hours and the average distance driven within the delivery area (excluding stem mileage) was 11.9 km (7.4 miles) with a mean speed of 7 kph (4.35 mph) (and 8.9 kph (5.5 mph) including stem mileage). Of interest was the fact that 62% of the total round time was spent with the vehicle parked while the driver unloaded and sorted on average 126 parcels and delivered these on foot to 72 establishments from 37 stopping places. The average distance walked per vehicle round was 7.9 km (4.9 miles) which accounted for 28% of the total distance travelled from the depot (i.e., including distance driven), with 95% of vehicle stops taking place on street at the curbside. On average, the driver delivered/collected 3.8 parcels from 2.1 establishments per vehicle stop, with establishments receiving/dispatching 1.9 parcels per delivery/collection.

The mean drive time between stopping locations was 3.7 min, with an average 8.1 min dwell time observed at each vehicle stop, which was comparable with previous studies (*15*, *16*). Mean driving and parking times per parcel were 1.5 and 2.3 min respectively, with associated driving and walking distance of 202 m (221 yards) and 72 m (79 yards). The walking distance per establishment served was 105 m on average. The findings suggest that last-mile parcel operations are characterized by walking with the vehicle left stationary, often conflicting with a curbside infrastructure legislated in favor of passenger transportation (*17*). In these circumstances, carriers are increasingly facing fines with UPS receiving penalty charge notices totaling over $17 m from servicing clients in New York alone during 2016 (*18*). With the growth in parcel delivery set to continue (*19*), carriers are becoming increasingly interested in exploring new ways of working.

The Concept of "Portering" to Reduce Last-Mile Vehicle Impacts

Human carriage of goods has been an important means of commercial freight transport in our cities for centuries (*20, 21*). The advent of the railways largely resulted in the demise of the City of London porter and the use of barrows and hand carts for goods movement (*22, 23*) but this concept could be viable once again for parcel logistics in London and other dense urban areas where curbside parking is problematic. *Espace de livraison de proximité* (ELP) or in English, "nearby delivery areas," have been operated in several French cities in which consignee goods are offloaded at a reserved, centrally located unloading space and delivered by ELP staff using trolley, carts, bicycles and electric vans (*24–26*). A similar approach was implemented in Paris in which goods were unloaded from motorized vehicles at a "virtual exchange point" and then delivered locally on foot using a trolley (*27*). In Brussels, the parcel carrier TNT experimented with a "mobile city hub" from which cargo bike deliveries were made (*28*). Meanwhile, in New York's lower Manhattan financial district, DHL operates a "walking courier" facility from which deliveries of packages and documents are made on foot (*29*). In this paper, we suggest that portering could take two operational forms, both with the specific objective of reducing vehicle stopping times at curbside:

Scenario 1— In this case the van stops at the curbside in the driver's preferred location, where a pre-notified porter is waiting to receive the parcels from the driver for local delivery on foot (this could be referred to as "drop-and-drive"). In this sense, the driver would still be making the same number of stops as if he/she were making the deliveries on foot and the time taken to make the deliveries by the porter would be the same. No porter facilities are required in terms of dedicated curbside space or storage facilities but carriage provision for parcels would be necessary in terms of a hand cart. This scenario is akin to the notions of crowdshipping on a large scale (*30–32*) and would be most applicable in locations with extremely dense delivery networks or where substantial vehicle access and curbside parking restrictions exist.

Scenario 2—In this case, there would be a fixed number of portering reception points (substantially less than the current number of vehicle stopping locations per round) which could be reserved curbside spaces, small permanent facilities/buildings, existing retail stores that already provide ecommerce collection-point services or temporary mobile depots which are delivered to the location each day. These portering reception points would cover a greater delivery catchment area compared with Scenario 1, and therefore drivers would drop off a larger number of parcels destined for more consignees at each stop. The porters would make deliveries from these points either on foot, possibly using handling equipment, or using cargo cycles (depending on the size, weight and number of parcels to be conveyed and the distances involved).

The key benefit to carriers of adopting such portering services is the reduction in vehicle stopping time at the curbside and, in the case of scenario 2, reductions in the overall distance travelled and time taken on vehicle rounds. Portering would have the potential to make rounds more efficient and vehicles more productive in terms of their carrying capacity and utilization. These gains would be traded against the additional cost of the porters, the carrying equipment and the telematics systems needed to manage the last-100 m transaction to the consignee (however, the time that porters take to walk and distribute the parcels should match the current on-foot performance of drivers in scenario 1).

Understanding the Role Portering Might Play in Existing Carrier-Round Structures

Researchers have previously used several approaches in an attempt to gauge the intensity of parcel operations on street including individual business audits through "Delivery and Servicing Plans" (*33*), observational high street surveys (*34*), and driver activity studies (*35*). In this research, a new research approach was adopted in which manifest data from two major carriers were used in an attempt to understand the spatial intensity of last-mile delivery and collection activity within central London, and what role portering might play in current operations. Manifest data covering transactions in the WC1, WC2 and EC1-4 postcode areas between 1 October 2016 and 7 February 2017 were used in the analyses.

Approximately 90% of Carrier 1's work was B2C-related across a mixed land use profile including retail, commerce, and domestic customers while Carrier 2 specialized more in B2B parcel movements. A total of 396 and 112,785 unique consignors were observed in the two carrier datasets respectively, with major fashion, general retailers and on-line ticket companies (C2C) generating the greatest number of records (˜110,000 in the case of Carrier 1). The database comprised 894,136 and 394,551 records for carriers 1 and 2 respectively with each record corresponding to a delivery/collection attempt.

To better understand the spatial distribution of deliveries and how portering might operate at the local level, a smaller study area based around Oxford Street was chosen, representing around 2% and 3.2% of the overall datasets from carriers 1 and 2 respectively. The area is

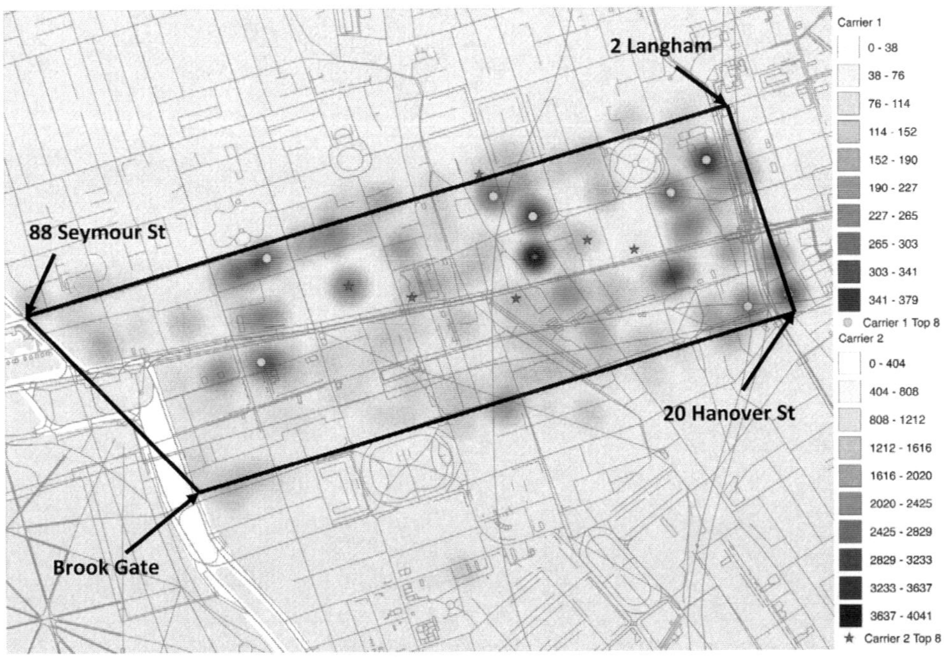

Figure 1. Total number of deliveries (Carrier 1 and Carrier 2) by location around Oxford Street between 1 October 2016 and 7 February 2017 (129 days); Carrier 2 (Primarily B2B) covers 28 August 2016 to 5 November 2016 (69 days).

approximately 1.3 km (0.8 miles) along the topmost edge (Seymour Street, A5204) by 0.4 km (0.25 miles) along the rightmost edge (Regent Street) and has a dense land use made up of shops, offices, and private addresses containing 1172 distinct postcodes. For spatial analyses, heat maps were generated using GIS software (QGIS) based on latitude and longitudes obtained for each postcode. These enable the numbers of parcels destined for particular postcodes to be displayed, with a radius of 50 m (55 yards) being drawn around each point to illustrate where overlap in delivery locations occurs.

Of the 1172 postcodes, 836 received successful deliveries with 336 postcodes recording a failed delivery attempt at some point over the analysis period, (Carrier 1 reported that 38% of failures occurred between 12:00 and 15:00). Aggregate deliveries from Carrier 1 and Carrier 2 were mapped to reveal the distribution and delivery hotspots in the area (Figure 2). The locations receiving the largest number of deliveries are of particular interest as these would be likely best served through direct van deliveries with porters rendezvousing with the van at those locations to then make deliveries to surrounding clusters of smaller consignees on foot (*36*).

Most of the activity hot spots appeared to be in areas of mixed land use with multi-tenanted offices, shops, restaurants and hotels, including those on Oxford Street, Regent Street and opposite Portman Square (Figure 2). Due to the data anonymization process, it was not possible to determine the extent to which personal deliveries were made to workplaces, but this is of interest to employers and transport authorities in London who would like to restrict such activity (*37*).

In designing portering rounds it would be important to identify the busiest locations in terms of consignee service requests. An analysis of the data suggested that the "top 8" (0.9%) postcodes, corresponding to the three "hottest" delivery activity bands identified in Figure 2 (i.e., those with over 522 aggregate deliveries from Carrier 1 and 2 over the period), accounted for 12.3 times the mean activity, or 29.1% of the total activity (Table 1). In addition, the "top 20" (2.4%) and "top 45" (5.4%) postcodes accounted for 42.4% and 58% of total activity, highlighting that a relatively small number of locations generate significant package volumes and could be the starting locations for portering rounds. This does however depend on the characteristics of the individual client base (e.g., B2C, B2B), (Figure 1).

In terms of potential workloads, Mondays and Tuesdays were the busiest days of the week for Carrier 1 and Carrier 2 respectively, with approximately 241 (Monday, Carrier 1) and 290 (Tuesday, Carrier 2) manifest entries per day in the Oxford Street area, 69.8% taking place between 11:00 and 16:00. For Carrier 1, the Monday peak was due to the very high proportion (49%) of failed first-time deliveries experienced on Saturdays that required subsequent redelivery on the

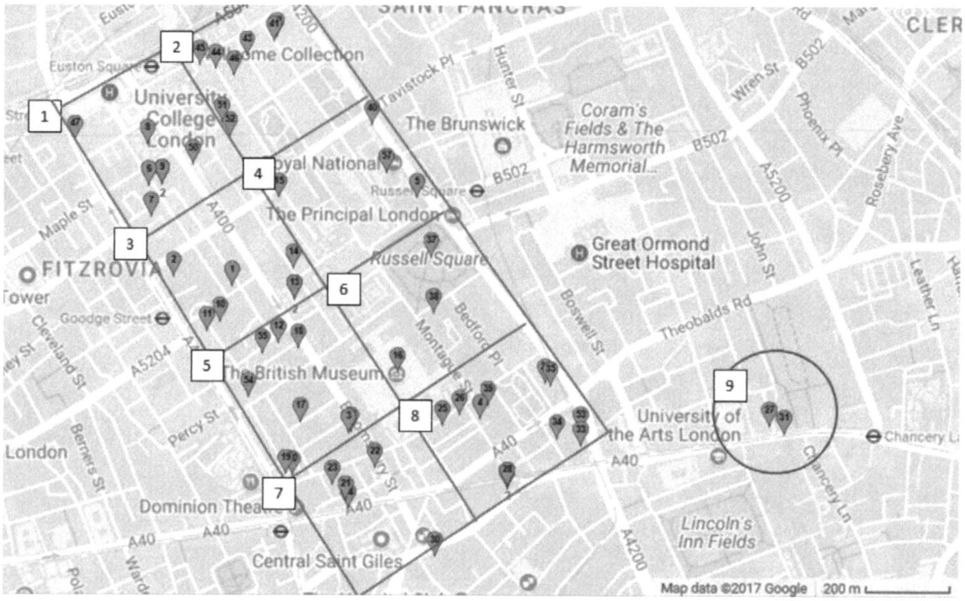

Figure 2. Customer locations on example round and proposed "drop-and-drive" delivery patches.

Table 1. Comparison of "Top 8" and All Postcode Areas in Terms of Delivered Items

Activity (days)	Number of deliveries—all 836 postcodes (top 8 postcodes)			
	Total	Average per postcode	Standard deviation	Maximum
Carrier 1 (129)	14009 (2348)	16.8 (293.5)	40 (56)	379
Carrier 2 (69)	19218 (8637)	23 (197.5)	158 (140)	4041
All Deliveries	33227 (9684)	39.8 (491)	169 (163)	4041

Note: Data for Carrier 1 (primarily B2C) covers 1 October 2016–7; February 2017 (129 days); and Carrier 2 (primarily B2B) covers 28 August 2016–5 November 2016 (69 days).

Monday, reflecting the number of offices closed on Saturdays. The majority of the activity was related to deliveries which outweighed collections by 18.6 to 1 in the case of Carrier 1. This is an important factor as portering would function best as a delivery-only activity with a one-way transaction of packages from the driver to the porter and through to the consignee. Collections would necessitate a porter having to reconvene with a carrier's vehicle at some point to offload packages. This process might also be necessary when dealing with failed first-time deliveries which ranged from 7.4% (Thursdays) to 14% (Mondays) for Carrier 1 and 2.3% (Monday) to 4.4% (Thursday) for Carrier 2, both in line with national averages, IMRG (2014).

Quantifying the Potential Benefits of a Portering Service

Using the data collected from the 25 vehicle rounds studied in detail, an attempt was made to understand the likely vehicle time and distance savings from both the drop-and-drive scenario (scenario 1) and the use of a reduced number of vehicle stopping points (scenario 2). For each of the rounds, estimated round times (T_{new}) for the drop-and-drive element were calculated (Table 2) as

$$T_{new} = T_{actual} - \text{Total parked timed(before)} + (\text{Number of stops} \times \text{Y minutes per stop})$$

In scenario 1 it was assumed that the same number of stops were made, using a conservative estimate of 3 mins per stop (Y) to unload, scan and transfer parcels from the driver to the porter based on surveyor observations. Replicating the same round orders using mapping software, the results suggested that an average time saving of approximately 4 hours per round (55% of the total round time) could be possible which would have significant implications on driver- and vehicle-utilization. The estimated time savings for each of the 25 individual rounds ranged from 2 to 6 hours, which reflected the

Table 2. Estimated Workload for Porters with Reduced Vehicle Stopping Locations (Scenario 2, Figure 2)

Delivery patch (no. consignees)	Parcels	Walking time (seconds)	Walking distance (m (yards))	Handover time for driver to porter (seconds)	Collections (no. consignors)
1 (6)	54	602	849 (928)	586	0
2 (8)	10	527	741 (810)	133	0
3 (6)	15	559	790 (864)	185	0
4 (4)	4	475	662 (724)	71	3
5 (9)	15	792	1107 (1211)	185	2
6 (3)	6	445	627 (686)	92	0
7 (5)	13	458	647 (708)	164	0
8 (9)	11	565	791 (865)	143	1
9 (2)	3	31	44 (48)	61	0
Total (52)	131	4454	6.26 km (3.89 miles)	1620	6

variability in observed total parking times per round (from 1.9 hours using 14 stopping locations to 6.3 hours using 72 stopping locations). Parking times were mainly influenced by the total workload in different "hot spot" areas and the individual driver's preference between moving the vehicle frequently to minimize walking or to walk between groups of customers to avoid driving and finding parking places.

To demonstrate the likely portering workload that could be involved with scenario 2, one of the surveyed vehicle rounds was studied in detail (Figure 2). This round involved 138 items being delivered to 54 consignees, (including 7 time-guaranteed deliveries and 6 collections) for which the driver used 52 stopping locations across the 1.3 km^2 area. The van covered 16.8 km over 7.3 hours during the round (excluding stem mileage), recording a mean speed of 8.8 km/hr. Sixty-one percent of the round involved the vehicle being parked (5.3 hours, 87% at on-street locations) with the driver making deliveries on foot.

To illustrate how portering "patches" might be allocated, the 54 consignees were separated into 9 defined delivery patches made up of approximate 350 m (383 yard) squares (Figure 2) with two outlying customers to the south East (patch 9). Previous relevant work focused on where to site "mini-hubs" in Seville based on 200 m (219 yards) radius circles of influence (*38*). Clearly, the size of the delivery patch has a direct influence on the amount of walking that may be entailed. This has been demonstrated in that the length of the optimal tour over a given patch is proportional to the square root of the size of the area (*39*), with implications for vehicle routing problems, (*40*). The geographical scale of walking patches would depend on the package generation characteristics of the surrounding land use and the consequential ability of the porter to physically handle the packages.

Within each delivery patch, a shortest-path walking tour between all the customers was devised and approximate walking times and distances quantified using mapping software (Table 2). Handover times in each patch were adjusted to reflect the number of parcels actually delivered. This was achieved by assuming 30 s to park the van, access packages for the specific patch and then book them over to the porter (10 s per parcel), being consistent with the average of 3 min per stop. This produced a range of van-to-porter handover times from 61 s to 586 s (Table 2) where delivery patch (*1*) received considerably more parcels (n = 54) than the others, mainly due to one customer receiving 32 parcels. The walking and handover times totaled 1.69 hours across all the patches with porter walking distances within each patch ranging between 44 m and 1107 m (48 to 1211 yards). The major benefit to the carrier is in the time and distance savings from only having to service one handover point in each patch. If in this example, the vehicle traversed patches 7-5-3-1-2-4-6-8-9 in order, stopping in each to drop packages to a porter, the vehicle driving distance within the delivery area (excluding stem distances) could be approximately 2.2 km (1.37 miles), a reduction of 14.6 km (9 miles) (86%) over the current system; the time saved would be approximately 6 hours (69%), comprising 5.3 hours spent making deliveries plus around 1 hour driving time savings less around 20 minutes spent with porters.

Any portering system would have to cater for instances where single consignees were receiving multiple parcels (Figure 1) as it would make logistical sense to service large receivers directly from the van, or to situate the drop site as close as possible to them where they featured in a given patch. There would also be the issue of how collections would be managed given the driver-porter transaction is one-way at each rendezvous point. It would be feasible for porters to work across multiple patches and hand back parcels to the driver at another location e.g., moving across after completing the 9 deliveries and picking up 2 collections in patch 5 and then moving to 6 to wait for the driver, hand over the 2 collections and pick up the 3 deliveries for that patch. To operate effectively, this concept would require careful consideration and optimization of both the driving and walking tours to account for things like dynamic collection requests during the round, failed deliveries and potential redelivery attempts, extended portering time associated with servicing high-rise buildings, carrying capacity limitations of the porters. Carrying capacity is a key issue which will differ between parcel carriers depending on their market specialism (e.g., Amazon states that 86% of its delivered products weigh 2.3 kg (5 lbs) or less (*41*) whereas 54% of Carrier 2 parcels weighed less than 5 kg (11 lbs)).

The Optimization Challenge Associated with Portering

Last-mile parcel delivery problems are generally studied under city logistics systems (*42*), where the corresponding optimization problems are modelled using two-tier distribution structures. The first tier usually involves vehicles with relatively large carrying capacities off-loading goods at rendezvous points, for the second tier to undertake the last-mile transactions.

The optimization of plans involves deciding on the routing and scheduling of vehicles across both tiers, the demand locations to be served, the locations of the van-porter rendezvous points, and the capacities of any reception facilities to be used. In our context, we envisage two cases: i) fixed cluster case, where the delivery patches have been identified (as in the case of the nine patches shown in Figure 2) prior to the routing and scheduling; ii) unknown clusters where the delivery or collection points have not been grouped into patches. In either case, vans would operate in the first tier and porters in the second. We discuss two cases below in further detail:

Fixed Cluster—The first case gives way to generalized vehicle routing problems where the aim is to route vehicles over a given set of clusters that correspond to the delivery patches. Although this class of optimization problems have been studied at a sufficient level of detail (*43*), the generalized vehicle routing problem and its variants do not explicitly consider the way in which intra-cluster deliveries are performed. Assuming Scenario 1, where there is no need for a reception facility for handing over of parcels, there are at least two ways in which deliveries within clusters can be done. Depending on the total weight and size of parcels handed over, the porters would perform direct deliveries in a so-called "hotelling" mode, back and forth from the van, or, assuming they have sufficient carrying capacity, would operate a smaller tour from consignee to consignee using consolidation. The combined use of hotelling and consolidation within a cluster is another possibility.

Deciding on the location of the hand-over point will be a key part of the optimisation problem, which may be limited to one of the delivery points. In addition, the consolidation poses an additional challenge of finding an optimal tour over the delivery points within a cluster. If there are no additional constraints present in the problem, then the optimal assignments for both the hotelling and the consolidation options can simply be precomputed for each possible selection of the rendezvous point without forming an integral part of the optimization problem. Such preprocessing will reduce the solution complexity, assuming that each cluster is feasible with respect to the porters being able to carry the parcels. However, if there are additional constraints on the time-sensitive nature of the parcels, then this preprocessing is no longer possible.

Unknown clusters—If the delivery patches were not predefined, then the optimization problem would have to involve decisions pertaining to the formation of clusters along with the routing and scheduling decisions. The interdependent nature of both sets of decisions means that they will have to be taken in conjunction. In the case of Scenario 1 where a porter is available to receive the parcels, the corresponding optimization problem would be akin to the truck and trailer routing problem (*44*) where, in our context, the trailer would correspond to the van performing first-tier deliveries and the porters would act as the trucks in the second tier for the last-mile deliveries. The problem will also involve additional constraints for time-sensitive deliveries as well as the capacity of the porter in terms of the total weight and size of the parcels they are able to carry. An additional set of constraints will also be needed to synchronize the timing between the van(s) and the porter(s) for a timely hand-over such that neither will stay idle waiting for the other at the rendezvous points.

Issues to Consider When Devising and Implementing Portering Systems

In devising and implementing a portering system for central urban areas, there are a range of issues that would require further consideration. These issues have emerged from a combination of reviewing the literature available on portering and related freight concepts (such as Urban Consolidation Centers) and original concepts about portering systems developed through dialogue between the authors, parcel carriers and policy makers as part of the research.

Geographical Coverage and the Influence of Major Consignees

The larger the catchment area for portering, the more likely the need for handling equipment such as trolleys or cargo cycles in addition to porters manually carrying parcels and packages. Understanding the major origins of demand across an area in terms of the number of deliveries, first-time delivery failures, returns, and collections is very important when devising the scale and applicability of portering patches and where the most optimal drop locations for vans would be.

The Location and Type of Portering Infrastructure Necessary

This will depend on the geographical area served, the variability in weight and size of packages handled, the portering infrastructure requirements associated with the land-use needs and the availability of space. Such infrastructure could include a reception facility with or without storage space for incoming and outgoing parcels, overnight storage for handling and transport equipment used by porters, scanning and computing equipment to track and trace goods, recharging facilities for any electric equipment such as cargo cycles, and off-street parking space for vehicles/drivers delivering to or making collections from the portering facility. Autonomous vehicles of varying types might be used in portering operations of the future (*45*) but given the complexity of crossing roads, climbing stairs, using lifts, and communicating with consignees, it is likely to remain far more efficient and cost effective to use humans to carry out this last leg of the supply chain in the short term.

Financing and Operating the Portering Service

Financing could be led by the private sector given the efficiency savings that parcel carriers would enjoy associated with vehicle and fuel use reduction. Reduced driver and vehicle costs are likely to offset the cost of porter labor but there are potential additional costs associated with portering equipment and the development of apps to support work allocation to porters. Consignees and consignors would benefit from having an improved customer environment at their premises as a result of reduced vehicle operations. Financial contributions by the public sector (i.e., city authorities) could also be justified on the basis of the traffic, environmental, and safety benefits associated with the portering approach. Aligning the costs and benefits of the portering scheme with the financial contributions is likely to be important in its success; as is the case with urban consolidation centers (UCCs), public sector financial support may well be necessary in terms of meeting the capital costs of any buildings and other infrastructure required (*46*).

In terms of running the portering system, the day-to-day operations may be best served by a private operator, selected through a tendering process. If the portering scheme is a private-sector-led initiative, this could be achieved through a single company (either a market entrant/start-up company specializing in providing this service or an established freight operator diversifying into this service) or it could be a joint venture formed by several collaborating parcel carriers who will each use and benefit from the scheme. One could envisage a last-mile crowd-sourced operator such as Deliveroo (www.deliveroo.co.uk) providing a porter smartphone-based interface to integrate with the carriers where they would in essence become a "freight traffic controller" over the last 100 m on behalf of a number of separate carriers. One concern is the potential security issue associated with handing over parcels to a 3rd party but given the tracking and proof-of-delivery functionality embedded in the existing systems used in last-mile fast food delivery, this may not be such a problem in reality.

Acknowledgments

This research reported in this paper was carried out as part of the UK EPSRC-funded Freight Traffic Control 2050 project (www.ftc2050.com).

Author Contributions

The authors confirm contribution to the paper as follows: study conception and design: all authors; data collection: Cherrett, T. J., Allen, J., McLeod, F. M.; analysis and interpretation of results: all authors; draft manuscript preparation: all authors. All authors reviewed the results and approved the final version of the manuscript.

References

1. Keynote. *Courier & Express Services: Market Report 2015*. Keynote, 2015.
2. Postal and Logistics Consulting Worldwide. *Review of the Impact of Competition in the Postal Market on Consumers*. Final Report to Citizens Advice. Postal and Logistics Consulting Worldwide, 2015.
3. Department for Transport. *Road Traffic Statistics 2016 Edition*. Department for Transport, 2016.
4. Department for Transport. *Vehicle Licensing Statistics*. Department for Transport, 2016.
5. Transport for London. *Travel in London: Report 9*. Transport for London, 2016.
6. Browne, M., C. Rizet, and J. Allen. A Comparative Assessment of the Light Goods Vehicle Fleet and the Scope to Reduce Its CO_2 Emissions in the UK and France. *Procedia - Social and Behavioral Sciences*, Vol. 125, 2014, pp. 334–344.

7. European Commission. *Roadmap to a Single European Transport Area - Towards a Competitive and Resource Efficient Transport System.* Transport White Paper, European Commission, 2011. http://eur-lex.europa.eu/legal-content/EN/ALL/?uri=CELEX:52011DC0144 Accessed July 24, 2017.
8. Royal Mail. *Full Prospectus.* Royal Mail, 2013.
9. Royal Mail. *Market Overview.* Royal Mail, 2016.
10. Ofcom. *Annual Monitoring Update on the Postal Market: Financial Year 2014–15.* Ofcom, 2015.
11. Royal Mail. *Response to Ofcom's July 2015 Discussion paper: Review of the regulation of Royal Mail.* Royal Mail, 2015.
12. Barry, J. *Head of Bus Network Development.* Transport for London, Personal Interview, 2014.
13. Transport for London. *Freight Forum.* London, 2015.
14. Westminster City Council. *The West End: Developing Westminster's Local Plan.* Westminster City Council, 2015. http://transact.westminster.gov.uk/docstores/publications_store/West%20End%20consultation%20booklet.pdf
15. Cherrett, T., G. McLay, and M. McDonald. *Effects of Freight Movements in Winchester.* Final Report. University of Southampton, Southampton, 2002.
16. Cherrett, T., J. Allen, F. McLeod, S. Maynard, A. Hickford, and M. Browne. Understanding Urban Freight Activity – Key Issues for Freight Planning. *Journal of Transport Geography*, Vol. 24, 2012, pp.22–32.
17. Allen, J., and M. Browne. *Success Factors of Past Initiatives and the Role of Public-Private Cooperation, Deliverable 2.3.* CITYLAB Project, 2016.
18. Jensen, T. F. Viewpoint from UPS. Presented at 96th Annual Meeting of the Transportation Research Board, Presentation 17-21812, Washington D.C., 2017.
19. Mintel. *Online Retailing.* Mintel, UK, 2016.
20. Bastien, G., P. Willems, B. Schepens, and N. Heglund. The Mechanics of Head-supported Load Carriage by Nepalese Porters. *Journal of Experimental Biology*, Vol. 219, 2016, pp. 3626–3634.
21. Gaurav, K., and M. Singhal. *Licensing and Livelihood: Railway Coolies, Internship Paper.* Center for Civil Society (CCS), India, 2003.
22. Allen, J., and M. Browne. Road Freight Transport To, From, and Within London. *The London Journal*, Vol. 39, No. 1, 2014, pp. 59–75.
23. Stern, W. *The Porters of London.* Longmans, 1960.
24. Browne, M., J. Allen, T. Nemoto, D. Patier, and J. Visser. Reducing Social and Environmental Impacts of Urban Freight Transport: A Review of Some Major Cities, The Seventh International Conference on City Logistics. *Procedia - Social and Behavioral Sciences*, Vol. 39, 2012, pp. 19–33.
25. Huschebeck, M. *Espace de Livraison de Proximité, Bordeaux, ELTIS Case Study*, 2012. http://www.eltis.org/index.php?id=13&lang1=en&study_id=1284
26. SUGAR. *City Logistics Best Practices: A Handbook for City Authorities.* SUGAR, 2011. http://www.eltis.org/index.php?ID1=6&id=62&list=&concept_id=3
27. Ducret, R. Parcel Deliveries and Urban Logistics: Changes and Challenges in the Courier Express and Parcel Sector in Europe-the French Case. *Research in Transportation Business & Management*, Vol. 11, 2014, pp. 15–22.
28. Verlinde, S., C. Macharis, L. Milan, and B. Kin. Does a Mobile Depot Make Urban Deliveries Faster, More Sustainable and More Economically Viable: Results of a Pilot Test in Brussels. *Paper Presented at mobil.TUM 2014 "Sustainable Mobility in Metropolitan Regions"*, Munich, Vol. 4, 19–20 May 2014, pp. 361–373. Transportation Research Procedia.
29. DC Velocity. *DHL Express Opens "Walking Courier" Facility in Manhattan Financial District*, 7 July 2016. http://www.dcvelocity.com/articles/20160707-dhl-express-opens-walking-courier-facility-in-manhattan-financial-district/
30. Allen, J., and M. Browne. *Success Factors of Past Initiatives and the Role of Public-private Cooperation, Deliverable 2.3*, CITYLAB Project, 2016.
31. Fung Business Intelligence Center. *Crowdsourced Delivery.* The Fung Business Intelligence Center, 2015.
32. McKinnon, A. *Crowdshipping: A Communal Approach to Reducing Urban Traffic Levels?* Logistics White Paper 1/2016. Alan McKinnon, 2016.
33. Transport for London. *Delivery and Servicing Plans: Making Freight Work for You.* Transport for London, 2014. https://www.tfl.gov.uk/cdn/static/cms/documents/delivery-and-servicing-plans.pdf Accessed July 24, 2017.
34. Transport for London. *TfL High Street Freight Survey Project, Stratford High Street: Case Study Summary.* Transport for London, 2015.
35. Transport for London. *Regent Street – Delivery and Servicing: Regent Street Site Survey.* Transport for London, 2009.
36. Allen, J., M. Piecyk, and M. Piotrowska. *An Analysis of Online Shopping and Home Delivery in the UK, Report Carried Out as Part of the Freight Traffic Control (FTC) 2050 Project.* University of Westminster, 2017.
37. Harris. *Online Shoppers could be Banned from Accepting Parcels at Work*, 2017. http://www.itv.com/news/london/2017-01-19/online-shoppers-could-be-banned-from-accepting-parcels-at-work/
38. Muñuzuri, J., P. Cortés, R. Grosso, and J. Guadix. Selecting the Location of Minihubs for Freight Delivery in Congested Downtown Areas. *Journal of Computational Science*, Vol. 3, 2012, pp. 228–237.
39. Beardwood, J., J. H. Halton, and J. M. Hammersley. The Shortest Path through Many Points. *Mathematical Proceedings of the Cambridge Philosophical Society*, Vol. 55, No. 4, 1959, pp. 299–327.
40. Chien, T. W. Operational Estimators for the Length of a Traveling Salesman Tour. *Computers & Operations Research*, Vol. 19, No. 6, 1992, pp. 469–478.
41. Guglielmo, C. *Turns Out Amazon, Touting Drone Delivery, Does Sell Lots of Products That Weigh Less Than 5 Pounds*, 2013. https://www.forbes.com/sites/connieguglielmo/2013/12/02/turns-out-amazon-touting-drone-delivery-does-sell-lots-of-products-that-weigh-less-than-5-pounds/#af3364c455ed. Accessed July 20, 2017.
42. Crainic, T. G., N. Ricciardi, and G. Storchi. Models for Evaluating and Planning City Logistics Systems. *Transportation Science*, Vol. 43, No. 4, 2009, pp. 432–454.

43. Bektas, T., G. Erdogan, and S. Ropke. Formulations and Branch-and-cut Algorithms for the Generalized Vehicle Routing Problem. *Transportation Science*, Vol. 45, No. 3, 2009, pp. 299–316.
44. Derigs, U., M. Pullmann, and U. Vogel. Truck and Trailer Routing – Problems, Heuristics and Computational Experience. *Computers & Operations Research*, Vol. 40, 2013, pp. 536–546.
45. Starship Technologies. Starship Technologies Launches Testing Program for Self-driving Delivery Robots with Major Industry Partners, *Press Release*, July 6, Starship Technologies, 2017. https://www.starship.xyz/starship-technologies-launches-testing-program-self-driving-delivery-robots-major-industry-partners/
46. Allen, J., M. Browne, A. Woodburn, and J. Leonardi. The Role of Urban Consolidation Centres in Sustainable Freight Transport. *Transport Reviews*, Vol. 32, No. 4, 2012, pp. 473–490.

The Standing Committee on Urban Freight Transportation (AT025) peer-reviewed this paper (18-01708).

Article

TRR
JOURNAL OF THE TRANSPORTATION RESEARCH BOARD

Transportation Research Record
2018, Vol. 2672(9) 194–203

DOI: 10.1177/0361198118796926
journals.sagepub.com/home/trr
SAGE

Potential of Implementing Urban Freight Strategies in the Accommodation and Food Services Sector

Iván Sánchez-Díaz[1]

Abstract

This paper studies freight demand patterns of establishments in the accommodation and food services sector, and the potential of implementing urban freight strategies in this sector to improve local and city traffic conditions. This research uses data from the City of Stockholm and assesses the potential implementation of two strategies (i.e., consolidation and off-peak hour deliveries) proposed in The Stockholm Freight Plan to foster safe, clean, and efficient freight deliveries. The results show that accommodation establishments have different freight generation patterns and opinions about the strategies proposed than other food services establishments. The results also show that the strategies proposed have good acceptance among 20% of the receivers surveyed, which would facilitate the implementation of the strategies and lead to significant benefits both at the local and city levels.

The increasing urban population and the congestion that this growth entails are consistently ranking very high among the issues concerning major cities around the world. Congestion has numerous negative effects including hampering accessibility for people and goods, producing environmental and health issues, diminishing livability, and distressing the competitiveness of local economies. Schrank, Eisele et al. (*1*) estimates that, in 2014, congestion led to USD$160 billion in extra cost only from increased travel times and fuel consumption in the US. It is noteworthy that although freight only represents 7% of the traffic, it accounts for 17% of this extra cost (*1*).

Congestion has major impacts on local commercial establishments that, in addition to assuming higher logistics costs due to increased transportation and inventory costs (caused by lower reliability of deliveries), must cope with the impacts on attractiveness for their customers. The latter is a particularly important issue for the accommodation and food services (AFS) sector for which the price of the service sold barely depends on the cost of the goods provided (e.g., food costs represent as low as 30% of the meal price), as other intangible elements, such as the establishment's environment, are the main source of added-value to their service (*2*). Hence, the interest of this sector in initiatives that decrease congestion and make their neighborhoods more attractive.

The AFS sector is defined by a common business purpose focused on service. It covers establishments with a wide range of services offered, i.e., basic/luxury accommodation, catering, café and bakery services, fast food or fancy dinner. Verlinden, Van de Voorde et al. (*3*) propose four different market typologies for this sector to characterize the flows and explain linkages: (i) small local specialists and cash-and-carry, (ii) big local and goods specialists, (iii) Ho.Re.Ca. supermarkets with deliveries, and (iv) franchises. AFS establishments can also be classified as independent or part of a chain. This distinction is relevant because it may influence the level of integration with suppliers and carriers. Chain establishments share a brand and standardized business practices with other establishments that are part of the same chain. In some cases, chain establishments are part of a franchise scheme where an existing business model is adopted and the supplies are bought from the franchisor or designated suppliers. Cherrett, Allen et al. (*4*) explain that independent retailers generate larger traffic impacts because of their decentralized distribution system, whereas larger chains given their centralized system generate less trips of a larger size.

[1]Department of Technology Management and Economics, Chalmers University of Technology, Göteborg, Sweden

Corresponding Author:
Address correspondence to Iván Sánchez-Díaz: ivan.sanchez@chalmers.se

AFS businesses do not only suffer the impacts of traffic, they are also a major contributor to it. AFS establishments are often located in central areas of cities (*5*) and generate a large amount of freight trips. Schoemaker, Allen et al. (*6*) estimate that this sector is responsible for about 31% of the freight moved in urban areas, and Holguín-Veras, Sánchez-Díaz et al. (*7*) estimate that this sector is responsible for 14%–17% of the total freight traffic in the main metropolitan areas of Minneapolis, Pennsylvania, New Jersey, and Florida, USA. These numbers reveal the important contribution of this sector to urban freight traffic and the need to study in detail the freight demand of AFS establishments. This knowledge can be instrumental for planning and implementing efficient strategies that reduce the impacts of deliveries and alleviate traffic conditions, thus benefiting commercial activity and at the same time enhancing the city's livability.

Developing and implementing urban freight plans is becoming an important priority for city authorities in recent years. The City of Stockholm has developed The Stockholm Freight Plan 2014–2017 to foster safe, clean, and efficient freight deliveries (*8*). The plan seeks to enable more reliable delivery times, to facilitate access for freight vehicles, to promote clean vehicles, and to advance freight delivery partnerships. To achieve those goals, the City proposes nine strategies including developing a freight consolidation project, studying the feasibility of off-peak hours deliveries (OPHD), and increasing the number of loading zones for large freight vehicles, among others. The plan also highlights the importance of collecting freight data to analyze the freight system and to support the implementation of the strategies proposed.

The consolidation strategy proposed seeks to promote consolidation centers, i.e., "staffed freight reception areas where distributors can unload freight which is then consolidated and dispatched with other freight destined for the same part of the city." While OPHD seeks to shift "freight deliveries to times when the city's road network is less congested. This does not necessarily mean night delivery, but can also be late-evening or early morning delivery." The increase in the number of loading zones can be achieved via local traffic regulation, adopted by the municipality or the county administrative board, to modify waiting and parking restrictions (*8*). Similar strategies have been successfully implemented in the AFS sector in various countries (*9–13*). Key findings from these experiences include the necessity to gain receivers' support for a successful implementation, as well as finding cost-effective schemes to implement OPHD and consolidation. This paper focuses on assessing receivers' support.

The purpose of this paper is twofold: it seeks (i) to analyze the freight demand of AFS receivers; and (ii) to investigate their opinion about freight deliveries in general (e.g., satisfaction, problems) and about two strategies aimed at alleviating the impacts of deliveries (i.e., consolidation and OPHD). To this effect, the paper is organized in four sections in addition to this introduction. Section 2 describes the method adopted. Section 3 and 4 present the results and practical implications. Section 5 discusses the main conclusions from this research.

Method

Data Collection Framework

This paper studies the AFS sector using data collected in the City of Stockholm during the fall of 2016. The geographical area of the study was selected based on the target of the strategies on consolidation and OPHD formulated in the Stockholm Freight Plan. It covers the central zone of the city that suffers from high congestion and where a congestion fee is currently being charged.

The sample size was designed taking into account past experience of mean and variance in the AFS sector as proposed by Ortúzar and Willumsen (*14*), and respecting a budget constraint to allow data collection for all commercial sectors of interest. An aspirational sample size was designed using the mean and variance of weekly deliveries from a past study (*15*) and assuming a response rate of about 80%, as

$$n = \frac{100}{80}\left(\frac{S}{0.051\mu}\right)^2 = 240 \quad (1)$$

where

n = sample size for the AFS sector, and

S and μ = sample variance and mean for weekly deliveries in (*15*), i.e., $S = 9.9$ and $\mu = 14.0$

0.051 assumes a desired 95% confidence level and a Normal distribution for the variable of interest.

The data was segmented in two strata using commercial sectors as designated by the Swedish National Industry (SNI) code 55 which includes accommodation services such as hotels, youth hostels, and bed and breakfast; and 56 which includes restaurants, cafés, pubs, bakeries, hotels, and catering services (*16*). A random sampling was used to select the establishments for the survey. No prior information was available about the variance in the accommodation sector. Therefore, the criterion for the partition of the sample between strata was to ensure a minimum of observations for each subsector (i.e., 40 observations).

The data collection combined internet surveys, computer-assisted telephone interviews and in-person interviews to ensure a high response rate and that the respondent was the right person to respond to questions.

In the case of accommodation, the study zone has 190 hotels, from which 60 (or 31.6% out of the universe) were sampled and 46 (or 76.7% out of the sample) provided complete data. For the food sector, the zone is home of 1,896 establishments from which 180 (or 9.5% out of the universe) were sampled and, after a first call, 28 establishments were excluded from the sample either because the business had closed or because the establishment was not in Stockholm, leading to an effective sample of 152 establishments. From this sample, only 43 (or 28.2% out of the sample) provided complete data. The main reasons for not participating in the study were confidentiality and lack of time. Although the sample size obtained (i.e., 89 observations) is less than the sample size expected based on Equation 1, it is still higher than the number of observations required for a 90% confidence level (i.e., 67 observations), and thus considered valid for this study.

The questionnaire was developed together with representatives from the public sector and was pilot tested with a few establishments before full deployment. It included a set of structured questions and an open question at the end where respondents could give their opinion about current deliveries and the strategies. More specifically, the first set of questions inquired about the type of commercial activity, the type of business (i.e., independent or part of a chain), the size of the establishment in number of employees and area, the number of weekly deliveries and shipments, estimates of the amount of cargo attracted, the number of carriers, the type of vehicles used for deliveries, the proportion of freight trips that included a service (i.e., defined as trips that include an activity or service performed by the carrier including cleaning, vending and coffee machine refill, paper refill, etc.), the type of receiving facility (e.g., main entrance, loading dock), and the time of the day when deliveries are received. The second set of questions inquired about their level of satisfaction with deliveries, the importance they give to personal contact with carriers, their interest in consolidating deliveries and receiving OPHD, as well as their ability and preference to receive deliveries during different hours of the day.

Data Analysis and Model Development

The variables were classified by nature (i.e., continuous, discrete, and categorical) and by attribute (i.e., freight quantity, establishment attributes, delivery characteristics, and attitudinal variables). Non-parametric tests were then performed to assess statistically significant relationships, chi^2 tests were used for assessing continuous vs. discrete/categorical variables, and Cramer's V tests were used to assess relationships between discrete/categorical variables. The t-stat, the Cramer's V values and probability values are reported to support the descriptive analysis.

The random sampling implemented enabled the use of regression models for inferential statistical analyses (i.e., the findings can be generalized to the overall AFS sector). Inferential analyses are used in this paper to relate establishments' attributes to freight trip generation (FTG) patterns and to the opinion of the person in charge of deliveries at the establishment towards the strategies proposed.

The terms used to explain FTG patterns are defined as follows. FTG can be divided into freight trip attraction (FTA) and freight trip production (FTP). FTA is defined as the number of deliveries attracted by an establishment to fulfill the freight needs that allows its commercial activity; while FTP is defined as the number of shipments produced by an establishment (*15*, *17*). The summation of FTA and FTP, as well as the freight trips that involve both a delivery and a shipment pick-up, is defined as FTG. As explained by Holguín-Veras, Jaller et al. (*17*), the estimation of FTG models requires an understanding of establishments' logistics costs explaining ordering patterns. In particular, the economic order quantity (EOQ) can provide a good basis to explain the optimal order quantity (or shipment size) and frequency as a function of demand, setup cost, and inventory cost.

The EOQ model shows that an increase in demand can be satisfied via an increase in the shipment size without necessarily increasing frequency, thus optimal delivery frequency (or FTA) is a nonlinear function of demand (which depends on business size) (*17*, *18*). In essence, the EOQ model shows that FTA, FTP, and FTG are better modeled without assuming direct proportionality between freight trips and business size. The regression models proposed in this paper follow the method explained in (*15*) to estimate FTA, FTP, and FTG as a function of establishments' attributes. The main methodological novelty in this paper is the focus on an explanatory purpose (in line with this paper's purpose) instead of on forecasting, and the focus on AFS establishments, which allows the assessment of unexplored variables. The estimation of nonlinear FTP and FTG models is also a contribution of this paper not present in others.

Following (*15*), both linear and nonlinear functional forms (lin-log and log-log) are assessed and the best specification is selected based on the lowest value of the Akaike's information criterion (AIC) (*19*). Area and employment are not used in the same model to prevent imperfect multicollinearity that could harm the precision of the parameters estimates. The parameters are estimated using sandwich estimators that are robust to specification errors and outliers (*20*). To ensure strong statistical relationships, only variables that are statistically significant at the 5% level are kept in the models. The overall form of the model can be expressed as

$$Y_n = f(\alpha, \boldsymbol{\lambda\delta}_n, \boldsymbol{\beta}\mathbf{X}_n, \boldsymbol{\theta\delta}_n\mathbf{X}_n) \quad (2)$$

where

Y_n = FTA, FTP or FTP which is a continuous dependent variable for establishment n;

α = intercept;

$\boldsymbol{\lambda}$ = vector of estimable parameters for the binary variables;

$\boldsymbol{\delta}_\mathbf{n}$ = vector of binary variables denoting an establishment attribute or the commercial sector of establishment n; for the latter case each binary variable denotes a sector and takes the value of 1 if establishment n belongs to the sector, 0 if not;

$\boldsymbol{\beta}$ = vector of estimable parameters;

$\mathbf{X}_\mathbf{n}$ = vector of continuous variables or attributes proper to establishment n;

$\boldsymbol{\theta}$ = vector of estimable parameters.

For the attitudinal analysis, in addition to the description analysis, a set of discrete choice models are estimated to analyze the opinion of respondents towards the strategies proposed in the Stockholm Freight Plan. The opinion of respondents was captured using a scale from one to six, where one denotes a very negative opinion about the action and six a very positive opinion. However, due to the limited number of observations, the scale was recoded to have only 3 levels, negative, neutral and positive. Given the ordinal nature of the data, an ordered logit model is proposed to model the respondents' opinion following (*21*, *22*). In this type of model, the dependent variable is estimated using a utility function and a set of thresholds to replicate the choice probabilities captured in the calibration data. The model uses the postulates from random utility theory, in which the respondents select a rate based on the level of utility that the action proposed represents for them. The utility function depends on the respondent's establishment attributes and the action proposed (*14*, *23*). The utility function can be expressed with

$$U_n = \alpha + \boldsymbol{\lambda\delta}_n + \boldsymbol{\beta}\mathbf{X}_n + \boldsymbol{\theta\delta}_n\mathbf{X}_n \quad (3)$$

where U_n = utility function for the respondent in establishment n.

The set of thresholds, $\mu(1) < \mu(2) < \mu(3)$, are estimated concurrently with the utility function. The parameters and the values of the thresholds are computed via a likelihood maximization approach, and the quality of the model can be measured through the highest likelihood ratio index, i.e., Pseudo R^2 (*23*). The criteria for including/excluding variables follow (*14*); variables with a t-stat higher than 1.96 are included; in cases where the variable is relevant for the purpose of the study it is included even if the t-stat is between 1 and 1.96. In all cases the conceptual validity of the model is the main selection criterion.

Results

The analysis of the data was conducted in three parts. The first part analyses current freight activity in AFS establishments, the second part analyses the opinion about deliveries and potential strategies, and the third, overall perceptions captured in the open question.

Freight Activity in AFS Establishments

The descriptive analysis in Table 1 provides an overall picture of freight activities in AFS establishments. Table 1 displays the number of observations with valid data, the mean, the coefficient of variation (i.e., standard deviation divided by the mean), and the minimum and maximum for each variable.

As shown, accommodation establishments generate on average more than twice as many trips than food services (18.4 vs. 8.5 trips/week). 80% of the accommodation establishments both produce (e.g., laundry) and receive trips, while only 30% of food services establishments do so. The average share of trips including a service is about 18%, with establishments in the accommodation sector having a slightly (but not statistically significant: t = 0.93, Pr = 0.35) higher share of service trips than other food services. Accommodation establishments attract more freight than food services establishments, with 36% and 19% of the establishments attracting more than 1 ton per week, respectively. Overall establishments in the AFS sector generate on average 4.4 tons per week.

Accommodation establishments are significantly larger than food services establishments both in terms of employees and area (t = 3.52, Pr = 0.0007 and t = 3.82, Pr = 0.003, respectively). Accommodation establishments are more likely to be part of a chain (50% vs. 30%, t = 2.21, Pr = 0.03) and also have more carriers (5.8 vs. 3.3, t = 2.42, Pr = 0.018). However, no statistical relationship was found between belonging to a chain and the number of carriers (corr = 0.15, t = 0.61, p = 0.542).

The share of establishments using a loading dock is similar for both sectors (30%–40%). More specifically, 52% of the establishments in the overall sample reported to receive their deliveries through the main entrance, another 38% reported to receive them through a loading dock and the remaining 10% through a different way. Although establishments generating a lower amount of freight tend to receive their goods through the main entrance, this difference is not statistically significant

Table 1. Descriptive Statistics

Variable	Unit	Obs.	Mean	CV	Min	Max
Accommodation						
FTG	Trips generated/week	46	18.4	109%	1	102.0
FTA	Deliveries attracted/week	46	14.4	121%	0.5	90.0
Intermediary	1 if yes, 0 if not	46	0.8	n.a.	0.0	1.0
Interm. FTP	Trips produced/week	37	5.0	104%	0.5	22.5
Service	Percentage	44	21.1%	150%	0	100%
Employment	Employees	46	24.1	136%	1.0	130.0
Area	100 m^2	37	37.3	153%	2.5	270.0
Chain	1 if yes, 0 if not	46	0.5	n.a.	0	1.0
Loading dock	1 if yes, 0 if not	46	0.4	n.a.	0	1.0
Carriers	# of carriers	40	5.8	101%	0	24.0
Food services (restaurants, cafés, pubs, etc.)						
FTG	Trips generated/week	43	8.5	81%	1	26.0
FTA	Deliveries attracted/week	43	7.2	81%	1.0	20.0
Intermediary	1 if yes, 0 if not	43	0.3	n.a.	0.0	1.0
Interm. FTP	Trips produced/week	14	3.9	60%	1.0	8.0
Service	Percentage	39	15.0%	197%	0	100%
Employment	Employees	43	6.2	109%	1.0	35.5
Area	100 m^2	39	2.4	189%	0.1	25.0
Chain	1 if yes, 0 if not	43	0.3	n.a.	0	1.0
Loading dock	1 if yes, 0 if not	43	0.3	n.a.	0	1.0
Carriers	# of carriers	41	3.3	95%	0	14.0

(Cramer's V = 0.22, Pr = 0.493). However, establishments that belong to a chain are significantly more likely to receive their deliveries through loading docks than independent establishments (Cramer's V = 0.33, Pr = 0.002). In terms of delivery units, half of the deliveries are made in roller cages, only 17% delivered as packages and another 8% delivered as pallets. Other types of units used included laundry bags, boxes, sacks, and hose for beer.

The questionnaire also inquired about the size of the vehicles delivering to the establishments and the ownership. Most of the trips (55%) are done by medium-heavy vehicles, i.e., gross vehicle weight over 3.5 tons, followed by light vehicles (38%), i.e., gross vehicle weight between 1 tons and 3.5 tons, cars (6%), and others including cargo bikes and walking. Establishments that are part of a chain have a significant larger share of medium-heavy vehicles deliveries than independent establishments (65% vs. 26%, t = 4.82, Pr = 0.000). The share of own vehicles is very low for most sub-sectors, except for cafés and bakeries where 14% of the trips are made using their own vehicles.

Several regression analyses were conducted to relate FTG patterns to establishments' attributes. The models assessed all the attributes captured in the questionnaire as independent variables. Table 2 shows the best employment and area models for FTA, FTP, and FTG (FTA + FTP).

As shown, the best model specification is a log-log function in which FTA, FTP, and FTG are estimated as a concave function of employment or area, i.e., FTA, FTP, and FTG increase at a diminishing marginal rate with business size. Both business size variables, area and employment, have equivalent effects on FTA, FTP, and FTG. Area models display a better AIC.

As this is a log-log model, the interpretation of the results requires an exponential transformation of the parameters. The models show that FTA, FTP, and FTG are influenced by other variables beside business size. The main finding is that belonging to a chain has an incremental effect on FTG (1.6–1.9 times larger). Another interesting finding is that the marginal effect of business size on FTA is larger for accommodation establishments than for food services establishments and even larger if it belongs to a chain. In most cases there is no statistically significant difference between sub-sectors in the food services sector (i.e., café, bakery, pubs), except for catering in the FTP model where the effect of business size is smaller than for other sub-sectors. The price paid for a night at the establishment or the price of food were not found to be statistically significant.

The application of employment-based FTG models to the 2,086 AFS establishments (190 accommodation and 1,896 food services establishments) registered in the area of study shows that this sector generates 24,648 freight trips every week which corresponds to 15% of the FTG (food services account for 12.5% and accommodation for 2.5%).

Opinion About Deliveries and Potential Strategies

The respondents were asked their opinion about consolidation, OPHD, and OPHD that involve a service in line

Table 2. Nonlinear Freight Trip Generation Models

Variable	FTA models		FTP models		FTG models	
	ln(FTA)	ln(FTA)	ln(FTP)	ln(FTP)	ln(FTG)	ln(FTG)
Constant	0.79 (4.88)	1.55 (10.13)		0.49 (2.15)	1.01 (7.15)	1.46 (10.11)
Chain	0.62 (2.92)				0.55 (2.68)	0.47 (2.01)
Accommodation	−0.78 (−2.28)	−0.88 (−2.74)				
ln(number of employees)	0.41 (5.28)		0.48 (9.61)		0.44 (6.52)	
ln(area [100 m^2])		0.32 (2.72)		0.24 (2.70)		0.27 (4.47)
Accommodation_chain* ln(area [100 m^2])		0.27 (2.81)				
Catering_ln(area [100 m^2])				−1.96 (−2.76)		
Observations	89	76	51	42	89	76
Akaike's information criterion (AIC)	230.27	206.04	129.61	108.06	223.33	203.08
RMSE	0.87	0.91	0.85	0.87	0.83	0.90

Note: t-stats = shown in parenthesis under the parameters.

Table 3. Opinion About Strategies, Importance of Personal Contact, and Satisfaction

Score	Consolidation		OPHD		OPHD with service		Personal		Satisfaction	
3	20%	18	23%	20	27%	20	40%	40	49%	43
2	15%	13	16%	14	19%	14	34%	34	40%	35
1	65%	57	61%	54	55%	41	26%	26	11%	10
Average	1.5		1.6		1.5		2.1		2.0	

with the strategies proposed in the Stockholm Freight Plan. Additionally, the respondents were inquired about the importance they give to personal contact with their vendors and their current level of satisfaction with deliveries. The results are shown in Table 3.

As shown, the opinion about the strategies proposed is mainly negative (>50%), but there is high heterogeneity among respondents. More than 20% of the respondents had a very positive opinion. OPHD rank slightly better than consolidation. Accommodation establishments have significantly more positive opinions about consolidation than food services (1.8 vs. 1.26, t = 3.30, Pr = 0.001), but there is no significant difference in terms of OPHD.

The importance given to personal contact is also heterogeneous. The overall score is 2.1 and two thirds of respondents gave medium-high importance to personal contact with their vendors, there is no significant difference between AFS establishments. The level of satisfaction ranks high, with about half of respondents rating it as good-very good, and accommodation establishments have a better satisfaction level than food services (2.3 vs. 1.63, t = 2.60, Pr = 0.01).

To shed light on the interaction between the opinion about strategies and the attitude towards current deliveries, a correlation analysis was performed. The results are shown in Table 4.

The results reveal a high positive correlation between OPHD and OPHD including a service, and a moderate positive correlation between OPHD and consolidation. There is also a low negative correlation between the importance given to personal contact and the opinion about the strategies. A similar result is found for satisfaction with current deliveries, for which the correlation with the opinion about the strategies is also very low. There is also a moderate positive correlation between personal contact importance and satisfaction.

The respondents were inquired about the times when they prefer to receive their deliveries, as well as about the time when they are currently receiving and when they could receive them. The results are summarized in Table 5.

The first salient result is that the time slot between 09:00 and 12:00 is the most popular and the slot between 18:00 and 06:00 the least popular, which is in line with the opinion about OPHD. However, 31% of the respondents said they can receive between 18:00 and 22:00 from which 24% are still not receiving at that time any delivery; and 21% of the respondents said they can receive between 22:00 and 06:00 from which 12% are still not receiving. It is also interesting to note that most respondents (more than 85%) receive their deliveries in the time they prefer to receive them.

Although the previous results provide an overall picture of the opinion of receivers, additional insight can be

Table 4. Correlation Between Interests in Initiatives and Personal Contact

	Consolidation	OPHD	OPHD with service	Personal contact
OPH	0.309	–	–	–
OPH service	0.183	0.527	–	–
Personal contact	−0.130	−0.271	−0.118	–
Satisfaction	−0.176	−0.196	−0.078	0.254

Table 5. Preferences of Delivery Time

Period of day	Hours	Prefer to receive	Receive today		Can receive		Overall
			Yes	No	Yes	No	
Peak hours	06:00–09:00	Yes	53%	2%	54%	1%	55%
		No	4%	40%	12%	33%	45%
	09:00–12:00	Yes	60%	7%	66%	0%	66%
		No	16%	18%	18%	16%	34%
	12:00–15:00	Yes	42%	9%	67%	0%	51%
		No	13%	36%	20%	13%	49%
	15:00–18:00	Yes	27%	15%	42%	0%	42%
		No	15%	44%	27%	31%	58%
Late evening/early morning off-peak hours	18:00–22:00	Yes	1%	7%	7%	1%	8%
		No	3%	89%	24%	69%	92%
	22:00–06:00	Yes	4%	6%	9%	1%	10%
		No	3%	87%	12%	78%	90%

gained from behavioral models that analytically relate establishments' attributes and the opinion about the strategies. The results are shown in Table 6.

These behavioral models show that accommodation establishments, and in general larger establishments, have a more positive opinion about consolidation and OPHD. Being part of a chain has more intricate implications. For accommodation establishments that are part of a chain, the tendency of large establishments to have a more favorable opinion about consolidation is less evident than for other establishments (effect of an extra employee on utility is 0.014 vs. 0.034 for others). However, these same types of establishments have a more positive opinion about OPHD, whereas restaurants that are part of a chain have less positive opinion about OPHD. For OPHD, establishments that give more relevance to personal contact with vendors have more negative opinion about OPHD. Although the same was expected for consolidation, the variable denoting relevance of personal contact was not significant so it was not included in the model.

Table 6. Ordered Logit Models: Opinion About Strategies

Variable	Consolidation	OHD
Accommodation	1.180	–
	(2.32)	–
Accommodation_chain	–	0.996
	–	(1.99)
Restaurant_chain	–	−2.000
	–	(−2.89)
Number of employees	0.034	0.013
	(2.15)	(1.17)
Employees* Accommodation_chain	−0.020	–
	(−1.13)	–
Personal contact relevance	–	−0.610
	–	(−1.84)
Threshold parameters		
$\mu(1)$	−3.8	−6.06
$\mu(2)$	1.62	−0.45
$\mu(3)$	2.48	0.447
Observations	89	89
Pseudo R^2	0.09	0.10
AIC	165.90	171.55

Overall Perception of Receivers

As mentioned before, the last part of the questionnaire included an open question where receivers could mention their topics of concern and/or opinion on matters related to their deliveries. The most common topic mentioned was about conflicts with other users, in particular related to loading zones (over 11 mentions), congestion, pedestrian zones, and noise.

The comments about loading zones included concerns about illegal parking from private vehicles, the use of loading zone for construction bins, the lack of loading zones or inappropriate size, a poor maintenance during winter and the lack of a correct demarcation that could allow more efficient use of these spaces. Other issues mentioned included the conflicts of deliveries with road works, garbage collection trucks (that also were judged too noisy) and taxi zones deemed as unnecessary.

The main concerns raised about OPHD are related to noise. In the case of hostels, establishments' owners are concerned about the impact OPHD could have on their guests' rest. Some establishments mentioned that they were already receiving OPHD, and some others stressed the need for changes in space planning and policy.

Another issue that arose a couple of times was the difficulty for some carriers to find addresses.

Practical Implications

The first finding is that AFS have similar business activities in which the added-value depends largely on intangible elements related to quality of service and location, and an important share of the goods are perishable. Although AFS establishments are often grouped under the same industry code and are often studied together, the results show that accommodation establishments have different freight generation patterns than food services establishments, which makes it more appropriate to study the two sectors separately.

Accommodation establishments have significantly larger business size, freight generation and FTG, a larger share of accommodation establishments belong to a chain, and the size of the vehicles serving accommodation establishments tend to be larger. This means that the implementation of the strategies in the accommodation sector can have a larger potential to improve local effects. However, applying the models to the whole city reveals that food services establishments generate 12.5% of all freight trips compared to 2.5% generated by accommodation establishments, meaning that targeting food services can provide larger benefits at the network level.

An interesting finding relates to the type of business. The results show that there is no correlation between being part of a chain and the number of carriers, meaning that being part of a chain does not necessarily imply being supplied by fewer carriers. This is in line with the findings from (*3*) who explain that some chains outsource logistics and thus deliveries are not necessarily optimized within the franchisors' network. In this case, the distribution system resembles a big local and goods specialist typology where owners of a variety of goods distribute directly to receivers. As the rationality for a consolidation strategy is that coordinating multiple carriers can result in fewer and more consolidated delivery trips, these findings suggest that consolidation has a potential for both establishments that are and are not part of a chain. When analyzing the effects of commercial sector and business size via econometric modeling, the results show that large accommodation establishments that are part of a chain receive more deliveries relative to business size than establishments not being part of a chain. However, these types of establishments show more interest in OPHD than in consolidation.

The attitudinal study shows that the opinion about the strategies proposed in the plan are homogeneous. More than half of the respondents have a negative-neutral opinion, and about 20% have a positive opinion, in line with the findings in other cities, such as New York City (*21*, *22*). This shows that the implementation of these strategies will work better when implemented in a voluntary program with some incentives, e.g., public recognition, rather than as a mandatory regulation. The comments from respondents to the open question also showed some concerns mainly related to loading zones and conflicts with other users. This insight can be used to design other incentives, such as more or better access to loading zones to the establishments that participate in the strategies proposed by the city.

As shown in the results about the time when establishments receive their deliveries, more than 85% of the respondents currently receive in the times they prefer, confirming the market power held by receivers as described by (*24*). Only 5% are currently receiving OPHD, which is normal given the current regulation prohibiting OPHD. Although the possibility to receive during certain times can be influenced with the introduction of unassisted deliveries technologies, 70%–80% of the establishments said that they cannot receive between 18:00 and 6:00. In essence, the potential market share for OPHD and consolidation in the AFS sector is around 20% which means that there is potential to influence about 5,000 freight trips every week.

The behavioral models show that, even when controlling for business size, accommodation establishments have a more positive opinion about consolidation and OPHD. Larger establishments have a more positive opinion about the strategies suggested, although this effect is more significant for consolidation than for OPHD. These results show that the strategies would have more acceptance among small accommodation establishments that are part of a chain, as well as among large AFS establishments that are not part of a chain.

Conclusions

In 2014, the City of Stockholm launched the Stockholm Freight Plan with the objective of fostering safe, clean, and efficient freight deliveries. Two of the main strategies proposed were implementing consolidation centers and OPHD. This paper studies the demand characteristics and opinion of establishments in the AFS sector about these initiatives. The findings from this study can be used as insight when designing urban freight strategies in other cities.

AFS establishments show an interest in improving their deliveries, decreasing congestion and making their neighborhoods more attractive. Studying accommodation establishments separately allowed some interesting conclusions to be drawn. Although they represent a lower share of establishments, they generate more deliveries than other food services establishments which leads to large local traffic impacts. Accommodation establishments also show more interest in taking part in the strategies proposed by the city which make them a good stakeholder to target.

Overall, more than half of the respondents have a negative-neutral opinion about consolidation and OPHD, while 20% have a positive opinion. Therefore, a smooth implementation of these strategies requires voluntary participation supported by incentives, e.g., public recognition and better access to loading zones.

The application of the FTG models to the whole city shows that the AFS sector generates 24,648 freight trips every week (about 15% of the city's FTG). Moreover, the behavioral study showed that the potential market share for consolidation and OPHD in the AFS sector is close to 20% (without offering any incentive or promoting an unassisted OPHD program), meaning that there is a potential to influence about 5,000 freight trips every week. The experience in other cities has shown that influencing the deliveries of 20% of the establishments can bring about major local and network benefits for the city (*21*), which clearly supports the objectives stated in the Stockholm Freight Plan.

The research design proposed in this paper is cross-sectional and focuses on the receivers' side which leads to certain limitations that should be addressed in further research. The main limitations of this design are the impossibility of studying in detail the supply chain structure and its impact on FTG and the opinion about the strategies, as well as the lack of study of the opinion of other stakeholders such as suppliers and carriers, and a detailed study of practical mechanism to implement the programs (e.g., type of commodities to target, incentives, noise control, consolidation facilities). These limitations can be addressed in the future with a complementary set of case studies.

Acknowledgments

The author would like to acknowledge the support from the Traffic and the Environmental Offices of the City of Stockholm through the project "Freight Trip Generation models for the City of Stockholm," from the Volvo Research and Educational Foundations (VREF) that fund the Urban Freight Platform (UFP) through its Future Urban Transport research program, and from the Area of Advance Transport at Chalmers University of Technology. The author would also like to acknowledge the contributions from Märta Brolinson and Elin Skogens formerly at the Traffic Office of the City of Stockholm, and from Chris Goodall and Nicole Bielak formerly at Atkins Sverige AB, whose inputs and contributions to the project were crucial.

Author Contributions

The author confirms sole responsibility for the following: study conception and design, data collection, analysis and interpretation of results, and manuscript preparation.

References

1. Schrank, D., B. Eisele, T. Lomax, and J. Bak. *2015 Urban Mobility Scorecard.* Texas A&M Transportation Institute and INRIX, Kirkland, WA, 2015.
2. Wirtz, J., and C. Lovelock. *Services Marketing: People, Technology, Strategy.* World Scientific Publishing Company, Hackensack, NJ, 2016.
3. Verlinden, T., E. Van de Voorde, and W. Dewulf. Ho. Re. Ca. Logistics and Medieval Structured Cities: A Market Analysis and Typology. In *Commercial Transport*, Springer, New York City, NY, 2016, pp. 369–384.
4. Cherrett, T., J. Allen, F. McLeod, S. Maynard, A. Hickford, and M. Browne. Understanding Urban Freight Activity: Key Issues for Freight Planning. *Journal of Tranport Geography*, Vol. 24, No. 1, 2012, pp. 22–32.
5. Sanchez-Diaz, I., and J. Gil. Exploring the Role of Urban Form on Freight Trip Generation. *Proc., VREF Conference on Urban Freight*, Gothenburg, SE, 2016.
6. Schoemaker, J., J. Allen, M. Huschebek, and J. Monigl. *Quantification of Urban Freight Transport Effects I.* BESTUFS Consortium, NEA, University of Westminster, PTV, Transman, Warszawa, Poland, 2006.
7. Holguín-Veras, J., I. Sánchez-Díaz, and M. Browne. Freight Demand Management: Role in Sustainable Urban Freight Systems. *Proc., 9th International Conference on City Logistics*, Springer, Tenerife, Spain, 2015.
8. City of Stockholm. *The Stockholm Freight Plan 2014-2017.* Stockholm Traffic Department, Stockholm, Sweden, 2015.
9. Morganti, E., and J. Gonzalez-Feliu. City Logistics for Perishable Products: The Case of the Parma's Food Hub. *Case Studies on Transport Policy*, Vol. 3, No. 2, 2015, pp. 120–128.
10. Sanchez-Diaz, I., P. Georen, and M. Brolinson. Off-Peak Hour Deliveries: A Review of Theory and Practice. *Transport Reviews*, Vol. 37, No. 4, 2015, pp. 521–543.
11. Holguín-Veras, J., J. M. Wojtowicz, C. Wang, M. Jaller, X. J. Ban, F. Aros-Vera, S. Campbell, X. Yang, I. Sanchez-Diaz, J. Amaya, C. González-Calderón, R. Marquis, S. D. Hodge, T. Maguire, M. Marsico, S. Zhang, S.

Rothbard, K. Ozbay, E. F. Morgul, S. Iyer, K. Xie, and E. E. Ozguven. *Integrative Freight Demand Management in the New York City Metropolitan Area: Implementation Phase*. Rensselaer Polytechnic Institute, New York City Department of Transportation, Rutgers University, New York City, NY, 2013.

12. BESTUFS. *Good Practice Guide on Urban Freight Transport*. BESTUFS, University of Westminster, London, United Kingdom, 2007.
13. Verlinde, S., C. Macharis, and F. Witlox. How to Consolidate Urban Flows of Goods Without Setting Up an Urban Consolidation Centre? *Procedia-Social and Behavioral Sciences*, Vol. 39, No. 2012, 2012, pp. 687–701.
14. Ortúzar, J. D., and L. G. Willumsen. *Modelling Transport*. John Wiley & Sons, Hoboken, NJ, 2011.
15. Sánchez-Díaz, I., J. Holguín-Veras, and X. Wang. An Exploratory Analysis of Spatial Effects on Freight Trip Attraction. *Transportation*, Vol. 43, No. 1, 2016, pp. 177–196.
16. Statistiska Centralbyrån. *SNI 2007*. http://www.sni2007.scb.se/snisokeng.asp. Accessed October 15, 2007.
17. Holguín-Veras, J., M. Jaller, L. Destro, X. Ban, C. Lawson, and H. S. Levinson. Freight Generation, Freight Trip Generation, and Perils of Using Constant Trip Rates. *Transportation Research Record: Journal of the Transportation Research Board*, 2011. 2224: 68–81.
18. Harris, F. *Operations and Costs*. A. W. Shaw Co., Chicago, IL, 1915.
19. Akaike, H. Factor Analysis and AIC. *Psychometrika*, Vol. 52, No. 3, 1987, pp. 317–332.
20. Freedman, D. A. On the So-Called "Huber Sandwich Estimator" and "Robust Standard Errors". *The American Statistician*, Vol. 60, No. 4, 2006, pp. 299–302.
21. Holguín-Veras, J., C. Wang, I. Sánchez-Díaz, S. Campbell, S. D. Hodge, M. Jaller, and J. Wojtowicz. Fostering Unassisted Off-Hour Deliveries: The Role of Incentives. *Tranportation Research Part A: Policy and Practice*, Vol. 102, No. 2017, pp. 172–187.
22. Holguín-Veras, J., and I. Sánchez-Díaz. Freight Demand Management and the Potential of Receiver-Led Consolidation Programs. *Transportation Research Part A: Policy and Practice*, Vol. 84, No. 2016, pp. 109–130.
23. Domencich, T. A., and D. McFadden. *Urban Travel Demand: A Behavioural Analysis*. North-Holland Publishing Co., Amsterdam, Netherlands, 1975.
24. Holguín-Veras, J. Necessary Conditions for Off-Hour Deliveries and the Effectiveness of Urban Freight Road Pricing and Alternative Financial Policies in Competitive Markets. *Transportation Research Part A: Policy and Practice*, Vol. 42, No. 2, 2008, pp. 392–413.

The Standing Committee on Urban Freight Transportation (AT025) peer-reviewed this paper (18-04282).

The content of this paper does not represent the official opinion of agencies involved in the funding of this research.

Article

TRR
JOURNAL OF THE TRANSPORTATION RESEARCH BOARD

Transportation Research Record
2018, Vol. 2672(9) 204–215

DOI: 10.1177/0361198118783161
journals.sagepub.com/home/trr
SAGE

Parking Conditions for Residential Delivery in New York City: A Case Study Analysis

Pierre Escand[1], Quanquan Chen[2,3], and Alison Conway[3]

Abstract
This study employed basic demand estimation, field observation, text analysis, and spatial analysis methods to examine the adequacy of the existing supply of commercial dedicated parking space in high-density areas of New York City to accommodate expected demand for direct-to-home deliveries. The study also examined the proximity of available commercial dedicated parking space to end delivery locations. The study estimated and mapped two performance metrics: (1) the share of on-street commercial dedicated parking demanded for expected U.S. Postal Service residential freight deliveries, and (2) the share of these package deliveries expected to occur within a reasonable walking distance of a commercial dedicated parking space. The study relies on a variety of open data sources and on limited field observations; owing to data limitations, and resulting assumptions for baseline analysis, sensitivity analysis was also conducted. Results suggest that there is currently both a spatial and temporal mismatch between the commercial dedicated parking supply and expected residential delivery demand, and that shifts toward express deliveries may exacerbate this mismatch. Future research needs are also discussed.

Continuing and rapid growth in e-commerce-driven home delivery, and increasing consumer expectations to control speed and arrival time, are transforming last-mile urban goods movements. Different from commercial truck trips to commercial locations, e-commerce home deliveries usually consist of smaller parcel sizes; these deliveries are mostly completed using smaller delivery vehicles such as vans that make many stops spread throughout the city and at all times of the day, often destined to residential homes. Consequently, neighborhoods are being inundated with new goods movement activity. The buildings to which goods are destined, the streets on which the trucks operate, and the curb regulations at delivery locations were not designed to accommodate this type of demand, resulting in congestion, emissions, and safety impacts. However, quantifying these impacts is extremely difficult because of a lack of available data on local freight vehicle activity and limited tools available to understand the relationship between direct-to-home delivery frequencies and associated vehicle trips. This study aimed to investigate one emerging challenge—new parking activity associated with residential deliveries—by employing literature review, basic demand estimation, field observation, and spatial analysis methods. Spatial analysis examined two aspects of the residential freight parking problem: the adequacy of the existing supply of commercial dedicated parking space to accommodate expected demand for direct-to-home deliveries; and the proximity of this space to end delivery locations.

Literature Review

Goods purchased online are in most cases distributed by a logistics operator to a destination determined by the consumer. Not all e-commerce deliveries end up at a home address (*1*), as alternative solutions are available such as pick-up points and automatic locker systems (*2*). However, most online purchases are still delivered to residential addresses; UPS's 2016 *Pulse of the Online Shopper* report revealed that an estimated 65% of individuals surveyed preferred delivery to their home, and another 6% to a friend or family member's home (*3*).

[1]ENTPE–Ecole Nationale des Travaux Publics de l'Etat (National Graduate School of Sustainable Civil Engineering, Transportation, and Planning), Vaulx-en-Velin Cedex, France
[2]Freight Planning, North Jersey Transportation Planning Authority, Newark, NJ
[3]Department of Civil Engineering, The City College of New York, New York, NY

Corresponding Author:
Address correspondence to Quanquan Chen: qchen@njtpa.org

These home-based deliveries present a major challenge for carriers; frequent, small deliveries to spatially dispersed residences are inefficient, and frequently result in failed deliveries (*4*, *5*).

In urban areas, high-demand density, limited road and curb capacity, and outdated regulations often exacerbate these challenges. Previous freight parking-related research studies have primarily focused on delivery vehicle parking problems in areas where high levels of commercial activity exist (*6*, *7*). For traditional commercial freight delivery, trips are often concentrated in brief time frames, leading to overcrowding situations (*7*). Muñuzuri et al. categorized four parking choices for delivery vehicles: normal (spaces available for all types of vehicles), loading/ unloading zone, double parking, and parking on a sidewalk (*8*). Based on driver interviews, Nourinejad et al. established a binary logit model for parking location choice (*9*). The results showed that drivers are prone to choose a parking location if it is closer to a destination, or if it is a loading bay. Silva and Alho used structural equation modeling to examine the relationship between establishment characteristics and distribution channels versus parking challenges (*10*). Allen et al. note that from a parcel carriers' perspective, extremely limited curb parking space is one of the important factors that result in longer walking distance and longer time spent on foot making deliveries, potentially resulting in unreliable service or even financial penalties such as parking fines (*11*). Having recognized the parking challenges of business deliveries, major cities have implemented a series of management strategies for improving truck parking accessibility, as previously detailed by Chen et al. (*12*). For example, the New York City Department of Transportation (NYCDOT) has tested several solutions, including but not limited to dedicated delivery windows, paid commercial parking, peak-hour parking pricing, and the off-hour delivery program (*13*, *14*).

An important and difficult step in evaluating the adequacy of current infrastructure to accommodate emerging residential delivery trips is estimation of demand. This demand is expected to vary as a function of the socioeconomic characteristics of residents; however, these specific relationships are difficult to quantify as a result of a lack of comprehensive and timely local data and because of the rapid pace of change in consumer shopping behavior. Previous studies have employed survey data to examine the relationship between expected delivery frequencies and socioeconomic characteristics. Clarke et al. found the geo-demographic characteristics of age, income, rural/urban location, and physical store accessibility were important factors affecting online shopping frequency (*15*). Wang and Zhou found that several demographic factors were highly correlated to the number of online purchase delivered to home: web use frequency, education level, age, gender, self-employed status, race, household income, households owning more vehicles, and households with child (*16*). In addition, the geographic factors of the household such as housing unit density and rural area also were tested to be significant but with limited influence. The U.S. Postal Service (USPS)'s annual Household Diary Study survey also estimates package delivery frequencies as a function of socioeconomics; their most recent report provided package frequency rates with different groups of household income, age, education, size of the household, number of adults, and internet access (*17*). However, as the e-commerce market matures and as technology penetration increases, these relationships may not remain constant; for example, UPS's recent 2017 *Pulse of the Online Shopper, Vol. 1* concluded that age and place of residence have little impact on online shopping behavior (*18*). The other method of estimating household parcel delivery frequencies is observation of delivery events; Rodrigue used building records to characterize parcel delivery to a large residential building in Fort Lee, NJ, a suburb of New York City (NYC) (*19*).

Of the above-referenced studies, only Wang and Zhou converted package frequencies to vehicle trips, and their study relied on a simple conversion factor assuming delivery truck stops were inversely proportional to the number of residential units in a building (*16*). The other method employed to estimate truck trips to residential addresses is use of carrier tour information to directly observe delivery activity. This method requires access to carrier data, which is difficulty to obtain given industry privacy concerns. Cárdenas et al. utilized a sample dataset provided by a single carrier to project delivery trips of the whole market (*20*).

This study aims to address an existing gap in the literature by examining the parking demand associated with residential delivery activity, and evaluating the relationships between this demand and the existing dedicated parking supply in the study area.

Methodology

The study area for this project includes high-density neighborhoods in three NYC boroughs—Manhattan, Brooklyn, and Queens. In each census tract, a baseline demand for USPS package deliveries was estimated as a function of household income. Using conversion factors obtained both from recent literature and from field observation, this package demand was converted to a demand for curb space during four distinct time windows. As this analysis relied on several assumptions and field-observed values, in addition to a baseline scenario, ten sensitivity scenarios were also examined. A supply of

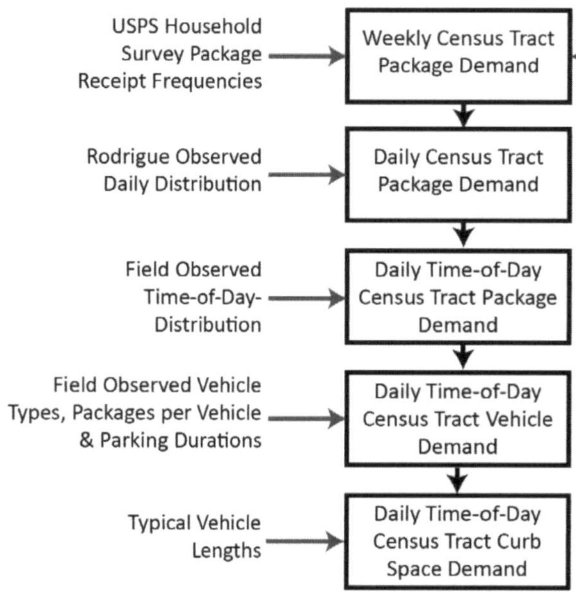

Figure 1. Residential freight parking demand estimation method.

dedicated curb space for commercial deliveries during specific time windows was estimated by development of an algorithm to read and classify signage text from a geocoded NYCDOT database. The ratio of estimated demand to available supply was then calculated to compare expected demand for curb space from USPS deliveries with available dedicated parking capacities in each census tract during each time window. Finally, to assess the proximity of available on-street parking supply to expected demand locations, a buffer analysis was conducted.

Residential Freight Parking Demand Estimation

As discussed in the literature review above, there is no one comprehensive data source available to assess demand for package delivery directly to residential homes. As a result, this study relied on an approximation method using a variety of inputs (Figure 1).

First, an overall expected USPS demand for package delivery within each census tract in NYC was estimated using the Postal Service's *The Household Diary Study: Mail Use and Attitudes in FY 2015* (*17*). Based on a survey of more than 5,000 households, this study estimated average package receipt frequencies based on three variables: head-of-household education level, head-of-household age, and household income category. As only the last of these applies to the entire household, only income-based package receipt rates were applied in this study (Table 1).

Table 1. USPS Package Frequencies

Household income ($k)	Weekly package receipts
0–35	0.37
35–65	0.58
65–100	0.82
100	0.98

These rates were applied at the census tract scale to estimate total weekly package demand. The number of households in each income category was estimated from U.S. Census Bureau 2015 American Community Survey (ACS) 5-year Household Income Estimates. As a result of incongruity in ACS and USPS income categories, some households could fall within one of two package frequency categories; for these cases, the lower rate was applied. Estimated weekly package frequencies were then converted to an expected weekday package demand. In analysis of an apartment building in suburban NYC, Rodrigue (*19*) found that the average weekday demand was about 17% of total weekly demand; this study did recognize differences in demand across different weekdays, but did not definitively identify the drivers of this variation. As a result, for the purpose of this study, the estimated weekly demand was divided by a factor of six to estimate daily package deliveries (Equation 1).

$$n_i = \sum_k \frac{PR_k * H_{i,k}}{6} \qquad (1)$$

where:

n_i: daily number of packages generated by households in census tract i

PR_k: USPS weekly package receipt rate for kth income level

$H_{i,k}$: number of households of kth income level in census tract i

In order to convert daily package demands to time-of-day vehicle space demand, several unknown conversion factors were needed: expected time-of-day trip distributions, expected packages per vehicle, and expected vehicle types (Equation 2). These variables were estimated using data from direct field observation.

In November and December 2016, 72 hours of curbside observation were performed at eight large residential buildings in four boroughs of NYC. All deliveries to these buildings were recorded by an observer between the weekday hours of 9 a.m. and 9 p.m., when the vast majority of residential parcel deliveries were expected to occur. Among other variables, the observers recorded the carrier, arrival time, departure time, vehicle type, and number of total parcels during each delivery event. These shares were applied directly to estimate the total number of packages delivered during each of four time windows: morning (9 a.m.–12 pm), afternoon (12 p.m.–4 p.m.), evening off-peak (4 p.m.–7 p.m.), and evening off-peak (7 p.m.–9 p.m.).

In the observation, 123 delivery attempts were observed carrying 1,036 parcels from all carriers. Normalizing for the exact number of minutes observed during each period, shares of total 12-hour (9 a.m.–9 p.m.) demand were expected to be about 8.8% per hour between 9 a.m. and noon, 12.2% per hour between noon and 4 p.m., 6.3% per hour between 4 p.m. and 7 p.m., and 3.1% per hour between 7 p.m. and 9 p.m.

To assign these packages to individual vehicles, the observed average number of packages/vehicle for each delivery event was 8.7. To convert vehicle demand to curb space demand, an expected vehicle length was estimated as the weighted average vehicle length. For 98 delivery events, the vehicle origin was observable; for the remainder, a delivery-person arrived carrying goods by hand or cart from a vehicle parked at another location. Of these 98 events, 47 occurred by step van (6.9 m), 25 by cargo van (18.7 m), 12 by passenger car (19 m), 12 by single unit truck (9.1 m), and two by semi-trailer (13.9 m), resulting in a weighted average vehicle length of 6.9 m, which is equivalent to a typical step van. Finally, to estimate the total required time-space, this value was multiplied by the mean observed parking duration of 13.2 min.

$$LH^D_{i,w} = \frac{n_i * s^w}{p} * l^V * t^V \qquad (2)$$

Table 2. Sensitivity Analysis Scenarios

	Assumptions		
Scenario	Parking duration (t^V)	Packages/ vehicle (p)	Vehicle length (l^V)
1 (baseline)	Mean	Mean	Mean
2	Median	Median	Mean
3	1st Qu.	Mean	Mean
4	3rd Qu.	Mean	Mean
5	Mean	1st Qu.	Mean
6	Mean	3rd Qu.	Mean
7	Mean	Mean	Mean − 25%
8	Mean	Mean	Mean + 25%
9 (best)	1st Qu.	3rd Qu.	Mean − 25%
10 (worst)	3rd Qu.	1st Qu.	Mean + 25%

where:

$LH^D_{i,w}$: parking demand in census tract i during time window w (meter*hour)

n_i: daily number of packages generated by households in census tract i

s^w: observed share of packages delivered during time window w

p: observed average number of packages delivered per vehicle (packages/veh)

l^V: observed weighted average length of delivery vehicle (meters)

t^V: observed average parking duration per vehicle (hours)

A future scenario was also examined. This scenario applied the 17% annual growth rate observed by Rodrigue (*19*) over 5 years, assuming no change in vehicle loading or weekly or daily distributions.

Sensitivity Analysis

As baseline demand analysis relied on mean field-observed values, ten sensitivity analysis scenarios were also evaluated to examine the impacts on expected demand from individual and compounded changes in three variables: parking duration, packages/vehicle, and vehicle length (Table 2). These scenarios examined the estimated demand under median, 25th percentile, and 75th percentile observed values for parking duration and packages/vehicle, and with +/− 25% in average vehicle size.

Commercial Dedicated Parking Supply Estimation

Next, the supply of on-street commercial dedicated parking space was estimated using NYCDOT's STATUS database, which is a geocoded dataset containing the text of all signage in NYC. Posted signage along the curb regulates a time-sensitive parking inventory; spaces are allocated to different users during different time periods. For example, while many midtown streets allow only commercial vehicle parking from 7 a.m. to 7 p.m., most of these spaces are available for other vehicle types in the evening and overnight. Similarly, on many commercial corridors, parking is available only during a fixed delivery period; for example, spaced dedicated for commercial loading may become metered parking later in the day, or space dedicated as a curbside travel lane during peak hours may become commercial loading space during off-peak periods.

Algorithms were developed in R and Visual Basic for Application (VBA) in Excel to extract a parking space availability matrix. After the algorithms were applied, the availability matrix included the length of curb space covered and the type of regulation provided by the sign during 30-minute time segments. The length of space it covered (meters) was estimated using the curb length of the street, the distance between signs or between a sign and the block-end, and the sign's arrow direction. Signs were classified into four types: no parking/no standing, free parking (spaces where all types of vehicles can park without any time or pricing restriction), restricted parking (spaces where all types of vehicles can park, but only for a limited time period, may or may not be metered), and commercial dedicated parking (spaces where only commercial vehicles can park for loading and unloading for a limited time period, may or may not be metered). Each regulation type was identified as applicable or not applicable during each 30-minute segment for a 24-hour period. Information from this matrix was utilized to calculate the total available on-street time-space of commercial dedicated parking during each time window (Equation 3).

$$LH^S_{i,w} = \sum_{s \in W} \sum_{j \in I} l^P_j * d_s * A_{j,s} \quad (3)$$

$LH^S_{i,w}$: total time-length of commercial dedicated parking in census tract i during time window w (meter*hour)

l^P_j: curb length of commercial dedicated parking provided by sign j (meters)

d_s: duration of time segment $s = .5$ (hours)

$A_{j,s}$: active sign index for sign j during time segment s →(1 if sign is active, 0 if sign is inactive)

W: set of all time segments belonging to time window w

I: set of all signs in census tract i

Census Tract Demand versus Supply Ratio Estimation

Following demand and supply estimation for each census tract, a parking demand/supply ratio was calculated for each scenario during each time window (Equation 4). For census tracts with zero commercial dedicated parking supply, an artificial supply of 0.01 m*hrs was assigned for estimation of this ratio. The resulting ratio represents the share of commercial dedicated parking supply that would be required to accommodate the expected USPS residential delivery demand. Resulting ratios were mapped to compare results in different neighborhoods and during different periods of the day.

$$R_{i,w} = \frac{LH^D_{i,w}}{LH^S_{i,w}} \quad (4)$$

where:

$R_{i,w}$: demand to supply ratio in census tract i during time window w

$LH^D_{i,w}$: parking demand in census tract i during time window w (meter*hour)

$LH^S_{i,w}$: commercial vehicle dedicate parking supply in census tract i during time window w (meter*hour)

Parking Proximity Analysis

In the census tract analysis, parking demand and supply were estimated and compared at the aggregate level; however, neither the demand nor the available parking are evenly distributed within a census tract. As noted in the literature review, drivers will frequently choose to double park rather than travel a long distance to complete a delivery. To provide insights on how the available supply serves the demand within each census tract, a tax lot-level analysis was also conducted. While demand will vary according to the demographic characteristics of a building's residents, this variability cannot be quantified with the data available; as a result, the estimated census tract package demand was distributed proportionally to each tax lot based on the number of residential units in buildings located on that lot (Equation 5). The number of residential units in each building was determined using the NYC Department of City Planning's geocoded Primary Land Use Tax Lot Output (PLUTO) database.

$$n^L_j = n_i * \frac{u_j}{\sum_{k \in I} u_k} \quad (5)$$

where:

n^L_j : daily number of packages generated by tax lot j in census tract i

n_i: daily number of packages generated by census tract i

u_k: number of residential units on tax lot k;

I: set of all tax lots in census tract i

Both the tax lots and each commercial dedicated parking sign were mapped in ArcGIS. A buffer analysis was then conducted to determine the share of demand falling within a variable catchment area for each commercial dedicated parking sign. Four distances were considered: 25 m, 50 m, 61 m, and 75 m. The distance 25 m is considered the average walking distance by delivery drivers, 50

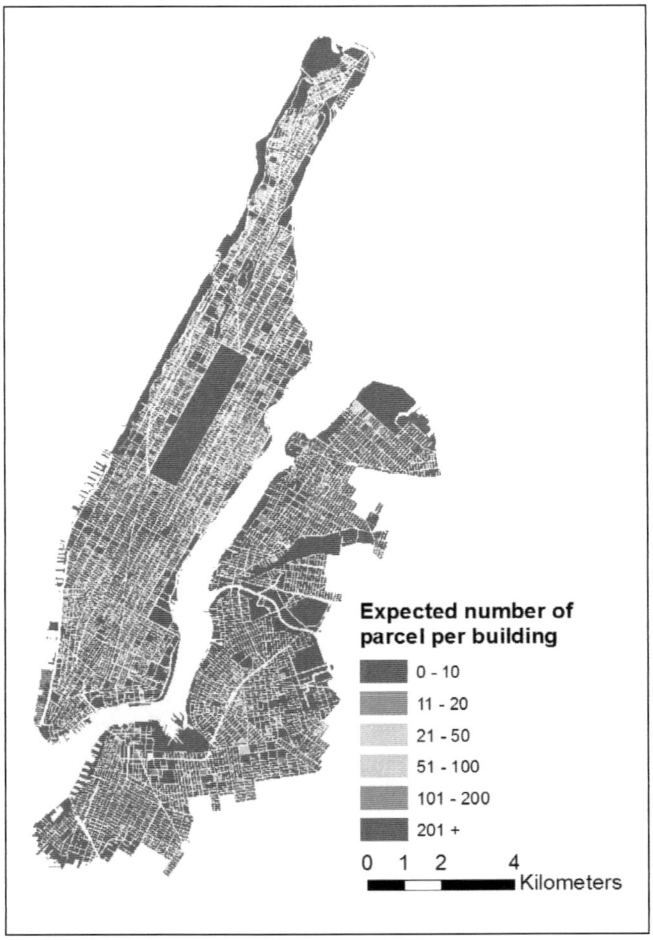

Figure 2. Expected weekly package demand by tax lot.

m is a commonly assumed reasonable walking distance, and 75 m is a maximum acceptable walking distance (*10*). The distance 61 m (200 ft) is a typical block length in Manhattan. The share of overall package demand not covered by each buffer during each time period was estimated.

Results

Demand/Supply Analysis

Approximately 1.7 million USPS packages are expected to be delivered each week in NYC, which represents 195,000 delivery stops. Individual tax lots have expected package demand as high as 7,800 packages per week. Figure 2 shows the spatial distribution of expected package demand across the selected study area, which represents only 13% of the city's land area but generates half of the total package demand. In Manhattan, demand is distributed throughout many mixed-use and residential neighborhoods, with particular concentration in high-density areas of Upper Manhattan, the Upper East and West Sides, and the Lower East Side. In the outer boroughs, demand is more concentrated in high-density areas such as Williamsburg, Bedford-Stuyvesant, and Downtown in Brooklyn, and Astoria in Queens.

In the total study area, between 9 a.m. and 9 p.m. on a typical weekday there are 11, 274 km*hours (7005 mile*hours) of on-street parking available; of this, 85.9% is open parking, 7.7% is restricted parking, and 6.4% is commercial dedicated parking. Commercial dedicated parking is not evenly distributed throughout the study area, but is concentrated in midtown Manhattan and along commercial corridors in other areas of the city (Figure 3). Commercial dedicated parking also varies throughout the day. While only minor variations in supply occur during daytime hours, very limited dedicated parking is available in off-peak evening hours, when much of the on-street space is made available for passenger car parking.

Comparing supply and expected demand, trends can be observed throughout the day and in different neighborhoods (Figure 4). Despite it being the lowest demand period, there appears to be a parking shortage in the evening off-peak, when 362 out of 474 census tracts have a demand/supply ratio greater than one. During this time frame, heavily commercial areas—such as parts of Midtown, the Financial District, and the Westside Highway corridor in Manhattan, East Williamsburg and Red Hook in Brooklyn, and Long Island City and Sunnyside in Queens—have ratios less than one. These census tracts have very few residents and some dedicated commercial parking that is available even in the off-peak. In the morning, parking access is much better; 308 out of 474 census tracts have demand less than .5 supply, and 332 less than the supply. Comparing the afternoon and evening peak periods with the morning, there is a clear increase in demand/supply ratios in some neighborhoods. Seventy-one new census tracts reach demand of .5 supply during this period, and 43 new census tracts exceed the available supply. These changes are particularly visible in moderately dense residential areas such as the Upper East and Upper West sides in Manhattan and South Brooklyn. In several neighborhoods—including the northern part of Central Harlem and Washington Heights in Manhattan, much of Steinway and Astoria in Queens, and Bedford-Stuyvesant in Brooklyn—there is little variation in the demand/supply ratio throughout the day. In general, these areas have ratios much larger than one, indicating there is little to no parking supply in these areas at any time.

Demand for residential parking is expected to continue to grow at a rapid pace. Applying Rodrigue's (*19*) observed annual growth rate of 17% and assuming growth in vehicle trips is proportional with package demand, new demand/supply ratios were estimated for a

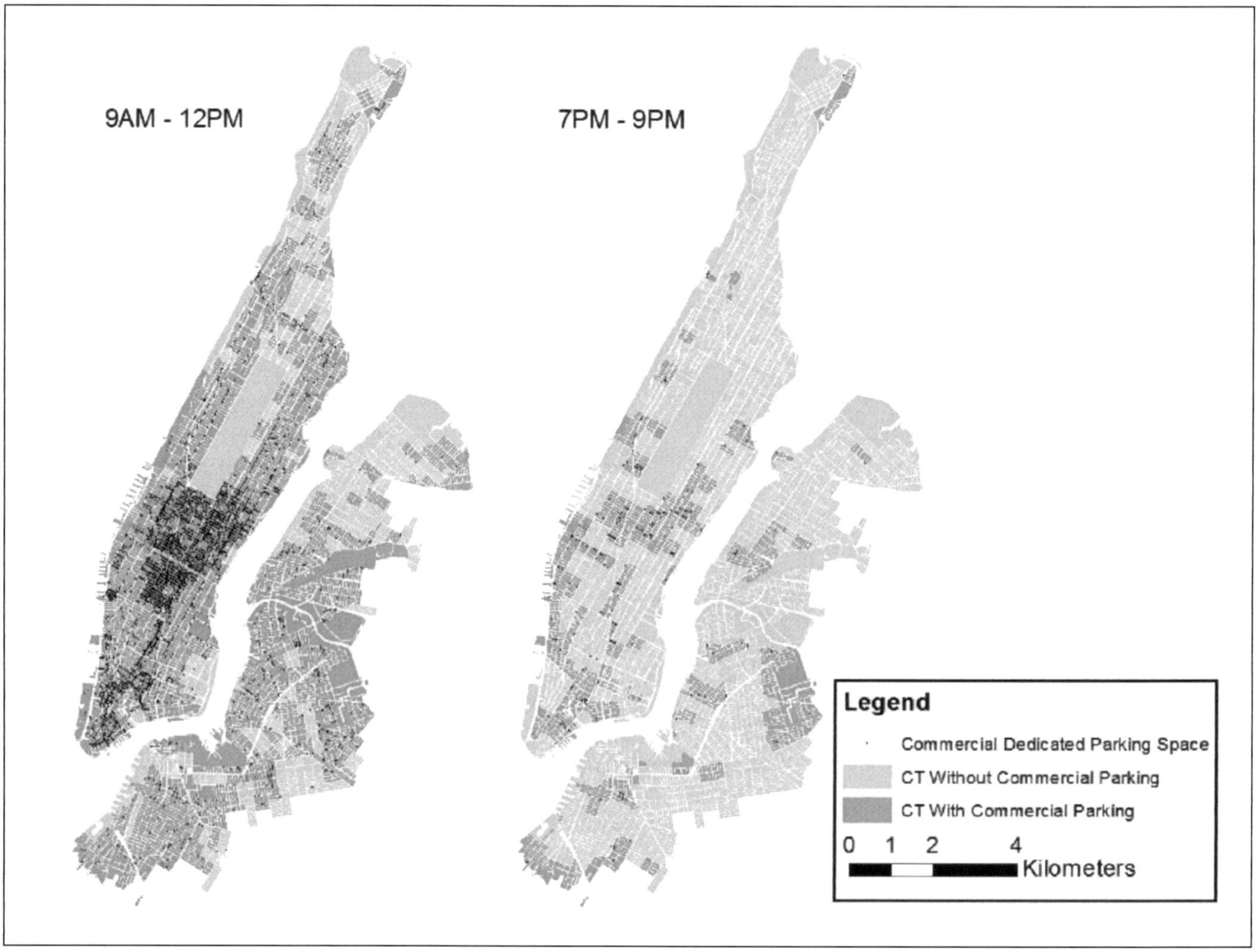

Figure 3. Commercial dedicated parking space.

5-year future projection. With this growth, USPS demand in 74 new census tracts exceeds the currently available supply; again this change occurs in moderately dense residential areas of the Upper East and West Sides, Upper Manhattan, and the Lower East Side, as well as in much of the Brooklyn study area (Figure 5).

Sensitivity Analysis

Table 3 summarizes the results from the sensitivity analysis scenarios. While minor changes are observed for all three variables, the only variable having an extreme impact on demand estimation, and ultimately on the demand/supply ratio, is the number of packages per vehicle. The 25th percentile observed number of parcels per vehicle was 2.0 compared with a mean value of 8.7; with this type of vehicle loading, the share of census tracts with a deficient parking supply nearly doubled compared with the baseline scenario.

Parking Proximity Analysis

Figure 6 summarizes the share of demand that falls outside the preferred (25 m), average (50 m), one-block (61 m), and maximum (75 m) walking distance buffers during each time period. Even during the morning, when supply is greatest, 55% of package demand falls outside even the maximum walking distance buffer. In each subsequent time period throughout the day, the share of deliveries falling outside the buffer areas increases; following a slight increase from morning to afternoon, a 6% increase is observed from afternoon to evening peak, and an even greater increase of 25% is observed from evening peak to off-peak. During the evening off-peak, 93% of package demand is outside of even the maximum

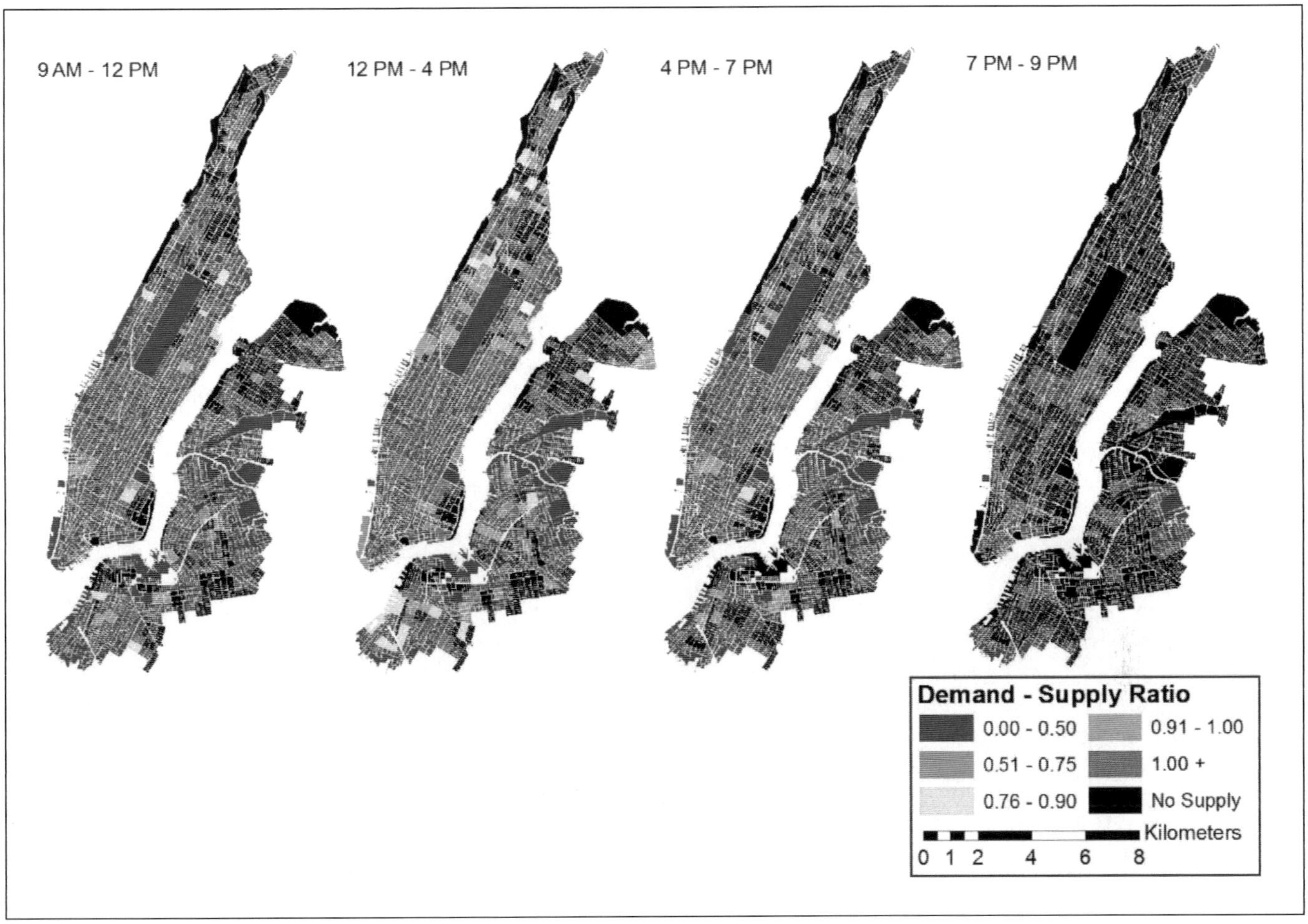

Figure 4. Ratios of demand to supply during each time period.

walking distance buffer area, and 97% outside the preferred walking distance buffer.

Research Limitations and Future Research

This study compares the expected demand for USPS parking with the theoretically available supply of commercial dedicated on-street parking to provide insights on how supply compares with growing residential demand. These values do not represent either the total demand or the total available parking supply in the study area. The demand estimated in this paper accounts for only deliveries made by the USPS applying national average rates from a 2015 survey. Even for USPS deliveries, these rates are likely conservative when applied in 2017 to NYC, as e-commerce is growing rapidly and the city's many carless residents rely on delivery of difficult-to-carry household goods. Field observation conducted during this study suggests that USPS demand accounts for only about 20% of total delivery trips generated by residences. In this study, the daily parcel demand estimation represents an average weekday. Within a typical weekday, parcel demand distribution across different times of day was built on field observation. Recognizing that there is variation across weekdays and in delivery times at different locations, the accuracy of both distributions can be improved by performing more comprehensive citywide data collection. Additional research is also needed to better understand the supply chain factors driving changes in the observed changes in spatial and temporal distribution of deliveries and in the vehicle types used for these movements. This will better inform the potential impacts that growth in on-demand and shared economy services may have on these factors.

In practice, commercial dedicated parking space will be used not only for residential deliveries, but also for commercial deliveries and for service vehicle parking. In many of the areas where little residential demand exists (e.g., the midtown Manhattan central business district), commercial demand is tremendous (*6*). In many mixed-use areas, both commercial and residential demand are considerable. Even where space is dedicated for commercial use, the space may also be illegally occupied by passenger cars or other non-commercial users. While this

Figure 5. Current versus projected 5-year demand/supply ratios.

Table 3. Sensitivity Analysis Results

Scenario	Assumed values: Parking duration (min)	Assumed values: Packages/vehicle	Assumed values: Vehicle length (m)	Demand > supply (% of census tracts)
1 (baseline)	13.2	8.7	6.92	37.85
2	7	4	6.92	39.14
3	4	8.7	6.92	29.25
4	14.3	8.7	6.92	38.06
5	13.2	2	6.92	65.38
6	13.2	10	6.92	36.77
7	13.2	8.7	5.19	35.48
8	13.2	8.7	8.65	40.43
9 (best)	4	10	5.19	28.6
10 (worst)	14.3	2	8.65	68.39

study reveals the need for residential delivery space, future research is needed to study how to balance this demand with: (1) the needs of other commercial operators, for example, commercial deliveries and service vehicles; and (2) the needs of other street users, including transit, passenger cars, and bicycles.

The supply estimated in this study represents total time-space dedicated for all commercial vehicles assuming no non-commercial vehicle use of this space. The locations of fire hydrants and driveway entrances were not considered; as a result, this supply is slightly overestimated. The estimated supply also does not include

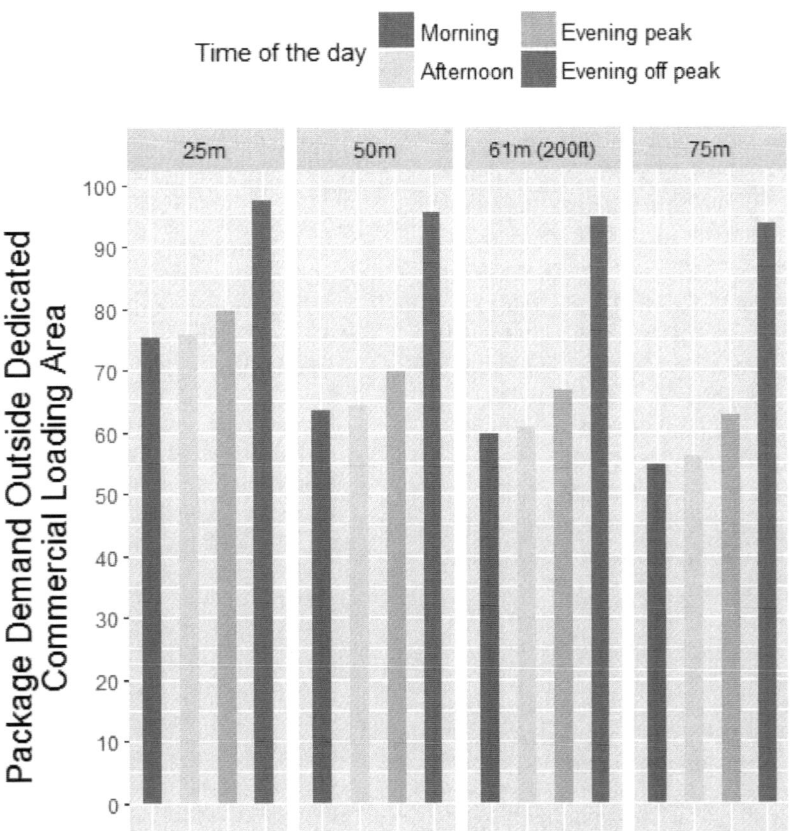

Figure 6. Share of package demand outside buffer area during each time period.

the other types of parking space theoretically available to delivery vehicles, including free and restricted on-street parking as well as loading docks. Current zoning regulations do not mandate loading docks for residential land uses, so it is expected that very few buildings have off-street space in these high-density areas. Real-time availability of free parking is also expected to be very low, as passenger cars frequently park for long durations with little turnover. However, it is likely that some residential freight demand could be accommodated in restricted, free, or off-street space. Determining the real-time share of parking space available for use by residential delivery vehicles would require an inventory of off-street space as well as comprehensive analysis of parking usage by user type and of parking turnover rates not possible with current data. However, emerging technologies such as video analytics may help to address this gap in the future.

Conclusion

Results from this analysis suggest that there are mismatches between the times and locations of NYC's commercial dedicated parking supply and expected residential freight demand. Time-of-day analysis suggests that the current time-variable parking supply of commercial dedicated space is not aligned with the distribution of demand; there is less space available in high-demand afternoon hours than in the morning, and there is very little space available in the low-demand, but non-negligible, evening off-peak. Results from the future projection suggests that growth in demand will be particularly problematic for parking in moderately dense residential areas. In several neighborhoods, there is little to no dedicated commercial parking at any time of day despite demand for residential deliveries. Even in parts of the city where commercial dedicated parking is provided, this space is not necessarily available within a reasonable walking distance to residential delivery locations. Even at times of day when the parking supply is relatively high, more than half of residential freight demand destinations are located more than a block away from the nearest dedicated commercial parking space. Results from the sensitivity analysis also suggest that increased dispersion of deliveries into smaller load sizes, as is expected to occur with growth in on-demand services, can have a severe impact on the demand for parking.

Overall, these results suggest that curb regulations and land-use regulations must be revisited. Space is needed to accommodate demand within a reasonable proximity to residential addresses. If this space is not available, double parking and other types of illegal parking are likely to increase. Space must be provided not only during morning hours, but at all times of day. Given that expected USPS demand alone is greater than the total available supply for all commercial parking in some census tracts, the necessary supply may be difficult to accommodate with limited on-street space. Off-street space may be needed to supplement on-street space. While changes in zoning regulations will do little to address the available supply in established neighborhoods, new loading space requirements for residential buildings could help to alleviate the deficiency in developing areas. In developed areas, demand management strategies such as delivery consolidation will likely be needed to limit the number of small-size deliveries made by individual vehicles despite increasingly specific customer demands.

Acknowledgments

The field observation utilized in this study was conducted as part of a related project supported by the Volvo Research and Education Foundations through the MetroFreight Center. Student work on this project was also supported by the French Ministry of Ecology, Sustainable Development and Energy.

Author Contributions

The authors confirm contribution to the paper as follows: study conception and design: Alison Conway; data collection: Alison Conway, Quanquan Chen; analysis and interpretation of results: Pierre Escand, Quanquan Chen; draft manuscript preparation: Quanquan Chen, Alison Conway, Pierre Escand. All authors reviewed the results and approved the final version of the manuscript.

References

1. Gardrat, M., F. Toilier, D. Patier, and J. Routhier. How Can the Impact of New Practices for Supplying Households be Quantified in Urban Goods Movements? *Proc., 6th International Conference on Information Systems Logistics and Supply Chain*, Bordeaux, France, 2016, pp. 1–9.
2. Morganti, E., L. Dablanc, and F. Fortin. Final Deliveries for Online Shopping: The Deployment of Pickup Point Networks in Urban and Suburban Areas. *Research in Transportation Business & Management*, Vol. 11, 2014, pp. 23–31.
3. United Problem Solvers™. *UPS Pulse of the Online Shopper: Tech-savvy Shoppers Transforming Retail*, 2016. https://pressroom.ups.com/assets/pdf/pressroom/white paper/2016_UPS_Pulse of the Online Shopper_white paper final.pdf. Accessed May, 2017.
4. Gevaers, R., T. Vanelslander, and E. Van de Voorde. Characteristics and Typology of Last-mile Logistics from an Innovation Perspective in an Urban Context. *City Distribution and Urban Freight Transport: Multiple Perspective*, Edward Elgar Publishing, 2011, pp. 56–71.
5. Andrea, M., and D. Meyer. *City Logistics Research: A Transatlantic Perspective*. Transportation Research Board, Washington, D.C., 2013.
6. Jaller, M., J. Holguín-Veras, and S. D. Hodge. Parking in the City: Challenges for Freight Traffic. *Transportation Research Record: Journal of the Transportation Research Board*, 2013. 2379: 46–56.
7. Malik, L., I. Sánchez-Díaz, G. Tiwari, and J. Woxenius. Urban Freight-parking Practices: The Cases of Gothenburg (Sweden) and Delhi (India). *Research in Transportation Business & Management*, Vol. 24, 2017, pp. 37–48.
8. Muñuzuri, J., J. Racero, and J. Larrañeta. Parking Search Modelling in Freight Transport and Private Traffic Simulation. *Advances in Transport*, Vol. 60, 2002, pp. 335–344.
9. Nourinejad, M., A. Wenneman, K. N. Habib, and M. J. Roorda. Truck Parking in Urban Areas: Application of Choice Modelling Within Traffic Microsimulation. *Transportation Research Part A: Policy Practice*, 2014, pp. 54–64.
10. Silva, J., and A. Alho. Using Structural Equations Modeling to Explore Perceived Urban Freight Deliveries Parking Issues. Vol. 102, 2017; pp. 18–32.
11. Allen, J., M. Piecyk, and M. Piotrowska. *An Analysis of Online Shopping and Home Delivery in the UK*, 2017. http://www.ftc2050.com/reports/Online_shopping_and_home_delivery_in_the_UK_final_version_Feb_2017.pdf. Accessed June 16, 2017.
12. Chen, Q., A. Conway, and J. Cheng. Parking for Residential Delivery in New York City: Regulations and Behavior. *Transport Policy*, Vol. 54, 2017, pp. 53–60.
13. Schaller, B., T. Maguire, D. Stein, W. Ng, and M. Blakeley. *Parking Pricing and Curbside Management in New York City*. Transportation Research Board of the National Academies, Washington, D.C., 2011.
14. New York City Department of Transportation. *Urban Freight Initiatives*. New York City Department of Transportation, 2015. http://www.nyc.gov/html/dot/downloads/pdf/2015-09-14-urban-freight-initiatives.pdf. Accessed June 10, 2017.
15. Clarke, G., C. Thompson, and M. Birkin. The Emerging Geography of E-commerce in British Retailing. *Regional Studies, Regional Science*, Vol. 2, No. 1, 2015, pp. 371–391.
16. Wang, X., and Y. Zhou. Deliveries to Residential Units: A Rising Form of Freight Transportation in the U.S. *Transportation Research Part C: Emerging Technologies*, Vol. 58, 2015, pp. 46–55.
17. Mazzone, J., and S. Rehman. *The Household Diary Study Mail Use & Attitudes in FY 2015*. United States Postal Service, Washington, D.C., 2016. https://www.prc.gov/docs/96/96795/Household Diary 2015_2.pdf
18. United Problem Solvers™. *UPS Pulse of the Online Shopper*, 2017. https://www.ups.com/media/en/2014-UPS-Pulse-of-the-Online-Shopper.pdf. Accessed June 5, 2017.
19. Rodrigue, J-P. Residential Parcel Deliveries: Evidence from a Large Apartment Complex. *MetroFreight*, 2017. https://

www.metrans.org/sites/default/files/MF 5.1a_Residential Parcel Deliveries_Final Report_030717.pdf

20. Cárdenas, I., J. Beckers, T. Vanelslander, and A. Verhetsel. Spatial Characteristics of Failed and Successful E-commerce Deliveries in Belgian Cities. *Proc., 6th International Conference on Information Systems Logistics and Supply Chain*, Bordeaux, France, Kedge Business School, Bordeaux, 2016.

The Standing Committee on Urban Freight Transportation (AT025) peer-reviewed this paper (18-05388).

Article

TRR
JOURNAL OF THE TRANSPORTATION RESEARCH BOARD

Transportation Research Record
2018, Vol. 2672(9) 216–228

DOI: 10.1177/0361198118776783
journals.sagepub.com/home/trr
SAGE

Testing the "Freight Landscape" Concept for Paris

Takanori Sakai[1], Adrien Beziat[2], Adeline Heitz[2], and Laetitia Dablanc[3]

Abstract

The concept of "freight landscape," the basis for a modeling approach for urban freight traffic estimation using commonly available datasets, was proposed in 2017 with a case study applying it to the Los Angeles metropolitan area. To extend the scope of that research, we conduct another case study using data from the Paris region, France. We estimate spatial lag models using population, employment, or establishment transportation accessibilities as explanatory variables and network-based truck traffic as the dependent variable, modifying the approach used in the Los Angeles study. We identify differences in the characteristics of the variables and the models between the Los Angeles and Paris cases, each having a distinctively different urban structure. While the models estimated for the Paris region provide beneficial insights into the relationships between freight landscape indicators and urban freight traffic, the complex correlation structure among indicators, as well as the limitation of the models for specifying the areas of very high truck traffic, underlines the need for further research on the modeling framework and for more case studies.

The scarcity of data for urban freight is a long-standing issue that hampers the analysis of urban freight flows and impacts. The complexity and heterogeneity of urban freight activities make data collection costly and challenging to implement. Only a limited number of cities implement surveys for urban freight. Even where survey data exists, its scope is often limited. On the other hand, growth in population and economic activities in cities around the world results in an increase of urban freight activities. Development of policies that address various issues associated with such increase in freight activities requires urban freight traffic analysis. As such, finding robust yet easily obtainable secondary data to estimate urban freight traffic demand is an important step toward overcoming the lack of resource for data collection. The methodology for using the secondary data to characterize urban freight traffic is an important research subject.

The primary focus of this research is methodological. Following the research by Giuliano et al. (*1*), we explore the use of available secondary data to estimate urban freight traffic. The objective of their work is to determine whether it is possible to accurately predict the spatial patterns of urban freight flows "using simple measures of population, employment and transport access" (*1*). The authors define the concept of "freight landscape" as "a description of freight activity imputed from population, employment and transport network characteristics," with the hypothesis that "freight flows depend systematically on the spatial organization of freight suppliers and demanders as well as on the transportation facilities within the metropolitan areas." To test the hypothesis, they analyze the relationship between the distributions of population, employment, and transportation supply and truck flow using data from the Los Angeles metropolitan area. The estimation of spatial regression models indicates the systematic relationships between those factors and truck traffic. In this line of work, we add another case study from the Paris region, which is the largest urban cluster of population and business activities in France, and one of the largest in Europe.

The rest of this paper consists of the following five sections: a brief literature review that focuses on the urban form—urban freight relationship; freight features and freight data collection in the Paris region; analytical

[1]Singapore-MIT Alliance for Research and Technology (SMART), CREATE Tower, Singapore
[2]IFSTTAR-SPLOTT – University of Paris-East, Marne-la-Vallée, France
[3]IFSTTAR – University of Paris-East, Marne-la-Vallée, France

Corresponding Author:
Address correspondence to Takanori Sakai: takanori@smart.mit.edu

framework and data sources; the discussion on the estimated models; and the conclusions drawn from this research.

Literature Review: Relationship between Urban Forms and Urban Freight

"Freight landscape," i.e., a systematic relationship between easily accessible socio-economic indicators and goods transport activity, is an important concept, as the relationship between freight and urban forms, a broad category which encompasses the distribution of various indicators of human activities in an area (*2*), is generally understudied in the literature (*3*, *4*). Past efforts of urban freight modeling, since at least the 1970s, have developed various types of freight (or freight trip) generation methods (*5*). Studies, including Gonzalez-Feliu and Routhier (*6*), Taniguchi et al. (*7*), and Holguín-Veras et al. (*8*), have attempted to inventory the variety of urban freight modeling efforts. Most models try to estimate freight trips generated from the economic establishments of an urban area. Even though spatial distribution plays no part in the generation of freight trips at the establishment level, other studies (*9*, *10*) emphasize the possibility that space can explain operational characteristics of deliveries/pick-ups, such as the type of vehicle, type of parking, and duration of parking.

Studies of the relationship between freight flows and urban forms are emerging at a theoretical level. Researchers recognize, for example, the links between density and inefficiency of deliveries. Dense urban centers bring about efficiency in goods transport activity, namely shorter distances between delivery points and, therefore, the higher potential for consolidated urban freight tours. On the other hand, high density can lead to inefficiency because of narrow streets (*11*), the use of small vehicles due to local or national regulations (*12*), traffic congestion (*13*), and the lack of supply of parking spaces for freight vehicles (*14–16*).

Freight Transportation and Data Collection in the Paris Region

Freight Transportation in the Paris Region

The Paris region (i.e., Ile-de-France) is by far the richest and most productive French region, concentrating about one fifth of the nation's population and employment as well as roughly 30% of its GDP for 2012 in only 2% of the surface area. The region is also a highly attractive tourist destination. As for freight tonnage, the Paris region displays a slight imbalance between inflows and outflows: 55% of the tonnage of transported goods is internal, 25% is inflow, and 20% is outflow (*17*); 31% of the tonnage of goods moved in the Paris region is finished products, and 9% is food products (*17*). The Paris region requires a very efficient logistics network of freight terminals to provide high frequencies and variety of small deliveries. That is why, alongside its national economic prominence, the Paris region holds about 17% of the total surface area of warehouses in France (*18*) and 26% of the French logistics sector workforce (*19*), while it represents just 10.5% in tonnage of the nation's transported goods (*17*). It is estimated that through-traffic represents about 20% of the total truck traffic in the Paris region (*17*), even though it is the most congested French region (*20*) and one of the most congested urban areas in Europe (*21*).

The Paris region relies mostly on trucks to transport goods (*17*); in total, 88.6% of the tonnage of goods transported within, to, or from the Paris region is transported on roads. About 6.3% of tonnage is transported through waterway system, 60% of which is construction materials, and about 4.5% is transported by rail, which mainly carries long-distance shipments of construction materials and finished products. Despite policy attempts designed to reduce the truck-kilometers traveled (TKT), road transportation has maintained its share for moving products over the last decade (*17*). Last-mile freight movements rely heavily on small vehicles: about 60% of the deliveries/pick-ups are made using light goods vehicles of 3.5 tons or less in gross vehicle weight (*22*).

Collection of Truck Traffic Data

French local authorities have jurisdiction over large portions of the road network and frequently conduct counts and cordon surveys, although at irregular intervals. The national government has jurisdiction over the highway network (autoroutes), and produces yearly data on vehicle flows obtained through automatic counts (most of the national network is equipped with traffic sensors that can distinguish between small and large vehicles). The results of these surveys are often aggregated at the regional level. Other data collection efforts focus on freight flows to produce OD matrices, which can then be used in traffic assignment models. The national government conducts a yearly mandatory survey of heavy goods vehicles' users (vehicles over 3.5 tons in gross weight) called the TRM survey (Transport Routier de Marchandises, Road Freight Transport), which is used to produce freight flow matrices at the departmental and regional levels (*23*). Those data on vehicle counts and flows allows regional agencies to produce regular reports on the state of the Paris region's mobility and traffic conditions, including truck traffic (*17*, *24*), which can be used for policy implementation (*25*). This paper relies heavily on the data from the counts (automatic or not), which are calibrated using the data from the TRM survey, to obtain the truck traffic in the Paris region.

Methodology and Data

Methodology

The Paris region consists of 1,300 local municipalities, consisting of 20 *arrondissements* in the City of Paris and 1,280 *communes* outside of Paris, with an average size of 9.3 km^2. A "municipality" is the most detailed unit in the data of population, employment, and establishments available to us. This spatial unit also matches the availability of the road network and truck traffic data, which do not include the municipal road network. It is still necessary to exclude 262 municipalities from the analysis, which have no truck traffic data as the road network and truck traffic data do not cover those municipalities.

The purpose of this research is to test whether widely available socio-economic indicators, such as population, employment, and establishments, and transportation accessibility, can describe freight traffic with the level of accuracy that warrants the use of such indicators for urban freight traffic analysis. We develop regression models for estimating TKT per km^2, which represents truck traffic demand. The model development is conducted in four steps, using the following groups of indicators for independent variables: (a) population and employment indicators, (b) population and establishment indicators, (c) transportation accessibility indicators, and (d) population, employment, or establishment, and transportation accessibility indicators. Employment and establishment indicators are not used at the same time for regression analysis due to high correlations between the two indicators.

As the spatial auto-correlation becomes an issue in this type of analysis, in which the effects from neighboring municipalities influence truck traffic demand (*1*, *26–29*), we use a spatial lag model of which the formula is shown below.

$$y_m = \alpha + \sum_i \beta_i x_{m,i} + \rho \sum_{n \neq m} w_{mn} y_n + \varepsilon_m \quad (1)$$

where:

y_m: TKT per km^2 (dependent variable) in municipality m

$x_{m,i}$: an indicator (independent variable) i for municipality m

w_{mn}: the spatial weight for the pair of municipality m and n

ε_m: a randomly distributed unobserved component

α, β, ρ: parameters to be estimated; ρ is the autoregressive lag coefficient

The spatial lag model requires spatial weight matrix. We tested various formulas including radial distance weights, power distance weights, exponential distance weights, and queen contiguity weights, and found that the power distance weights provide the best fits for the tested models. Weight matrix is calculated using the formula below. We tested $\gamma = 0.5, 1, 2, 3, 4$ and $\gamma = 3$ gives the best results.

$$w_{mn} = d_{mn}^{-\gamma} \quad (2)$$

where:

d_{mn}: network distance between municipality m and n

γ: a positive exponent

All variables, including independent and dependent variables, are log-transformed as it gives higher Pseudo R^2 and AIC. "Spdep" package (*30*) in R, the software environment for statistical computing, is used for estimating models, applying the maximum-likelihood method. The model estimations are conducted so as to find statistically significant independent variables and minimize AIC.

Data

We use population, employment, and the number of establishments at the municipal level in the Paris region as the independent variables. To measure population density and employment density, we use the data provided by the French National Institute for Statistics (INSEE) for the year 2014. The Paris region is a monocentric metropolitan area. Population, employment, and establishments (especially the latter two) are concentrated in the center of the region, the City of Paris, and the surrounding suburban municipalities (Figure 1).

The source for detailed employment and establishment patterns is the "Local Knowledge of the Productive System" (CLAP) database provided by INSEE. This database uses the European NACE classification (Statistical Classification of Economic Activities in the European Community), which provides similar categories as the North American Industry Classification System (NAICS). The NACE classification only considers the main purpose of each building so that other functions in the establishment may not be recognized. In this analysis, we use an aggregated version of the CLAP database that groups the NACE classes into 17 categories (Table 1).

In the Paris region, economic activities of various types are located close to one another. As for the log-transformed employment densities, all pairs of the 17 industry categories excluding AZ (agriculture, forestry, and fishing), C2 (coke and refined petroleum) and C4 (manufacture of transport equipment) have strong correlations, with Pearson's correlation coefficients of 0.58 or

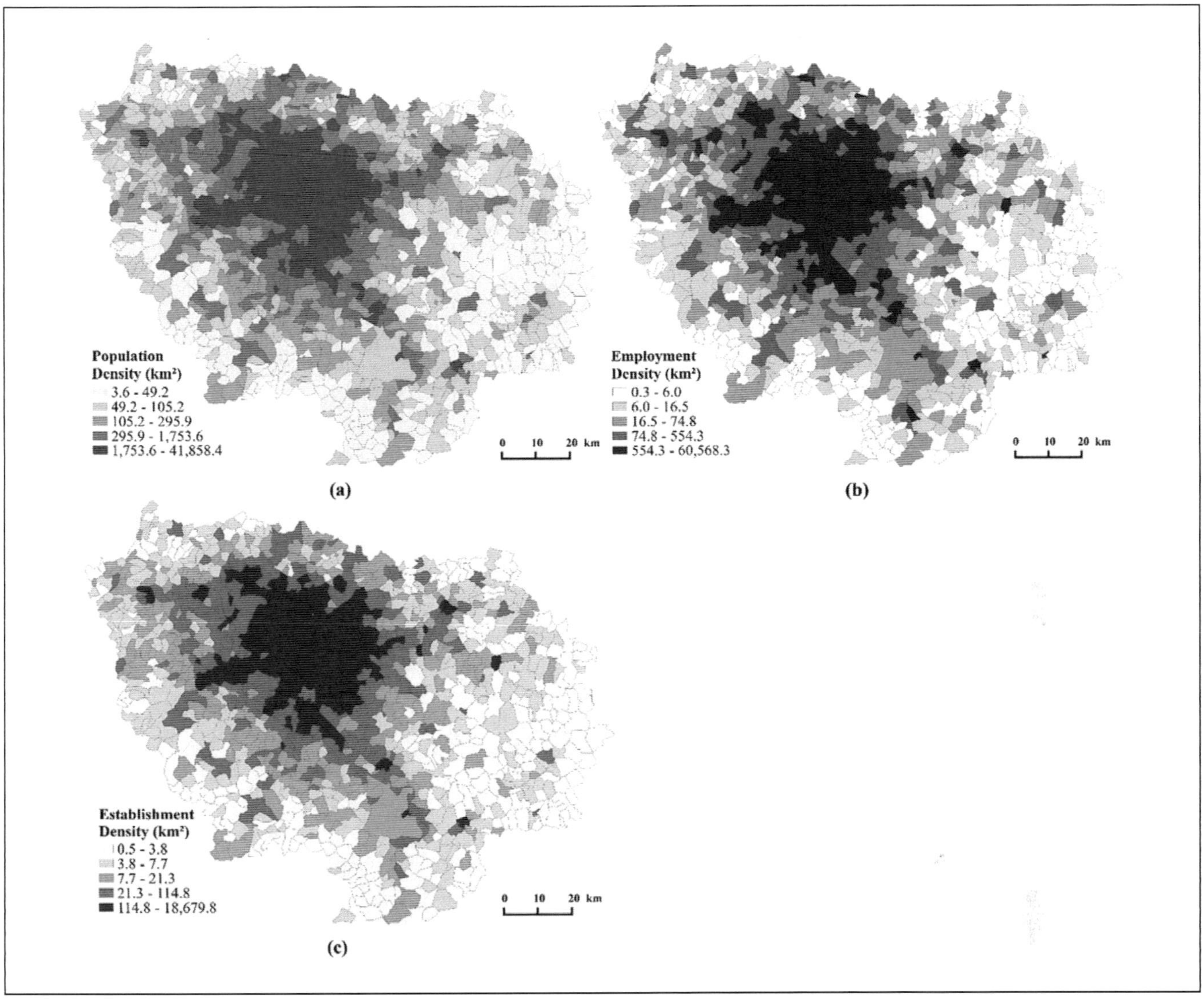

Figure 1. Density of population, employment, and establishments in the Paris region: (*a*) population density, (*b*) employment density, (*c*) establishment density.
Source: French National Institute for Statistics and French Institute for Geography.

higher. Such strong correlations indicate high potential for the multicollinearity problem, leading to biased estimations of the effects, even showing the opposite signs from the actual ones. To avoid the problem, we consider only statistically significant variables in the final models.

The list of road infrastructures in the Paris region was provided by the French Institute for Geography (IGN). We use the Route 120® database which gives us all the roads and highways as well as road nodes and intersections to calculate the road accessibilities to logistics infrastructures, for year 2015. The database of logistics infrastructures is the product of our own compilation from HAROPA (Ports of Paris) and SNCF (French National Railway Company) information. Those infrastructures are widespread, with some located within the inner suburban areas, while others stretch from the north to the east, and in the southeast (Figure 2). This has led to the development of the areas close to the infrastructures and the ports of Gennevilliers and Bonneuil or the Charles de Gaulle and Orly airports. The spatial distributions of some logistics infrastructures are similar and, therefore, there are strong correlations among the associated accessibility indicators like those of employment/establishments. Pearson's correlation coefficient of the log-transformed distances to rail–road and river–road intermodal terminals is as high as 0.96. Those two distance measures are also highly correlated with the distance to airport (log-transformed), with the coefficients of 0.82 and 0.78, respectively.

Table 1. Industry Category Based on NACE Classification (17 Categories)

Code	Industry	% of employment in the Paris region (total = 100)
AZ	Agriculture, forestry, and fishing	0.08
BE	Mining, energy, water, waste management, and remediation activities	1.57
C1	Manufacture of food products, beverages, and tobacco products	0.88
C2	Coke and refined petroleum	0.02
C3	Manufacture of electrical equipment, electronic, computer, and manufacturing machines	1.39
C4	Manufacture of transport equipment	1.43
C5	Manufacture of other industrial products	3.33
FZ	Construction	5.08
GZ	Trade and automobile and motorcycle repairs	12.90
HZ	Transportation and storage	6.80
IZ	Accommodation and food	5.17
JZ	Information and communication	6.99
KZ	Financial and insurance activities	6.14
LZ	Real estate activities	1.44
MN	Scientific and technical activities and administrative and support services	16.76
OQ	Public administration, education, human health, and social work	25.44
RU	Other service activities	4.41

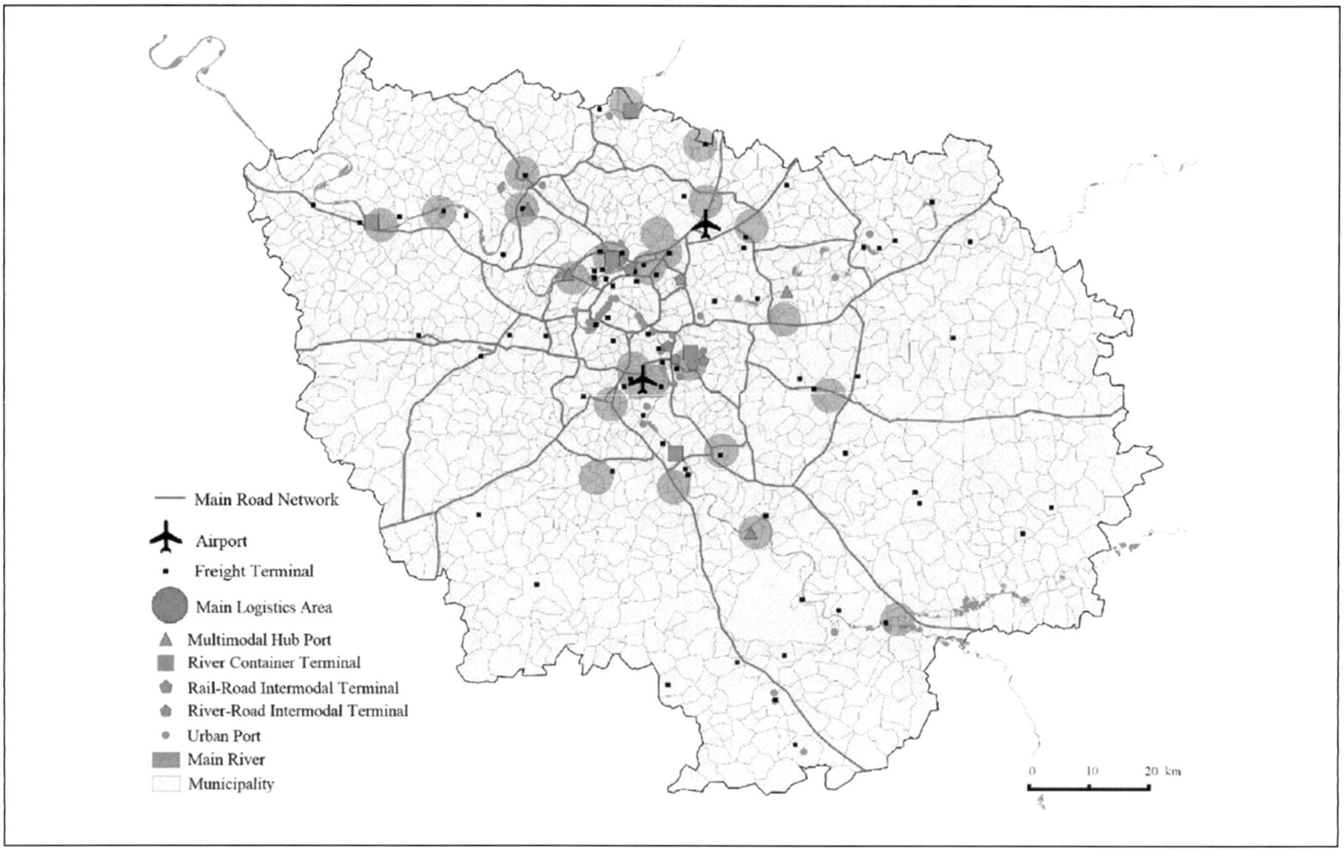

Figure 2. Main logistics infrastructures in the Paris region.
Source: Regional bureau of infrastructures and urban planning for the Paris region, French Institute for Geography and others.

Truck traffic data were provided by the regional bureau of infrastructures and urban planning for the Paris region (DRIEA). We use the hourly average truck traffic volume of morning rush hour on the road network in the Paris region, 2009, which were estimated based on the samples collected by each department (comparable to

Table 2. Summary of Variables (Before Log-Transformation)

Variables	Annotation	Mean	Median	SD	Min	Max
TKT per km^2	–	180.7	48.4	363.6	0.0	4089.2
Population density (per km^2)	Pop.	2094.8	263.9	4688.6	3.6	41858.5
Employment density (per km^2)	Emp.	1110.0	56.4	4528.4	0.2	68238.4
Establishment density (per km^2)	Estt.	262.3	19.1	1187.4	0.5	18679.8
Distance to *autoroute* (km)	Dist. to autoroute	9.0	6.6	8.2	0.0	51.4
Distance to airport (km)	Dist. to airport	36.9	33.7	20.0	0.9	94.4
Distance to freight terminal (km)	Dist. to freight terminal	7.9	7.1	5.0	0.0	29.7
Distance to main logistics area (km)	Dist. to main logistics area	15.0	11.2	10.9	0.0	51.3
Distance to multimodal hub port (km)	Dist. to multimodal hub port	25.8	23.7	13.6	0.9	69.4
Distance to river container terminal (km)	Dist. to river container terminal	28.6	24.6	16.7	0.0	83.9
Distance to rail–road intermodal terminal (km)	Dist. to rail–road intermodal terminal	34.6	32.7	19.6	0.8	85.0
Distance to river–road intermodal terminal (km)	Dist. to river–road intermodal terminal	34.5	32.5	19.0	1.8	84.2
Distance to urban port (km)	Dist. to urban port	15.4	12.5	11.4	0.0	54.4

Note: Breakdowns of employment density and establishment density by industry category are omitted from the table.

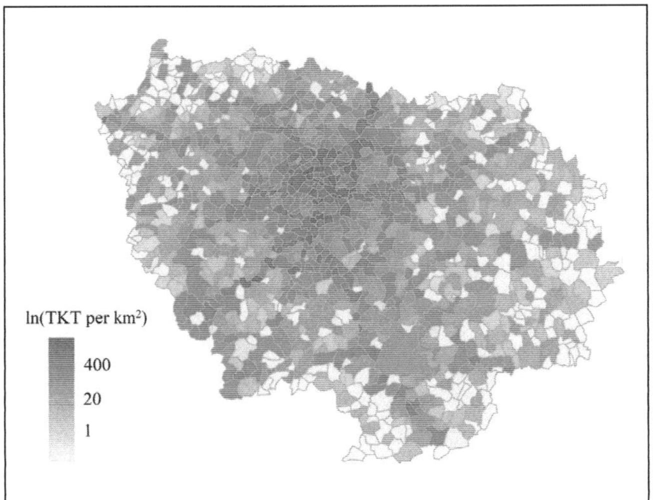

Figure 3. Density of truck-kilometers traveled (TKT) (log-transformed).

county in the U.S.A.) in the Paris region (Figure 3). It should be noted that the data do not include light commercial vehicles (i.e., commercial vehicles of 3.5 tons or less in gross weight), which account for 57% of freight flows for deliveries, according to the urban freight survey conducted in 2012 for the Paris region (*22*). The summary of the variables is shown in Table 2.

Analysis and Results

In the following subsections, the results are discussed for the models with (a) population and employment variables, (b) population and establishment variables, (c) accessibility variables, and (d) the combinations of population, employment/establishment, and accessibility variables. These models underline the set of indicators that contribute to the truck traffic estimation in each category or among all available variables.

Models with Population and Employment/Establishment Data

The three models that apply population and employment variables are shown in Table 3. In the first model (left in Table 3), only total population and total employment are included; the second model (center in Table 3) is estimated using total population and employment for 17 industrial categories; in the third model (right in Table 3), independent variables are reduced to population and only the statistically significant employment variables that show the p-values of less than .10.

"Population" is insignificant for all three models; population is highly correlated with employment (Pearson's correlation coefficient is 0.96) so that its apparent contribution for estimating truck traffic demand is limited. The second model highlights the heterogeneity of the employment effect by industry type. Coefficients of some employment variables are negative, which is likely due to the multicollinearity problem, that is, the correlations among independent variables result in the biased estimated coefficients.

To find the variables that are highly related to truck traffic demand, the insignificant employment variables and those with negative coefficients are removed and, thus, the third model is obtained. "Emp. IZ" and "Emp. KZ" are omitted in the process as these variables turn out to be insignificant when some other variables are omitted. Among various industrial categories, GZ (trade and automobile and motorcycle repair) shows the largest

Table 3. Estimated Spatial Lag Models with Population and Employment Variables

	1. Total pop. and total emp.			2. Total pop. and emp. by industry group			3. Total pop. and significant emp. variables		
	Coef.	S.E.	*p*-value	Coef.	S.E.	*p*-value	Coef.	S.E.	*p*-value
Intercept	1.453	0.214	.000**	1.869	0.329	.000**	2.122	0.230	.000**
Pop.	−0.080	0.065	.220	0.017	0.087	.842	−0.073	0.051	.154
Emp.	0.386	0.051	.000**						
Emp. AZ				−0.036	0.101	.725			
Emp. BE				−0.036	0.043	.402			
Emp. C1				−0.129	0.075	.087*			
Emp. C2				−0.130	0.182	.475			
Emp. C3				0.075	0.048	.120	0.083	0.043	.054*
Emp. C4				0.053	0.044	.235			
Emp. C5				0.010	0.050	.847			
Emp. FZ				0.036	0.065	.584			
Emp. GZ				0.256	0.063	.000**	0.314	0.053	.000**
Emp. HZ				0.095	0.047	.042**	0.117	0.044	.007**
Emp. IZ				0.152	0.075	.042**			
Emp. JZ				−0.092	0.055	.094*			
Emp. KZ				0.181	0.076	.017**			
Emp. LZ				0.043	0.081	.599			
Emp. MN				0.054	0.057	.345			
Emp. OQ				−0.117	0.069	.091*			
Emp. RU				−0.136	0.083	.102			
Autoregressive lag coefficient	0.316	0.037	.000**	0.286	0.037	.000**	0.288	0.037	.000**
Pseudo R^2	0.403			0.440			0.429		
AIC	3611.2			3576.0			3568.9		
AIC for linear regression model	3678.5			3630.2			3624.9		

*Significant at 90% confidence level; **significant at 95% confidence level.

Note: Dependent variable is a log of TKT per km^2; all independent variables are in log-transformed density (per km^2); "pop." stands for population and "emp." stands for employment; refer to Table 1 for the codes of industrial category (AZ~RU).

effect on truck traffic demand, followed by HZ (transportation and storage), and C3 (manufacture of electrical equipment, electronic, computer, and manufacturing machines). Autoregressive lag coefficients are significant for all three models and the comparison between AIC and AIC for linear regression model indicates that the spatial autoregressive effects are non-trivial.

Similarly, three models that apply population and establishment, instead of employment, variables are shown in Table 4. Compared with the models in Table 3, Pseudo R^2 is slightly lower and AIC is higher for each step, which indicates that employment variables are likely to be better indicators for predicting truck traffic demand.

Only "Estt. GZ" (trade and automobile and motorcycle repair) is the significant establishment variable that remains in the third model, and autoregressive lag coefficient is, again, significant in all three models.

Models with Accessibility Indicators

Next, the models are estimated using nine accessibility variables. All variables are used in the first model and, in the second model, non-significant variables and the variables with positive coefficients are excluded. Among all accessibility variables considered, "Dist. to autoroute" is most important for truck traffic prediction. This is likely due to the simple fact that a large number of trucks use autoroutes. Also, the locations near autoroute interchanges and exits are efficient for logistics activities.

"Dist. to rail–road intermodal terminal" is also significant and its effect on truck traffic demand is strong. Two municipalities, Gennevilliers and Bonneuil, accommodate two of the region's main rail–road intermodal terminals, as well as the two main river ports in the Paris region. The other two main intermodal terminals are in Valenton and Noisy. They are all located in the fringes of the Paris urban area. The other variable which is statistically significant is "Dist. to freight terminal," though the effect is smaller than the two accessibility variables mentioned above.

Comparison of the models in Table 5 with Tables 3 and 4 indicates that accessibility indicators tend to produce more reliable models than the models with only population and employment/establishment variables. The autoregressive lag coefficient is significant, but its effects are smaller in the models with accessibility variables, which means that the accessibility variables could capture a part of the spatial auto-correlation effect.

Models with Population, Employment/Establishment, and Accessibility Indicators

The first two models (1 and 2) in Table 6 are those using population, employment, and accessibility variables. We use the three accessibility variables that are statistically significant in the earlier analysis ("Dist. to autoroute," "Dist. to freight terminal," and "Dist. to rail–road combined terminal") (see the second model in Table 5). The first model in Table 6 uses the total employment and the second model uses, instead, the categories of the employment that are significant in the earlier model estimation (see the third model in Table 3).

When population, employment, and accessibility are included in the model, the contribution of autoregressive lag coefficient to AIC is small (i.e., the gap between AIC and AIC for linear regression model is small), although autoregressive lag coefficient is still statistically significant. Interestingly, the effect of population becomes significant with the inclusion of accessibility variables. The effect of population density is masked if accessibility variables are not adequately considered. All employment categories considered ("Emp. C3," "Emp. GZ," and "Emp. HZ") remain significant in the second model. On the other hand, "Dist. to freight terminals" becomes insignificant in both models and therefore is removed. The second model generates the highest Pseudo R^2 and the lowest AIC, which indicates the model is superior to the other models tested in this section.

The next two models (3 and 4) in Table 6 are those that apply the number of establishments instead of employment. Again, "Dist. to freight terminal" is insignificant in the models. Also, when "Estt. GZ" (trade and automobile and motorcycle repair) is included, "Dist. to rail–road terminals" is insignificant, because the two variables are negatively correlated to each other (Pearson's correlation coefficient is −0.76).

We also test the models using income level (i.e., median income) as an independent variable; however, the income level is always insignificant, unlike the case of Los Angeles metropolitan region (*1*). As discussed by De Lara (*31*), logistics activities may be associated with low-income neighborhoods in the Los Angeles but that is not the case in the Paris region (*32*).

Further Discussion

The analysis in this section highlights the socio-economic and accessibility indicators that best characterize truck traffic density in the Paris region. Our approach also clarifies the challenging issues for estimating truck traffic using those indicators. First, for the cities where mixed land use is common, like the Paris region, high correlations among indicators of urban forms make it difficult to specify accurately the effects of the indicators. As for accessibility indicators, the distances to rail–road intermodal terminal, river–road intermodal terminal, and airport, are highly correlated with one another and, therefore, the use of one of those three indicators returns the combined effect.

Table 4. Estimated Spatial Lag Models with Population and Establishment Variables

	1. Total pop. and total estt.			2. Total pop. and estt. by industry group			3. Total pop. and significant estt. variables		
	Coef.	S.E.	*p*-value	Coef.	S.E.	*p*-value	Coef.	S.E.	*p*-value
Intercept	1.935	0.293	.000**	1.892	0.456	.000**	2.387	0.317	.000**
Pop.	−0.304	0.105	.004**	0.012	0.126	.924	−0.182	0.077	.017**
Estt.	0.728	0.111	.000**						
Estt. AZ				−0.305	0.147	.038**			
Estt. BE				−0.063	0.139	.650			
Estt. C1				−0.522	0.236	.027**			
Estt. C2				−0.224	0.753	.766			
Estt. C3				0.498	0.207	.016**			
Estt. C4				−0.204	0.374	.586			
Estt. C5				−0.090	0.179	.615			
Estt. FZ				−0.236	0.157	.134			
Estt. GZ				0.699	0.176	.000**	0.683	0.090	.000**
Estt. HZ				0.266	0.140	.058*			
Estt. IZ				0.344	0.202	.089*			
Estt. JZ				−0.363	0.172	.035**			
Estt. KZ				0.408	0.190	.031**			
Estt. LZ				−0.122	0.202	.546			
Estt. MN				0.418	0.185	.024**			
Estt. OQ				−0.624	0.191	.001**			
Estt. RU				−0.046	0.233	.844			
Autoregressive lag coefficient	0.316	0.037	.000**	0.286	0.038	.000**	0.309	0.037	.000**
Pseudo R^2	0.396			0.439			0.404		
AIC	3622.8			3578.1			3609.0		
AIC for linear regression model	3688.8			3630.8			3672.7		

*Significant at 90% confidence level; **significant at 95% confidence level.

Note: Dependent variable is a log of TKT per km^2; all independent variables are in log-transformed density (per km^2); "pop." stands for population and "estt." stands for number of establishments; refer to Table 1 for the codes of industrial category (AZ~RU).

Table 5. Estimated Spatial Lag Models with Accessibility Variables

	1. All accessibility variables			2. Significant accessibility variables		
	Coef.	S.E.	*p*-value	Coef.	S.E.	*p*-value
Intercept	6.481	0.480	.000**	6.732	0.386	.000**
Dist. to autoroute	−0.791	0.071	.000**	−0.732	0.066	.000**
Dist. to airport	−0.077	0.122	.531			
Dist. to freight terminal	−0.226	0.079	.004**	−0.193	0.073	.008**
Dist. to main logistics area	−0.071	0.095	.455			
Dist. to multimodal hub port	0.173	0.101	.087*			
Dist. to river container terminal	0.146	0.089	.103			
Dist. to rail–road intermodal terminal	−0.450	0.227	.048**	−0.553	0.078	.000**
Dist. to river–road intermodal terminal	−0.212	0.232	.359			
Dist. to urban port	0.111	0.078	.157			
Autoregressive lag coefficient	0.204	0.041	.000**	0.214	0.041	.000**
Pseudo R^2	0.446			0.441		
AIC	3547.7			3545.6		
AIC for linear regression model	3571.2			3572.2		

*Significant at 90% confidence level; **significant at 95% confidence level.
Note: Dependent variable is a log of TKT per km^2; all independent variables are in log-transformed.

Secondly, key factors should be jointly and comprehensively considered. Our analysis shows that adding accessibilities to the models with population and employment/establishment densities is essential to capture the effect of population density on truck traffic. Two suburban municipalities, Cregy-les-Meaux and Limours, exemplify this mechanism. They have similar employment densities and truck traffic densities, although their population densities are different ["Pop." (log-transformed) are 7.1 and 6.1, respectively]. The effect of population density becomes apparent only when the distance to autoroute, which is negatively associated with truck traffic, is considered; "Dist. to autoroute" (log-transformed) is 1.5 and 2.1, respectively. In this case, the effects of population density and distance to autoroute counteract each other. Missing accessibility indicators lead to the negation of the population density effect.

Taking one step further, we compare the actual TKT per km^2 (non-log-transformed) and those estimated using the best model (the second model in Table 6). The root-mean-square error (RMSE) is 297 and the adjusted R^2 is 0.76. The correlation between actual and estimated TKT per km^2 is 0.64, which indicates the difficulty in differentiating the areas with very high truck traffic densities from those with moderately high truck traffic densities (and, the areas with very low traffic densities from those with moderately low traffic densities). While the modeling approach used for Los Angeles and the Paris region works for identifying the factors for truck traffic and the locations where truck traffic concentrates, the approach still has difficulty in estimating truck traffic volume with high accuracy. Incorporating data of other dimensions, such as network-based characteristics (e.g., centrality), land use regulations, and spatial design (road width, plot size, parking capacity, etc.) is worth testing for both accurate specifications of indicator effects and prediction.

Conclusion

In this paper, we use the concept of "freight landscape" and apply the modeling approach of Giuliano et al. (*1*) with some refinements for the Paris region. The concept hypothesizes that urban freight patterns can be predicted from simple indicators of population, employment, and accessibility to transportation infrastructure, which are generally available. For the Los Angeles case, the results mostly validate the hypothesis, with nuances. Giuliano et al. find that there is a systematic relationship between density and truck traffic. In Los Angeles, truck volumes are associated with employment density, especially for services, manufacturing, and trade. Their analysis also shows that transportation supply and highway access are good indicators explaining truck traffic, although accessibility to major freight generators (airports and seaports) is not. The intensity of truck activity is strongly and negatively associated with population density and household income.

We develop and test models using TKT per km^2 on the Paris region's road network as a dependent variable for finding the approach to estimate urban freight traffic. The main results are the following:

(i) In the Paris region, the distributions of the residential population and employment are monocentric and quite similar. Thus, the contribution of population for estimating truck

Table 6. Estimated Spatial Lag Models with Total Population, Employment/Establishment and Accessibility Variables

	1. Total pop. and emp. and accessibility variables			2. Total pop. and significant emp. and accessibility variables			3. Total pop. and estt. and accessibility variables			4. Total pop. and significant estt. and accessibility variables		
	Coef.	S.E.	*p*-value	Coef.	S.E.	*p*-value	Coef.	S.E.	*p*-value	Coef.	S.E.	*p*-value
Intercept	5.017	0.561	.00**	5.305	0.556	.00**	5.714	0.577	.00**	5.226	0.384	.00**
Pop.	−0.169	0.065	.01**	−0.169	0.052	.00**	−0.491	0.099	.00**	−0.323	0.073	.00**
Emp.	0.320	0.049	.00**									
Emp. C3				0.078	0.041	.06*						
Emp. GZ				0.279	0.051	.00**						
Emp. HZ				0.101	0.042	.02**						
Estt.							0.758	0.105	.00**			
Estt. GZ										0.716	0.084	.00**
Dist. to autoroute	−0.673	0.064	.00**	−0.670	0.063	.00**	−0.740	0.064	.00**	−0.749	0.063	.00**
Dist. to rail–road intermodal terminal	−0.246	0.096	.01**	−0.176	0.096	.07*	−0.189	0.096	.05**			
Autoregressive lag coefficient	0.164	0.041	.00**	0.142	0.041	.00**	0.146	0.041	.00**	0.159	0.040	.00**
Pseudo R^2	0.471			0.493			0.476			0.481		
AIC	3489.7			3448.7			3480.0			3468.6		
AIC for linear regression model	3504.5			3459.3			3491.0			3483.3		

*Significant at 90% confidence level; **Significant at 95% confidence level.

Note: Dependent variable is a log of TKT per km^2; all independent variables are in log-transformed; "pop." stands for population, "emp." stands for employment in density (per km^2) and "estt." stands for number of establishments in density (per km^2); refer to Table 1 for the codes of industrial category.

traffic demand is limited when employment variables are also in the model. However, the addition of accessibility variables highlights the significant negative effect of population on truck traffic.

(ii) Employment is a better business activity indicator for estimating truck traffic than establishments.

(iii) The following activities: trade and automobile and motorcycle repair, transportation and storage, and manufacture of electrical equipment, electronic, computer, and manufacturing machines, show the largest effects on truck traffic demand.

(iv) Among all accessibility variables, distance to autoroute is the most important for prediction of truck traffic demand.

(v) Distance to rail–road intermodal terminal is also significant and its effect on truck traffic demand is strong. The effect of this accessibility indicator is enhanced due to its high correlations with the distances to river–road terminal and airport.

(vi) The best fit of the model is achieved by the set of variables including population, employment in the three industrial categories mentioned earlier, distance to autoroute, and to rail–road intermodal terminal; and

(vii) The income level is always insignificant.

We identify the relationship between population, employment and accessibilities, and truck traffic demand in the Paris region, similarly to the Los Angeles case, but with many differences. This is not surprising, as the urban structure of the Paris region is monocentric and different types of activities are located close to one another, while in Los Angeles the urban area spreads widely with a more evenly distributed road network. The analysis presented in this research has shortcomings. One of them is that we did not consider light commercial vehicles (i.e., vans), which represent more than half of the vehicles used for deliveries and pick-ups in the Paris region. The updating of the analysis using comprehensive freight traffic flow data is a future research task.

For pushing the freight landscape concept forward and making the proposed approach of use for urban freight planning, more case studies are required, especially those for cities that have similar urban structure to the Paris region (or to Los Angeles) for comparison. Also, as discussed earlier, other commonly available indicators that can characterize urban truck traffic must be explored. It should be noted that rapidly emerging practices in logistics, such as night-time deliveries, consolidations, and crowdsourced deliveries, may be changing the relationships between urban forms and truck traffic over time and the validity of the concept of "freight landscape" must be reviewed accordingly.

Acknowledgments

This research was made possible with the support of the MetroFreight/VREF center of excellence, financing Heitz and Beziat PhD works as well as an academic visit of Sakai at IFSTTAR during Fall 2015.

Author Contributions

The authors confirm contribution to the paper as follows: study conception and design: Laetitia Dablanc, Takanori Sakai; data collection: Adrien Beziat, Adeline Heitz; analysis and interpretation of results: Takanori Sakai; draft manuscript preparation: Takanori Sakai, Adrien Beziat, Adeline Heitz. All authors reviewed the results and approved the final version of the manuscript.

References

1. Giuliano, G., S. Kang, and Q. Yuan. Using Proxies to Describe the Metropolitan Freight Landscape. *Urban Studies*, Vol. 55, No. 6, 2018, pp. 1346–1363. https://doi.org/10.1177/0042098017691438
2. Schwarz, N. Urban Forms Revisited—Selecting Indicators for Characterizing European Cities. *Landscape and Urban Planning*, Vol. 96, No. 1, 2010, pp. 29–47.
3. Allen, J., M. Browne, and T. Cherrett. Investigating Relationships Between Road Freight Transport, Facility Location, Logistics Management and Urban Form. *Journal of Transport Geography*, Vol. 24, 2012, pp. 45–57.
4. Anand, N., R. Van Dujn, L. Tavasszy, and H. Quak. City Logistics Modeling Efforts: Trends and Gaps—A Review. *Procedia—Social and Behavioral Sciences*, Vol. 39, 2012, pp. 101–115.
5. Ogden, K. W. Modeling Urban Freight Generation. In *Traffic Engineering and Control*, Transportation Research Board of the National Academies, Washington, D.C., Vol. 18, 1977, pp. 106–109.
6. Gonzalez-Feliu, J., and J. L. Routhier. Modeling Urban Goods Movements: How to Be Oriented with So Many Approaches? *Procedia—Social and Behavioral Science*, Vol. 39, 2012, pp. 89–100.
7. Taniguchi, E., R. Thompson, and T. Yamada. Emerging Techniques for Enhancing the Practical Application of City Logistics Models. *Procedia—Social and Behavioral Sciences*, Vol. 39, 2012, pp. 3–18.
8. Holguín-Veras, J., C. Lawson, C. Wang, .M. Jaller, C. González-Calderón, S. Campbell, L. Kalahashti, J. Wojtowicz, and D. Ramirez. *NCFRP Research Report 37: Using Commodity Flow Survey Microdata and Other Establishment Data to Estimate the Generation of Freight, Freight Trips, and Service Trips—Guidebook*. Transportation Research Board of the National Academies, Washington, D.C., 2016.
9. Routhier, J. L., and F. Toilier. Freturb V3, A Policy Oriented Software of Modeling Urban Goods Movements.

Presented at 11th World Conference on Transport Research, Berkeley, California, 2007.
10. Beziat, A., and F. Toilier. Using Urban Forms to Characterize Logistics Profiles: The Geography of Freight Movements in the City of Paris. Presented at 14th World Conference on Transport Research, Shanghai, China, 2016.
11. Organization for Economic Co-operation and Development. *Delivering the Goods: 21st Century Challenges to Urban Goods Transport*. Organization for Economic Co-operation and Development, France, 2003.
12. Dablanc, L. Goods Transport in Large European Cities: Difficult to Organize, Difficult to Modernize. *Transportation Research Part A: Policy and Practice*, Vol. 41, No. 3, 2007, pp. 280–285.
13. Beziat, A., and M. Koning. Marginal Congestion Costs with Multi-class Traffic: Some Tentative Estimates for the Paris Region. Presented at 6th International Conference on Urban Freight, Long Beach, USA, 2015.
14. Habib, P. A., and P. W. Crowley. Economic Approach to Allocating Curb Space for Urban Goods Movement. *Transportation Research Record: Journal of the Transportation Research Board*, 1976. 591: 18–24.
15. Jaller, M., J. Holguin-Veras, and S. D. Hodge. Parking in the City: Challenges for Freight Traffic. *Transportation Research Record: Journal of the Transportation Research Board*, 2013. 2379: 46–56.
16. Beziat, A. Parking for Freight Vehicles in Dense Urban Centers: The Issue of Delivery Area in Paris. Presented at 94th Annual Meeting of the Transportation Research Board, Washington, D.C., 2015.
17. *Document d'orientations stratégiques pour le fret en Ile-de-France à l'horizon 2025*. Direction Régionale et Interdépartementale de l'Equipement et de l'Aménagement d'Ile-de-France, France, 2012.
18. *Entrepôts et plateformes logistiques*. Commissariat Général au Développement Durable, France, 2015.
19. INSEE Analyses No. 13. *375 000 emplois salariés dans la logistique en Ile-de-France*. Institut National de la Statistique et des Etudes Economiques, France, 2015.
20. *Faits et chiffres 2013: Statistiques des transports*. Union Routière Française, France, 2013.
21. Report for INRIX. *The Future Economic and Environmental Costs of Gridlock in 2030: An Assessment of the Direct and Indirect Economic and Environmental Costs of Idling in Road Traffic Congestion to Households in the UK, France, Germany and the USA*. INRIX, France, 2014.
22. *Enquête Transport de Marchandises en Ville—Méthodologie et premiers résultats*. Laboratoire Aménagement Economie et Transports, France, 2015.
23. *Enquête Transport Routier de Marchandises (TRM)—Métadonnées*. Commissariat Général au Développement Durable, France, 2013.
24. *La circulation routière en Ile-de-France en 2010*. Institut d'Aménagement et d'Urbanisme de la Région Ile-de-France, France, 2013.
25. Regional Plan for Transportation. *Plan de Déplacements Urbains en Ile-de-France*. Région Ile-de-France, France, 2014.
26. Miyamoto, K., V. Vichiensan, N. Shimomura, and A. Paez. Discrete Choice Model with Structuralized Spatial Effects for Location Analysis. *Transportation Research Record: Journal of the Transportation Research Board*, 2004. 1898: 183–190.
27. Nguyen, C. Y., and K. Sano. Location Choice Model for Logistic Firms with Consideration of Spatial Effects. *Transportation Research Record: Journal of the Transportation Research Board*, 2010. 2168: 17–23.
28. Mohammadian, A., M. Haider, and P. S. Kanaroglou. Incorporating Spatial Dependencies in Random Parameter Discrete Choice Models. Presented at 84th Annual Meeting of the Transportation Research Board, Washington, D.C., 2005.
29. Woudsma, C., J. F. Jensen, P. Kanaroglou, and H. Maoh. Logistics Land Use and the City: A Spatial–Temporal Modeling Approach. *Transportation Research Part E: Logistics and Transportation Review*, Vol. 44, No. 2, 2008, pp. 277–297.
30. Bivand, R., and G. Piras. Comparing Implementations of Estimation Methods for Spatial Econometrics. *Journal of Statistical Software*, Vol. 63, No. 18, 2015, pp. 1–36.
31. De Lara, J. Goods Movement and Metropolitan Inequality: Global Restructuring, Commodity Flows and Metropolitan Development. In *Cities, Regions and Flows* (P. Hall, and M. Hesse, eds.), Routledge, New York, 2013.
32. Raimbault, N. *Gouverner le développement logistique de la métropole : périurbanisation, planification et compétition métropolitaine—Le cas du bassin parisien et éclairages étrangers*. PhD thesis. University of Paris-Est, Paris, France, 2014.

The Standing Committee on Urban Freight Transportation (AT025) peer-reviewed this paper (18-01910).

TRR
JOURNAL OF THE TRANSPORTATION RESEARCH BOARD

Article

Transportation Research Record
2018, Vol. 2672(9) 229–241

DOI: 10.1177/0361198118793243
journals.sagepub.com/home/trr
SAGE

Statistical Distributions of Pavement Damage Associated with Overweight Vehicles: Methodology and Case Study

Dario Batioja-Alvarez[1], Seyed-Farzan Kazemi[2], Elie Y. Hajj[2], Raj V. Siddharthan[3], and Adam J. T. Hand[4]

Abstract

The trucking industry is the primary mode of transporting for goods and commodities in the United States. Currently, there is an increasing trend in the use of overweight (OW) vehicles on the highway network. State highway agencies (SHAs) are challenged to address this increase, particularly relative to associated pavement damage. In this study, a probabilistic method was developed to evaluate rutting and fatigue cracking damage caused by OW vehicles under different road and environmental conditions. The influential input parameters in this analysis included gross vehicle weight (GVW), axle configuration, axle weight, pavement temperature, and vehicle-miles traveled (VMT). Necessary information was obtained by analyzing more than 170,000 entries of a Nevada Department of Transportation (NDOT) OW permit database. The developed model was based on mechanistic-empirical (ME) approach and considered asphalt concrete (AC) viscoelastic characterization. The results of this study were distributions of AC critical responses, load equivalency factors (LEFs), and relative damage factors (RDFs). The analysis showed that load equivalent factor (LEF) distributions could be incorporated in pavement design methods to account for OW vehicles. Furthermore, the damage induced by specific OW vehicles could be assessed using the relative damage factor (RDF) concept and may be efficiently used by SHAs during the permit application process. A case study was presented illustrating the impact of an OW axle configuration on pavement damage. Finally, a method was suggested for developing RDF tables with damage ranges corresponding to different axle configurations and the GVW that could be a tool for SHAs to evaluate and understand pavement damage induced by OW vehicles.

The trucking industry is a fundamental pillar in the U.S. economy as it is the primary mode of transporting goods and commodities. It carries 85% of all domestic cargo by value and 70% by weight (*1*). Freight volumes have been growing over recent years and are expected to increase significantly in the future (*2*). State Highway Agencies (SHAs) spend millions of dollars annually to maintain highway infrastructure at an acceptable level of service, while simultaneously attempting to reduce traffic congestion and travel time. Due to the significant and continuous growth of freight transportation, SHAs are currently challenged in accomplishing these tasks, especially relative to overweight (OW) vehicles.

As per federal law, the commercial vehicle gross vehicle weight (GVW) standard limit is currently 80,000 lb for the interstate highway network. The GVW limit and other axle loading statutes regulate highway traffic loading to prevent premature deterioration which could drastically increase associated costs of pavement maintenance and rehabilitation. One approach to reduce the number of commercial vehicles on the highway network is the allowance of multi-trailer vehicles. Multi-trailer vehicles make it possible for shippers to accommodate larger and heavier cargo in a single haul that would otherwise require multiple shipments. However, multi-trailer units cannot be used for large non-divisible loads, thus freight companies usually request an OW permit to transport heavier-than-standard loads. An OW vehicle generally has a GVW exceeding 80,000 lb (*3*). SHAs issue special permits for OW vehicles and collect a nominal fee, making the

[1]Lyles School of Civil Engineering, Purdue University, West Lafayette, IN
[2]Pavement Engineering and Science Program, Department of Civil Engineering, University of Nevada Reno, Reno, NV
[3]Geotechnical Engineering Program, Department of Civil Engineering, University of Nevada Reno, Reno, NV
[4]Pavement Engineering and Science Program, Department of Civil and Environmental Engineering, University of Nevada Reno, Reno, NV

Corresponding Author:
Address correspondence to Dario Batioja-Alvarez: dbatioja@Purdue.edu

operation of them legal on the highway network (*4*). It is common practice for SHAs to use funds collected for OW permits to partially cover costs associated with pavement damage induced by passing of OW vehicles (*5*).

Quantifying pavement damage caused by OW vehicles has historically been challenging. Therefore, this damage has not typically been appropriately considered in permit fees and/or pavement design and rehabilitation. One of the difficulties associated with determining OW vehicle damage is properly accounting for the viscoelastic behavior of asphalt concrete (AC) layers in the analysis process. This is critically important as it influences estimated pavement damage associated with GVW, axle loads, vehicle speed, and environmental conditions, such as pavement temperature. For example, pavement damage caused by an OW vehicle operating at lower than posted speeds on a summer day may be significantly greater than the damage caused by the same vehicle operating at night just due to the change in pavement temperature. As heavier axle loads of OW vehicles (GVW > 80,000 lb) introduce greater stresses and strains compared to those from a standard vehicle (GVW < 80,000 lb), a single OW vehicle pass could induce the same damage as multiple passes of a standard vehicle. OW to standard vehicle load equivalencies are known to be dependent on multiple factors including pavement structural capacity and environmental conditions at the time of OW vehicle operation (*6*). As a result, engineers and transportation officials need tools to evaluate and understand pavement damage induced by OW vehicles operating under different loading and environmental conditions. Once an OW permit is requested, appropriate information should be available to staff at SHAs during the permit issuance/review process to assess pavement damage before and under the proposed OW vehicle pass in a timely manner.

The objective of this paper is to present a probabilistic methodology to assess additional pavement damage caused by OW vehicles. The methodology is based on mechanistic-empirical (ME) analysis which relies on the well-established concept of load equivalency factors to assess relative pavement damage potential caused by OW vehicles. The developed method uses Monte Carlo simulations to mimic OW vehicle operation under different conditions. It considers multiple factors influencing pavement damage caused by a single OW vehicle pass under different road and environmental conditions existing in Nevada. The method also mimics the operation of a reference vehicle under the same set of conditions, enabling a relative comparison and determination of pavement damage scenarios. SHAs could use the presented method to obtain necessary information for regulating OW vehicle operations considering OW vehicle characteristics (e.g., axle configuration and load distribution). It could also be used to develop guidelines to limit certain OW vehicle types operating on specific roadways. Trucking companies could use the method to optimize vehicle axle configurations and axle loading distributions to minimize pavement damage and reduce permit cost. The developed method is based on the information that is typically collected by SHAs during the permit application process.

Background

The impact of OW vehicles on pavements has been the topic of several research studies. Titi et al. investigated pavement deterioration caused by heavy vehicles in Wisconsin (*7*). The authors analyzed an OW permit database with over 95,000 entries and conducted a routing analysis to identify highway segments that received significant OW traffic. Additionally, visual condition surveys were performed to determine related pavement conditions on the identified segments. A strong correlation between OW traffic level and observed pavement distress was found. In another recent study, the impact of OW traffic on pavement life was investigated (*8*). The predicted pavement life of different pavement structures was investigated using various OW traffic scenarios. It was reported that a 1% increase in OW vehicle traffic led to a 1.8% reduction in pavement life based on ME analysis.

Estimation of pavement damage has historically been assessed using load equivalency factors (LEFs). The LEF concept was an outcome of the AASHO road test in 1960's (*9*). The concept behind LEFs is the conversion of any axle load and configuration to an equivalent standard axle configuration (18,000 lb single axle) for use in pavement design. Several researchers have applied this concept to investigate the impact of heavy vehicles on pavement damage. Sebaaly et al. evaluated the impact of agricultural vehicles on AC pavement rutting for low volume roads using LEFs (*10*). The authors determined the ratio of the number of repetitions to failure for an 18,000 lb single axle to the number of repetitions to failure for a given agricultural vehicle axle configuration relative to a surface rutting threshold.

Researchers have also investigated the effect of environmental conditions on pavement performance. In particular, pavement mechanical responses are significantly affected by pavement temperature. Sebaaly et al. concluded that rutting-based LEFs were not constant from season to season because of temperature variation (*10*). In another study, the effect of environmental conditions on pavement damage induced by OW bus rapid transit (BRT) was investigated (*6*). The authors determined LEFs for several BRT vehicles using pavement temperature distributions representing different seasons of the year. It was found that pavement damage from BRT vehicles in Nevada is significantly influenced by the variability in passenger ridership and more importantly the corresponding climatic conditions at the time of the trips.

Banerjee and Prozzi documented a framework to determine load equivalencies for individual axle configurations using a ME approach (*5*). The concept of LEF was extended to incorporate multiple distresses and to account for vehicle dynamic loading. The equivalency concept was employed to determine the pavement service life consumption and associated costs (using a marginal cost allocation approach) attributable to the operation of OW vehicles. It was concluded that LEF were significantly affected by axle configuration and the distribution of loads on an axle. As the LEF method has been used as a measure of OW vehicle damage potential to the pavement, it is essential to understand how different factors influence LEF. This understanding will be helpful to modify axle configuration and load distribution so that potential pavement damage can be minimized. Batioja-Alvarez et al. focused on pavement damage costs associated with multi-trip scenarios of a few selected OW vehicles in Nevada with single yearly permits (*11*). The NDOT database on past OW vehicle passes was used to determined pavement repair and maintenance associated costs. As multiple factors contribute to pavement damage caused by OW vehicles, a high degree of uncertainty is associated with damage estimation caused by OW vehicle passes. Furthermore, the deterministic analysis only provides a time-shot of the potential pavement damage caused by a single OW vehicle pass. Therefore, a probabilistic analysis was employed in this study to account for the uncertainty associated with OW traffic induced pavement damage. The advantage of using probabilistic analysis is obtaining statistical distributions for decision parameters. In the developed method, distributions of GVW, axle loads and configurations, pavement structures, and pavement temperature were used to generate distributions of critical AC pavement responses, LEFs and relative damage factors (RDFs). A database containing 340,000 permit entries issued by Nevada Department of Transportation (NDOT) over 10 years was analyzed for this purpose. Nearly 170,000 of the entries were classified as OW with necessary information for the developed analysis. Key information such as vehicle configuration (i.e., type and number of axles), GVW, and axle loads were analyzed and converted into empirical probability distributions. The distributions were then used to generate output distributions for important parameters (e.g., LEF and RDF) for a wide range of environmental conditions existing in Nevada. Additionally, LEFs for multiple vehicle configurations and various load levels were determined for annual average pavement temperature (AAPT).

Methodology

The Monte Carlo simulation method was employed to obtain the distribution of output parameters such as AC critical responses and LEF. This simulation method uses random sampling of input parameters based on their distribution to process these inputs and obtain the distribution of output parameters (*12*, *13*). In this study, influential simulation input parameters were GVW, axle configuration, axle weight, pavement structure, and pavement temperature. In each simulation step, the number of repetitions to rutting and fatigue-cracking failure were calculated for OW and reference vehicles. A fairly large number of simulation scenarios (10,000 simulation steps) was considered to obtain adequate points for simulation output parameters. Parallel processing was used to accelerate simulation time because simulation steps could be executed independently (*14*). In this study, the reference vehicle was a typical 80,000 lb 18-wheeler truck with one steering axle (12,000 lb) and two tandem axles (34,000 lb each).

Data Collection

The NDOT OW database was scrutinized to extract the necessary influential input parameters. It was found that over 50% of OW loads have a GVW within the 100,000–150,000 lb category. There are also a considerable number of OW permit entries in the 80,000–100,000 lb and 200,001–250,000 lb categories. The least frequent category corresponds to OW permits with a GVW greater than 250,000 lb. These OW loads are generally regarded as superheavy loads and usually require a detailed engineering analysis that is either performed by SHAs or an independent engineering firm. Concerns about superheavy load passes are not only confined to failure of AC layer(s) but most importantly to the shear failure of unbound layer(s). As a very small percentage of OW permits correspond to the superheavy load category (GVW >250,000), the developed method excludes them. The following sections present the probabilistic method developed for damage analysis associated with OW vehicle passes.

After scrutinizing the NDOT OW database, information on the GVW, axle configuration, axle weight distribution, and routing were assembled. NDOT has three districts under its jurisdiction: District I (representing southern Nevada), District II (representing northwest Nevada), and District III (representing northeast Nevada). Environmental conditions vary significantly between these districts requiring integration of this variation in the simulations.

NDOT categorizes roads into three categories. Namely, highways and freeways (NDOT road category 1), major arterials (NDOT road category 2), and collectors (NDOT road category 3). Based on the most recent Nevada cost allocation study, highways and freeways are exposed to 59% of overall OW vehicle-miles traveled (VMT) making road category 1 the most common road type in which OW vehicles operate in Nevada (*15*). Furthermore, OW traffic VMT is almost equally divided between the other road categories with 20% and 21% on road categories 2 and 3, respectively. It was assumed that OW vehicles do not operate on roads with lower structural capacity (e.g., local roads). VMT distribution was used to randomly assign the road category in each simulation step.

For the simulations, pavement structures consisting of 10 inch, 8 inch, and 6 inch of AC over a 10 inch crushed aggregate base (CAB) over a semi-infinite subgrade (SG) were considered for road categories 1, 2, and 3, respectively. For NDOT District I, a dynamic modulus of PG76-22NV Type II mixture was used to characterize AC viscoelastic properties, and a dynamic modulus of PG64-28NV Type II was used for NDOT Districts II and III (*16*). CAB and SG were modeled as linear elastic with modulus of E_{CAB} = 30,000 psi and E_{SG} = 15,000 psi, respectively.

As the pavement temperature affects the AC dynamic modulus, it influences AC critical responses. Thus, AC temperature variation was considered in this study. The model developed by Alavi et al. was used to estimate pavement temperature (*17*). This one-dimensional model is based on the finite-volume control method (FVCM) and requires inputs of climatic data (solar radiation, air temperature, and wind speed), material thermal properties (density, specific heat capacity, and thermal conductivity), and surface characteristics (albedo, emissivity, and absorption). Climatic data were obtained from different weather stations located in NDOT districts. The output of the prediction model was spatial pavement temperature as a function of depth on an hourly basis for NDOT districts. Mid-depth AC temperature was selected as the representative AC temperature in the simulations after appropriately shifting the AC dynamic modulus.

Distribution of Influential Input Parameters

Figure 1 presents distributions of influential input parameters included in this study. For each input, the corresponding data was translated into histograms. Probability distribution functions (PDFs) were calculated based on the frequency of histogram bins. Subsequently, cumulative distribution functions (CDFs) were derived by integrating the PDFs.

As presented in Figure 1*a*, OW vehicles were divided into seven bins based on GVW: 80,000–90,000 lb, 90,000–100,000 lb, 100,000–110,000 lb, 110,000–130,000 lb, 130,000–150,000 lb, 150,000–200,000 lb, and 200,000–250,000 lb. The frequency of each bin is presented based on information retrieved from the NDOT OW database. Consequently, CDF was generated based on the GVW frequency distribution. Figure 1*b* presents the distribution of axle configuration, for the GVW bin of 80,000–90,000 lb as an example, as well as its respective CDF. In the last step of determining the OW vehicle axle load and configuration, the axle load was randomly assigned based on the empirical axle weight distribution for the particular axle configuration derived from the database. For instance, Figure 1*c* presents axle weight CDFs for a tandem axle of an OW vehicle within the GVW category of 80,000–90,000 lb.

Pavement temperature was another influential factor used for viscoelastic characterization of AC. Figure 1*d* presents the mid-depth AC temperature distribution for District I as an example. Furthermore, Figure 1*e* presents the VMT frequency distribution and CDFs for different NDOT road categories which were used to randomly select pavement structure.

In each simulation step, inputs were randomly sampled from their respective distributions. AC critical responses were calculated for OW and reference vehicles using 3D-Move analysis software (*18*). 3D-Move Analysis software employs a finite-layer approach to calculate pavement responses. The software is also capable of handling the viscoelastic properties of an AC layer, as well as the influence of a moving surface load which made it suitable for the integration with the Monte Carlo simulations. The number of repetitions to rutting and fatigue cracking failure were calculated for OW and reference vehicles using AC critical responses obtained from 3D-Move Analysis software. Subsequently, damage indicators (LEF and RDF) were calculated based on the number of repetitions to failure. Figure 2 provides a flowchart representation of the steps associated with the use of Monte Carlo simulations including generating input parameters, simulation process, and generating output parameters for a single simulation step. The distribution of output parameters (i.e., LEF and RDF) was obtained by running the simulation for 10,000 steps. Details of the calculation of these output parameters are presented next.

ME Damage Prediction

This study focused on AC rutting and AC bottom-up fatigue cracking which are two of the most prominent distresses observed in Nevada. The locally calibrated mechanistic-empirical pavement design guide (MEPDG) performance models and transfer functions were used in this study (*19*). The necessary critical pavement responses were assembled to estimate pavement damage for each axle. Equations 1 and 2

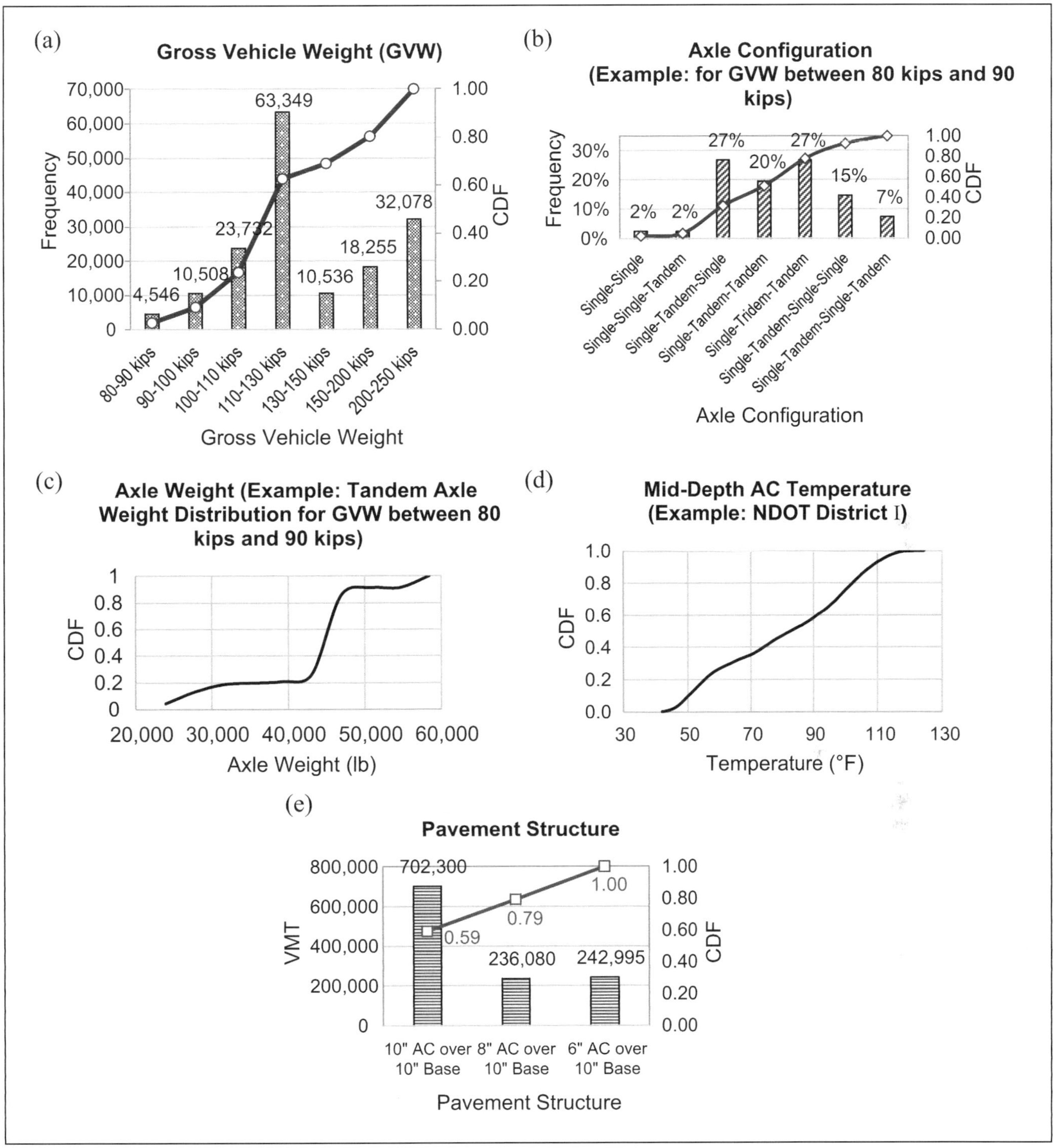

Figure 1. Sample input parameters for Monte Carlo simulation: (*a*) GVW, (*b*) axle configuration, (*c*) axle weight, (*d*) mid-depth AC temperature, and (*e*) pavement structure.

represent general MEPDG models for rutting and fatigue cracking, respectively (*20*). These equations were used to obtain the number of repetitions of OW or reference axles to certain rehabilitation failure criteria. NDOT failure criteria for AC rutting and AC bottom-up fatigue cracking are 0.15 inch and a 5% area cracking, respectively (*19*). The local calibration factors corresponding to each of the three NDOT districts were used in calculating the number of repetitions to failure (*19*).

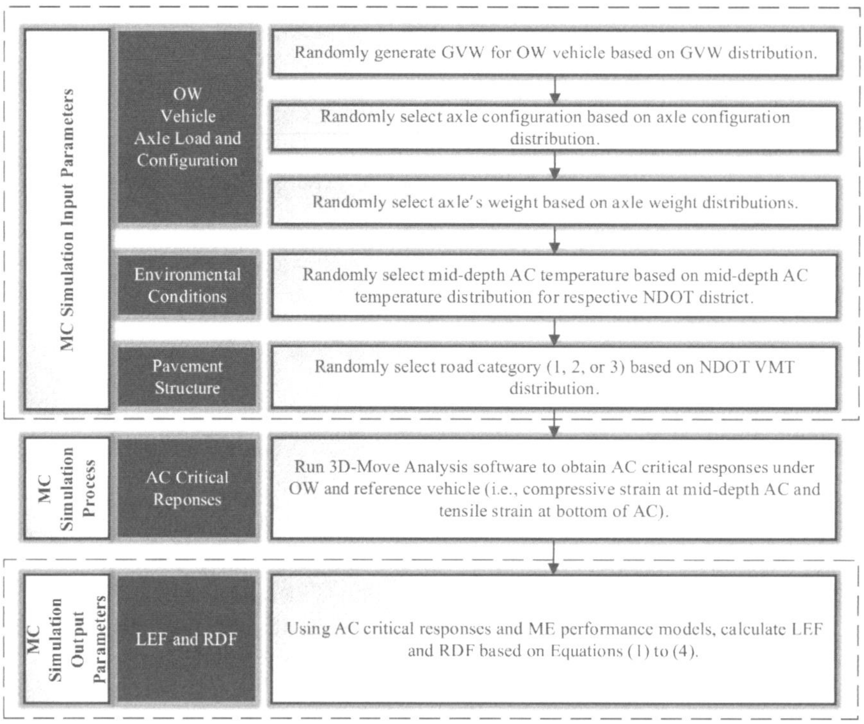

Figure 2. Monte Carlo simulation flowchart presenting inputs, process, and outputs.

$$\frac{\varepsilon_p}{\varepsilon_r} = B_{r1} 10^{K_1} T^{K_2 B_{r2}} N_r^{K_3 B_{r3}} \tag{1}$$

where:

ε_p = plastic strain (inch/inch),
ε_r = resilient strain obtained from 3D-Move Analysis software (inch/inch),
T = mid-depth AC temperature (°F),
N_r = number of load repetitions to rutting failure,
K_1, K_2, K_3, = −2.9708, 1.7435, 0.3547 (NDOT District I),
−3.2605, 2.0055, 0.3161 (NDOT District II),
−3.4717, 2.0258, 0.3946 (NDOT District III); and
B_{r1}, B_{r2}, B_{r3} = 0.10451, 1.0, 1.0 (NDOT District I),
0.16981, 1.0, 0.9 (NDOT District II),
0.13654, 0.9, 0.8 (NDOT District III).

The Fatigue cracking MEPDG performance model uses the following equation:

$$N_f = B_{f1} K_1 \left(\frac{1}{\varepsilon_t}\right)^{k_2 B_{f2}} \left(\frac{1}{E}\right)^{k_3 B_{f3}} \tag{2}$$

where:

N_f = number of load repetitions to fatigue cracking failure,
ε_t = tensile strain at the bottom of AC (inch/inch),
K_1, K_2, K_3, = 214.18, 5.0284, 2.3072 (NDOT District I),
30.08, 5.0537, 2.8904 (NDOT District II and III).
B_{f1}, B_{f2}, B_{f3} = 0.005, 1.0, 1.0 (NDOT District I),
50, 1.0, 1.0 (NDOT District II and III).

Relative Axle and Vehicle Pavement Damage. As mentioned earlier, the LEF is a routinely-used parameter in pavement design and analysis to compare pavement damage caused by a given axle group (single, tandem, tridem, or quad) to that caused by the reference dual tire 18,000 lb single-axle. The concept of LEF was expanded in this study to develop ME based LEFs. The presented analysis specifically develops equivalencies solely based on AC rutting and fatigue cracking distresses. This was done for each and every axle of the OW and reference vehicles using the following relationship (*5, 6, 10*):

$$LEF = \frac{N_{18}}{N_{axle}} \tag{3}$$

in which N_{18} and N_{axle} represent the number of repetitions to rutting or fatigue cracking failure corresponding to the reference axle and specific OW vehicle axles, respectively. The LEF for an OW vehicle was determined by summing the LEFs of its individual axles.

The RDF for an OW vehicle is defined as the ratio of LEF for the entire OW vehicle (LEF_{OW}) to that of reference vehicle (LEF_{Ref}), given by

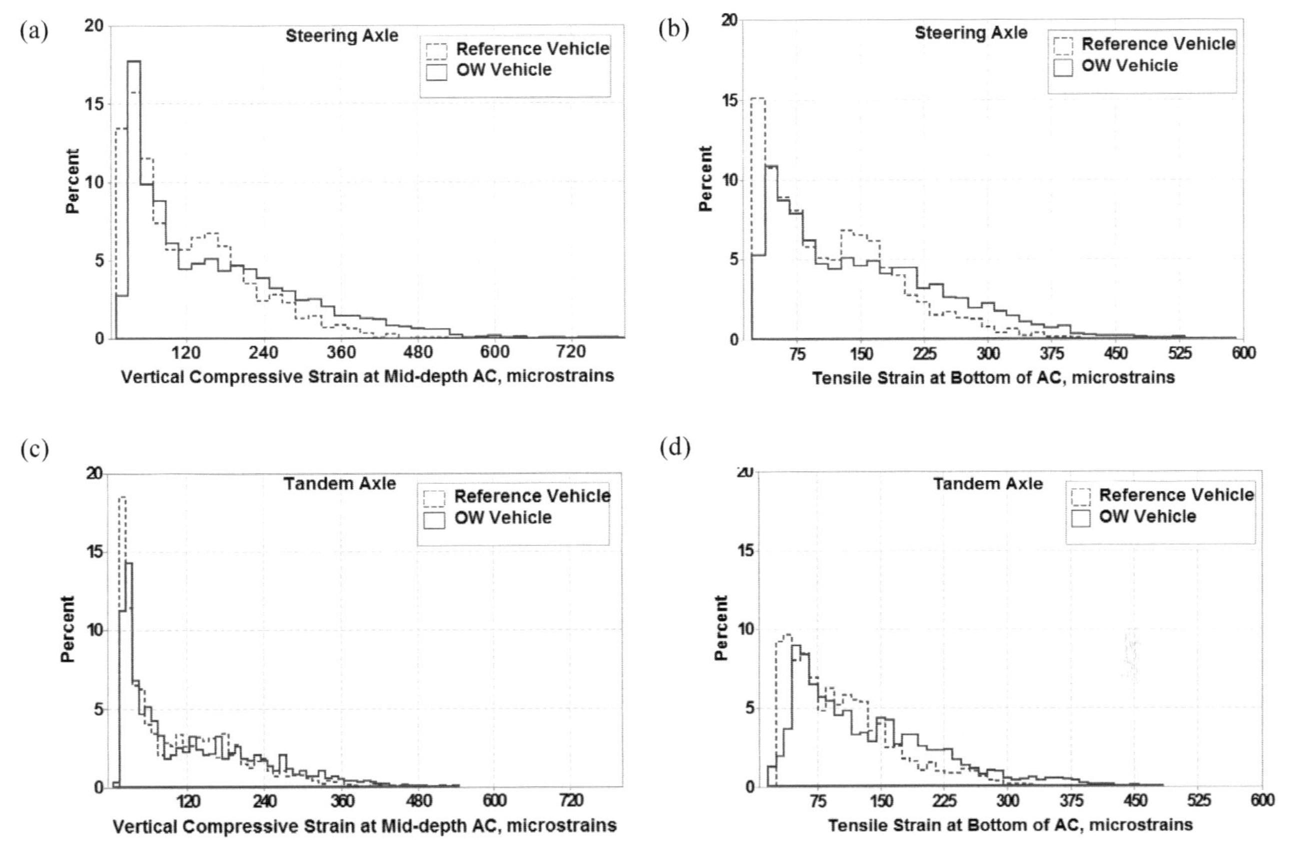

Figure 3. Distribution of AC critical responses: (*a*) vertical compressive strain at mid-depth AC under pass of steering axle, (*b*) horizontal tensile strain at bottom of AC under pass of steering axle, (*c*) vertical compressive strain at mid-depth AC under pass of tandem axle, and (*d*) horizontal tensile strain at bottom of AC under pass of tandem axle.

$$RDF = \frac{LEF_{OW}}{LEF_{Ref}} \tag{4}$$

For example, an RDF greater than 1.0 indicates that fewer passes of an OW vehicle cause the same damage compared to the reference vehicle. Thus, RDF can be used as a measure of potential pavement damage induced by permitted OW vehicles relative to the reference vehicle.

Results

AC Critical Response Distributions

Vertical compressive strain at mid-depth (ε_r) of AC layers and maximum horizontal tensile strain at bottom of AC layers (ε_t) were calculated using 3D-Move Analysis software (*18*). The reason for calculating ε_r and ε_t was to estimate the number of repetitions to failure for rutting and fatigue cracking, respectively, using locally calibrated MEPDG performance models. AC critical responses were determined for each and every axle of OW and reference vehicles in each simulation step. Figure 3, *a–d* present distributions of ε_r and ε_t for OW and reference vehicles for NDOT District I. The presented distributions in Figure 3 report critical strains caused by steering and tandem axle passes. It is clear in all plots that the critical responses corresponding OW vehicles were significantly higher than those of the reference vehicles. For instance, those distributions corresponding to OW vehicle's axles present righter-skewness, suggesting higher strain levels.

LEF Distributions

Figure 4 presents the LEF distribution information for aggregated axles of OW vehicles based on simulation results. Figure 4*a* is the box plot representation of the LEF corresponding to AC rutting for different NDOT districts. Rutting failure criteria was assumed to be 0.15 in an AC rut depth. The rutting-based LEF ranged from 1.5 to 12.0 for District I, 2.4 to 9.8 for District II, and 2.0 to 6.0 for District III. The median, a representative

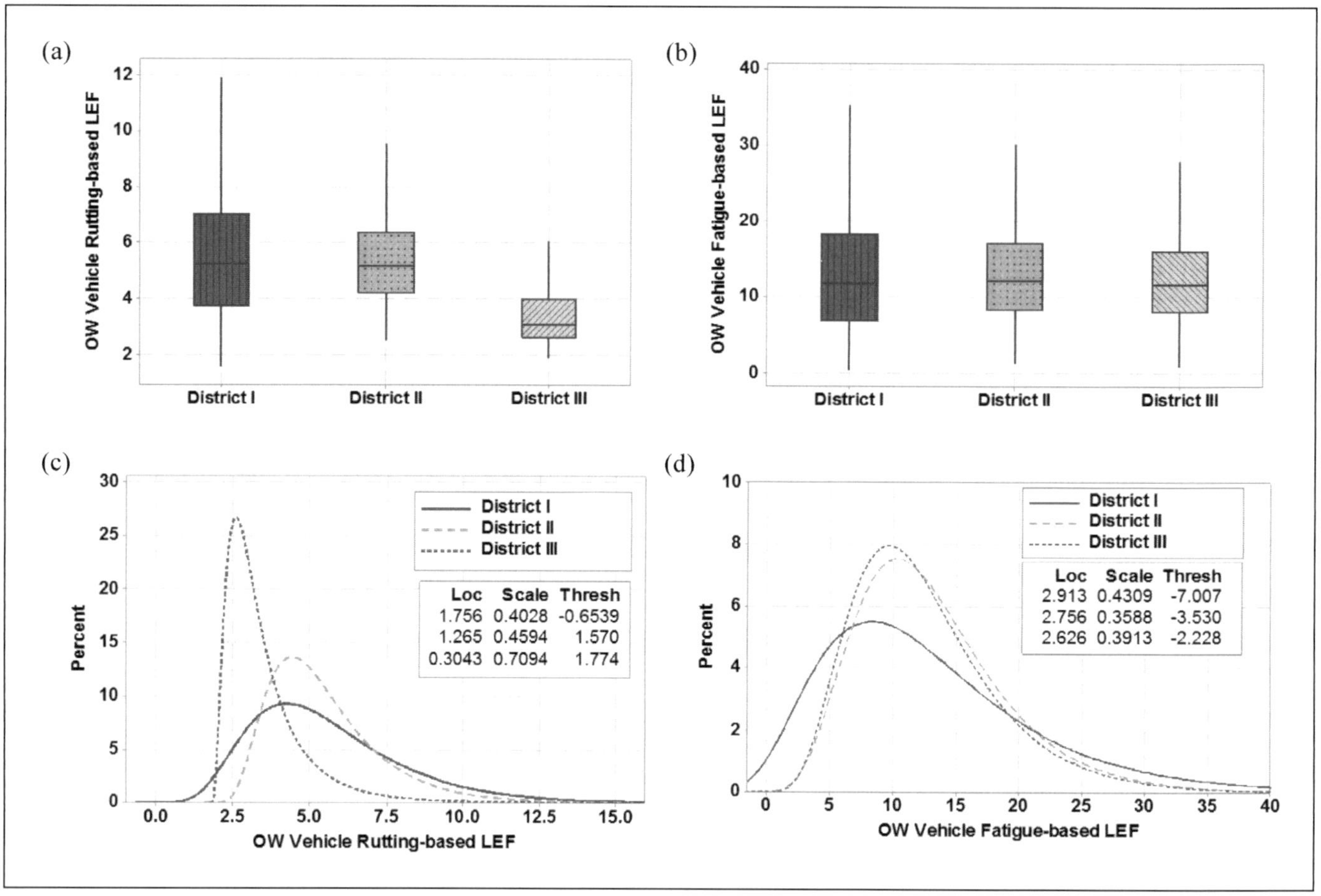

Figure 4. LEF distributions for OW vehicles: (*a*) boxplot representation of rutting-based LEF, (*b*) boxplot representation of fatigue-based LEF, (*c*) log-normal distribution fitted to rutting-based LEF, and (*d*) log-normal distribution fitted to fatigue-based LEF.

of central tendency, is also presented on this plot for each district. For instance, District III had the lowest rutting-based median (LEF = 3.13) suggesting that half of OW vehicle axles in this district had a rutting-based LEF greater than 3.13.

A box plot presentation of fatigue-based LEFs is provided in Figure 4*b*. Comparing Figure 4*a* with Figure 4*b*, one can conclude that fatigue-based LEFs were clearly higher than the rutting-based LEFs. Therefore, a single OW vehicle pass induces greater fatigue cracking damage than rutting damage compared to the damage caused by a pass of the reference vehicle. These findings are in agreement with conclusions reported by Hajj et al. where it was observed that fatigue cracking damage was the dominant distress caused by OW BRT vehicles in Nevada (*6*). A review of fatigue-based LEF distributions reveals that the highest LEFs corresponded to OW vehicles with a GVW over 150,000 lb with more than three axle groups (i.e., tandem, tridem, and quads).

Best fit probability distribution plots of LEFs are presented in Figure 4, *c* and *d* for rutting-based and fatigue-based LEFs, respectively. These distributions resemble a bell-shaped curve with positive skewness. A three-parameter log-normal distribution was fitted to the LEF data as this distribution generally matches the skewed data. The three parameters of the log-normal probability distribution function (i.e., threshold (γ), scale (μ), and shape (σ)) are also presented in the figures. These three parameters determine the overall shape of log-normal distributions and can be used to calculate the mean, median, and other quartiles. Equation (5) presents the probability density function (PDF) of the three-parameter log-normal distributions.

$$f(x;\mu,\sigma,\gamma) = \frac{1}{(x-\gamma)\sigma\sqrt{2\pi}} exp\left\{-\frac{[ln(x-\gamma)-\mu]^2}{2\sigma^2}\right\} \quad (5)$$

From a pavement design perspective, LEFs have been used to characterize traffic, and it is a critical design factor. With the introduction of MEPDG, full axle load spectra are required to characterize traffic loads. However, OW traffic loads are generally not represented on these load distributions. Thus, probability-based LEF

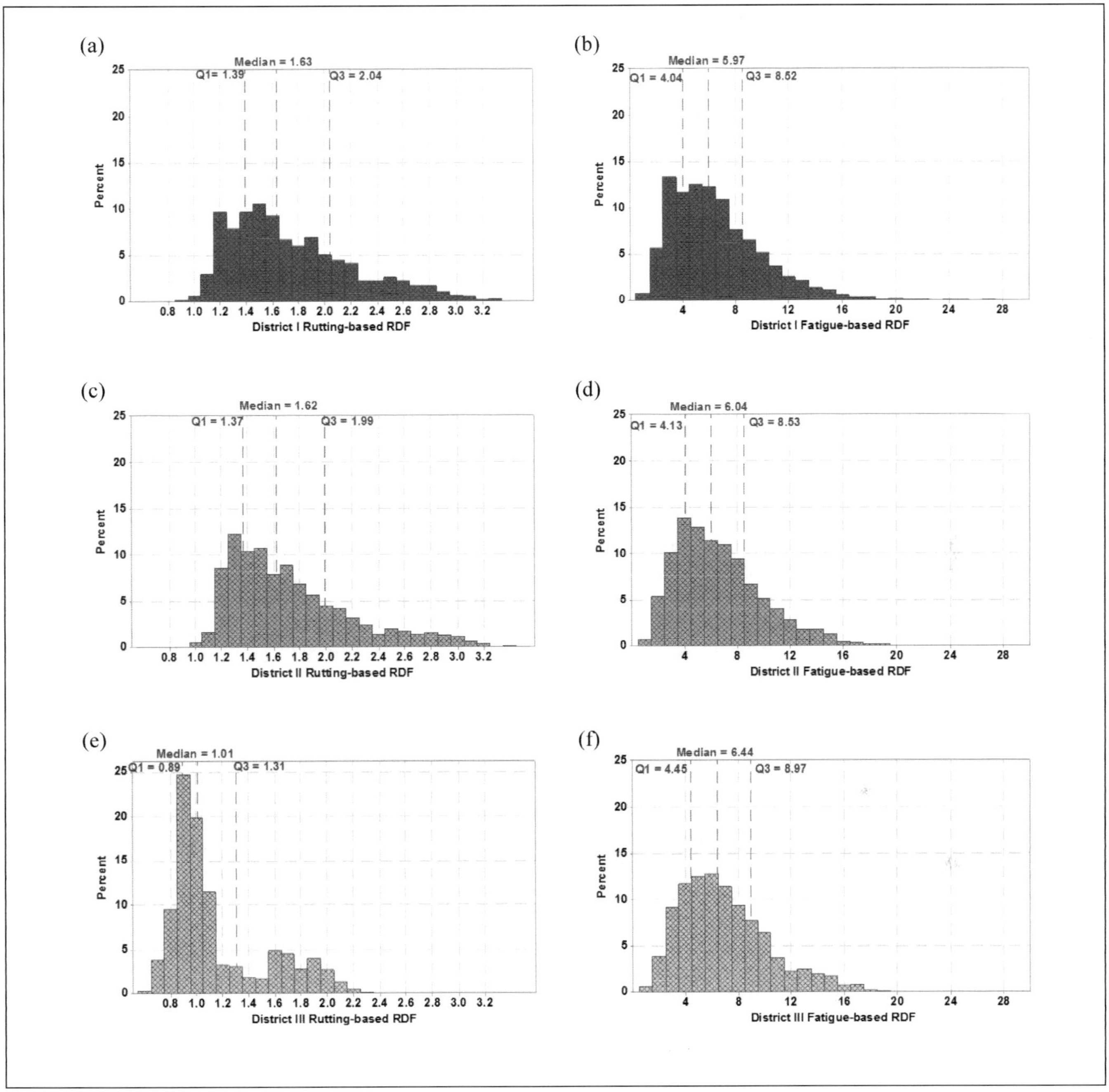

Figure 5. RDF distributions for full vehicles: (*a*) rutting-based RDF for District I, (*b*) fatigue-based RDF for District I, (*c*) rutting-based RDF for District II, (*d*) fatigue-based RDF for District II, (*e*) rutting-based RDF for District III, and (*f*) fatigue-based RDF for District III.

modeling of OW traffic would be useful for SHAs as it provides an estimation of pavement damage induced by different OW vehicles having different axle loads and configurations. The results obtained by using the developed method showed that there is significant variation in rutting-based and fatigue-based LEFs for different NDOT districts. This variation is mainly caused by different axle types, axle loads and configurations, and environmental conditions. When these variations are not accounted for in pavement design, a high potential for premature failure exists.

RDF Distributions

RDF distributions are presented in Figure 5 for different NDOT districts and distress types. Figure 5, *a, c,* and *e* present rutting-based RDF distributions for NDOT Districts I, II, and III, respectively. Figure 5, *b, d,* and *f* illustrate fatigue-based RDF distributions for NDOT Districts I, II, and III, respectively. The values for the 25th quartile (Q_1), 50th quartile (Median), and 75th quartile (Q_3) are also presented on the figures. In general, fatigue-based RDFs were considerably higher than rutting-based RDFs. In fact, the highest median for rutting-based RDF distributions was 1.63 which occurred in District I. Based on Figure 5, *a* and *c*, District I and II had similar rutting-based RDF distributions when considering the shape and range of RDF, which varied between 0.9 and 3.5 for these districts. Conversely, District III rutting-based RDF distributions ranged between 0.6 and 2.3 based on Figure 5*e*. As a result, OW vehicles can potentially cause up to 3.5 times more rutting damage than the reference vehicles in District I and II, and up to 2.3 times more rutting damage in District III compared with rutting damage caused by the reference vehicle. Results were inspected for instances in which an OW vehicle caused the same or less rutting or fatigue cracking damage than the reference vehicle (i.e., cases with RDF $\leq$ 1). Few cases were found, and in all of them, the OW vehicle included at least a tridem or a quad axle in its axle configuration. In addition, OW vehicles in those cases had a GVW close to 80,000 lb.

Fatigue-based RDF values were significantly higher (between 1.0 and 20.0 for aggregated RDF values). In the most severe cases, a single pass of an OW vehicle could induce the same fatigue cracking damage as 20 passes of the reference vehicle. These extreme cases are not very common in the NDOT database and correspond to cases where OW vehicles had a GVW greater than 200,000 lb with multiple axle groups passing on a pavement section with relatively low structural capacity (i.e., 6 inch AC over 10 inch CAB). Comparison of fatigue-based RDF results for the three districts shows that all districts had similar distributions with similar median and quartile values.

The RDF distributions can also be used for optimization purposes. As mentioned before, the probabilistic approach uses material properties, environmental conditions and performance model calibration factors that are district-specific when determining pavement performance associated with OW vehicles. However, the same representative pavement structures (e.g., layer thicknesses) were implemented in these analyses. The RDF distributions are district specific and present variability from district to district. Agencies could use a probabilistic analysis to determine the required pavement layer thicknesses to obtain equivalent RDF for common OW configurations through the regions.

The RDF distributions presented in Figure 5, *a–f* provide practical information to evaluate potential pavement damage caused by OW vehicles during a permit review process. If an OW permit is requested, the respective district's RDF distributions can be used as a baseline for evaluating pavement damage potential of the OW vehicle of interest. If the RDFs for the OW vehicle are significantly higher than those of the district distributions, SHAs could request a different axle configuration or another type of mitigation procedure. This type of analysis could quickly provide an indication of relative pavement damage before an OW permit is issued.

Case Study

To illustrate an application of the presented probabilistic method, a specific OW vehicle traveling in northern Nevada (NDOT District II) was evaluated. The following presents information provided by freight companies during the permit application process. The evaluated vehicle contained a single axle single tire steering axle and four tandem axles with eight tires each. LEF values for each and every axle of the studied OW were then derived based on the information presented. In addition, simulation outputs included LEFs for the reference vehicle traveling under the same road and environmental conditions as the OW vehicle. Subsequently, the RDF was calculated for the OW vehicle based on Equation (4) and is presented below.

- Location of OW Pass = Northern Nevada (District II)
- GVW = 202,687 lb
- Axle Loads/configurations = 14,915 lb (Single), 46,943 (Tandem), 46,943 (Tandem), 46,943 (Tandem), 46,943 (Tandem)
- Determined OW RDF
  - AC Rutting: 2.7 > District II Q_3: 1.9
  - AC Fatigue Cracking: 5.8 < District II Median: 6.0

The rutting-based and fatigue-based RDF for the OW vehicle were 2.7 and 5.8, respectively. These values were compared to the District II rutting-based and fatigue-based RDF distribution parameters (i.e., median and third quartile, Q_3). It was observed that rutting-based RDF for the OW vehicle ($RDF_{OW,rutting}$ = 2.7) was greater than the third quartile of the District II rutting-based RDF (Q_3 = 1.99) (see Figure 5*c*). However, fatigue-based RDF ($RDF_{OW,fatigue}$ = 5.80) was lower than the median District II value (50th quartile = 6.04) (see Figure 5*d*). This information could raise concerns as the actual axle configuration presents considerable AC rutting damage risk for the State highway agency. After this evaluation, the agency could ask for a different OW vehicle axle configuration. Further analysis showed that if three out of the four tandem axles were replaced by

Table 1. Relative Pavement Damage Potential for Specific OW Vehicle Configurations

	OW Vehicle Axle Configuration					
	S-T-T-T		S-T-T-T-T		S-TR-TR-TR	
	RDF					
OW vehicle load levels, GVW (lb)	Rutting-based	Fatigue-based	Rutting-based	Fatigue-based	Rutting-based	Fatigue-based
90,000	1.11	1.01	-	-	-	-
100,000	1.21	1.35	1.17	0.79	-	-
110,000	1.32	1.79	1.27	1.04	1.09	1.44
120,000	1.44	2.34	1.41	1.65	1.16	1.49
130,000	1.55	3.03	1.51	2.13	1.25	1.94
140,000	1.67	4.20	1.65	3.19	1.32	2.03
150,000	1.82	6.53	1.78	4.64	1.42	2.61
160,000	1.99	8.62	1.91	6.48	1.49	2.74
170,000	2.07	10.09	2.04	8.65	1.60	3.51
180,000	2.20	11.90	2.13	10.00	1.71	4.63
190,000	-	-	2.22	10.37	1.81	5.54
200,000	-	-	2.34	10.90	1.92	6.86
220,000	-	-	2.52	12.04	2.09	7.65
240,000	-	-	2.70	13.70	2.29	9.91

Note: OW = over weight; GVW = gross vehicle weight; S-T-T-T = single-tandem-tandem-tandem, S-T-T-T-T = single-tandem-tandem-tandem-tandem; S-TR-TR-TR = single-tridem-tridem-tridem; RDF = relative damage factor. Numbers shown indicate RDF.

tridem axles (12 tires each), rutting-based and fatigue-based RDFs would be reduced to 1.80 and 3.04, respectively. Consequently, potential pavement damage would be reduced significantly. There might also be other axle configurations that could be suggested by the agency or freight company to minimize potential pavement damage.

Clearly, the probabilistic aspect of the analysis provided yields useful information for SHAs to evaluate potential pavement damage, and at the same time useful information for freight companies to optimize vehicles and permit fees. As SHAs migrate to the implementation of ME pavement design, permit fee methodology might adopt weight-distance or axle/damage-distance schemes. Thus, exploring axle configurations that induce the minimum potential pavement damage will become critical for freight companies to minimize pavement damage associated costs.

Developing Tables for Potential Pavement Damage

Table 1 shows the average RDF values for three selected OW vehicle axle configurations obtained from simulation outputs. These axle configurations are single–tandem–tandem–tandem, single–tandem–tandem–tandem–tandem, and single–tridem–tridem–tridem (represented as S-T-T-T, S-T-T-T-T, and S-TR-TR-TR in Table 1). To develop the comparisons presented in Table 1, the analysis was conducted at a single temperature, 82 °F, which is the average annual pavement temperature in Reno, Nevada (NDOT District II). Additionally, the simulation results were filtered to only include the evaluated axle configurations. It was assumed that if the RDF for the OW vehicle was less than the median of the District II RDF distribution, then a low damage risk is associated with a pass of the OW vehicle (green zone). Similarly, if the RDF for the OW vehicle is greater than the third quartile (Q_3) of the District II RDF distribution, then a high risk of damage is associated with a pass of the OW vehicle exists (red zone). If the RDF for the OW vehicle lies between the median and third quartile (Q_3) of the District II RDF distribution, then it was assumed a medium risk of damage would be associated with a pass of the OW vehicle (yellow zone).

Based on the information presented in Table 1, there is a higher potential for fatigue cracking damage than rutting for a single pass of the OW vehicles. For instance, it is estimated that an OW vehicle with a 130,000 lb GVW and S-T-T-T axle configuration would induce 55% more AC rutting damage than the reference vehicle. Interestingly, the same OW vehicle would cause 300% more fatigue cracking damage than the reference vehicle. Additionally, the S-T-T-T axle configuration could potentially induce excessive pavement damage if GVW is more than 160,000 lb. Conversely, if one more tandem axle group is added (S-T-T-T-T axle configuration), high AC rutting and fatigue cracking damage levels would be observed for vehicles with a GVW greater than 170,000 lb. OW configurations containing tandem axles in Table 1 also present limited yellow zones (medium damage risk) indicating that once the OW vehicle axle configurations exceed certain GVW thresholds, pavement damage potential increases significantly. On the other hand, significantly lower RDF values were observed for the S-TR-TR-TR configuration along with a wider yellow zone. In general, the RDF values shown in Table 1 suggest that when considering similar GVW and axle load levels, the use of tridem axles instead of tandems have a significant impact on RDF values. RDF values are also impacted by pavement temperature. Rutting-based RDF values are expected to increase, and fatigue-based RDF values to decrease with the increase in temperature.

Conclusion

In this study, potential pavement damage induced by OW vehicle passes was investigated using a Monte Carlo simulation. The distributions of influential parameters including GVW, axle configuration, axle weight, pavement temperature, and VMT were obtained by analysis of more than 170,000 entries from the NDOT OW permit database. In each simulation step, the number of repetitions to rutting and fatigue cracking failure were calculated for OW and reference vehicles based on locally calibrated MEPDG performance models. The simulation output was a distribution of the AC critical responses, LEFs, and RDFs. Although the NDOT OW database did not include illegal OW data, the methodology presented in this paper could be implemented to account for illegal OW vehicles. The influence of illegal OW traffic on pavement damage could be estimated once a solid distribution is obtained. The methodology presented in this paper allows for the implementation of multiple distributions during the Monte Carlo simulation. For instance, other distributions, such as the axle spacing and tire spacing could also be considered as part of the presented approach.

The following conclusions can be drawn from this study:

- A new application of the LEF concept was established to assess OW vehicle rutting and fatigue cracking damage potential based on ME principles.
- Rutting-based and fatigue-based RDF distributions may be used to assess the relative pavement damage of an OW vehicle with respect to a reference vehicle.
- Observed fatigue-based LEFs were significantly greater than rutting-based LEFs. This suggests that for Nevada conditions OW vehicles have the potential to induce greater fatigue cracking damage than rutting damage.
- A skewed three-parameter log-normal distribution was effectively fitted to rutting-based and fatigue-based LEFs. These distributions could be used for pavement design procedures to include the effect of OW vehicles.
- A case study was presented illustrating the effective application of the probabilistic method. It showed how the developed distributions can be employed to assess the impact of specific OW vehicle configurations on pavement damage.
- RDF probabilistic modeling of OW traffic would be useful for SHAs as it provides an estimation of pavement damage induced by different OW vehicles having different axle loads and configurations. The RDF methodology presented can be used as the basis for permit fee cost allocation for both single- and multi-trip OW vehicle scenarios.

Acknowledgments

The content of this paper is part of the overall effort in the NDOT P609-13-803.

Author Contributions

The authors confirm contribution to the paper as follows: study conception and design: Elie Y. Hajj, Raj V. Siddharthan, Dario D. Batioja-Alvarez; data collection: Dario D. Batioja-Alvarez, Seyed-Farzan Kazemi; analysis and interpretation of results: Dario D. Batioja-Alvarez, Seyed-Farzan Kazemi; draft manuscript preparation: Dario D. Batioja-Alvarez, Seyed-Farzan Kazemi, Elie Y. Hajj, Adam J.T. Hand, Raj, V. Siddharthan. All authors reviewed the results and approved the final version of the manuscript.

References

1. Mallett, W., B. Sedor, and J. Sedor. *Freight Transportation Improvements and the Economy*. Federal Highway Administration, Washington, D.C., 2004.
2. Ostria, S. J. *Evaluation of U.S. Commercial Motor Carrier Industry Challenges and Opportunities*. Final Report. ICF Consulting, Fairfax, VA, 2003.
3. Bilal, M. K., M. Irfan, A. Ahmed, S. Labi, and K. C. Sinha. *A Synthesis of Overweight Truck Permitting*. Publication FHWA/IN/JTRP-2010/12. Joint Transportation Research Program, Indiana Department of Transportation and Purdue University, West Lafayette, 2010.
4. FHWA. *Oversize/Overweight Load Permits*. FHWA, Office of Freight Management and Operations, Washington, D.C., 2017.
5. Banerjee, A., and J. A. Prozzi. Practical Approach for Determining Permit Fees for Overweight Trucks.

Transportation Research Record: Journal of the Transportation Research Board, 2015. 2478: 93–102.
6. Hajj, E. Y., D. Batioja-Alvarez, and R. Siddharthan. Assessment of Pavement Damage from Bus Rapid Transit: Case Study for Nevada. *Transportation Research Record: Journal of the Transportation Research Board*, 2016. 2591: 70–79.
7. Titi, H., N. Coley, V. Latifi, and M. Matar. Characterization of Overweight Permitted Truck Routes and Loads in Wisconsin. *Transportation Research Record: Journal of the Transportation Research Board*, 2014. 2411: 72–81.
8. Wang, H., J. Zhao, and Z. Wang. Impact of Overweight Traffic on Pavement Life Using Weigh-In-Motion Data and Mechanistic-Empirical Pavement Analysis. *Proc., 9th International Conference on Managing Pavement Assets*, Virginia Tech Transportation Institute, Washington, D.C., 2015.
9. Huang, Y. H. *Pavement Analysis and Design*. Pearson College Division, 2003.
10. Sebaaly, P., R. Siddharthan, and D. Huft. Impact of Heavy Vehicles on Low-Volume Roads. *Transportation Research Record: Journal of the Transportation Research Board*, 2003. 1819: 228–235.
11. Batioja-Alvarez, D., S.-F. Kazemi, E. Y. Hajj, R. V. Siddharthan, and A. J. T. Hand. Probabilistic Mechanistic-Based Pavement Damage Costs for Multi-Trip Overweight Vehicles. *Journal of Transportation Engineering, Part B: Pavements*, Vol. 144, No. 2, 2017, p. 04018004.
12. Coleri, E., and J. T. Harvey. Evaluation of Laboratory, Construction, and Performance Variability by Bootstrapping and Monte Carlo Methods for Rutting Performance Prediction of Heavy Vehicle Simulator Test Sections. *Journal of Transportation Engineering*, Vol. 137, No. 12, 2011, pp. 897–906.
13. Hand, A., and A. Epps. Effects of Test Variability on Mixture Volumetrics and Mix Design Verification. *Asphalt Paving Technology*, Vol. 69, 2000, pp. 635–674.
14. Kazemi, S. F., and Y. Shafahi. An Integrated Model of Parallel Processing and PSO Algorithm For Solving Optimum Highway Alignment Problem. *ECMS*, 2013, pp. 551–557.
15. Balducci, P., J. Stowers, R. Mingo, H. Cohen, and H. Wolff. *Nevada Highway Cost Allocation Study*. Compilation Report, Battelle, Pacific Northwest Division, Richland, Wash., 2009.
16. Malfabon, R., P. Frost, and R. Hurley. *Standard Specifications for Road and Bridge Construction*. Nevada Department of Transportation, 2014.
17. Alavi, M., M. Pouranian, and E. Hajj. Prediction of asphalt pavement temperature profile with finite control volume method. *Transportation Research Record: Journal of the Transportation Research Board*, 2014. 2456: 96–106.
18. ARC. 3D-Move Analysis Software (Version 2.1). 2013.
19. Hajj, E., P. Sebaaly, and P. Nabhan. *Manual for Designing Flexible Pavements in Nevada Using AASHTOWare Pavement-ME Design*. Nevada Department of Transportation, 2015.
20. *NCHRP: Guide for Mechanistic-Empirical Design of New and Rehabilitated Pavement Structures*. Transportation Research Board of the National Academies, Washington, D.C., 2004.

The Standing Committee on Truck Size and Weight (AT055) peer-reviewed this paper (18-05657).

The contents of this paper reflect the views of the authors and do not necessarily reflect the official views and policies of the NDOT.

Article

TRR
JOURNAL OF THE TRANSPORTATION RESEARCH BOARD

Transportation Research Record
2018, Vol. 2672(9) 242–248

DOI: 10.1177/0361198118775869
journals.sagepub.com/home/trr
SAGE

A Comparative Analysis of Truck Parking Travel Diary Data

Caroline Boris[1] and Rebecca Brewster[2]

Abstract

Critical information on the effects that truck parking shortages have on crucial stakeholders, truck drivers in particular, has historically been obtained through survey methods. However, survey methodologies rely heavily on generalizations, respondents' memory, the recency effect in memory recall, or experiences that evoked strong feelings. To investigate detailed truck parking shortage impacts using a method less reliant on historical recollections, the researchers developed and implemented a 14-day truck parking travel diary study. A total of 148 U.S. drivers participated, providing detailed data for 2,035 days of parking activity and 4,763 unique stops. The analysis focused on truck parking topics that have a nexus to previous survey findings that relate to drivers, state departments of transportation and enforcement activities, and other stakeholders, to facilitate the comparison of truck parking survey data with the diary data. Generally, the truck parking diary findings corroborated survey results. Almost 90% of drivers parked in an unauthorized location at least once a week – supporting survey findings that a majority of drivers find parking is difficult at least once a week. Data on unauthorized parking demonstrated that time-of-day can affect the ease of finding available parking, with peak difficulty occurring during evening hours. Drivers primarily choose parking locations for the mandated 10-h hours-of-service breaks based on practical needs – proximity to route, restroom and shower access, and expected parking availability. The research findings can provide policy, regulatory, and infrastructure guidance to both private industry and public agencies.

Truck parking shortages have been documented for over 20 years – in the late 1990s a national survey of truck drivers revealed that commercial drivers had difficulty locating truck parking at least once a week (*1*). The survey results were supported by a supply/demand model that estimated a parking shortage of 52,700 spaces nationally (*2*). Subsequent analyses of truck parking adequacy have shown truck parking shortages persist (*3*, *4*). Additionally, the truck parking shortage has been aggravated by the closure of public rest areas, changes in supply chains and industry operations that have shifted preferred parking locations toward urban areas, and difficulties in building new parking facilities caused by zoning and land values (*5*). Existing truck parking facilities seeking to expand often encounter similar obstacles (*5*).

Recent truck driver surveys have shown a majority of drivers experience difficulty finding parking at least once a week (*5*). Yet access to legal truck parking is critical to roadway safety and ensuring compliance with federal hours-of-service (HOS) regulations. Drivers may continue to drive while fatigued when searching for available parking, or park on roadway shoulders and ramps if parking is not available (*6*). The electronic logging device (ELD) mandate may aggravate existing parking shortage issues, as change-of-duty status will be automatically recorded. If available drive time is completely exhausted, drivers may be forced to decide between violating HOS and parking in an unsafe location.

Truck driver surveys have been the primary method of assessing parking shortage impacts on commercial drivers. Although this approach is effective for collecting general information on truck parking issues, survey responses may be skewed toward recent experiences or experiences in which strong emotional responses occurred. Due to this, survey results may not accurately represent truck parking shortage impacts. Additionally, survey methods may omit causes and nuances of truck parking shortage impacts.

To address the shortcomings of survey data collection methods, the American Transportation Research Institute developed and deployed 14-day truck parking travel diaries. This method collected continuous and

[1]American Transportation Research Institute, Roseville, MN
[2]American Transportation Research Institute, Marietta, GA

Corresponding Author:
Address correspondence to Caroline Boris: cboris@trucking.org

more granular data on the effects of parking shortages on truck drivers. The truck parking travel diary analysis will inform public agencies and private industry of truck driver needs, behaviors, and issues related to parking. This additional, detailed information can guide public agency and private industry policy, and regulatory and infrastructure decisions related to the parking shortage.

Methodology

Driver participants were recruited through satellite radio programs targeted at commercial drivers, the Mid-America Trucking Show, and trucking industry trade press. First, drivers completed a short online survey asking for information on operating region(s), the number of nights spent away from home each week, general information on industry sector, and contact information. Drivers who spent zero nights away from home each week were excluded from consideration, as one goal of the data collection was to better understand parking behavior surrounding 10-h required HOS breaks.

Qualifying drivers were then contacted by email for a shipping address. A package containing the 14-day diary, a short survey on parking and planning behaviors, and a postage-paid envelope were then sent to drivers. An incentive of US$50 was offered to the first 100 drivers to return a completed diary.

The diaries were developed with the goals of gauging the amount of time drivers spend searching for parking, estimating the prevalence of unauthorized truck parking, exploring how reasons for stopping influence location choices, and what factors affect stop location choices. Drivers were asked to provide the following information for up to four stops per day:

- Arrival/departure time,
- Stop reason,
- Location (urban/suburban/rural, city, and state),
- Location type (rest area, ramp/shoulder, truck stop, other),
- Officially designated parking space,
- Number of locations searched, and,
- Total non-productive drive time spent searching for parking.

The diary also requested information specific to the 10-h required HOS break. These questions related to remaining drive time, parking reservations, and the top five factors influencing stop location choice.

A total of 148 drivers returned completed parking diaries between June and September of 2016, representing 2,035 days of parking activity and 4,763 unique stops. Diary data included all states except Hawaii.

Findings

Driver Demographics and Behaviors

Driver demographics were collected to assess the representativeness of the sample compared with the overall industry. The sample underrepresented males (88.4%) relative to the industry-at-large, of which 94.4% are male (*7*). Most participating drivers were 45 to 64 years of age (71.4%) or 25 to 44 years of age (25.9%). Compared with the industry overall, the sample overrepresented drivers 45 to 64 years of age (*8*).

Participants were primarily employee drivers (72.3%) operating in the truckload sector (56.1%) in large fleets of more than 1,000 power units (50.0%) that are paid on a per-mile basis (74.1%). Vehicle configurations operated in the sample were primarily dry van (41.2%), flatbed (30.4%), or refrigerated trailers (16.2%). The majority of participants (91.2%) drove irregular routes. HOS compliance was recorded by ELDs for the majority of drivers in the sample (85.0%).

The diary package sent to drivers also included a survey to gather information on participants' planning and parking practices. Drivers identified parking locations primarily through websites/applications (55.5%) or global positioning systems (53.4%). Nearly half of the drivers had previously used a parking reservation system (45.6%); 7.0% of participants had been ticketed for parking on a road shoulder or ramp.

Free Responses

The supplemental survey in the diary package also included a free response question for additional input from drivers. Responses most frequently referenced the following topic areas:

- Time-of-day impacts on parking availability (61.6%).
- Public rest area challenges were referenced in 24.4% of responses – in particular, the closing of public rest areas and the distance between rest areas. Greater distances between parking facilities are associated with elevated crash risk (*9*). Drivers also noted the conflict between public rest area parking time limits, which typically range from 2 to 10 h (*5*), and HOS compliance.
- Drivers described specific regions where finding available parking is difficult (20.1%).

Unauthorized Parking

Unauthorized parking is defined here as parking in a location where parking is illegal, such as a ramp or road shoulder, or "creating" an ad hoc parking space in a facility where parking may be permitted but not where the parking occurred. Parking in unauthorized locations creates safety risks by exposing parked trucks to risky traffic interactions. Most road shoulder crashes are the result of driver inattention, but there is an imperative to prevent these crashes by not parking on a road shoulder (*10*). "Created" parking spaces also pose hazards, as created spaces do not consider the space requirements to safely maneuver a truck, increasing the risk of property damage. Surveys of truck stop operators found that roughly one-half to three-quarters of truck stop operators are aware of unauthorized parking at or near their facility (*5, 11*). Unauthorized parking is primarily the result of no available nearby parking, lack of awareness of available nearby parking, or HOS constraints (*1, 3, 11, 12*).

In surveys, drivers report that HOS compliance and fatigue are top reasons for seeking parking (*13–15*). However, state departments of transportation (DOTs) and state troopers typically approach unauthorized parking as an issue for the rest of the motoring public, without consideration for why commercial drivers park in unauthorized locations. Nearly half of all state DOTs reported problems with trucks parking on freeway interchange ramps (47.1%), freeway shoulders (45.1%), and highway roadsides (35.3%) (*5*). A survey of state troopers found that the majority ask drivers to move if they are parked on a road shoulder or ramp (*11*). Asking drivers to move from unauthorized parking locations will likely present other safety or compliance issues if a driver parked in a location for fatigue-related reasons, or if the driver was out of available HOS.

In the parking diaries, unauthorized parking was identified by respondents when "unauthorized space" was marked or if the location was a road shoulder or ramp. In the truck parking diary dataset, 3% of total parking stops were unauthorized. This may underestimate the total prevalence of unauthorized parking, as identifying unauthorized parking in the diaries often relied on the respondent indicating a space was unauthorized. For example, if a driver parked in a public rest area, but not in a marked space and did not indicate it was an unauthorized location, this instance of unauthorized parking would be missed. In a survey of drivers operating in North Carolina, drivers estimated they parked on road ramps/shoulders for 15.4% of total parking stops in North Carolina (*16*). The disparity between the frequency of unauthorized parking reported in the North Carolina survey and the 3% observed in the diary data is significant. The observed disparity may have a variety of causations, such as the survey drivers overestimated the frequency of unauthorized parking in North Carolina, driver experiences in North Carolina are not characteristic of other regions, or the diary sample underestimated the frequency of unauthorized parking nationally.

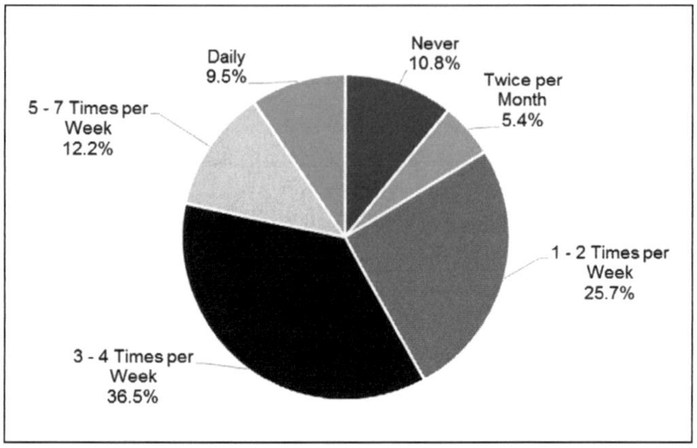

Figure 1. Frequency of drivers parking in an unauthorized location.

Although the percentage of total unauthorized parking stops is low (3%), it is easier to understand the issues drivers face through the frequency that each respondent engaged in unauthorized parking over the 14-day period (Figure 1). Nearly 11% of drivers never parked in an unauthorized location over the 2-week period, possibly due to company policy prohibiting roadside parking or an individual's risk aversion to legal consequences, such as being ticketed or asked to move in conflict with HOS requirements. Drivers most frequently reported parking in an unauthorized location one to two times per week (25.7%) or 3 to 4 times per week (36.5%). Over 20% of drivers parked in an unauthorized location five times or more per week. The frequency that most drivers in the sample utilize unauthorized locations supports survey findings that a majority of drivers experience difficulty locating available parking at least one to two times a week, if not more (*5*).

Unauthorized parking peaked during evening (7:00 p.m. to midnight) and early morning (midnight to 5:00 a.m.) hours. The relationship between time-of-day and parking scarcity is also reflected in truck driver survey results (*5, 16*). Both private truck stop operators and drivers report that time-of-day significantly impacts parking availability; available truck parking is scarce during evening and early morning hours (*5, 16*). The time-of-day impacts draw attention to the importance of addressing truck parking issues at the supply chain level; peak parking demand periods are not solely the result of driver choices and preferences. Parking shortage issues may be improved if alternative solutions, such as more flexibility in HOS regimen or shipper/receiver appointments were possible.

Unauthorized parking also varied by whether a location was urban, suburban, or rural. Drivers parked in an unauthorized location 26.5% of the time in urban locations, 21.3% of the time in suburban locations and 16.5% of the time in rural locations. This reflects survey

Table 1. Driver-Reported Search Times by Region (*13, 16, 17*)

Search time	Kansas	Mid America Association of State Transportation Officials States	North Carolina
Less than 30 minutes	21.9%	39.9%	13.2%
30 minutes to 1 hour	46.6%	42.1%	46.7%
More than 1 hour	31.4%	18.0%	40.2%

findings that locating available parking is more difficult in urban areas (*5, 14*). State DOT representatives have noted that the highest demand for truck parking is typically within a 20-mi radius of urban areas, but increasing urban parking capacity is difficult due to land use issues and the scarcity of land parcels near major business centers. Expanding truck parking availability near residential land use is even more problematic (*5*). The shortage of parking in urban areas likely explains the high prevalence of unauthorized parking.

Overall, unauthorized parking reported in the diaries reinforces driver parking survey findings. The frequency drivers have difficulty locating parking, time-of-day impacts, and issues in urban areas reinforce and provide more detail to driver survey findings.

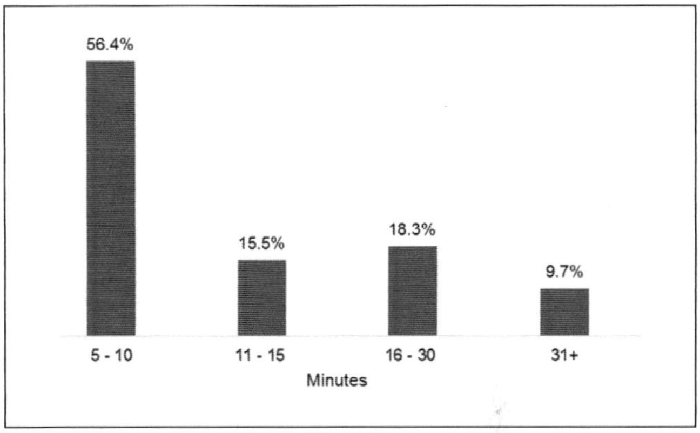

Figure 2. Daily search time.

Productivity Impacts: Search Time

The diaries were designed to collect details on the productivity impacts of truck parking shortages. Numerous surveys have asked drivers about how much time they spend searching for parking. Table 1 displays driver-reported search times in U.S. regions and states. A majority of drivers reported spending at least 30 min searching for parking in surveys of drivers operating in Kansas (80.0%) (*13*), states represented by the Mid America Association of State Transportation Officials (60.2%) (*17*), and North Carolina (86.9%) (*16*). As most drivers in the sample and in the industry are compensated on a per-mile or per-load basis, devoting revenue time to locating available parking has a tangible effect on driver compensation.

However, driver diary entries for the "search time" field – the drive time drivers devoted to locating available parking rather than drive revenue-earning miles – were frequently zero minutes. This disparity suggests that search time may be overestimated in surveys, search time was underestimated in the diaries, or drivers conceptualize the sacrificed drive time differently when asked in the two separate methods. Alternatively, drivers may sacrifice available drive time by stopping early to avoid uncertainty (e.g., a destination has no available parking and they must keep searching with limited HOS). Given the interest drivers have in truck parking – the third highest ranked industry issue by drivers in 2016 (*18*) – it is unlikely that drivers would systematically underestimate the lost productivity resulting from inadequate parking. Therefore, in line with findings from multiple, robust driver surveys, search times of less than 5 min were excluded from consideration here.

Figure 2 displays the total daily search time reported by drivers in the diaries. The search times from the driver diaries are low compared with the typical search times of 30 min or more reported by drivers in the surveys.

Individual parking search times were used to assess location, day of week, and time-of-day relationships to search times. Drivers were more likely to spend 15 min or more searching for parking from 7:00 p.m. to midnight (34.8%) or from 4:00 p.m. to 7:00 p.m. (28.6%). This reflects the findings of other surveys related to time-of-day impacts on parking availability (*5, 16*).

The reason for a stop also affected driver search times. When drivers stopped for 10-h HOS breaks, search times exceeded 15 min 8.6% of the time, whereas search times for other stops exceeded 15 min only 1.3% of the time. Drivers may dedicate more time to find parking for 10-h HOS breaks for a number of reasons, such as finding parking at a facility with specific amenities or to find a parking space with adequate room to easily exit the space after the required 10 h.

Additionally, the willingness to use reserved parking spaces, by typical search times was examined. Drivers willing to reserve parking were more likely to spend more than 15 min searching for parking (16.2%) relative to drivers that do not use reserved parking (11.9%). The observation that parking reservations do not have lower

Table 2. Driver Willingness to Pay for Parking Reservations (*19*)

Search time	Percent of responses
I would not be willing to pay any amount	48.2%
US$1–$5	20.0%
US$6–$10	19.8%
US$11–$15	9.0%
US$16–$20	2.1%
US$21 +	0.9%

search times suggests that parking reservations are used when no other parking is available. Individual driver preferences, such as willingness to park in an unauthorized location if no free parking is available, may also play a significant role in the use of parking reservation systems. Use of parking reservation systems peaked during the hours of 7:00 p.m. to 4:59 a.m., suggesting that the reason for reservation system use relates primarily to the lack of available free parking. In a survey focused on parking reservations (Table 2), 48.2% of drivers were not willing to pay for parking reservations (*19*); 57.1% of drivers indicated they were willing to pay US$1 to US$10 for a reservation, however, this amount is less than the average reservation fee of $12.86 at TA Petro (*19*, *20*). Other surveys have found similar results (*6*).

The search times reported in the driver diaries were generally lower than the typical search time of 30 min or more reported in driver surveys. The reason for this disparity is unknown. However, despite differences in average search times, the diary search time data corroborates survey data in confirming that the time-of-day can affect parking availability, more time is devoted to finding parking for 10-h required HOS breaks, and that reserving parking does not result in lower search times.

Productivity Impacts: Remaining Drive Time

Another approach to assessing parking impacts on roductivity was to examine the amount of remaining revenue drive time in relation to variables of interest (see Figure 3). Drivers park prior to exhausting available drive time if they anticipate difficulty locating available parking if they continue driving. Parking before drive time is exhausted decreases revenue-earning miles, negatively affecting driver compensation. Motor carriers are also adversely affected if trucks do not maximize their revenue-earning miles and may have to hire additional drivers or purchase additional equipment to meet customer demands. Remaining drive time figures do not include stops where the location was marked as a shipper or receiver, as it was assumed drivers had reached their destination if they were parked at a shipper or receiver.

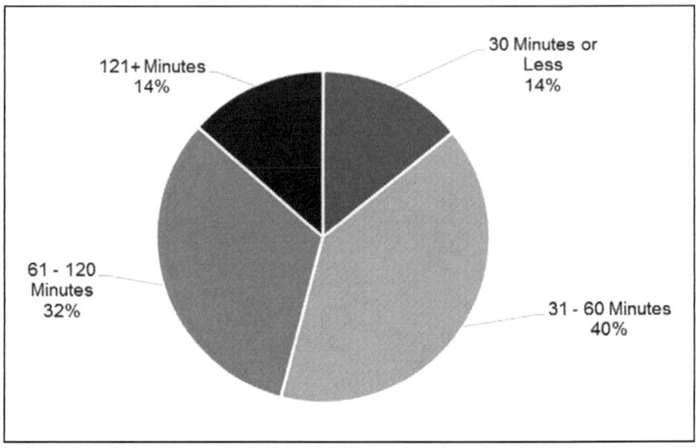

Figure 3. Participant average remaining drive time.

In the diary sample, drivers averaged 31 to 60 min (39.9%) of remaining drive time per day, followed by 61 to 120 min (32.4%) of remaining drive time. A median of 56 min of remaining drive time was observed in the diary sample. Using the median remaining drive time of 56 min per day and assuming an average speed of 39.98 mph and 250 days worked per year, over 9,300 additional revenue-earning miles could be driven per driver per year (*21*). As most drivers are compensated on a per-mile or per-load basis, stopping before available drive time is exhausted has a significant and tangible effect on driver compensation.

Other factors, such as ELD-use and parking reservation system use were investigated in relation to remaining drive time, and no relationships were observed.

Driver Location and Amenity Preferences

Location and amenity preferences for 10-h HOS breaks were also investigated through the diary data. Drivers predominantly spent 10-h HOS breaks at private truck stops (71.4%), followed by public rest areas (9.6%), and customer facilities (8.9%). More amenities and the large supply of private truck stop parking spaces relative to public rest area parking spaces likely explains the prevalence of parking at a private truck stop for the 10-h HOS breaks – private truck stops have an average of 7.63 spaces for every parking space at a public rest area (*5*). Drivers have also indicated a preference for private truck stops for long breaks in numerous surveys (*6*, *22*).

Drivers were asked to indicate what factors were most important in their stop location choice (Table 3). Drivers placed highest priority on proximity to their route (96.5%), restrooms/showers (79.8%), and expected parking availability (75.5%). Most top amenities influencing stop choice are typically available at both private truck stops and public rest areas. Drivers also prioritized parking space width and ease of access (31.9%), and

Table 3. Factors Influencing Stop Location Choice

Important factor	Percent of responses
Proximity to route/destination	96.5%
Restroom/showers	79.8%
Expected parking availability	75.5%
Width of parking space/ease of access	31.9%
Restaurant	30.5%
Security	20.3%
Company policy/loyalty program	18.1%
Internet	6.9%
Laundry	4.0%
Maintenance/service center	3.7%
Weather conditions	3.6%

restaurants (30.5%). Driver amenity preferences for showers and restaurants supplement the supply explanation for why private truck stops are the preferred location for 10-h HOS breaks. Public rest areas are prohibited from offering commercial services (*23*) and parking space design may not be suited to current vehicle sizes/configurations (*24*).

Conclusions

Truck parking travel diaries provide more detailed information about the effects of the truck parking shortage on the trucking industry. The diary data can inform the decisions of public agencies and private industry to address truck parking shortages through numerous areas including policy, regulatory, and infrastructure. For example, states can examine existing caps on how long drivers are allowed to park in public rest areas to more closely align with federal HOS rules requiring that drivers take a 10-h break.

Most diary-derived information corroborated existing knowledge obtained from survey data collection methods. However, unauthorized parking was estimated to occur in 3% of all parking stops in the diaries, which contrasts with an estimated prevalence of 15.4% from a North Carolina driver survey (*16*). This disparity may be attributable to numerous causes. However, the use of survey and diary methods provides a baseline range in the frequency of unauthorized parking.

Most drivers parked in an unauthorized location regularly – three to four times per week (26.5%) or one to two times per week (25.7%). The driver-level prevalence of unauthorized parking demonstrates that difficulty finding safe and legal parking presents an issue for most drivers at least once a week (*5*). However, some drivers were able to avoid unauthorized parking entirely (10.8%). Drivers may have used a variety of methods to avoid having to park in an unauthorized location, such as reserving parking or extensive knowledge of a geographic region.

Unauthorized parking was most prevalent in the late evening (7 p.m. to midnight) hours, followed by early morning (midnight to 5 a.m.), similar to survey results (*5*, *16*). Unauthorized parking was more prevalent in or near highly populated areas – unauthorized parking comprised 26.5% of all parking in urban areas, 21.3% of parking in suburban areas, and 16.5% of all parking in rural areas. This reflects state DOT survey findings that the highest parking demand is within 20 mi of an urban area, but these are also areas that are difficult to develop or expand truck parking in (*5*).

Productivity impacts related to truck parking scarcity were examined in the areas of remaining drive time and search time. Drivers had a median of 56 min of remaining revenue drive time left per day, which represents an additional 9,300 revenue-earning miles that could be driven per year, assuming an average speed of 39.98 mph and 250 days worked per year. This has a tangible impact on driver compensation, as many drivers are compensated on a per-mile or per-load basis.

Productivity with regard to search time, related to multiple survey findings that drivers, on average, spend at least 30 min per day searching for parking in a variety of regions (*14*, *16*, *17*). However, the diaries varied from the surveys in that stops frequently involved no search time. As the survey results are supported by thousands of truck drivers rather than a sample of 148 drivers, researchers focused on non-zero responses. The pattern of greater difficulty finding parking in the evening was reflected in search times peaking in the evening (7 p.m. to midnight) hours (*5*). Drivers who used reservation systems typically spent more time searching for parking than those who did not use reservation systems, suggesting that reservation systems are a "last-ditch" solution when there is no available free parking. Surveys specifically about reservation systems have found drivers are generally unwilling to pay for parking, which is similar to the unwillingness to pay for parking observed in the diary sample (*6*, *19*). Finally, drivers tended to devote the most search time to finding parking for 10-h HOS breaks relative to stops made for other reasons.

The diaries validated the findings of driver surveys through gathering information about amenity and location preferences of drivers when searching for parking, focusing on 10-h HOS breaks in particular. Drivers in the sample predominantly used private truck stops (71.4%) for 10-h HOS breaks. This reflects the composition of facilities at truck stops, which have 7.63 parking spaces for every parking space at a public rest area (*5*). Drivers in the sample primarily parked at locations for practical reasons, such as proximity to route (96.5%), restrooms/showers (79.8%), and expected parking availability

(75.5%). Parking space width and ease of access (31.9%) and restaurants (30.5%) were often noted to be important factors.

Author Contributions

The authors confirm contribution to the paper as follows: study conception and design: R.B., C.B.; data collection: R.B., C.B.; analysis and interpretation of results: R.B., C.B.; draft manuscript preparation: R.B., C.B. All authors reviewed the results and approved the final version of the manuscript.

References

1. *Highway Special Investigation Report Truck Parking Areas*. Report NTSB/SIR-00/01. National Transportation Safety Board, Washington, D.C., 2000. http://ntl.bts.gov/lib/17000/17100/17171/PB2000917001.pdf.
2. Trucking Research Institute, Apogee Research Inc., and Wilbur Smith Associates. *Commercial Driver Rest and Parking Requirements: Making Space for Safety*. FHWA-MC-96-0010. Federal Highway Administration, U.S. Department of Transportation, Washington, D.C., 1996.
3. Fleger, S. A., R. P. Haas, J. W. Trombly, R. H. Cross III, J. E. Noltenius, K. K. Pécheux, and K. J. Chen. *Study of Adequacy of Commercial Truck Parking Facilities*. Technical Report, Report FHWA-RD-01-158, Federal Highway Administration, McLean, VA, 2002.
4. Federal Highway Administration. *Commercial Motor Vehicle Parking Shortage*. Federal Highway Administration, Washington, D.C., 2012.
5. Federal Highway Administration and Department of Transportation. *Jason's Law Truck Parking Survey Results and Comparative Analysis*. http://www.ops.fhwa.dot.gov/freight/infrastructure/truck_parking/jasons_law/truckparkingsurvey/index.htm. Federal Highway Administration, Washington, D.C. 2015.
6. Washington State Department of Transportation. *Washington State Truck Parking Study*. Washington State Department of Transportation, Olympia, WA, 2016.
7. American Trucking Associations. *American Trucking Trends 2016*. American Trucking Associations, Arlington, VA, 2016.
8. Short, J. *White Paper: Analysis of Truck Driver Age Demographics across Two Decades*. American Transportation Research Institute, Arlington, VA, 2014.
9. Banerjee, I., J. Lee, K. Jang, S. Pande, and D. Ragland. *Rest Areas—Reducing Accidents Involving Driver Fatigue*. University of California Berkeley Traffic Safety Center and California Department of Transportation, Berkeley. http://www.dot.ca.gov/newtech/researchreports/reports/2009/2009-05_design_and_construction.pdf. Accessed November 23, 2016.
10. Roberts, G. L., and C. W. Lynn. *Passenger Vehicle Crashes Into Stationary Large Trucks: Incidence and Possible Countermeasures*. Project No.: 63408. Virginia Department of Transportation, Richmond, VA, 2003.
11. *Virginia Truck Parking Study*. Kimley Horn, 2015. http://www.virginiadot.org/projects/resources/VirginiaTruckParkingStudy_FinalReport_July2015.pdf.
12. Trombly, J. W. *NCHRP Synthesis 317 Dealing with Truck Parking Demands: A Synthesis of Highway Practice*. Transportation Research Board, National Academies of Science, Washington, D.C., 2003.
13. American Transportation Research Institute. *I-95 Corridor Coalition Truck Parking Availability Notification System User Requirements/Preferences Survey*. American Transportation Research Institute, Arlington, VA, 2010.
14. Kansas Truck Parking Plan. *Kansas Statewide Freight Network Truck Parking Plan*. Kansas Turnpike Authority, 2016. https://www.ksdot.org/Assets/wwwksdotorg/bureaus/burRail/Rail/Documents/Kansas_Statewide_Freight_Network_Truck_Parking_Plan_2015_2016.pdf.
15. Morris, T., V. Morellas, N. Pananikolopolous, D. Cook, D. Murray, K. Fender, and A. Weber. *A Comprehensive System for Assessing Truck Parking Availability*. University of Minnesota Department of Computer Science, American Transportation Research Institute, Arlington, VA, 2017.
16. American Transportation Research Institute. *North Carolina Truck Parking Survey Analysis*. American Transportation Research Institute, Arlington, VA, 2016.
17. American Transportation Research Institute. *MAASTO Truck Parking Survey Analysis*. American Transportation Research Institute, Arlington, VA, 2016.
18. American Transportation Research Institute. *Critical Issues in the Trucking Industry – 2016*. American Transportation Research Institute, Arlington, VA, 2016.
19. Boris, C. R., and M. A. Johnson. *Managing Critical Truck Parking Tech Memo #1: Commercial Driver Perspectives on Truck Parking*. American Transportation Research Institute, Arlington, VA, 2015.
20. Overdrive. *The Rise of Pay-to-Park*. http://www.overdriveonline.com/the-rise-of-pay-to-park/, Accessed July 12, 2017.
21. Torrey, W. F., and D. Murray. *Analysis of the Operational Costs of Trucking: 2016 Update*. American Transportation Research Institute, Arlington, VA, 2016.
22. Wang, H., and N. J. Garber. *Final report of ITS Center project: Truck Stop Information*. Center for Transportation Studies, University of Virginia, 2003.
23. 23 U.S.C. § 111. https://www.gpo.gov/fdsys/granule/USCODE-2011-title23/USCODE-2011-title23-chap1-sec111.
24. Phelan, T., V. Mantero, J. Purdy, and T. Kearney. National Coalition on Truck Parking: Activity Report, 2015-2016. FHWA-HOP-17-026. https://ops.fhwa.dot.gov/publications/fhwahop17026/index.htm. Federal Highway Administration, Washington, D.C., 2017. Accessed July 21, 2017.

The Standing Committee on Trucking Industry Research (AT060) peer-reviewed this paper (18-00538).

Article

TRR
JOURNAL OF THE TRANSPORTATION RESEARCH BOARD

Transportation Research Record
2018, Vol. 2672(9) 249–260

DOI: 10.1177/0361198118801337
journals.sagepub.com/home/trr
SAGE

Train-Level and Car-Level Modeling of Hazardous Materials Release in Railroad Incidents

Amirfarrokh Iranitalab[1] and Aemal Khattak[1]

Abstract

This paper quantifies the impacts of incident, railroad, environment. and train/car characteristics on the probability of hazardous material (hazmat) release in a hazmat-carrying train incident and provides a prediction tool for hazmat release. Two sets of models utilized the Federal Railroad Administration 2012–2016 rail equipment incident dataset. The units of analyses for these two sets were trains and hazmat cars. Binary logit and binary mixed logit models were investigated to account for hazmat release and potential single-level and two-level grouping in the data (owing to possible hazmat release interdependence among hazmat cars belonging to a train and trains belonging to an incident). Development of receiver operating characteristics curves improved the prediction performance of the models by defining an appropriate cutoff point. Results showed that derailment increased hazmat release probability more than other incident types. Incidents resulting from signal and communication causes were most likely to result in hazmat release. Higher proportion of damaged/derailed hazmat cars and proportion of hazmat cars in a train, track classes 2 and 3, higher train speed, and train gross tonnage were the other important factors. Results of mixed models showed hazmat release from cars belonging to a train were interdependent and hazmat release from trains belonging to an incident were independent. Although models at both levels led to useful results, car-level models had better prediction performance.

North American railroads carry substantial volumes of hazardous materials (hazmat), exceeding 2.6 million carloads per year in 2016 (*1*). Rail transportation of hazmat exposes people and the environment in proximity of the tracks to the potentially severe consequences of a hazmat release. Identification of operational, environmental, and technical factors affecting the probability of hazmat release from trains is important for making decisions toward decreasing the probability of hazmat release in train incidents; it can also be useful in risk-based methods designed to improve the safety of rail transportation of hazmat.

This paper presents analysis of hazmat-carrying train incidents to fulfill two objectives: (1) quantifying the impacts of incident, railroad, environment, and train/car characteristics on conditional probability of hazmat release (given a train incident), and (2) developing a prediction tool for this conditional probability. This research considered two sets of models; trains were the unit of analysis for the first set of models, and hazmat cars were the unit of analysis for the second set. For both sets, binary logit models and binary mixed logit models were estimated using the Federal Railroad Administration (FRA) 2012–2016 rail equipment incident dataset. Single-level and two-level groupings in the train-level and hazmat car-level models (owing to possible hazmat release interdependence among cars belonging to a train and trains belonging to an incident) were considered, and significant factors associated with hazmat release identified. Moreover, receiver operating characteristics (ROC) curves were developed to improve the prediction performance of the models, by defining an appropriate cutoff point.

A review of pertinent literature on safety aspects of rail transportation of hazmat follows this introduction. The next section introduces the statistical methods used in this paper. The ensuing section introduces the dataset and variables, followed by the estimation results of the models and interpretation. The last section presents the conclusions of this paper.

[1]Department of Civil Engineering and Nebraska Transportation Center, University of Nebraska-Lincoln, Lincoln, NE

Corresponding Author:
Address correspondence to Aemal Khattak: khattak@unl.edu

Literature Review

In rail transportation of hazmat literature the usual definition of risk is the multiplication of a set of components that include occurrence probability of a hazmat train incident, traffic (e.g., train-miles or ton-miles), measures for exposure to danger (population, environment, etc.), probability of hazmat release in an incident, and consequences of a release. Many studies focused on quantification of these components or utilizing risk in frameworks. Some of the proposed countermeasures regarding hazmat train risk reduction include tank car safety design enhancement (*2–5*), alternatives for hazmat train routing (*6*, *7*), train speed reduction (*8*, *9*), and emergency response improvement (*6*, *10*).

Zografos and Androutsopoulos presented a system for assessing alternative routes for hazmat shipment in terms of travel time, risk, and evacuation implications (*6*). They used a logit model to estimate the relative probabilities of occurrence of hazmat accidents throughout the network. Glickman et al. also considered an alternate routing strategy based on accident risk (*7*). Rail transport risk was quantified by estimating the expected population that resides within a radius of the location of a train accident, and then a weighted combination of cost and risk was used to generate alternate routes.

Saccomanno and El-hage developed models for predicting derailments using the position of tank cars carrying hazmat in a train, operating speed, the cause of derailment, and the number of tank cars following the point of derailment (residual train length) (*11*, *12*). Liu et al. analyzed severity of tank car derailments in freight-train accidents, in terms of number of derailed tank cars using regression models (*13*). They reported residual train length, derailment speed, distribution of train power, and proportion of loaded cars affecting derailment severity. Liu et al. evaluated the effects of railroad track characteristics and upgrades on derailments (*14*). Using track class as a proxy for track quality, they reported a higher likelihood of track-related derailments on lower track classes. Bearing failures resulted in less severe derailments but broken rails or welds led to more severe consequences. Speed affected the average number of derailed cars and track upgradation reduced track-related derailment risk.

Liu et al. developed an optimization method combining two risk-reduction strategies: broken rail prevention and tank car safety design (*5*). They estimated regression models for broken-rail-caused derailment rate and the probability that a derailment is caused by broken rails. Barkan et al. identified the speed of derailment and number of derailed tank cars as proxy variables highly associated with hazmat release given a train accident (*15*). Treichel et al. estimated probabilities of lading loss (given derailment) for a variety of tank car specifications (*16*). Head shield type, head thickness, tank insulation, shell thickness, tank car pressure, and yard/mainline affected probability of release from head, shell, top fittings, and bottom fittings of tank cars.

Besides use for modeling hazmat release in several studies (*6*, *16*), a common use of this study's methods in transportation safety is in crash severity modeling in which the crash severity variable is categorical with two levels of severity (*17*, *18*), or multiple levels of severity (*19*, *20*). Logit models are popular for modeling crash severity (*19–21*), as are mixed logit models for addressing different types of correlations in crash data (*20*, *22*, *23*).

Methodology

There is a possibility of interdependence in hazmat release from trains involved in an incident and hazmat cars that belong to a train, which may lead to a single-level correlation in the train-level models and a two-level correlation in the car-level models. Binary logit models do not consider multi-level correlation among observations but mixed binary logit models have the capability of addressing it. Therefore, this study took both binary logit and mixed binary logit models into account.

The response variable in the train-level binary logit and mixed binary logit models was a variable that indicated the occurrence of hazmat release from a train, given an incident. The response variable in the car-level binary logit and mixed binary logit models was the occurrence of hazmat release from a hazmat car (including tank cars, covered hoppers, gondolas, etc.), given an incident. The explanatory variables for the two sets of models included train/car, railroad, operation, environment, and incident characteristics.

Binary Logit Models

The binary logit model is a special case of multinomial logit model, which is commonly used for modeling choice behavior as a categorical outcome in economics and transportation (*24*). It is also useful for modeling non-choice categorical response variables, such as traffic crash severity (*19*), hit-and-run crash occurrence (*25*), and real-time crash occurrence (*26*). Assuming the nth observation with a categorical outcome i, the utility function for this outcome is:

$$U_{in} = V_{in} + \varepsilon_{in} = \alpha_i + \beta_i X_{in} + \varepsilon_{in} \quad (1)$$

In this equation V_{in} is the observable part of the utility function for nth observation with a categorical outcome i, α_i is a constant parameter, β_i is a vector of model parameters, X_{in} is a vector of observable factors that influence the outcome, and ε_{in} is an error term that accounts for unobserved effects. An assumption of the error terms independent and identically distributed with a generalized extreme value distribution (*27*, *28*) results in the multinomial model:

$$P_n(i) = \frac{e^{V_{in}}}{\sum_j e^{V_{jn}}} = \frac{e^{\alpha_i + \beta_i X_{in}}}{\sum_j e^{\alpha_j + \beta_j X_{jn}}} \tag{2}$$

In this equation $P_n(i)$ is the probability of occurrence of outcome i for observation n and X_{in} is the vector of explanatory variables. In case of binary outcome, the multinomial logit model turns into binary logit model. The probability of occurrence of each outcome, i and j, for the nth observation is:

$$P_n(i) = \frac{e^{\alpha_i + \beta_i X_{in}}}{e^{\alpha_i + \beta_i X_{in}} + e^{\alpha_j + \beta_j X_{jn}}}, P_n(j) = 1 - P_n(i) \tag{3}$$

Maximum likelihood (ML) is the common estimation method, but rarity of either outcome of a response variable may bias this estimator. Where appropriate, this paper utilized the bias-reduction method developed by Firth (*29*) because of the relative paucity of hazmat release cases in the dataset (denoted by FE).

Binary Mixed Logit Models

A binary logit model assumes uncorrelated observations in the dataset. However, single-level or a multi-level grouping might exist in a dataset causing correlation among the observations. As is shown in Figure 1, a single-level grouping existed in the train-level models, while a two-level grouping was present in the car-level models. Neglecting these correlations might result in smaller estimated variances, leading to model misinterpretation (*30*). Generalized linear mixed models (in this study binary mixed logit model) relax the uncorrelated observation assumption by inclusion of random effects in models. In a binary mixed logit model, Equation 4 defines the observable part of the utility function. In this equation, a_{in} and b_{in} are the vectors of random parameters (respectively for the model constant and explanatory variables) for outcome i and observation n and they are assumed to follow a normal distribution with mean of 0 and unknown variance (which will be estimated). A consideration of random parameters for all explanatory variables or the model constant is not necessary:

$$V_{in} = \alpha_i + \beta_i X_{in} + a_{in} + b_{in} X_{in} \tag{4}$$

This study utilized the Gaussian quadrature to estimate the binary mixed logit models. Interested readers may refer to Bilder and Loughin, and McCulloch and Neuhaus (*30*, *31*) for more details.

Interpretation Tools

This study utilized the odds ratio for model interpretation (odds were obtained by dividing probability of hazmat release by probability of no release). For a c-unit increase in a continuous explanatory variable, x, Equation 5 gives the odds ratio. The interpretation is that "the odds of hazmat release change by OR times for every c-unit increase in x, holding other variables constant." If x is a categorical explanatory variable, the value of c is 1, and the interpretation changes to "the odds of hazmat release change by OR times as x changes from 0 to 1, holding other variables constant" (*30*):

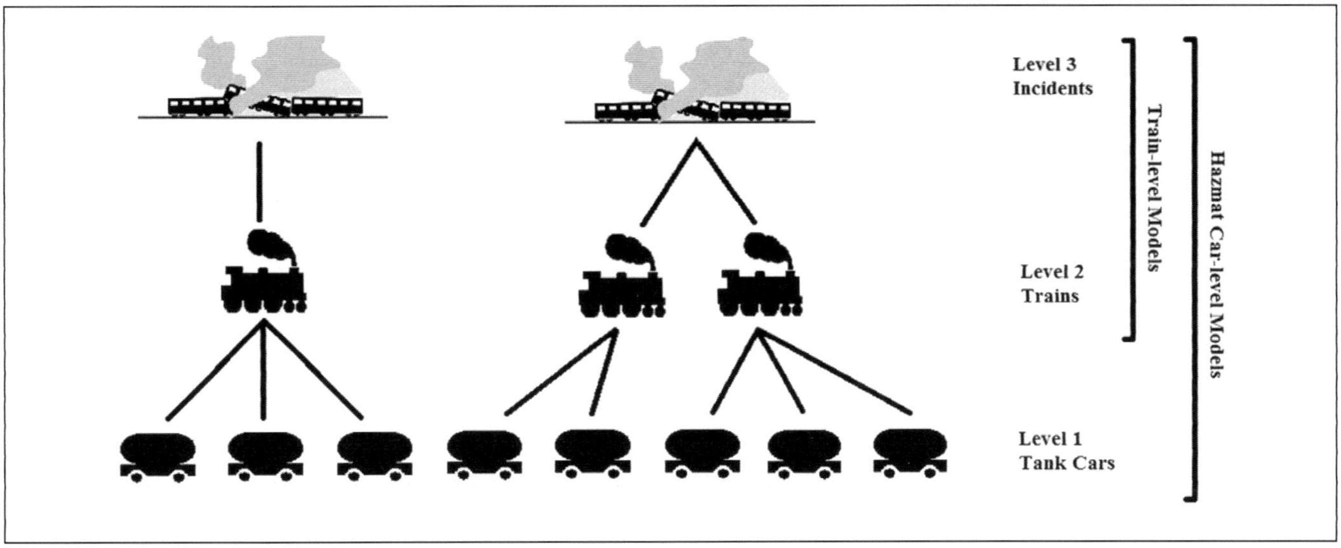

Figure 1. Multi-level structure of the mixed binary logit models.
Note: Icons were obtained from www.iconfinder.com, www.clipartfest.com and www.thenounproject.com.

$$OR = \frac{Odds_{x+c}}{Odds_x} = e^{c\beta_i} \quad (5)$$

Likelihood ratio (LR) test and Akaike information criteria (AIC) were the basis for variable selection, and profile LR confidence intervals were calculated for interpreting the odds ratios (*30*). An estimated odds ratios confidence interval that includes 1.0 for an explanatory variable indicates lack of sufficient evidence toward a significant effect on the response variable. On the contrary, exclusion of 1.0 in the confidence interval is indicative of an effect on the response variable.

Prediction and ROC Curves

The binary logit and binary mixed logit models can serve as prediction tools. ROC curves in this study aided visualizing, organizing, and selecting prediction models based on their performance. In the context of this paper, the definition of "sensitivity" was the proportion of actual incidents with hazmat release correctly predicted, whereas the "false positive rate" was actual incidents without hazmat release incorrectly predicted as hazmat release. "Cutoff probabilities" is the threshold for the estimated probabilities the model uses to predict "release" or "no release" for each incident. ROC curves plot sensitivity versus false positive rate for various cutoff probabilities used with a prediction model. These curves are helpful in choosing a cutoff probability that corresponds to an appropriate level of sensitivity and false positive rate. The area under these curves is a general criterion for performance evaluation in prediction (*32*).

Data and Variables

All railroad reported incidents including derailments, collisions, crossing incidents, and other incidents involving a hazmat-carrying train were extracted from the

Table 1. Statistics for Train-Level Variables

Variable	Variable name	Values and statistics
Response variable		
Hazmat release	hazrel	0 = No (96.73%), 1 = Yes (3.26%)
Explanatory variables		
Incident characteristics		
Type of incident	typinc	1 = Derailment (62.68%), 2 = Collision (12.77%), 3 = Crossing (8.50%), Others (16.04%)
Proportion of damaged/derailed hazmat cars to all hazmat cars	hazdamrate	Mean = 0.2504, Variance = 0.1540
Locomotive(s) derailed	locder	0 = No (91.96%), 1 = Yes (8.04%)
Proportion of damaged/derailed cars to all cars	derrate	Mean = 0.0989, Variance = 0.0331
Cause of incident	cause	E = Mechanical and Electrical Failures (12.16%), H = Human Factors (39.25%), M = Miscellaneous (20.99%), S= Signal and Communication (3.01%), T= Track, Roadbed and Structures (24.58%)
Railroad characteristics		
Type of railroad (Interstate Commerce Commission)	typrr	1 = Class I (83.05%), 2 = Class II (0.90%), 3 = Class III (16.05%)
Method of operation	mopera	1 = Signal indication (24.26%), 2 = Direct train control (6.71%), 3 = Yard/restricted limits (2.08%), 4 = Block register territory (0.47%), 5 = Other than main track rules (66.49%)
Track class	trkcls	1 = Classes 1 and X (67.60%), 2 = Class 2 (7.61%), 3 = Class 3 (6.71%), 4 = Class 4 (14.46%), 5 = Classes 5 to 9 (3.62%)
Type of track	typtrk	1 = Main (32.44%), 2 = Yard (59.78%), 3 = Siding (2.37%), 4 = Industry (5.42%)
Environmental characteristics		
Temperature	temp	Mean = 58.62, Variance = 496.55
Visibility	visibility	1 = Dawn (7.86%), 2 = Day (42.59%), 3 = Dusk (7.39%), 4 = Dark (42.16%)
Weather	weather	1 = Clear (66.49%), 2 = Cloudy (22.53%), 3 = Rain (7.14%), 4 = Fog (1.15%), 5 = Sleet (0.25%), 6 = Snow (2.44%)
Train characteristics		
Train speed (mph)	trnspd	Mean = 12.37, Variance = 211.40
Train gross tonnage (ton)	tonnage	Mean = 4404, Variance = 21787396
Proportion of hazmat tank cars to all tank cars	hazcarrate	Mean = 0.2947, Variance = 0.0958
Remote control locomotive	rclmod	0 = No (80.19%), 1 = Yes (19.81%)

Note: Data obtained from the FRA safety database (*33*).

Table 2. Statistics for Hazmat Car-Level Variables

Variable	Variable name	Values and statistics
Response variable		
Hazmat release	hazrel	0 = No (99.38%), 1 = Yes (0.62%)
Explanatory variables		
Incident characteristics		
Type of incident	typinc	1 = Derailment (67.46%), 2 = Collision (11.89%), 3 = Crossing (10.14%), Others (10.51%)
Locomotive(s) derailed	locder	0 = No (89.93%), 1 = Yes (10.07%)
Cause of incident	cause	E = Mechanical and Electrical Failures (14.38%), H = Human Factors (35.03%), M = Miscellaneous (20.01%), S= Signal and Communication (1.65%), T= Track, Roadbed and Structures (28.93%)
Railroad characteristics		
Method of operation	mopera	1 = Signal indication (32.67%), 2 = Direct train control (8.25%), 3 = Yard/ restricted limits (2.75%), 4 = Block register territory (0.60%), 5 = Other than main track rules (55.72%)
Track class	trkcls	1 = Classes 1 and X (57.83%), 2 = Class 2 (9.75%), 3 = Class 3 (10.03%), 4 = Class 4 (20.18%), 5 = Classes 5 to 9 (2.21%)
Type of track	typtrk	1 = Main (43.11%), 2 = Yard (48.92%), 3 = Siding (2.80%), 4 = Industry (5.17%)
Environmental characteristics		
Temperature	temp	Mean = 57.30, Variance = 509.56
Visibility	visibility	1 = Dawn (8.15%), 2 = Day (43.68%), 3 = Dusk (6.81%), 4 = Dark (41.37%)
Weather	weather	1 = Clear (65.81%), 2 = Cloudy (22.60%), 3 = Rain (6.84%), 4 = Fog (2.06%), 5 = Sleet (0.25%), 6 = Snow (2.45%)
Train/car characteristics		
Train speed (mph)	trnspd	Mean = 13.89, Variance = 213.38
Tank car tonnage (ton)	cartonnage	Mean = 77.30, Variance = 5356.59
Remote control locomotive	rclmod	0 = No (88.37%), 1 = Yes (11.63%)

Note: Data obtained from the FRA safety database (*33*).

2012–2016 U.S. rail equipment incident database (*33*). The extracted dataset consisted of 2,581 incidents, 2,787 trains, and 39,162 hazmat cars. Car-level data were generated based on the original train-level dataset using information on number of hazmat-carrying cars, and number of cars that released hazmat. Tonnage of train cars was approximated by dividing the gross tonnage of trains (excluding the power units) by the number of cars in each train.

Tables 1 and 2 present the variables of train-level and car-level datasets, respectively. Car-level models did not utilize train-level variables (e.g., *hazdamrate, derrate, typrr, tonnage*, and *hazcarrate*), as their possible impacts on hazmat release were at the train level; train-level models did not utilize the car-level variable *cartonnage*. All the categorical explanatory variables were used in the models as sets of dummy (indicator) variables, with the base level set to the first level (alphabetical order), with the exception of *typinc*, in which the third level (crossing incidents) was chosen as the base level. Track classes 1 and X were aggregated in one level for the variable *trkcls*, as they represented the same maximum speed for freight trains (10 mph). Also, track classes 5–9 were aggregated into one level (they were infrequent in the dataset).

Modeling Results

This section presents the estimation results, models' interpretations, and prediction considerations. Model coding and statistical calculations were in R version 3.4.0 (R Foundation for Statistical Computing).

Train-Level Models

This set of models provides the probability of hazmat release from one or more cars of each train, given an incident. The five binary response models (Table 3) include two binary logit models estimated by ML (A1 and A2), one binary logit model estimated by FE (A3), and two binary mixed logit models (A4 and A5). The LR test provided information on the impacts of each variable on probability of hazmat release and variable selection (along with AIC). A comparison of models A2 and A3 shows a slight difference in the estimations, indicating possible biased estimation of A2; thus, A3 is preferable. The AIC values and the *p*-values of the LR test of the random parameter show that a mixed effects model is not necessary in train-level models. So, A3 was chosen as the best model for further interpretation (in terms of cross-validation, the performance of models was similar).

Table 3. Estimated Train-Level Models

Variable		A1) Binary logit (ML)	A2) Binary logit (ML)	A3) Binary logit (FE)	A4)Binary mixed logit	A5) Binary mixed logit
		LR test *p*-value	LR test *p*-value	LR test *p*-value	LR test *p*-value	LR test *p*-value
typinc		0.01219*	0.00778**	0.00893**	0.01219*	0.00778**
hazdamrate		0.00000***	0.00000***	0.00000***	0.00000***	0.00000***
temp		0.79757	—	—	0.79757	—
visibility		0.62895	—	—	0.62895	—
weather		0.81558	—	—	0.81558	—
trnspd		0.00003***	0.00001***	0.00001***	0.00003***	0.00001***
tonnage		0.00470**	0.00547**	0.00564**	0.00470**	0.00547**
trkcls		0.01460*	0.01326*	0.01483*	0.01460*	0.01326*
typtrk		0.56286	—	—	0.56286	—
locder		0.88518	—	—	0.88518	—
derrate		0.27772	—	—	0.27772	—
hazcarrate		0.00000***	0.00000***	0.00000***	0.00000***	0.00000***
cause		0.00040***	0.00022***	0.00026***	0.00040***	0.00022***
typrr		0.95962	—	—	0.95962	—
rclmod		0.12957	—	—	0.12957	—
mopera		0.99563	0.03931*	0.04360*	0.99563	0.03931*
Random effects	Variables	NA	NA	NA	Intercept	Intercept
	Variance	NA	NA	NA	0.00000	0.00000
	Variance LR test comparison with	NA	NA	NA	Model A1	Model A2
	Variance LR test p-value	NA	NA	NA	0.49962	0.50000
Cross-validation	Release	20.88%	20.88%	20.88%	20.88%	20.88%
	Overall	97.24%	97.00%	97.20%	97.24%	97.20%
AIC		613.47	584.64	585.52	615.5	586.6

Note: NA = not applicable; — = variable not used.
$^{*}p < .05$. $^{**}p < .01$. $^{***}p < .001$.

Table 4. Estimated Coefficients and Odds Ratios for Model A3

	Coefficients		Odds ratios			
Variable	Estimate	Std. error	c	Point estimate	Lower bound of C.I.	Upper bound of C.I.
Intercept	−8.6420	0.9631	NA	NA	NA	NA
typinc1	1.8090	0.5819	1	6.1055	1.9992	24.3916
typinc2	1.4260	0.7245	1	4.1634	0.9403	20.0359
typinc4	1.0590	0.6620	1	2.8822	0.7167	12.1323
hazdamrate	1.9220	0.3043	0.1	1.2119	1.1401	1.2955
trnspd	0.0515	0.0124	5	1.2934	1.1467	1.4924
tonnage	0.0001	0.0000	1000	1.0632	1.0207	1.1049
trkcls2	1.1100	0.4647	1	3.0346	1.1827	7.8901
trkcls3	1.0140	0.5390	1	2.7559	0.9264	8.3848
trkcls4	0.0606	0.6027	1	1.0625	0.2987	3.5781
trkcls5	0.2340	0.7896	1	1.2637	0.2241	5.9801
hazcarrate	2.5200	0.3450	0.1	1.2867	1.2016	1.3918
causeh	1.2340	0.5194	1	3.4358	1.2622	11.6063
causem	1.1060	0.5278	1	3.0208	1.0801	10.0642
causes	2.0230	0.7551	1	7.5599	1.3317	34.6251
casuet	1.7910	0.4570	1	5.9956	2.5679	18.7879
mopera2	−0.0712	0.3487	1	0.9313	0.4435	1.8342
mopera3	0.1922	0.6187	1	1.2119	0.279	3.8557
mopera4	0.4817	0.8202	1	1.6189	0.2403	7.2008
mopera5	−1.4510	0.5132	1	0.2343	0.0819	0.6649

Note: NA = not applicable.

Table 4 presents estimated coefficients and standard errors, the values of *c* for odds ratios, estimated odds ratios and 95% profile LR confidence intervals for odds ratios for model A3. The model interpretation is as follows (all the statements in the next two paragraphs are subject to "95% confidence" and "conditional on keeping all the other variables constant"): Derailment incidents increased the odds of hazmat release by 2.0–24.4 times compared with a crossing incident, whereas sufficient evidence was not available to show that collisions and other types of incidents changed the odds of hazmat release, compared with crossing incidents. A 10% increase in the ratio of damaged or derailed cars resulted in 14–30% increase in the odds of hazmat release, and a 10% increase in the ratio of hazardous materials-carrying cars on a train increased these odds by 20–39%. A 5 mph increase in train speed was associated with 15–49% increase in the odds of hazmat release, and a 1000-ton increase in the gross tonnage of the train resulted in 2–11% increase in these odds.

FRA track class 2 increased the odds of hazmat release (given an incident) by 1.18–7.89 times, compared with FRA track class 1 and X. Other FRA track classes did not show statistically significant evidence of affecting the probability of hazmat release. Incidents caused by track, roadbed and structures, signal and communication, human factors and miscellaneous causes compared with mechanical and electrical issues increased the odds of release by 2.57–18.79 times, 1.33–34.63 times, 1.26–11.61 times, and 1.08–10.06 times, respectively.

Hazmat Car-Level Models

This set of five binary response models provides the probability of hazmat release from each hazmat car, given an incident. A number of cars carried hazmat on each train with some releasing hazmat, leading to a potential two-level grouping in the dataset: hazmat cars belonging to the same train, and hazmat cars that belonged to the same incident from the same or different trains. Table 5 presents the five estimated models including two binary logit models with ML (B1 and B2), two single-level binary mixed logit models (B3 and B5) with different explanatory variables and a 2-level binary mixed logit (B4) to account for the two possible levels of grouping (there is no FE model as it was similar to the ML models).

AIC values and the LR test for variances of the random parameters in the binary mixed logit models showed that grouping in the car level was statistically significant but the incident level grouping could be ignored. All mixed logit models showed the same cross-validation performance and were superior to the binary logit models; Model B5 was selected for interpretation.

Table 6 presents odds ratios and their 95% profile LR confidence intervals for Model B5 along with estimated coefficients, standard errors and the values for the parameter *c*. All interpretations are subject to "95% confidence" and "conditional on keeping all the other variables constant." The model interpretation is as follows: Derailment incidents, collisions, and other types of incident increased the odds of hazmat release from a car

Table 5. Estimated Hazmat Car-Level Models

Models		B1) Binary logit (ML)	B2) Binary logit (ML)	B3) Binary mixed logit	B4) 2-level binary mixed logit	B5) Binary mixed logit
Variable		LR test p-value	LR test p-value	LR test p-value	LR test p-value	LR test p-value
typinc		0.00000***	0.00000***	0.00033***	0.00033***	0.00055***
temp		0.00000***	0.00000***	0.20616	0.20524	0.14155
visibility		0.00002***	0.00001***	0.44398	0.44358	—
weather		0.00804**	0.00433**	0.89123	0.89141	—
trnspd		0.00000***	0.00000***	0.00000***	0.00000***	0.00000***
cartonnage		0.12937	0.06547.	0.89327	0.89327	—
trkclas		0.00006***	0.00000***	0.06485.	0.06488.	0.00001***
typtrk		0.87620	—	0.90482	0.90495	—
locder		0.01132*	0.00718**	0.87437	0.87437	—
cause		0.00000***	0.00000***	0.00000***	0.00000***	0.00000***
rclmod		0.04997*	—	0.04045*	0.04040*	0.10905
mopera		0.38831	—	0.80503	0.80434	—
Random effects	Levels	NA	NA	1 Level (trains)	2 Levels (trains, incidents)	1 Level (trains)
	Variables	NA	NA	Intercept	Intercept	Intercept
	Variance	NA	NA	4.0200	(4.0140, 0.0000)	4.2820
	Variance LR test comparison with	NA	NA	Model B1	Model B3	A binary logit with similar variables
	LRT p-value	NA	NA	0.00000***	0.50000	0.00000 ***
Cross-validation	Release	0.00%	0.00%	4.98%	4.98%	4.98%
	Overall	99.38%	99.38%	99.41%	99.41%	99.41%
AIC		2356.5	2354.8	2101.3	2103.3	2080.3

Note: NA = not applicable.
$^{*}p < .05$. $^{**}p < .01$. $^{***}p < .001$.

Table 6. Estimated Coefficients and Odds Ratios for Model B5

Variable	Coefficients		Odds ratios			
	Estimate	Std. error	c	Point estimate	Lower bound of C.I.	Upper bound of C.I.
Intercept	−11.1806	0.9912	NA	NA	NA	NA
typinc1	2.3070	0.7003	1	10.0447	2.5459	39.6298
typinc2	2.0391	0.8566	1	7.6837	1.4335	41.1857
typinc4	2.0841	0.7568	1	8.0376	1.8235	35.4276
temp	−0.0071	0.0059	10	0.9317	0.8305	1.0452
trnspd	0.0593	0.0145	5	1.3451	1.1671	1.5502
trkcls2	1.6586	0.4454	1	5.2517	2.1937	12.573
trkcls3	1.4134	0.5152	1	4.1098	1.4973	11.281
trkcls4	0.5925	0.5848	1	1.8085	0.5748	5.6904
trkcls5	0.4597	0.9016	1	1.5836	0.2705	9.2701
causeh	1.5382	0.5839	1	4.656	1.4825	14.6227
causem	1.5662	0.6045	1	4.7885	1.4644	15.6576
causes	3.3066	0.8892	1	27.2922	4.7768	155.934
casuet	2.0961	0.5125	1	8.1344	2.9793	22.2088
rclmod	0.7857	0.4812	1	2.1939	0.8543	5.634

Note: NA = not applicable.

by 2.55–39.63 times, 1.43–41.19 times, and 1.82–35.43 times, respectively, compared with a crossing incident. The odds of release given an incident also increased by 17–55% for each 5 mph increase in train speed. FRA track classes 2 and 3 increased the odds of hazmat release by 2.19–12.57 times and 1.50–11.28 times compared with FRA tack class 1 and X, respectively. Incidents caused by track, roadbed and structures, signal and communication, human factors and miscellaneous compared with mechanical and electrical issues increased the odds of release by amounts between 2.97–22.21 times, 4.78–155.93 times, 1.48–14.62 times, and 1.46–15.66 times, respectively.

Prediction

For both train-level and hazmat car-level models, the datasets were randomly divided to model estimation (80%) and model validation (20%) subsets. Models A3 and B5 were re-estimated using the estimation data subsets. Figure 2 presents the ROC curves for both models along with sensitivity versus cutoff probabilities plots. The larger area under ROC curves showed that the car-level model provided better overall predictions.

To achieve higher sensitivity without a large increase in false positive rates, using the ROC curves and sensitivity versus cutoff probabilities plots, new cutoff points were selected as 0.05 and 0.01 for the train-level and car-level models, respectively. Based on these curves, the value of 0.05 relative to the default value of 0.5 corresponds to increase in sensitivity and false positive rate by approximately 50% and 15% in the train-level model. These values for choosing 0.01 over 0.5 for the hazmat car-level model were approximately 90% and 10%. Table 7 shows the prediction results of both models for the validation subsets for the cutoff point of 0.5 and the new cutoff points of 0.05 and 0.01.

The results showed that both models, with the new cutoff probabilities, correctly predicted approximately 80% of hazmat release occurrences. However, the proportion of "no release" incidents predicted as "release" was twice as large for the train-level models. Overall, the new cutoff probabilities improved the prediction performance of the models, and the car-level models were preferred over train-level models.

Conclusions

This research showed that derailment type incidents increased the likelihood of hazmat release more than the other types. This finding strengthens the existing emphasis of researchers and policy makers on derailment incidents regarding safety improvement of hazmat shipment by rail. Higher proportion of damaged/derailed hazmat cars in a train increased hazmat release probability, emphasizing the need and use of countermeasures aimed at decreasing the number of damaged/derailed cars in incidents. Although all causes of incidents increased hazmat release probability, relative to the base level (mechanical and electrical failures), prioritization of the corresponding countermeasures and policies is suggested to be based on this descending order: signal and communication; track, roadbed and structures; human factors; and miscellaneous.

Track class 2 in both models and track class 3 in the car-level models were associated with higher probability

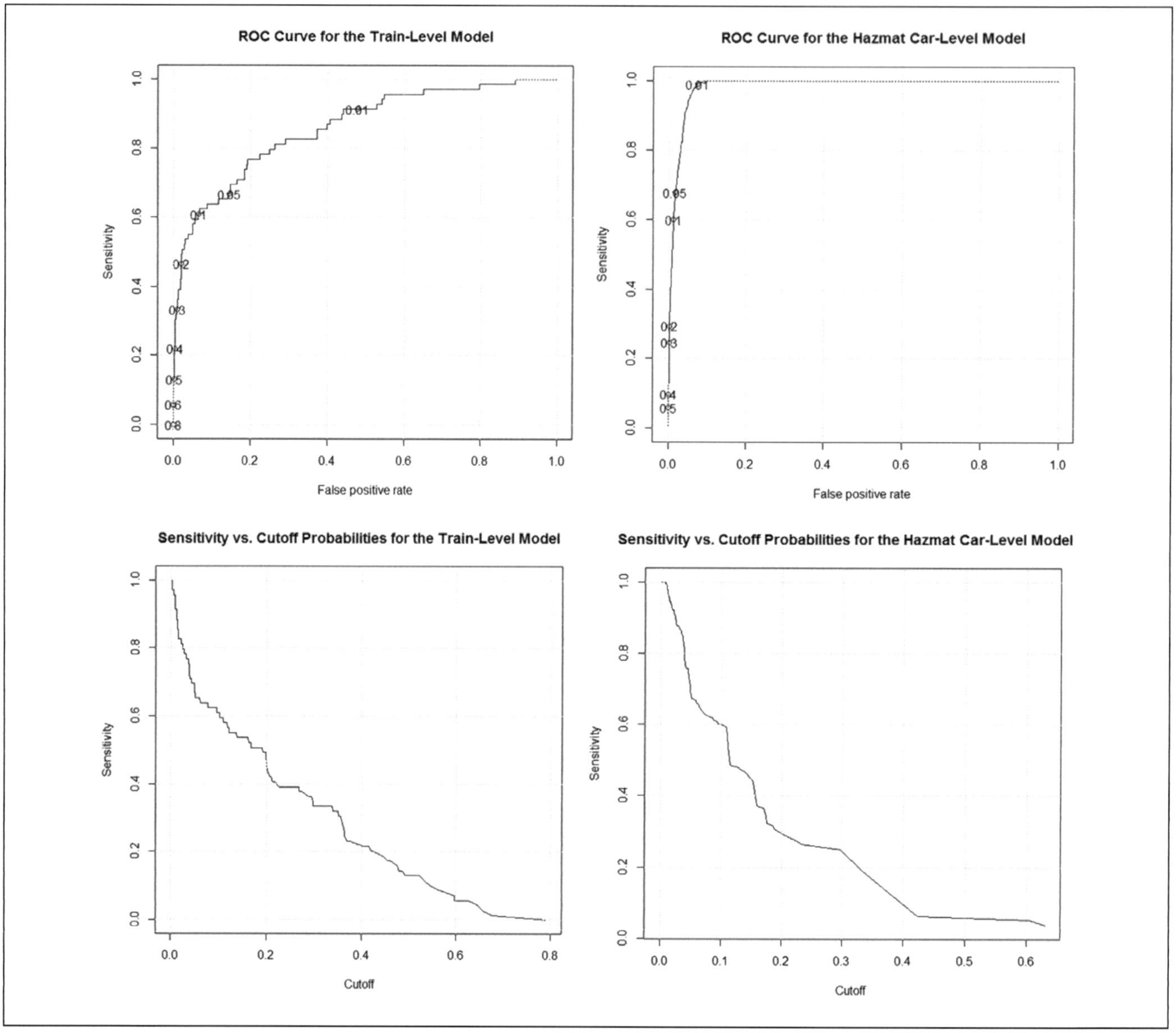

Figure 2. ROC curves and sensitivity vs. cutoff probabilities plots for train-level and hazmat car-level models.

Table 7. Prediction Results of Train-Level and Hazmat Car-Level Models

	Train-level model				Hazmat car-level model			
Cutoff probability	0.5		0.05		0.5		0.01	
	No release	Release	No release	Release	No release	Release	No release	Release
No release	534	1	456	79	7779	1	7206	574
Release	15	7	5	17	50	2	11	41
No release	99.81%	0.19%	85.23%	14.77%	99.99%	0.01%	92.62%	7.38%
Release	68.18%	31.82%	22.73%	77.27%	96.15%	3.85%	21.15%	78.85%
Overall	97.13%		84.92%		99.35%		92.53%	

of hazmat release. This may be a consideration in routing of hazmat-carrying trains. In the train-level model, other than main track rules as a method of operation was associated with a decrease in hazmat release probability, relative to signal indication. Statistically significant evidence was not available with respect to the effects of environmental characteristics on hazmat release probability. These characteristics might have indirect effects captured in the model through other variables and may be assessed in a future study. Higher train speed, train gross tonnage, and proportion of hazmat tank cars on trains increased the hazmat release probability. These variables are useful in developing policies aimed at requiring railroad companies to decrease train speed, gross tonnage, and proportion of hazmat cars in hazmat-carrying trains.

The results of mixed models showed hazmat release from hazmat cars belonging to a train were interdependent and hazmat release from trains belonging to an incident were independent. Analyzing the incidents in train-level and car-level gave relatively consistent results. The train-level model accounted for variables such as *hazdamrate, derrate*, and *hazcarrate*, leading to useful insights. The car-level model captured the effect of the number of cars released by having cars as units of analysis. Characteristics of hazmat cars were not available in the FRA dataset and therefore not considered in this study (which is a limitation of this study). In addition, the car-level model had better performance in prediction, suggesting the implementation of risk-based analyses at the car level.

As reported in the literature review, Liu et al. found track-related derailments more likely on lower class tracks (*14*). They also reported higher speed increased the average number of cars derailed in an incident, while Barkan et al. (*15*) reported that train speed increased the probability of hazmat release. These results are consistent with the findings of this study (reported here). Barkan et al. also identified the incident cause "broken rails or welds" as the most significant incident cause on the number of derailed cars (*15*). Although, the current study accounted for grouped incident causes (because of limitations of the number of variables usable in the models), the group that contained this incident cause (Track, Roadbed and Structures) increased hazmat release probability significantly. Other causes that were in the same group had different effects on release probability in (*15*), which could not be addressed in the current study.

Future research may investigate the effects of other variables, such as hazmat car specification and safety design, and type of hazmat on the hazmat release probability. The effects of incident causes on hazmat release at a more detailed level in train incidents can be the emphasis of a future study. Future research may evaluate other techniques, such as machine learning methods, and compare the results with those reported here.

Author Contributions

The authors confirm contributions to the paper as follows: study conception and design: AI and AK; data collection: AI; analysis and interpretation of results: AI and AK; manuscript preparation: AI and AK. All authors reviewed the results and approved the final version of the manuscript.

References

1. Association of American Railroads: Bureau of Explosives. *Annual Report of Hazardous Materials Transported by Rail: 2016*. Washington, D.C., 2017.
2. Saat, M. R., and C. P. L. Barkan. Generalized Railway Tank Car Safety Design Optimization for Hazardous Materials Transport: Addressing the Trade-Off between Transportation Efficiency and Safety. *Journal of Hazardous Materials*, Vol. 189, No. 1, 2011, pp. 62–68.
3. Barkan, C. P. L., S. V. Ukkusuri, and S. T. Waller. Optimizing the Design of Railway Tank Cars to Minimize Accident-Caused Releases. *Computers & Operations Research*, Vol. 34, No. 5, 2007, pp. 1266–1286.
4. Barkan, C. P. L. Improving the Design of Higher-Capacity Railway Tank Cars for Hazardous Materials Transport: Optimizing the Trade-Off between Weight and Safety. *Journal of Hazardous Materials*, Vol. 160, No. 1, 2008, pp. 122–134.
5. Liu, X., M. R. Saat, and C. P. L. Barkan. Integrated Risk Reduction Framework to Improve Railway Hazardous Materials Transportation Safety. *Journal of Hazardous Materials*, Vol. 260, 2013, pp. 131–140.
6. Zografos, K. G., and K. N. Androutsopoulos. A Decision Support System for Integrated Hazardous Materials Routing and Emergency Response Decisions. *Transportation Research Part C: Emerging Technologies*, Vol. 16, No. 6, 2008, pp. 684–703.
7. Glickman, T. S., E. Erkut, and M. S. Zschocke. The Cost and Risk Impacts of Rerouting Railroad Shipments of Hazardous Materials. *Accident Analysis and Prevention*, Vol. 39, No. 5, 2007, pp. 1015–1025.
8. Kawprasert, A., and C. Barkan. Effect of Train Speed on Risk Analysis of Transporting Hazardous Materials by Rail. *Transportation Research Record: Journal of the Transportation Research Board*, 2010. 2159: 59–68.
9. Nayak, P. R., D. B. Rosenfield, J. H. Hagopian, A. D. Lillie, and A. Park. *Event Probabilities and Impact Zones for Hazardous Materials Accidents on Railroads*. Report No. DOT-TSC-FRA-83-5. Federal Railroad Administration, Washington, D.C., 1983.
10. Berman, O., V. Verter, and B. Y. Kara. Designing Emergency Response Networks for Hazardous Materials Transportation. *Computers & Operations Research*, Vol. 34, No. 5, 2007, pp. 1374–1388.
11. Saccomanno, F. F., and S. M. El-Hage. Minimizing Derailments of Railcars Carrying Dangerous Commodities through Effective Marshaling Strategies. *Transportation*

Research Record: Journal of the Transportation Research Board, 1989. 1245: 34–51.
12. Saccomanno, F. F., and S. M. El-Hage. Establishing Derailment Profiles by Position for Corridor Shipments of Dangerous Goods. *Canadian Journal of Civil Engineering*, Vol. 18, No. 1, 1991, pp. 67–75.
13. Liu, X., M. R. Saat, X. Qin, and C. P. L. Barkan. Analysis of U.S. Freight-Train Derailment Severity using Zero-Truncated Negative Binomial Regression and Quantile Regression. *Accident Analysis and Prevention*, Vol. 59, 2013, pp. 87–93.
14. Liu, X., C. Barkan, and M. Saat. Analysis of Derailments by Accident Cause: Evaluating Railroad Track Upgrades to Reduce Transportation Risk. *Transportation Research Record: Journal of the Transportation Research Board*, 2011. 2261: 178–185.
15. Barkan, C. P. L., C. T. Dick, and R. T. Anderson. Railroad Derailment Factors Affecting Hazardous Materials Transportation Risk. *Transportation Research Record: Journal of the Transportation Research Board*, 2003. 1825: 64–74.
16. Treichel, T. T., J. P. Hughes, C. P. L. Barkan, R. D. Sims, E. A. Phillips, M. R. Saat, Y. K. Wen, and D. G. Simpson. Safety Performance of Tank Cars in Accidents: Probability of Lading Loss. *RSI-AAR Railroad Tank Car Safety Research and Test Project*. Association of American Railroads, Washington, D.C., 2006.
17. Khattak, A., P. Kantor, and F. Council. Role of Adverse Weather in Key Crash Types on Limited-Access: Roadways Implications for Advanced Weather Systems. *Transportation Research Record: Journal of the Transportation Research Board*, 1998. 1621: 10–19.
18. Zhao, S., A. Iranitalab, and A. Khattak. *Report No. 16-3697: Investigation of Pedestrian Injury Severity in Crashes at Highway-Rail Grade Crossings using Latent Class Analysis*. Transportation Research Board of the National Academies, Washington, D.C., 2016.
19. Iranitalab, A., and A. Khattak. Comparison of Four Statistical and Machine Learning Methods for Crash Severity Prediction. *Accident Analysis & Prevention*, Vol. 108, 2017, pp. 27–36.
20. Ye, F., and D. Lord. Comparing Three Commonly Used Crash Severity Models on Sample Size Requirements: Multinomial Logit, Ordered Probit and Mixed Logit Models. *Analytic Methods in Accident Research*, Vol. 1, 2014, pp. 72–85.
21. Shibata, A., and K. Fukuda. Risk Factors of Fatality in Motor Vehicle Traffic Accidents. *Accident Analysis & Prevention*, Vol. 26, No. 3, 1994, pp. 391–397.
22. Zhao, S., and A. Khattak. Motor Vehicle Drivers' Injuries in Train–Motor Vehicle Crashes. *Accident Analysis & Prevention*, Vol. 74, 2015, pp. 162–168.
23. Yasmin, S., and N. Eluru. Evaluating Alternate Discrete Outcome Frameworks for Modeling Crash Injury Severity. *Accident Analysis and Prevention*, Vol. 59, 2013, pp. 506–21.
24. McFadden, D. Econometric Models for Probabilistic Choice among Products. *The Journal of Business*, Vol. 53, No. 3, 1980, pp. S13–S29.
25. Xie, M., W. Cheng, G. S. Gill, J. Zhou, X. Jia, and S. Choi. Investigation of Hit-and-Run Crash Occurrence and Severity using Real-Time Loop Detector Data and Hierarchical Bayesian Binary Logit Model with Random Effects. *Traffic Injury Prevention*, Vol. 19, No. 2, 2018, pp. 207–213.
26. Huang, Z., Z. Gao, R. Yu, X. Wang, and K. Yang. Utilizing Latent Class Logit Model to Predict Crash. *Proc., 16th International Conference on Computer and Information Science (ICIS)*, Wuhan, China, IEEE, New York, NY, 2017.
27. Train, K. *Discrete Choice Methods with Simulation*. Cambridge University Press, Cambridge, UK, 2002.
28. Ben-Akiva, M. E., and S. R. Lerman. *Discrete Choice Analysis: Theory and Application to Travel Demand*. MIT Press, Cambridge, MA, 1985.
29. Firth, D. Bias Reduction of Maximum Likelihood Estimates. *Biometrika*, Vol. 80, No. 1, 1993, pp. 27–38.
30. Bilder, C. R., and T. M. Loughin. *Analysis of Categorical Data with R*. CRC Press, Boca Raton, FL, 2014.
31. McCulloch, C. E., and J. M. Neuhaus. *Generalized Linear Mixed Models*. Wiley Online Library, 2005. https://doi.org/10.1002/0470011815.b2a10021.
32. Fawcett, T. An Introduction to ROC Analysis. *Pattern Recognition Letters*, Vol. 27, No. 8, 2006, pp. 861–874.
33. Federal Railroad Adminsitration Office of Safety Analysis. *Railroad Safety Data*. http://safetydata.fra.dot.gov/OfficeofSafety/publicsite/downloads/downloads.aspx. Accessed May 1, 2017.

The Standing Committee on Transportation of Hazardous Materials (AT040) peer-reviewed this paper (18-00246).

TRR
JOURNAL OF THE TRANSPORTATION RESEARCH BOARD

Article

Transportation Research Record
2018, Vol. 2672(9) 261–272

DOI: 10.1177/0361198118801350
journals.sagepub.com/home/trr
SAGE

Developing an Empirical Pipeline and Rail Crude Oil Mode Split and Route Assignment Model

Adam Morrison[1], Chris Bachmann[1], and Frank Saccomanno[1]

Abstract

In recent years, improvements in pipeline capacities and connectivity have been inhibited by environmental and political concerns (e.g., the Dakota and Keystone XL pipeline expansion projects). This has resulted in a greater dependency on rail transport, and this modal shift of crude oil from pipeline to rail is likely to become more pronounced in the future. Therefore, there is a need to assess the impacts of future changes in pipeline/rail network connectivity, modal attributes, and shipment protocols, on the expected pattern of crude oil shipments. This paper presents a rule-based mode split and route assignment model that reflects real-world allocation, assignment, and apportionment rules. The decision-making process underlying this model is the shipper, who is prioritized by the carrier when there is limited pipeline capacity available. The proposed algorithm allows for the inclusion of crude oil shipments and the pipeline network into conventional freight demand models by capturing the complex interactions of crude shippers, pipeline carriers, and regulatory bodies. This paper demonstrates how the model can be used to predict changes in Canadian crude oil flow patterns and mode shares subject to changes in specific transportation network attributes or crude oil demands.

The majority of bulk crude oil is shipped by pipeline in Canada, with most of the remaining quantity being shipped by rail tankers. Rail tanker shipments become feasible where pipeline capacity is limited, or where access to pipelines is unavailable because of poor connectivity at the origin and/or destination points. In recent years, improvements in pipeline capacities and connectivity have been inhibited in a large part by environmental and political concerns, namely unintended releases in which detection and containment is difficult for pipelines buried several meters below ground. Recent moratoria on the Dakota and Keystone XL pipeline expansion projects are two examples of the uphill battle faced by industry proponents to receive government approval for new pipeline construction (*1*, *2*). This has resulted in a greater dependency on rail tank cars (*3*). The Canadian Association of Petroleum Producers (CAPP) note that without expansion of the pipeline network, every new barrel of oil will be transported by rail (*4*).

The mode shift of crude oil from pipelines to rail is likely to become more pronounced in the future, as the demand for crude increases subject to expanding global markets (especially in Asia). According to CAPP (*5*), almost 100,000 barrels per day (b/d) of western Canadian crude oil were transported to market by rail in 2016, and almost 140,000 b/d were transported by rail in the first quarter of 2017. Moreover, the current rail loading capacity originating in western Canada is 754,000 b/d. This rail capacity is expected to meet the 1 million b/d deficit between western Canadian production forecasted for 2025 (5 million b/d) and existing pipeline capacity (4 million b/d). Western Canadian production is further forecast to reach 5.4 million b/d by 2030, representing a 39% increase from 2016. The network-wide freight and risk implications of potential pipeline to rail modal shifts are not well understood and need further scientific study, serving as the main motivating principle underlying this research.

Research Objectives

The primary objective of this research is to develop an empirical rail and crude oil freight demand model. Modeling crude oil shipments allows for an assessment

[1]Department of Civil and Environmental Engineering, University of Waterloo, Waterloo, ON, Canada

Corresponding Author:
Address correspondence to Chris Bachmann:
chris.bachmann@uwaterloo.ca

of the impacts of future changes in pipeline/rail network connectivity, modal attributes, and shipment protocols, on the expected pattern of crude oil shipments. For example, policy makers could use a crude oil freight demand model to estimate the impact of implementing the Keystone XL pipeline expansion project on the distribution and costs of shipments by mode in the U.S. and Canada. Alternatively, historic or forecasted growth rates could be applied to existing demands to estimate crude oil shipments on a mode and route-specific basis in future years.

This research involves two major tasks:

1. Develop a rule-based mode choice and route assignment algorithm for shipping crude oil by pipeline and rail. The rule-based algorithm should reflect current regulatory requirements, and assumed shipper–carrier interactions and prioritization rules in allocating limited pipeline capacities.
2. Apply the proposed algorithm to the Canadian pipeline and rail crude oil distribution network, and demonstrate how it can be used to predict changes in crude oil flow patterns and mode shares subject to changes in specific transportation network attributes or crude oil demands.

Organization of Paper

The remainder of this paper is organized as follows. The next section provides a brief literature review on freight demand models, particularly on mode split and assignment techniques. Subsequently, a rule-based mode split and route assignment model is presented that combines crude oil origin–destination (OD) demands with generalized cost functions for rail and pipeline to determine the tonnage of crude oil shipped by each mode and route. A hypothetical example demonstrates the logic and unique properties of the proposed model. A full-scale Canadian crude oil freight demand model is presented and applied to changing pipeline capacities and crude oil demands. The final section of the paper provides a summary of key findings and directions for future research.

Literature Review

Many concepts in freight demand modeling have been borrowed largely from travel demand modeling. As noted by de Jong et al., important differences between modeling freight and personal travel demand include the diversity of decision-makers (shippers, carriers, intermediaries, drivers, operators), and items being transported (such as, parcel deliveries with many stops to single bulk shipments comprising hundreds of tonnes) (*6*). Despite the differences between freight and personal travel, researchers have successfully applied the four-stage model paradigm to predict and describe shipper decision-making patterns, where OD commodity demands by mode are assigned to a multimodal transportation network utilizing shortest-path, user equilibrium, or system optimal traffic assignment procedures. The literature to date has focused primarily on rail and truck, as they are the two dominant modes of freight transportation (*7*). There do not appear to be any studies dealing specifically with assigning shipments to pipeline networks.

In terms of mode split, the conventional approach to modeling choice at the aggregate and disaggregate levels is based on random utility maximization theory, particularly the multinomial logit model and aggregate logit model, estimated on aggregate and disaggregate data, respectively (*8*). In recent years, several activity-based and rule-based models have been adopted for mode choice and route assignment applications. For example, Ben-Akiva and de Jong propose an aggregate–disaggregate–aggregate model (*9*), in which commodity generation, commodity distribution, and route assignment are carried out at an aggregate level, whereas the mode choice process is at a disaggregate (activity-based) level. Arentze and Timmermans' developed a model referred to as ALBATROSS, for **A** **L**earning-**Ba**sed, **Tr**ansportation-**O**riented **S**imulation **S**ystem, which uses activity-based logic and decision-trees to model trip making behavior (*10*). Along similar lines to ALBATROSS, a rule-based **I**ntelligent **N**etwork **Si**mulation **M**odel (INSIM) expert system was developed to simulate real-time mode choice decision-making in the presence of multimodal information (*11*). Neither conventional nor contemporary approaches to mode split modeling have been applied to model the behavior of crude oil shippers, in particular using pipelines.

Crude oil forecasts are typically done at an aggregate level without considering transportation modes or routes. For example, CAPP forecasts crude oil production to 2030 by Canadian region (eastern, western) using "historical trends, expected drilling activity, as well as discussions with industry stakeholders and government agencies" (*5*). The U.S. Energy Information Administration (EIA) forecasts aggregate U.S. crude oil to 2040 using the National Energy Modeling System—"an integrated model that aims to capture various interactions of economic changes and energy supply, demand, and prices" (*12*). The Organization of the Petroleum Exporting Countries forecasts world oil supply and demand by country/region to 2040 based on assumptions about population, gross domestic product, policy, technology, and the oil price (*13*).

The remainder of this paper develops a crude oil freight demand model. This research contributes a new mode choice and route assignment algorithm specifically for pipeline and rail networks. Crude oil OD demands are estimated by traditional commodity generation and distribution methods (e.g., trade flows, gravity models, etc.), as in the Canadian model presented toward the end of this paper.

Methodology

This section presents a mode split and route assignment model that reflects real-world Canadian allocation, assignment, and apportionment rules for crude oil pipeline and rail shipments (*14*). The decision-making process underlying this model is the shipper, who is prioritized by the carrier when there is limited pipeline capacity available. Priority is established on the basis of contractual arrangements between shippers and carriers (e.g., such as whether the producer is a term shipper or uncommitted), other attributes of the destination points (e.g., such as whether the destination is accessible by other modes), and sometimes by the shippers willingness to pay. In Canada, the term "carriers" refers to common transportation companies (e.g., for pipelines: Enbridge, or TransCanada Pipelines; and for rail: Canadian National (CN) or Pacific (CP) railways). "Shippers" refers to the oil production companies, who are normally members of CAPP in Canada.

From the perspective of mode choice and routing decisions, shippers are assumed to bear the full brunt of transportation costs and tolls, and thus they are assumed to adjust their decisions in order to minimize shipment costs on a given route and mode. Shipper costs for crude oil transport are obtained for each mode by empirically based cost functions:

$$c_i = f(\mathbf{X}_i) \tag{1}$$

where c_i is the cost of route i, and $\mathbf{X}_i$ is a vector of variables that affect the costs along route segment i (such as distance, capacity, throughput, etc.). This study assumes that shippers will always first select pipelines (given their lower costs) and then, when pipeline capacity is reached, the remaining crude oil demand is diverted to rail at a higher cost to the shipper. The capacity of pipeline segments is determined based on their diameters. In this paper, rail capacity is assumed to be unrestricted (or uncapacitated). No transfers are permitted between modes en-route.

For real-world applications, shippers can be aggregated into representative zones. Zonal production totals are the sum of individual shipper production levels in the zone. Similarly, all shipments to a destination zone are assumed to concentrate on a common attraction centroid within the zone.

Shipper Types

Two types of shippers are considered: term shippers and uncommitted shippers.

- *Term shippers* are defined as shippers who are subject to legal contracts with the carriers (pipeline operators or railways), and so are prioritized in the order of allocated modal capacities over uncommitted shippers who do not have any binding agreement with the carriers.
- *Uncommitted shippers* are either not subject to contract with the carriers, or they are designated as term shippers, but have desired shipment volumes exceeding the amounts specified in their monthly maximum contracted volumes.

The contract refers to the "Term Service Transportation Service Agreement" entered into between the carrier and the term shipper. It also refers to any other agreement for service between the carrier and a term Shipper (e.g., as per Kinder Morgan introduction of a service fee to Westridge Marine Terminal shipments in Canada).

Destination Types

Different delivery points have different shipment priorities, as established by the regulatory agencies. In Canada, these agencies are the National Energy Board (NEB) for pipelines and Transport Canada for rail. Three types of destination points are considered:

- *Designated delivery points* are designated by the shippers. These destinations can be either crude oil refineries (e.g., Edmonton), industrial user (e.g., petroleum user), or a port terminal/transshipment point (e.g., Westridge Marine Terminal in Vancouver).
- *Priority destinations* refers to refineries, marketing terminals, or other facilities connected to and capable of receiving petroleum from facilities of a given carrier (e.g., Trans Mountain Pipeline to Puget Sound) and not capable of being supplied economically from alternative sources (e.g., no rail connection).
- *Alternate delivery point* refers to a demand by a shipper to a destination other than the designated delivery points.

Prioritization Rules

Priority rules are commonly adopted by the pipeline industry to allocate limited capacity on specific routes (or route segments) and to specific points of disposition. In

Table 1. Hypothetical Pipeline Link and Rail Path Attributes

Pipeline			Rail		
Link	Cost ($/unit)	Capacity	Path	Cost ($/unit)	Capacity
a.	10	12000	h.	200	Unlimited
b.	20	12000	i.	150	Unlimited
c.	12	12000	j.	100	Unlimited
d.	16	12000	k.	250	Unlimited
e.	18	12000			
f.	14	5000			
g.	8	5000			

the model, it is assumed that all shipments are assigned to the minimum (i.e., shortest) paths for pipeline and rail.

The default priority ranking used in this model is:

1. Term shipper to priority destinations
2. Uncommitted shipper to priority destinations
3. Term shipper to designated destinations
4. Uncommitted shipper to designated destinations (based on bid premiums)
5. Term shipper to alternative delivery points (based on proportionate allocation)
6. Uncommitted shipper to alternative delivery points

As noted previously, shipments begin with assignment to pipelines, as pipelines have lower transport costs than rail. Once a pipeline segment has reached capacity, any remaining shipments seek an alternative pipeline route. If there is no available pipeline route, the shipment is diverted to the rail network.

Allocation Rules

When multiple shippers of the same type compete for limited pipeline capacity to the same destination point (i.e., the shippers have the same priority in the above ranking), the carrier allocates capacity either proportionally or according to bid premiums:

- *Proportional allocation*: All shippers demands $(d_1, d_2, \ldots, d_n)$ are $_n$reduced proportionally (i.e., divided by $d_{\text{total}} = \sum d_i$).
- *Bid premium allocation*: All shippers bid on the capacity by submitting a bid premium, where the bid premium is the multiplication of their bid price ($/m^3) and volume (m^3). For the purpose of this model, it was assumed that the bid price is the difference in cost between shipping by pipeline and rail, as any higher value for the bid premium would make rail cheaper than pipeline, and any lower value would result in a potential transfer to rail at a higher cost. The bid price (difference in cost) between origin r and destination s is denoted as Δp_{rs}.

The following hypothetical example illustrates how mode split and route assignment are jointly determined by shipper types, destination types, prioritization rules, and allocation rules.

Numerical Example

Consider a hypothetical region consisting of two crude oil production sources (zones O_1 and O_2) to three destinations (zones D_3, D_4, and D_5). Origin zone O_1 comprises 3 individual shippers [1 term (T) and 2 uncommitted (U)] and zone O_2 has 2 shippers (1T and 1U). The pipeline and railway networks connecting these zones are illustrated in Figure 1*a*, with a few key link/node attributes summarized in Table 1. The costs given in Table 1 include tolls and other fees charged by the pipeline and rail carriers, which are assumed to be some function of link distance, congestion, and other financing and contractual arrangements established between the carrier, shipper, and regulatory agencies (e.g., toll charges and service fees).

Table 2 provides a summary of the assumed total crude oil volumes or tonnages which are desired to be shipped from a given origin to a given destination zone, by individual shippers and in total for each zonal pair. Each volume can be represented by V_{rs}^{mn}, where r is the origin zone, s is the destination zone, m is the shipper type (T or U), and n is the shipment number.

Mode Split and Route Assignment

The mode split and route assignment algorithm proceeds by assigning shipments in order of their priority.

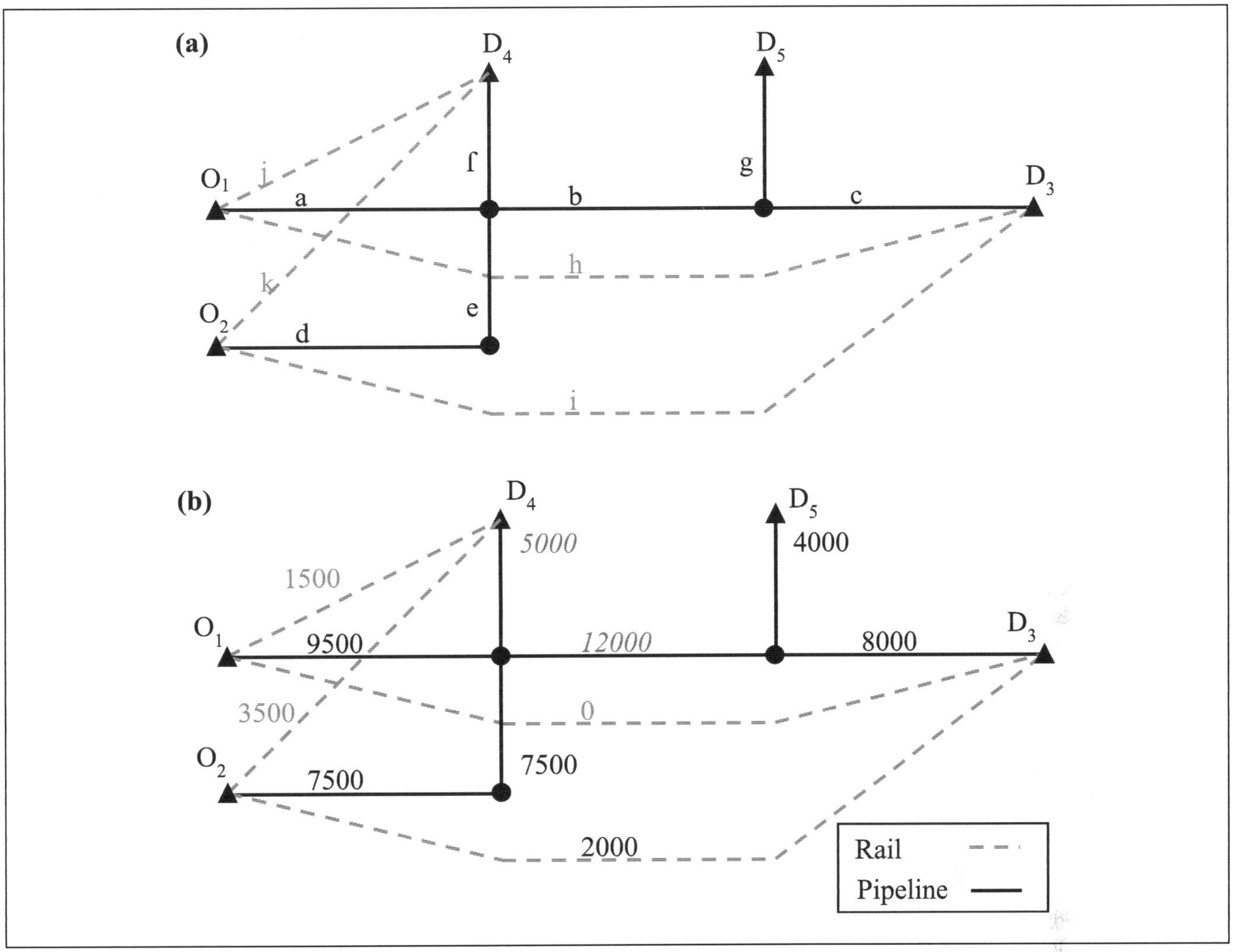

Figure 1. (*a*) Hypothetical pipeline and rail network and (*b*) hypothetical pipeline and rail network flows (after mode split and route assignment).

Table 2. Hypothetical Origin–Destination Matrix of Shipment Tonnages

Origin / destination	D3 (designated)	D4 (alternate)	D5 (priority)	Total
O1	1000 (T)	1000 (U)	1000 (T)	
	2000 (U)	1000 (U)	1000 (U)	
	3000 (U)	1000 (U)		
	6000	3000	2000	11, 000
O2	2000 (T)	3000 (U)	1000 (T)	
	2000 (U)	4000 (U)	1000 (U)	
	4000	7000	2000	13,000
Total	10,000	10,000	4000	24,000

Step 1: Assign term shipper flows to priority destination

$$V_{15}^{T1} = 1000; \quad V_{25}^{T1} = 1000.$$

Step 2: Assign uncommitted shipper flows to priority destination

$$V_{15}^{U1} = 1000; \quad V_{25}^{U1} = 1000.$$

Step 3: Assign term shippers to designated delivery point

$$V_{13}^{T1} = 1000; \quad V_{23}^{T1} = 2000.$$

Step 4: Assign uncommitted shippers to designated delivery point (based on bid premiums)

$$\text{For } V_{13}^{U1}: \Delta p_{13} x V_{13}^{U1} = [200-(10+20+12)]x2000 = 158x2000 = \$316,000$$

$$\text{For } V_{13}^{U2}: \Delta p_{13} x V_{13}^{U2} = [200-(10+20+12)]x3000 = 158x3000 = \$474,000$$

$$\text{For } V_{23}^{U1}: \Delta p_{23} x V_{23}^{U1} = [150-(16+18+20+12]x2000 = 84x2000 = \$168,000$$

Therefore, priorities are: 1) V_{13}^{U2}; 2) V_{13}^{U1}; and 3) V_{23}^{U1}.

Assigning all uncommitted shipments from highest to lowest priority, one shipper at a time:

$$V_{13}^{U2} = 3000(1^{st}\text{rank}); V_{13}^{U1} = 2000(2^{nd}\text{rank}).$$

Pipeline segment (b.) is now at capacity. Note assigning V_{23}^{U1} to pipeline, will exceed the pipeline capacity of 12,000 tonnes for route segment (b.). Thus, the decision is made to divert this shipment (V_{23}^{U1}) to rail.

Step 5: Assign uncommitted shippers to alternative land destinations (based on proportionate allocation)

We need to assign the following uncommitted volumes to D_4 from O_1: $V_{14}^{U1} = 1000$; $V_{14}^{U2} = 1000$; $V_{14}^{U3} = 1000$. We need to assign following uncommitted volumes to D_4 from O_2: $V_{24}^{U1} = 3000$; $V_{24}^{U2} = 4000$. If all shipments were assigned to pipeline, cumulative flow on links (a.), (d.), and (e.) would be below capacity (no apportioning required). However, cumulative flow on link (f.) would be 10000, which is double the 5000 capacity for this segment (requiring apportioning). Therefore, capacity of pipeline segment (f.) is allocated proportionately to shippers at origins *1* and *2*, and the remaining flows are assigned to rail: V_{14}^{U1}: $1000 \times 5/10 = 500$; V_{14}^{U2}: $1000 \times 5/10 = 500$; V_{14}^{U3}: $1000 \times 5/10 = 500$; V_{24}^{U1}: $3000 \times 5/10 = 1500$; V_{24}^{U2}: $4000 \times 5/10 = 2000$.

The final assignment of crude oil flow to the hypothetical network is given in Figure 1*b*, with pipeline bottlenecks shown in *red*. This assignment approach is similar to a traditional incremental assignment, in which each increment has been prioritized by shipper type, destination type, prioritization rules, and allocation rules.

In this example it is assumed that rail capacity is unrestricted (to reduce complexity). In reality, the above results represent the *demand* for crude oil rail transport, but these demands may not materialize because of rail capacity restrictions. For example, the Fair Rail for Grain Farmers Act of 2014 (Bill C-30) was passed in Canada to mandate grain shipment levels because of competition for rail transport. Therefore, the crude oil rail flows that ultimately materialize in real-world networks may be the result of complex market interactions. To properly address these issues is considered to be outside the scope of this paper.

Case Study

OD Demands

For a real-world Canadian implementation, OD demands are based on statistical trade data. In particular, crude oil shipments beginning and ending in Canada are available from CANSIM, Statistics Canada's key socioeconomic database. CANSIM Table 126-0001 describes the historical supply and disposition of crude oil and equivalent in Canada (cubic meters) (*15*). Canadian exports are also included in CANSIM Table 126-000, including exports from Canadian provinces to U.S. Petroleum Administration for Defense Districts (PADDs), as well as total exports from each province to all countries other than the U.S. Canada's total shipment of crude to individual PADDs is also available quarterly from the NEB, broken down by type (Conventional Light; Conventional Medium; Conventional Heavy; Synthetic; and Bitumen & Blended Bitumen) (*16*).

In terms of Canadian imports, CANSIM Table 126-0001 only provides total imports from all sources to Canada. However, Canada's total import of crude from the U.S. is available monthly from 1993 to 2016 from the U.S. EIA (*17*), and therefore total imports from other countries can be deduced. Data for Canadian imports from U.S. PADDs are not available. However, the EIA also disseminates data on "Movements of Crude Oil and Selected Products by Rail" (*18*), which includes data on rail movements between Canada and individual U.S. PADDs. The table suggests rail imports to Canada are non-existent or small (e.g., below the threshold for recordkeeping—volume of less than 0.5 thousand barrels per day), whereas some export rail shipments from Canada to U.S. PADDs take place. Other complementary data sources are described by Bachmann and Saccomanno (*19*).

As shown in Figure 2, these OD demands are linked to 17 North American production/disposition zones: 12 for Canada, and 5 (PADDs) for the continental U.S. Alberta was structured into four zones, and British Columbia into two zones (one in the north and one in the south). In addition, there is one zone for each province: Saskatchewan, Manitoba, Ontario, and Quebec. The Northwest Territories was also assigned a separate zone. Finally, all four Atlantic Provinces are represented

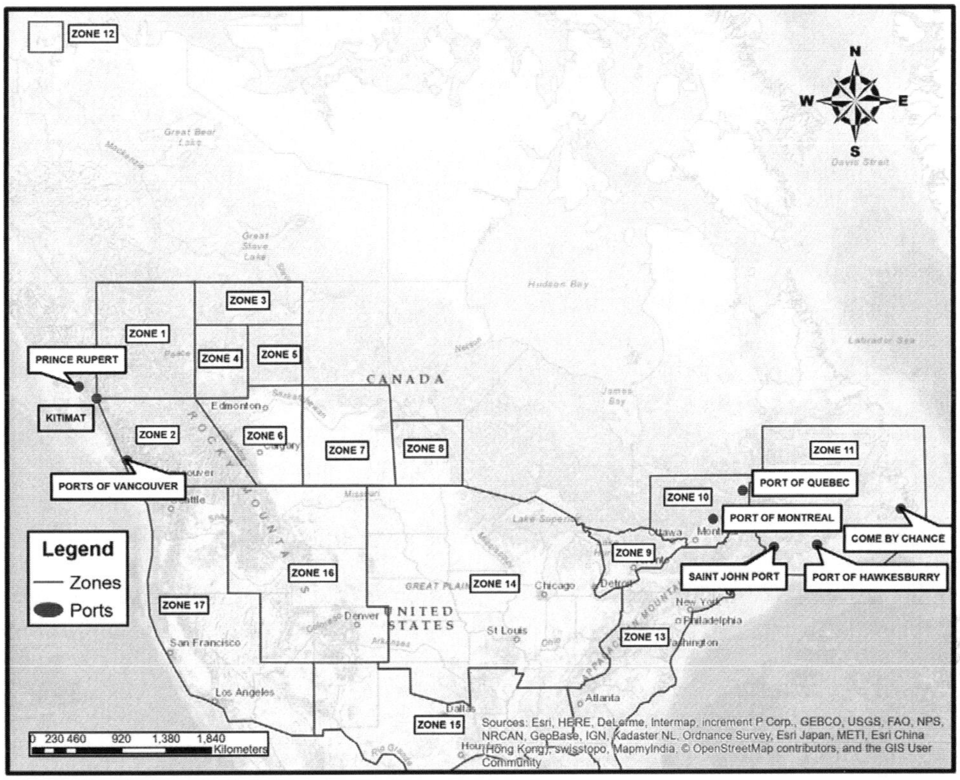

Figure 2. Zone system for modeling Canadian crude oil shipments.

by one zone. The resolution of zones was chosen to reflect the density of shippers and to properly identify usage of the Canadian rail and pipeline networks. As more disaggregate OD data are not available, shipments originating from British Columbia or Alberta were split based on total production data. Similarly, shipments destined for British Columbia or Alberta were split based on total attraction data. These estimates are based on apportioning values determined by data from the CAPP (*20*), Alberta Energy (*21*), and the Canadian Fuels Association (*22*).

Foreign sources require additional external zones (port locations shown on Figure 2) so as to accurately represent the usage of the Canadian pipeline and rail networks. On the West coast, crude is shipped through the Ports of Vancouver, Prince Rupert, and Kitimat. On the East coast, crude is predominantly shipped to the ports of Come by Chance, Port Hawkesbury, Saint John, Port of Quebec, and Port of Montreal (*23*).

Cost Functions for Shipping Crude by Pipeline or Rail in Canada

Numerous data sources were compiled to estimate cost functions for shipping crude by pipeline and rail in Canada. The NEB compiles data on the tariffs/tolls charged by Canadian pipeline companies, including the fees charged to crude oil shippers, which consist of usage fees as well as surcharges (*24*). For rail, CP railways publically disseminates a full tariff schedule for OD pairs of crude oil shipments (*25*). NRCan (*26*) has Geographic Information System (GIS) data for both the Canadian rail and pipeline transportation networks. These data were used for determining distances between the OD pairs given by the NEB toll and tariff documents. For some pipelines not covered in the NRCan GIS data, the Canadian Energy Pipeline Association liquids pipeline maps were used (*27*). Other data were collected from pipeline company documents. These companies consisted of Enbridge Inc., Kinder Morgan, Plains Midstream Canada, Spectra Energy Corp., and TransCanada Corp (*28*). These data include pipeline specifications, such as diameters.

Using these data an empirical crude oil pipeline cost function was established of the form:

$$C_p = 6.92 + 0.005d - 0.142\phi + 4.26D_T + 1.70D_M + 3.00D_H \tag{2}$$

where C_p is the pipeline cost ($CND/m^3), d is the pipeline distance between the origin and destination (km), ϕ

is the maximum pipeline diameter per segment (inches), D_T is an interval (dummy) variable for the toll type ($D_T = 0$ for a Canadian destination and $D_T = 1$ for a U.S. destination), D_M is a dummy variable for medium crude ($D_M = 1$ if the crude type is medium $D_M = 0$ for all others), and D_H is a binary value for heavy crude ($D_H = 1$ if the crude type is heavy crude and $D_H = 0$ for all others).

The above expression was estimated based on a sample of 323 OD observed pairs. The model yielded an R^2 value of 0.85 and an adjusted R^2 value of 0.84. All parameter estimates were found to be statistically significant at the 5% level. The average pipeline cost in the underlying data was \$14.87/m^3, with a range of \$ 0.58/ m^3 to \$42.16/m^3, and a standard deviation of \$8.32/m^3. The average distance in the underlying data was 1599 km, with a range of 0 km to 3589 km, and a standard deviation of 1054 km.

All parameter estimates have expected signs and magnitudes. The longer the distance (d) shipped, the higher the cost charged by the carrier. Larger pipelines (ϕ) have lower costs per individual shipment, as a larger pool of shippers split the total base rate (i.e., the total capital costs and operations and maintenance of the pipeline). Shipments to the U.S. (D_T) have higher cost because they are subject to an additional tariff. Finally, medium (D_M) and heavy (D_H) crude are more expensive to ship compared with condensate and light crude because of their lower viscosities. Further details and alternative specifications of the pipeline cost function are provided by Morrison et al. (*28*).

Note that term shippers are likely to incur lower shipment costs as part of their negotiated settlements. For example, the NEB regulates pricing and settlements between pipeline owners and term shippers for Canadian interprovincial and international pipelines (*29*). However, these cost functions estimate the additional costs borne by shippers when they choose to use rail because of pipelines bottlenecks. In these cases, it is assumed the shipments are uncommitted (i.e., that pipeline owners do not overcommit and the pipeline network can satisfy all of its committed shipments).

For rail, a crude oil cost function was obtained of the form:

$$C_r = 30.35 + 0.0123d + 13.29t \qquad (3)$$

where C_r is the rail cost (\$CND/m^3), d is the rail distance between the origin and destination (km), and t is an interval (dummy) variable for the toll type ($t = 0$ for a Canadian destination and $t = 1$ for a U.S. destination).

This model was estimated on a sample size of 917 OD pairs, resulting in an R^2 value of 0.74 and an adjusted R^2 value of 0.74. All parameter estimates are again statistically significant at the 5% level of significance and have expected signs and magnitudes (same rational for distance and toll type as for pipeline). The average rail cost in the underlying data was \$59.87/m^3, with a range of \$24.61/m^3 to \$126.35/m^3, and a standard deviation of \$19.24/m^3. The average distance in the underlying data was 1824 km, with a range of 0 km to 5139 km, and a standard deviation of 1209 km.

Comparing Equations 2 and 3 confirms that shipping crude by pipeline is cheaper than shipping by rail (as assumed previously in the mode split and route assignment algorithm). From the constants in Equations 2 and 3, the model estimates initial cost advantages for pipeline over rail of 4.4 for light crude (30.35/6.92), 3.5 for medium crude (30.35/[6.92 + 1.70]), and 3.1 for heavy crude (30.35/[6.92 + 3.00]). These findings are comparable with other estimates. For example, the U.S. Congressional Research Service found transporting crude oil by pipeline is cheaper than by rail, at a cost of under \$5 a barrel by pipeline compared with \$10 to \$15 a barrel for rail (*30*). The Association of Oil Pipelines presents the same statistic and a set of sample data showing that total pipeline costs range between \$2.71 and \$4.54 per barrel, while total rail costs range from \$10.38 to \$16.73 per barrel (*31*).

Base Case Network Model

Figure 3 illustrates the modeled base case (2015) crude oil volumes in Canada for the pipeline and rail networks. These networks consist of the majority of rail and pipeline infrastructure in Canada. For rail, all first-class railroads as defined by the U.S. Surface Transportation Board were included in the model. Crude oil rail shipments are typically sent only with first-class railroads because of restrictions surrounding the shipments of hazardous materials. Therefore, the rail network consists primarily of track owned by Canada's major rail line companies, CN and CP, as well some from smaller companies that are subcontracted for selected shipments. For pipelines, all existing pipelines in Canada that are meant to ship liquid petroleum products in the range of condensates to heavy crude oil were included in the model. Lines that ship only liquid petroleum gasses such as liquid ethane, butane, and natural gas liquids were excluded.

The Canadian pipeline and rail crude oil demand model uses the rule-based pipeline–rail mode split and route assignment algorithm presented in the previous section, with slight modifications to reflect data limitations. In particular:

- Alternative and designated destinations are not distinguished, and therefore the last four cases in the prioritization rules are collapsed into two (term shipper to designated/alternative destination; and

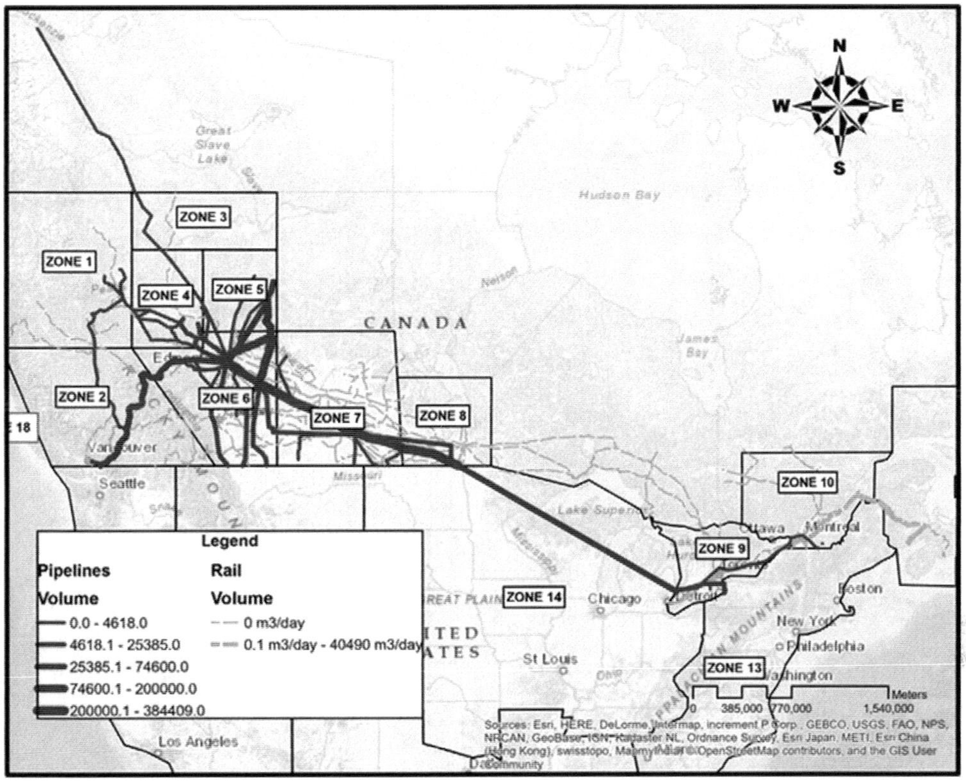

Figure 3. Base case crude oil volumes on pipeline and rail in Canada.

uncommitted shipper to designated/alternative destination). This aggregation is required because data on designated and alternate destinations were not available.

- As crude oil was not distinguished by type in all available OD data, an alternative cost function was used for pipeline costs:

$$C_p = 6.25 + 0.00345d - 0.104\phi + 9.56D_T \qquad (4)$$

where: d is the distance in (km), ϕ is the pipeline diameter (inches), and D_T is an interval (dummy) variable for the toll type ($D_T = 0$ for a Canadian destination and $D_T = 1$ for a U.S. destination). This cost function has an R^2 value of 0.79 (all variables statistically significant at 5% level of significance).

In other applications (e.g., smaller geographical scope, other areas), data availability may result in the use of a greater (or fewer) number of steps or other modifications in the proposed rule-based mode split and route assignment algorithm. Because of the size of this network, the steps of the mode split and route assignment algorithm were automated in Python, which interacted with the network coded in ArcGIS.

To validate the base case model, predicted throughputs were compared with those reported for Canada's group 1 and group 2 export pipelines: Enbridge Mainline, Keystone, Trans Mountain, Normal Wells, Express, Westspur, and Milk River. Their average throughputs in 2015 were 385305 m^3/day, 88360 m^3/day, 50085 m^3/day, 1221 m^3/day, 34710 m^3/day, 27945 m^3/day, and 15087 m^3/day, respectively (*29*). Their predicted throughputs were 411370 m^3/day, 8845 m^3/day, 43984 m^3/day, 0 m^3/day, 25383 m^3/day, 0 m^3/day, and 18800 m^3/day, respectively. These values have a correlation coefficient of $r = 0.98$ and a coefficient of determination of $r^2 = 0.96$. These statistics are high because of the much larger throughput on the Enbridge Mainline compared with other pipelines; the mean absolute percentage errors (MAPEs) are: 7%, 90%, 12%, 100%, 27%, 100%, and 25%. Note that Norman Wells and Westspur pipelines have predicted throughputs of zero (MAPEs of 100%) as a result of suppressed production data at their origin zones (i.e., not available). In general, suppressed OD data remain the largest contributor to prediction error. It was not possible to validate Canada's import pipelines because crude oil imports from the U.S. by PADD are not yet available. Therefore, the model is

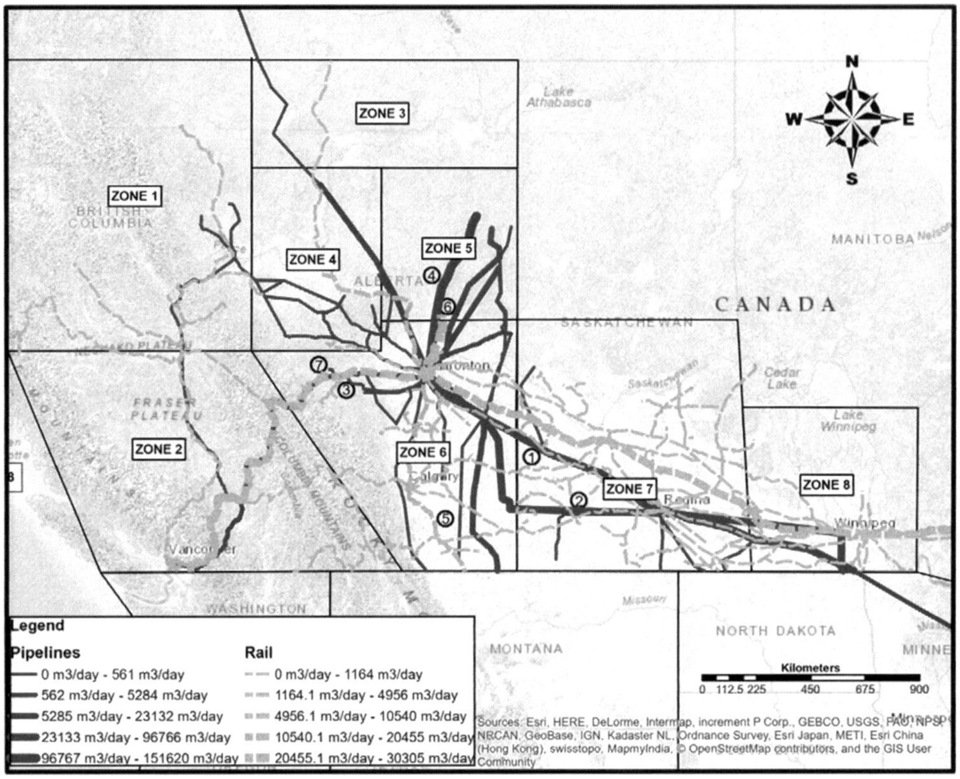

Figure 4. Modeled crude oil volume changes owing to a 39% increase in western Canadian crude oil production.

only suitable for export scenario analyses until import OD demands are refined.

Scenario Analysis: Crude Oil Forecasting

The developed crude oil freight demand model can be used to predict changes in crude oil flow patterns and mode shares subject to changes in specific pipeline or rail network attributes or crude oil demands. Figure 4 shows the results of a 39% increase in western Canadian crude oil production, forecasted by CAPP for 2030 (*5*). Their forecast is based on Canada's crude oil supply growing by 5% per year to 2020, and then slowing to 2% growth per year to 2030 (owing to many market uncertainties). The modeled scenario assumes a corresponding 39% increase in all Alberta production zones (zones 3–6), distributed to destinations in proportion to their base case demands.

The results summarized in Figure 4 highlight the usefulness of developing a complete crude oil freight demand model for forecasting and project/policy evaluation. The expected increases in pipeline volumes of crude oil shipments can be quantified. In general, increased volumes are forecasted on all pipeline links unless a link is already at capacity in the base case. For example, Figure 4 shows that volumes increase on several pipelines that have spare capacity in the base case, including: (1) Enbridge Mainline (maximum increase of 143,152 m^3/day); (2) TransCanada (increase of 85,154 m^3/day); (3) Kinder Morgan (increase of 3,718 m^3/day); and (4) Pembina Horizon/Syncrude Line (101,500 m^3/day). These increases cause the Enbridge Mainline and other pipelines serving the oil sands to reach their capacities.

In addition, the volume of crude oil shipments transferred to rail can be determined. Assuming no new pipeline infrastructure is constructed before 2030, several rail lines would carry increased amounts of crude oil. For example, Figure 4 shows that volumes increase on the following rail lines: (5) CP from Edmonton to PADD 4 (zone 16) (maximum increase of 4,956 m^3/day); (6) CN from the Athabasca and Cold Lake oil sands (zone 5) to Edmonton (increase of 30,305 m^3/day); (7) CN from Edmonton to PADD 5 (zone 17) (maximum increase of 12,382 m^3/day); and CN from Peace River oil sands (zone 4) to all remaining destinations (maximum increase of 15,499 m^3/day).

Conclusion

Conventional freight forecasting models exclude crude oil shippers and the pipeline network. Meanwhile, crude oil forecasts are typically done at an aggregate level without considering transportation modes or routes. This research found that the mode split and route assignment of crude oil shipments are jointly determined by shipper types, destination types, prioritization rules, and allocation rules. This paper proposes a rule-based mode choice and route assignment algorithm for shipping crude oil by pipeline and rail that captures the complex interactions of crude shippers, pipeline carriers, and regulatory bodies. The complete algorithm cannot be fully applied at the national level in Canada because of current data limitations. Nonetheless, a complete crude oil freight demand model was developed using the algorithm and two modifications to predict changes in crude oil flow patterns for the year 2030.

Although the current model reflects a set of Canadian rules, it can be extended to other geographic areas or contexts by modifying the shipper types, destination types, prioritization rules, and allocation rules. For example, it is possible that shipper allocations might increasingly be based on bid premiums (as opposed to proportional allocations) as a greater proportion of the network reaches capacity. The model may also require modifications depending on the scope of the application and data availability. In this light, the authors suggest that future researchers carefully review the rules and regulations applicable to their study area, available data sources, and their study purpose and scope, to make modifications as necessary. Once developed, the model can be used to predict changes in crude oil flow patterns and mode shares subject to changes in specific pipeline or rail network attributes or crude oil demands.

Future research is needed to develop an improved understanding of the risks associated with shipping crude oil by pipeline and rail. After a crude oil freight demand model is developed, a comprehensive quantitative risk assessment can be applied to a wide spectrum of shipment scenarios, shipping regulations and guidelines, as well as future changes in crude oil throughputs (OD). Such a model permits a complete objective assessment of the risk implication of transporting crude oil by pipeline as compared with rail. This assessment will assist in developing informed, safe, and cost-effective decisions on how best to transport crude oil.

Acknowledgments

Portions of this paper are based on work funded by Transport Canada, for the project titled "Modelling Crude Oil Rail and Pipeline Shipments in Canada for Quantitative Risk Assessment." The authors would like to thank Daniel Chisholm, who assisted in preparing the origin destination demands during his co-op term at the University of Waterloo.

Author Contributions

The authors confirm contribution to the paper as follows: study conception and design: FS, CB; data collection: AM, CB; analysis and interpretation of results: AM, CB, FS; draft manuscript preparation: CB, FS, AM. All authors reviewed the results and approved the final version of the manuscript.

References

1. CBC. *Redo Analysis for Dakota Access Pipeline, U.S. Judge Rules*. https://www.theguardian.com/commentisfree/2017/jun/16/dakota-access-pipeline-protesters-court-victory-fight-on. Accessed June 16, 2017.
2. CBC. *Indigenous Leaders Sign Opposition to Keystone XL in Calgary*. http://www.cbc.ca/news/canada/calgary/trans-canada-keystone-calgary-indigenous-1.4119301. Accessed June 16, 2017.
3. Green, K. P., and T. Jackson. *Safety in the Transportation of Oil and Gas: Pipelines or Rail?* Fraser Institute, 2015. https://www.fraserinstitute.org/sites/default/files/safety-in-the-transportation-of-oil-and-gas-pipelines-or-rail-rev2.pdf. Accessed June 16, 2017.
4. Canadian Association of Petroleum Producers. *Rail*. http://www.capp.ca/canadian-oil-and-natural-gas/infrastructure-and-transportation/rail. Accessed June 16, 2017.
5. Canadian Association of Petroleum Producers. *2017 CAPP Crude Oil Forecast, Markets & Transportation*. 2017. http://www.capp.ca/publications-and-statistics/publications/303440. Accessed June 16, 2017.
6. de Jong, G., H. Gunn, and W. Walker. National and International Freight Transport Models: An Overview and Ideas for Future Development. *Transport Reviews*, Vol. 24, No. 1, 2004, pp. 103–124.
7. Uddin, M. M., and N. Huynh. Freight Traffic Assignment Methodology for Large-Scale Road–Rail Intermodal Networks. *Transportation Research Record: Journal of the Transportation Research Board*, 2015. 2477: 50–57.
8. Tavasszy, L, and G. de Jong. *Modelling Freight Transport*. Elsevier Insights, London, UK, 2014.
9. Ben-Akiva, M., and G. de Jong. The Aggregate-Disaggregate-Aggregate (ADA) Freight Model System. In *Freight Transport Modelling*, Emerald Group Publishing Limited, Bingley, UK, 2013, pp. 69–90.
10. Arentze, T. A., and H. J. P. Timmermans. A Learning-Based Transportation Oriented Simulation System. *Transportation Research Part B: Methodological*, Vol. 38, No. 7, 2004, pp. 613–633.
11. Memon, A. A., M. Meng, Y. D. Wong, and S. H. Lam. Rule-Based Mode Choice Model: INSIM Expert System. *Journal of Transportation Engineering*, Vol. 141, No. 4, 2015, p. 04014088.
12. EIA. *Annual Energy Outlook 2017: With Projections to 2050*. https://www.eia.gov/outlooks/aeo/. Accessed October 20, 2017.
13. OPEC. *World Oil Outlook*. http://www.opec.org/opec_web/en/publications/340.htm. Accessed October 20, 2017.
14. Kinder Morgan. *Trans Mountain Pipeline ULC: Petroleum Tariff, Rules and Regulations Governing the Transportation of Petroleum*. 2015. https://www.kindermorgan.com/content/docs/NEB_Tariff_No_95.pdf. Accessed June 19, 2017.
15. Statistics Canada. *Table 126-0001 - Historical Supply and Disposition of Crude Oil and Equivalent, Monthly (Cubic*

Metres). http://www5.statcan.gc.ca/cansim/a26?lang=en-g&id=1260001. Accessed June 22, 2017.
16. NEB. *Commodity Statistics, Crude Oil and Refined Petroleum Products, Crude Oil – Summary by Type and Destination*. https://apps.neb-one.gc.ca/CommodityStatistics/Statistics.aspx. Accessed June 23, 2017.
17. EIA. *Exports by Destination*. https://www.eia.gov/dnav/pet/pet_move_expc_a_EP00_EEX_mbbl_m.htm. Accessed June 23, 2017.
18. EIA. *Movements of Crude Oil and Selected Products by Rail*. https://www.eia.gov/dnav/pet/PET_MOVE_RAIL-NA_A_EPC0_RAIL_MBBL_M.htm. Accessed June 23, 2017.
19. Bachmann, C., and F. Saccomanno. *Modelling Crude Oil Rail and Pipeline Shipments in Canada for Quantitative Risk Assessment*. University of Waterloo, Waterloo, ON, Canada, 2017.
20. CAPP. *Statistical Handbook*. http://www.capp.ca/publications-and-statistics/statistics/statistical-handbook. Accessed June 23, 2017.
21. Alberta Energy. *Oil Sands Production Profile: 2004–2014*. http://www.energy.alberta.ca/Org/pdfs/InitiativeOSPP.pdf. Accessed June 23, 2017.
22. Canadian Fuels Association. *Fuel Production*. http://www.canadianfuels.ca/The-Fuels-Industry/Fuel-Production/. Accessed June 23, 2017.
23. Transport Canada. *Get the Facts on Oil Tanker Safety in Canada*. https://www.tc.gc.ca/eng/marinesafety/facts-oil-tanker-safety-canada-4513.html#east-coast. Accessed June 23, 2017.
24. NEB. *National Energy Board: Traffic, Tolls & Tariffs*. https://apps.neb-one.gc.ca/REGDOCS/Item/View/90465. Accessed June 23, 2017.
25. CP. *Class Rate Tariffs: 00500 - Crude Products*. http://www.cpr.ca/en/customer-resources/pricing-and-tariffs/class-rate-tariffs. Accessed June 23, 2017.
26. NRCan. Natural Resources Canada. https://www.nrcan.gc.ca/home. Accessed June 23, 2017.
27. CEPA. *About Pipelines Map*. http://aboutpipelinesmap.com/#/NTYuMTMwMzY2Oy0xMDYuMzQ2NzcxOzU=. Accessed June 23, 2017.
28. Morrison, A., C. Bachmann, and F. Saccomanno. Generalized Cost Functions of Crude Oil Pipeline Shipments in Canada. Presented at the Canadian Transportation Research Forum (CTRF) 52nd Annual Conference, Winnipeg, MB, Canada, 2017.
29. NEB. *Canada's Pipeline Transportation System 2016*. https://www.neb-one.gc.ca/nrg/ntgrtd/trnsprttn/2016/pblctn-nfrmtn-dwnlds-eng.html. Accessed October 20, 2017.
30. Congressional Research Service. *U.S. Rail Transportation of Crude Oil Background and Issues for Congress*. https://fas.org/sgp/crs/misc/R43390.pdf. Accessed October 20, 2017.
31. Association of Oil Pipe Lines. *About Pipelines*. http://www.aopl.org/pipeline-basics/about-pipelines/. Accessed October 20, 2017.

The Standing Committee on Transportation of Hazardous Materials (AT040) peer-reviewed this paper (18-01844).

Opinions expressed in this paper are those of the authors and may not necessarily reflect the views and policies of Transport Canada.

Sage Publications USA
sagepub.com

ISBN 978-1-5108-8178-5

CURRAN ASSOCIATES INC.
proceedings.com